METHODS IN MOLECULAR BIOLOGY

Series Editor
John M. Walker
School of Life and Medical Sciences
University of Hertfordshire
Hatfield, Hertfordshire, AL10 9AB, UK

For further volumes:
http://www.springer.com/series/7651

Oligodendrocytes

Methods and Protocols

Edited by

David A. Lyons and Linde Kegel

Centre for Discovery Brain Sciences, University of Edinburgh, Edinburgh, UK

Humana Press

Editors
David A. Lyons
Centre for Discovery Brain Sciences
University of Edinburgh
Edinburgh, UK

Linde Kegel
Centre for Discovery Brain Sciences
University of Edinburgh
Edinburgh, UK

ISSN 1064-3745 ISSN 1940-6029 (electronic)
Methods in Molecular Biology
ISBN 978-1-4939-9070-2 ISBN 978-1-4939-9072-6 (eBook)
https://doi.org/10.1007/978-1-4939-9072-6

Library of Congress Control Number: 2019930142

Cover Caption: Single oligodendrocyte, labelled with GFP in a transgenic zebrafish. Image courtesy of Dr. Rafael Almeida, Lyons Lab, University of Edinburgh, Edinburgh, UK.

This Humana Press imprint is published by the registered company Springer Science+Business Media, LLC, part of Springer Nature.
The registered company address is: 233 Spring Street, New York, NY 10013, U.S.A.

Preface

We have known for nearly a century that the oligodendrocyte is the glial cell of the central nervous system that generates myelin. Similarly, we have known for decades that the presence of myelin on axons speeds up conduction velocities, by virtue of the fact that myelin sequesters key ion channels to the short nodes of Ranvier between consecutive myelin sheaths and due to the remarkable properties of the lipid-rich myelin sheath itself. The myelination of axons allows rapid information propagation throughout specific neural circuits, and as a consequence disruption to myelin has been known to underpin diseases of the central nervous system, such as multiple sclerosis, for a very long time.

However, it was not until the expansion of a myriad of advanced, molecular, cellular, and imaging techniques over the past two decades that our knowledge of how oligodendrocytes develop, mature, and make myelin has started to come together. We now have a very good understanding of the molecular genetic mechanisms that specify oligodendrocyte progenitor cells in distinct domains of the developing nervous system, and we are starting to understand how these cells migrate throughout the CNS. Indeed, it is now clear that oligodendrocyte progenitor cells reside throughout the gray and white matter of our CNS lifelong, and that they retain a capacity to generate mature oligodendrocytes, both in the healthy nervous system and in response to damage or loss of myelin in diseases, such as MS.

Our understanding of oligodendrocyte differentiation is now also being elucidated through complementary in vitro and in vivo molecular and cellular analyses. In addition, work over the past decade or so has also started to elucidate how oligodendrocytes subsequently select which axons to myelinate and how myelin sheath growth and remodeling occurs thereafter. An area of investigation that has undergone something of a renaissance in recent years is based around the fact that many stages of oligodendrocyte progenitor cell development, oligodendrocyte differentiation, and myelination are responsive to changes in neuronal activity. In-depth analyses of how neuronal activity regulates oligodendrocytes have been facilitated by numerous new techniques and have led to the concept that dynamic regulation of myelination may occur lifelong and may represent a form of functional nervous system plasticity. This concept is supported by reductionist molecular, cellular, and electrophysiological studies noted in this volume and by parallel system-level imaging of animal models and humans, which is also discussed within this volume.

Reflecting the idea that oligodendrocytes remain responsive to and active regulators of neural circuit function throughout life is the fact that new functions for oligodendrocytes and myelin have recently emerged, highlighting the fact that this cell lineage does more than converge on producing a static insulator to facilitate conduction. Oligodendrocytes regulate ion homeostasis and thus neuronal excitability; they provide metabolic support to axons and likely have an array of functions that we will begin to be able to unravel due to the approaches outlined in this volume and others. The prediction that oligodendrocytes play yet more functional roles than we currently understand is based on the fact that their disruption is now implicated in many, if not most, diseases of the nervous system, from neurodevelopmental disorders including autism, neuropsychiatric disorders including schizophrenia, neurodegenerative disorders including Alzheimer's disease, motor neuron disease, and others. In addition, oligodendrocytes and myelin exhibit dynamic changes across the life course that tightly correlate with the emergence, stability, and decline of cognitive performance.

In this volume, we bring together a series of methods that bridge in vitro and in vivo approaches, model organisms, analyses of health and disease, and scales from molecular through to system. We take a bottom-up approach and start with fundamental molecular analyses of oligodendrocytes and myelin and move through an array of in vitro and ex vivo molecular-cellular-electrophysiology-based approaches from mouse to man. In the second part of the volume, we collate methodologies for in vivo analyses of oligodendrocyte formation, homeostasis, and disruption, in zebrafish, *Xenopus*, mouse, and man. Additional methods of relevance not included in this volume are brought together in a separate volume on myelin, which we direct interested readers to.

We hope that this volume highlights the fast-paced and broad nature of research into the formation, health, and function of the oligodendrocyte and serves as a useful reference manual for the field.

Edinburgh, UK

David A. Lyons
Linde Kegel

Contents

Contributors

SYLVIA AGATHOU • *Department of Veterinary Medicine, Wellcome Trust-Medical Research Council Cambridge Stem Cell Institute, University of Cambridge, Cambridge, UK*

RAFAEL G. ALMEIDA • *Centre for Discovery Brain Sciences, The University of Edinburgh, Edinburgh, UK*

MARIE E. BECHLER • *MRC Centre for Regenerative Medicine, MS Society Edinburgh Centre for MS Research, The University of Edinburgh, Edinburgh, UK*

SILVIA BENITO • *Centre for Discovery Brain Sciences, The University of Edinburgh, Edinburgh, UK*

GRAHAM R. CAMPBELL • *Centre for Clinical Brain Sciences, University of Edinburgh, Edinburgh, UK*

GONÇALO CASTELO-BRANCO • *Laboratory of Molecular Neurobiology, Department of Medical Biochemistry and Biophysics, Karolinska Institutet, Stockholm, Sweden; Ming Wai Lau Centre for Reparative Medicine, Stockholm Node, Karolinska Institutet, Stockholm, Sweden*

SIDDHARTHAN CHANDRAN • *Centre for Clinical Brain Sciences, The University of Edinburgh, Edinburgh, UK; Euan MacDonald Centre, The University of Edinburgh, Edinburgh, UK*

ANNE COTLEUR • *Biogen, Cambridge, MA, USA*

CLAIRE L. DAVIES • *Centre for Reproductive Health, The Queen's Medical Research Institute, The University of Edinburgh, Edinburgh, UK*

LEDA DIMOU • *Molecular and Translational Neuroscience, Department of Neurology, Ulm University, Ulm, Germany; Graduate School of Systemic Neurosciences, Ludwig-Maximilians-University, Munich, Germany*

KATA DIOSI • *NDCN Department, Wellcome Centre for Integrative Neuroimaging, FMRIB Centre, University of Oxford, Oxford, UK*

MICHELLE S. ERWIG • *Department of Neurogenetics, Max Planck Institute of Experimental Medicine, Goettingen, Germany*

JAIME EUGENIN VON BERNHARDI • *Molecular and Translational Neuroscience, Department of Neurology, Ulm University, Ulm, Germany; Graduate School of Systemic Neurosciences, Ludwig-Maximilians-University, Munich, Germany*

CHARLES FFRENCH-CONSTANT • *MRC Centre for Regenerative Medicine, The Multiple Sclerosis Research Centre, The University of Edinburgh, Edinburgh, UK*

ROBIN J. M. FRANKLIN • *Wellcome Trust-MRC Cambridge Stem Cell Institute, University of Cambridge, Cambridge, UK*

SANDRA GOEBBELS • *Department of Neurogenetics, Max Planck Institute of Experimental Medicine, Göttingen, Germany*

STEVEN A. GOLDMAN • *Department of Neurology and the Center for Translational Neuromedicine, University of Rochester Medical Center, Rochester, NY, USA; The Neuroscience Center, Rigshospitalet, Copenhagen, Denmark; Center for Translational Neuromedicine, Faculty of Health and Medical Sciences, University of Copenhagen, Copenhagen, Denmark*

STEFKA GYONEVA • *Biogen, Cambridge, MA, USA*

DÖRTE HESSE • *Proteomics Group, Max Planck Institute of Experimental Medicine, Goettingen, Germany*

OLAF JAHN • *Proteomics Group, Max Planck Institute of Experimental Medicine, Goettingen, Germany*
ANDREW A. JARJOUR • *MRC Centre for Regenerative Medicine and MS Society Edinburgh Centre, Edinburgh bioQuarter, The University of Edinburgh, Edinburgh, UK*
HEIDI JOHANSEN-BERG • *NDCN Department, Wellcome Centre for Integrative Neuroimaging, FMRIB Centre, University of Oxford, Oxford, UK*
RAMONA B. JUNG • *Department of Neurogenetics, Max Planck Institute of Experimental Medicine, Goettingen, Germany*
RAGNHILDUR THÓRA KÁRADÓTTIR • *Department of Veterinary Medicine, Wellcome Trust-Medical Research Council Cambridge Stem Cell Institute, University of Cambridge, Cambridge, UK*
MARJA J. KARTTUNEN • *Centre for Discovery Brain Sciences, The University of Edinburgh, Edinburgh, UK*
LINDE KEGEL • *Centre for Discovery Brain Sciences, The University of Edinburgh, Edinburgh, UK*
ANNA KLINGSEISEN • *Centre for Discovery Brain Sciences, The University of Edinburgh, Edinburgh, UK*
WEN PING KUO-ELSNER • *Department of Biology, Molecular Cell Biology, Johannes Gutenberg University, Mainz, Germany*
KATHRIN KUSCH • *Department of Neurogenetics, Max Planck Institute of Experimental Medicine, Goettingen, Germany*
MATTHEW R. LIVESEY • *Centre for Discovery Brain Sciences, The University of Edinburgh, Edinburgh, UK*
DAVID A. LYONS • *Centre for Discovery Brain Sciences, The University of Edinburgh, Edinburgh, UK*
DARIO MAGNANI • *Centre for Clinical Brain Sciences, The University of Edinburgh, Edinburgh, UK; Euan MacDonald Centre, The University of Edinburgh, Edinburgh, UK*
DON J. MAHAD • *Centre for Clinical Brain Sciences, University of Edinburgh, Edinburgh, UK*
ABDELKRIM MANNIOUI • *Sorbonne Universités UPMC, Univ Paris 06, Inserm, CNRS, APHP, ICM-GH Pitié-Salpêtrière, Paris, France*
JOHN N. MARIANI • *Department of Neurology and the Center for Translational Neuromedicine, University of Rochester Medical Center, Rochester, NY, USA*
SUELI MARQUES • *Laboratory of Molecular Neurobiology, Department of Medical Biochemistry and Biophysics, Karolinska Institutet, Stockholm, Sweden*
CHRISTOPHER E. MCMURRAN • *Wellcome Trust-MRC Cambridge Stem Cell Institute, University of Cambridge, Cambridge, UK*
MARTIN MESCHKAT • *Max Planck Institute of Experimental Medicine, Department of Neurogenetics, Electron Microscopy Core Unit, Göttingen, Germany*
VERONIQUE E. MIRON • *Centre for Reproductive Health, The Queen's Medical Research Institute, The University of Edinburgh, Edinburgh, UK*
WIEBKE MÖBIUS • *Max Planck Institute of Experimental Medicine, Department of Neurogenetics, Electron Microscopy Core Unit, Goettingen, Germany; Center Nanoscale Microscopy and Molecular Physiology of the Brain (CNMPB), Göttingen, Germany*
CHRISTINA MÜLLER • *Institute of Physiology, University Medical Center, Johannes Gutenberg University, Mainz, Germany*

CHAITALI MUKHERJEE • *Institute of Neuronal Cell Biology, Technical University Munich, Munich, Germany; German Center for Neurodegenerative Disease (DZNE), Munich, Germany; Munich Cluster for Systems Neurology (SyNergy), Munich, Germany*

KLAUS-ARMIN NAVE • *Department of Neurogenetics, Max Planck Institute of Experimental Medicine, Göttingen, Germany*

RICHARD M. RANSOHOFF • *Third Rock Ventures, Boston, MA, USA; Department of Cell Biology, Harvard Medical School, Boston, MA, USA*

MARIA RUBIO • *Centre for Discovery Brain Sciences, The University of Edinburgh, Edinburgh, UK*

TORBEN RUHWEDEL • *Max Planck Institute of Experimental Medicine, Department of Neurogenetics, Electron Microscopy Core Unit, Göttingen, Germany*

BOGUSLAWA SADOWSKI • *Max Planck Institute of Experimental Medicine, Department of Neurogenetics, Electron Microscopy Core Unit, Göttingen, Germany; Center Nanoscale Microscopy and Molecular Physiology of the Brain (CNMPB), Göttingen, Germany*

CASSANDRA SAMPAIO-BAPTISTA • *NDCN Department, Wellcome Centre for Integrative Neuroimaging, FMRIB Centre, University of Oxford, Oxford, UK*

ISABELLE SCHÄFER • *Institute of Physiology, University Medical Center, Johannes Gutenberg University, Mainz, Germany*

GISELHEID SCHULZ-ËBERLIN • *Max Planck Institute of Experimental Medicine, Goettingen, Germany*

SOWMYA SEKIZAR • *MRC Centre for Regenerative Medicine and MS Society Edinburgh Centre, Edinburgh bioQuarter, The University of Edinburgh, Edinburgh, UK*

DIANE L. SHERMAN • *Centre for Discovery Brain Sciences, The Chancellor's Building, Edinburgh bioQuarter, The University of Edinburgh, Edinburgh, UK*

MIKAEL SIMONS • *Max Planck Institute of Experimental Medicine, Goettingen, Germany; Institute of Neuronal Cell Biology, Technical University Munich, Munich, Germany; German Center for Neurodegenerative Disease (DZNE), Munich, Germany; Munich Cluster for Systems Neurology (SyNergy), Munich, Germany*

MATTHEW SWIRE • *MRC Centre for Regenerative Medicine, The Multiple Sclerosis Research Centre, The University of Edinburgh, Edinburgh, UK*

STEFAN TENZER • *Institute of Immunology, University Medical Center, Johannes Gutenberg University, Mainz, Germany*

MARIA TRAKA • *Department of Anatomy, College of Graduate Studies, Midwestern University, Downers Grove, IL, USA*

MARINA UECKER • *Proteomics Group, Max Planck Institute of Experimental Medicine, Goettingen, Germany*

DAVID VAN BRUGGEN • *Laboratory of Molecular Neurobiology, Department of Medical Biochemistry and Biophysics, Karolinska Institutet, Stockholm, Sweden*

MARIE-THERES WEIL • *Max Planck Institute of Experimental Medicine, Department of Neurogenetics, Electron Microscopy Core Unit, Göttingen, Germany; Center Nanoscale Microscopy and Molecular Physiology of the brain (CNMPB), Göttingen, Germany; AbbVie Deutschland GmbH and Co. KG, Ludwigshafen, Germany*

HAUKE B. WERNER • *Department of Neurogenetics, Max Planck Institute of Experimental Medicine, Goettingen, Germany*

ANNA WILLIAMS • *MRC Centre for Regenerative Medicine and MS Society Edinburgh Centre, Edinburgh bioQuarter, The University of Edinburgh, Edinburgh, UK*

JILL M. WILLIAMSON • *Centre for Discovery Brain Sciences, The University of Edinburgh, Edinburgh, UK*

DAVID J. A. WYLLIE • *Centre for Discovery Brain Sciences, The University of Edinburgh, Edinburgh, UK*
BERNARD ZALC • *Sorbonne Universités UPMC, Univ Paris 06, Inserm, CNRS, APHP, ICM-GH Pitié-Salpêtrière, Paris, France*
CHAO ZHAO • *Wellcome Trust-MRC Cambridge Stem Cell Institute, University of Cambridge, Cambridge, UK*
LISA ZOU • *Department of Neurology and the Center for Translational Neuromedicine, University of Rochester Medical Center, Rochester, NY, USA*

Chapter 1

Single-Cell RNA Sequencing of Oligodendrocyte Lineage Cells from the Mouse Central Nervous System

Sueli Marques, David van Bruggen, and Gonçalo Castelo-Branco

Abstract

Single-cell RNA sequencing has emerged as a powerful technique for the identification of distinct cell states/populations in complex tissues. We have recently used this technology to investigate heterogeneity of cells of the oligodendrocyte lineage in the mouse central nervous system. In this chapter, we describe methods to perform single-cell RNA sequencing on this glial cell lineage, and discuss experimental and computational approaches to explore the potential and to tackle hurdles associated with this technology.

Key words Single-cell, RNA sequencing, Transcriptomics, Oligodendrocyte, Myelin, Central nervous system

1 Introduction

The advent in the last decade of transcriptomic technologies, coupling cDNA synthesis with next-generation sequencing (NGS), has allowed the analysis of millions of RNA transcripts from distinct cell populations and thus to determine key drivers in a plethora of biological processes. The original RNA-sequencing protocols were only compatible with pools of several thousand cells, resulting in a loss of any detectable heterogeneity within the analyzed sample. The first single-cell RNA sequencing (RNA-Seq) method was published in 2009 elucidating the transcriptome of dissociated mouse blastomeres [1]. Since then, microfluidic approaches such as the Fluidigm C1 system have been developed, enabling cDNA synthesis of hundreds of individual cells [2]. More recently, micro-droplet approaches such as Drop-seq have been described, as well as combinatorial barcoding allowing processing of many thousands of cells [3–5]. Recent studies have been focusing on cell lineage heterogeneity and deconstruction of tissues in populations and subpopulations, revealing new biomarkers and leading to profound insights in the cellular hierarchies existing within tissues [6–13]. We recently

David A. Lyons and Linde Kegel (eds.), *Oligodendrocytes: Methods and Protocols*, Methods in Molecular Biology, vol. 1936, https://doi.org/10.1007/978-1-4939-9072-6_1, © Springer Science+Business Media, LLC, part of Springer Nature 2019

investigated heterogeneity within the oligodendrocyte lineage in the mouse central nervous system in a similar manner [14, 15].

Oligodendrocytes (OLs) were first defined as an heterogeneous cell population by Rio Hortega and Penfield in the 1920s [16, 17]. Four types of oligodendrocytes were described, depending on the number of axons myelinated, the diameter of ensheathed axons and their orientations. Subsequent studies further supported and extended this view of morphological heterogeneity of oligodendrocyte populations in the central nervous system [18–23]. However, targeted ablation of oligodendrocyte precursor cell (OPC) populations from specific ventral/dorsal origins in the forebrain does not lead to behavior abnormalities in mice, with OPCs/oligodendrocytes derived from non-ablated regions functionally compensating the lost populations [24]. Furthermore, while myelinating preferentially specific tracks, dorsal and ventral derived oligodendrocytes present similar electrophysiological properties [25]. In order to determine whether there is an intrinsic heterogeneity of cells of the oligodendrocyte lineage, we used single-cell RNA sequencing with Fluidigm C1 technology to produce a comprehensive map of this lineage [6, 14]. Approximately 5000 cells expressing OL lineage markers, from ten distinct regions of the juvenile and adult mouse central nervous system (CNS), clustered in 12 cell populations/states [14]. We found a relatively homogeneous OPC population that followed a narrow path of differentiation through committed precursor cells (COP), newly formed OL (NFOL 1 and 2), myelin-forming OL (MFOL 1 and 2) and diversifying into 6 mature OLs (MOLs) (Fig. 1) [14]. The path of differentiation was ubiquitous throughout the CNS, while the mature OLs diversity was more region-specific, with distinct MOLs being over-represented in specific areas. We also observed a segregation of transcriptional profiles between four MOLs present in juvenile (MOL1-4) compared to the two main MOLs enriched in adult tissues (MOL5/6) [14]. The latter was enriched in genes involved in synapse and neurotransmitter regulation [14], which could indicate functional heterogeneity within the mature OLs. Using also single-cell RNA-Seq, we recently found that progenitors of the oligodendrocyte lineage transcriptionally converge during mouse CNS development, to generate postnatal OPCs with similar gene expression profiles regardless of the region of origin [15]. In this book chapter, we describe a detailed protocol for single-cell RNA-Seq of oligodendrocyte lineage cells in CNS tissues.

2 Materials

2.1 Tissue Collection

1. CNS tissue from wild-type mice or mice expressing eGFP reporter or another fluorescent marker in cells belonging to the oligodendrocyte lineage, such as Pdgfra-histone H2B-GFP [26] or Pdgfra-Cre-RCE [27] mice.

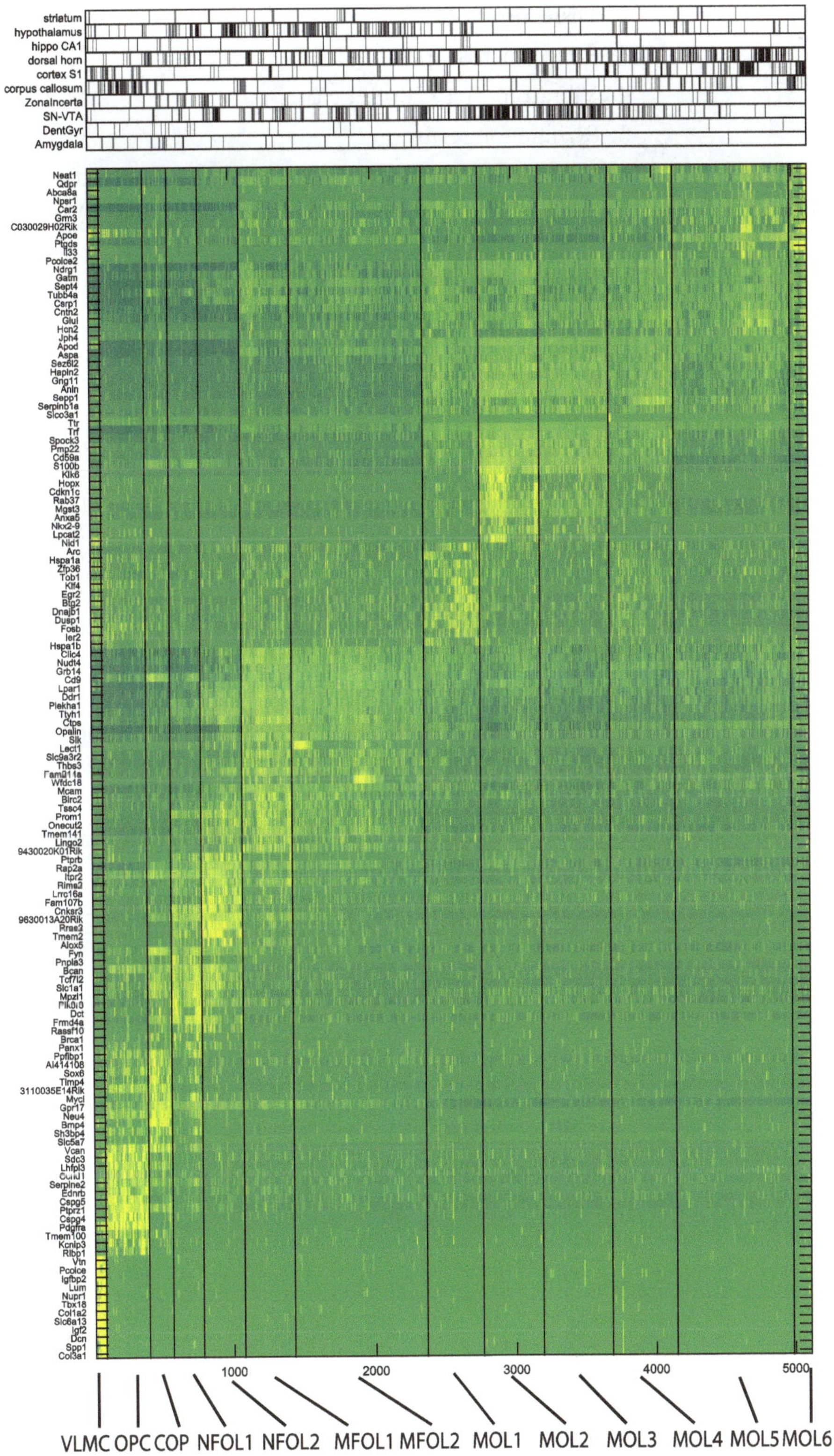

Fig. 1 Heatmap resulting from backspin algorithm of the 5072 cells from the OL lineage and showing markers specific for the subclusters

2. Dissection tools.
3. Stereomicroscopes.
4. Vibratome VT1200S, Leica.
5. Syringe for perfusion.
6. Cutting solution—87 mM NaCl, 2.5 mM KCl, 1.25 mM NaH_2PO_4, 26 mM $NaHCO_3$, 75 mM Saccharose, 20 mM Glucose, 1 mM $CaCl_2$, and 2 mM $MgSO_4$.
7. Oxygenation tank with 95% O_2/CO_2.

2.2 Tissue Dissociation

1. Pasteur glass pipettes.
2. HBSS with calcium and magnesium (no phenol red)—Thermo Fisher catalogue number: 14025050.
3. Neural tissue dissociation kit (P)—Miltenyi catalogue number: 130-092-628.
4. Adult brain dissociation kit, mouse and rat—Miltenyi catalogue number: 130-107-677.
5. Cell Trics® strainers 30 μm—Sysmex catalogue number: 25004-0042-2316.

2.3 FACS Sorting of OL

1. FACS tubes—Corning, catalogue number #352058.
2. HBSS, no magnesium, no calcium, and no phenol red—Thermo Fisher catalogue number: 14175095.
3. BSA 5%.
4. Nonsticky RNAse-free tubes—Thermo Fisher catalogue number: AM12450.
5. FACS machine (BD FACSAria III Cell Sorter B5/R3/V3 system).

2.4 Cell Capture and Imaging

1. C1™ Single cell Reagent for mRNA seq—Fluidigm, catalogue number #100-6201.
2. C1 Single cell mRNA seq IFC microfluidic chip-sizes 5–10 μm and 10–17 μm (catalogue numbers #100-5759 and #100-5760, respectively).
3. Fluidigm C1 instrument.
4. Nikon TE2000E microscope.
5. μManager software.

2.5 Lysis, Reverse Transcription, and PCR

1. Oligonucleotide sequences:

C1-P1-RNA-TSO	Bio-AAUGAUACGGCG ACCACCGAUNNNN NGGG	All-RNA oligo
C1-P1-T31	Bio-AATGATACGGCGACC ACCGATCGTTTTTTT TTTTTTTTTTTTTTT TTTTTTTTTT	
C1-P1-PCR-2	Bio-GAATGATACG GCGACCACCGAT	
C1-TN5-U	PHO-CTGTCTCTTATA CACATCTGACGC	Mix with C1-TN5-x to make adapter

2. Lysis buffer: 0.15% triton X-100, 1 U/μL Takara Rnase inhibitor, 4 μM reverse transcription primer C1-P1-T31, 5% C1 Loading Reagent, and 1:50,000 Life Technologies ERCC Spike-In Mix 1.
3. Reverse transcription mix: 1× SuperScript II First-Strand Buffer supplemented with 3 mM $MgCl_2$, 1.5 mM dNTP, 4 mM DTT, 3.3% C1 Loading Reagent, 1.8 μM template-switching oligo C1-P1-RNA-TSO, 1.5 U/μL TaKaRa RNase inhibitor, and 18 U/μL Life Technologies Superscript II reverse transcriptase.
4. PCR mix: 1.1× Clontech Advantage2 PCR buffer, 440 μM dNTP, 530 nM PCR primer C1-P1-PCR-2, 5% C1 Loading Reagent, and 2× Advantage2 Polymerase Mix.
5. C1 Harvesting Reagent.
6. Agilent Bioanalyzer.
7. Agilent High sensitivity DNA kit—catalogue number 5067-4626.

2.6 Bioinformatic Analysis, Quality Control, and Selection

1. MATLAB software—MathWorks.
2. R statistical computing and graphic language—R Foundation.

2.7 Tagmentation and Isolation of 5′ Fragments

1. 96 different 10× transposome stocks (6.25 μM barcoded adaptor C1-TN5-x, 40% glycerol, 6.25 μM recombinant Tn5 transposase, where x denotes a well-specific barcode). Barcode sequences can be found in Islam et al., 2014 [28].
2. Tagmentation buffer: 50 mM TAPS-NaOH, pH 8.5, 25 mM $MgCl_2$, and 50% DMF.
3. Dynabeads MyOne Streptavidin C1 beads.

4. BWT: 10 mM Tris–HCl, pH 7.5, 1 mM EDTA, 2 M NaCl, 0.02% Tween-20.
5. TNT: 20 mM Tris, pH 7.5, 50 mM NaCl, 0.02% Tween.
6. Qiagen Qiaquick PB.
7. Restriction mix: 1× NEB NEBuffer 4, 0.4 U/μL PvuI-HF enzyme.
8. NaOH 100 mM.
9. HCl 100 mM.
10. Neutralization buffer: 200 mM Tris, pH 7.5, 0.05% Tween-20.

2.8 Illumina High-Throughput Sequencing

1. KAPA library quantification kit—KAPA biosystems, catalogue number KK4824.
2. Illumina HiSeq 2000 instrument.

3 Methods

Single-cell RNA-Seq has been shown to be a powerful tool to profile cell-to-cell transcriptomic variability [6–13]. The general workflow for single-cell RNA-Seq is common to many other transcriptomic methods. First, a cell is lysed to obtain RNA, followed by first strand synthesis by a dedicated reverse transcriptase, then a PCR pre-amplification step is performed adding Illumina compatible adapters. An important factor in single-cell RNA-Seq is the manner (and efficiency) of single cell capture. Many existing dissociation methods are tissue specific, and new protocols are constantly being developed (*see* **Note 1**).

3.1 Tissue Collection

Tissues were collected in oxygenated 1× cutting solution.

1. For embryonic stage, excise brain and spinal cord under stereomicroscopes, after carefully removing meninges.
2. For postnatal and adult stages, intracardially perfuse mice with oxygenated 1× cutting solution to clear the tissue. For dissection of specific regions, section brain or spinal cord in 300 μm thick vibratome slices (kept in cold cutting solution with constant oxygenation).

3.2 Tissue Dissociation

Depending on the age of the mice, different dissociation protocols to purify oligodendrocytes are used.

For embryonic stage, cut the tissue in pieces (<1 mm of thickness) and perform manual dissociation using fire-polished Pasteur pipettes with decreasing diameter until the suspension looks homogeneous.

Myelin removal in postnatal and adult stages was required, involving extra steps of purification after papain-driven tissue dissociation, such as FACS sorting with strict gating parameters to exclude myelin debris, or using specific gradients (*see* **Note 2**).

For postnatal day zero (P0) to P30 stages, Neural dissociation kit with Papain from Miltenyi is used in a manual mode, adapting manufacturer's instructions:

1. Add 13.5 μL of 50 mM beta-mercaptoethanol to 10 mL of Buffer X.
2. Prepare enzyme mixes:

   Enzyme mix1—50 μL Enzyme P + 1900 μL Buffer X (complete).
   Enzyme mix2—10 μL Enzyme A + 20 μL Buffer Y.
   Volumes given are for up to 400 mg of starting tissue material. When working with less than 400 mg, use the same volumes as indicated.
3. Fire-polish three glass Pasteur pipettes so that decreasing tip diameters are achieved.
4. Preheat Enzyme mix1 at 37 °C for 10–15 min.
5. Determine the weight of tissue in 1 mL of cold HBSS (w/o) to make sure the 400 mg limit per digestion is not exceeded.
6. Cut brain into small pieces using a scalpel in a 35 mm diameter petri dish.
7. Using a 1 mL pipette tip, add 1 mL of HBSS without ions (w/o) and pipette pieces back into an appropriate-sized tube. Rinse with HBSS (w/o).
8. Centrifuge at 300 × *g* for 2 min at room temperature and aspirate the supernatant carefully.
9. Add Enzyme mix 1.
10. Incubate in closed tubes for 15 min at 37 °C, mixing gently every 5 min.
11. Add the Enzyme mix2. Invert gently to mix.
12. Dissociate tissue mechanically using the wide-tipped, fire-polished Pasteur pipette by pipetting up and down ten times slowly. Avoid forming air bubbles.
13. Incubate at 37 °C for 10 min, mixing gently every 5 min.
14. Dissociate tissue mechanically using the other two fire-polished pipettes in decreasing diameter. Pipette slowly up and down ten times with each pipette, or as long as tissue pieces are still observable. Be careful to avoid the formation of air bubbles.
15. Incubate at 37 °C for 10 min, mixing gently every 5 min.
16. Apply the cell suspension to a CellTric 30 μm, placed on a 15 mL tube.
17. Wash with 10 mL HBSS with ions.

18. Centrifuge cell suspension at 300 × *g* for 10 min at room temperature. Aspirate supernatant completely.
19. Resuspend in 1 mL cold cutting solution with 1% BSA.
20. Apply the cell suspension to another CellTric 30 μm, placed on a FACS tube.

For adult stages, the Adult brain dissociation kit was used in a manual mode, adapting manufacturer's instructions (Miltenyi). After dissociation was finished, a gradient was performed to clean myelin, according to the kit's protocol:

1. Prepare enzyme mixes:

   Enzyme mix1—50 μL Enzyme P + 1900 μL Buffer Z.
   Enzyme mix2—10 μL Enzyme A + 20 μL Buffer Y.
   Volumes given below are for one adult mouse brain (400–500 mg) in 1980 μL enzyme mix. When working with less than 400 mg, use the same volumes as indicated.
2. Fire-polish three glass Pasteur pipettes so that decreasing tip diameters are achieved.
3. Preheat Enzyme mix1 at 37 °C for 10–15 min.
4. Determine the weight of tissue in 1 mL of cold HBSS (w/o) to make sure the 400 mg limit per digestion is not exceeded.
5. Cut brain into small pieces using a scalpel in a 35 mm diameter petri dish.
6. Using a 1 mL pipette tip, add 1 mL of HBSS without ions (w/o) and pipette pieces back into an appropriate-sized tube. Rinse with HBSS (w/o).
7. Centrifuge at 300 × *g* for 2 min at room temperature and aspirate the supernatant carefully.
8. Add Enzyme mix 1.
9. Incubate in closed tubes for 15 min at 37 °C, mixing gently every 5 min.
10. Add the Enzyme mix2. Invert gently to mix.
11. Dissociate tissue mechanically using the wide-tipped, fire-polished Pasteur pipette by pipetting up and down ten times slowly. Avoid forming air bubbles.
12. Incubate at 37 °C for 10 min, mixing gently every 5 min.
13. Dissociate tissue mechanically using the other two fire-polished pipettes in decreasing diameter. Pipette slowly up and down ten times with each pipette, or as long as tissue pieces are still observable. Be careful to avoid the formation of air bubbles.
14. Incubate at 37 °C for 10 min, mixing gently every 5 min.

15. Apply the cell suspension to a MACS SmartStrainer 70 µm, placed on a 15 mL tube.
16. Apply 10 mL of cold PBS onto the strainer.
17. Centrifuge cell suspension at 300 × *g* for 10 min at 4 °C. Aspirate supernatant completely.
18. For 1 brain (400–500 mg), resuspend cell suspension carefully with 3100 µL cold PBS.
19. Transfer to a 15 mL tube.
20. Add 900 µL Debris Removal Solution. Mix well.
21. Overlay very gently with 4 mL of cold PBS (pipette very slowly to ensure that the PBS phase overlays the cell suspension and phases are not mixed).
22. Centrifuge at 4 °C and 3000 × *g* for 10 min with full acceleration and full brake.
23. Three phases are formed. Aspirate the two top phases completely and discard them.
24. Fill up with cold PBS to a final volume of 15 mL.
25. Gently invert the tube three times. Do not vortex!
26. Centrifuge at 4 °C and 1000 × *g* for 10 min with full acceleration and full brake. Aspirate supernatant completely.
27. Optional—remove red blood cells as kit's instructions.
28. Resuspend in 1 mL cold cutting solution with 1% BSA.
29. Apply the cell suspension to another CellTric 30 µm, placed on a FACS tube.

Independently of the dissociation protocol, the cells were at the end resuspended in oxygenated cutting solution with 1% BSA and filtered through a 30 µm filter into a FACS tube. Dissection of individual areas was performed according to the Allen Brain Mouse Reference Brain and Spinal Cord atlas—http://mouse.brain-map.org/static/atlas.

3.3 FACS Sorting

In the juvenile stage, when using mice lines where oligodendrocyte lineage cells express a reporter gene, FACS sorting the cells, using stringent gating to avoid collection of myelin debris, produces a fairly clean cell suspension.

1. FACS sort GFP+ cells using a BD FACSAria III Cell Sorter B5/R3/V3 system (*see* Fig. 2 for gate selection).
2. Collect cells in cutting solution with 1% BSA, on ice and quickly prepare for capture on the C1 Fluidigm system.

3.4 Cell Capture and Imaging

1. Use cell suspensions in a concentration of 600–1000 cells/µL.
2. Add C1 Suspension Reagent (all "C1" reagents were from Fluidigm, Inc.) in a ratio of 4 µL to every 7 µL cell suspension.

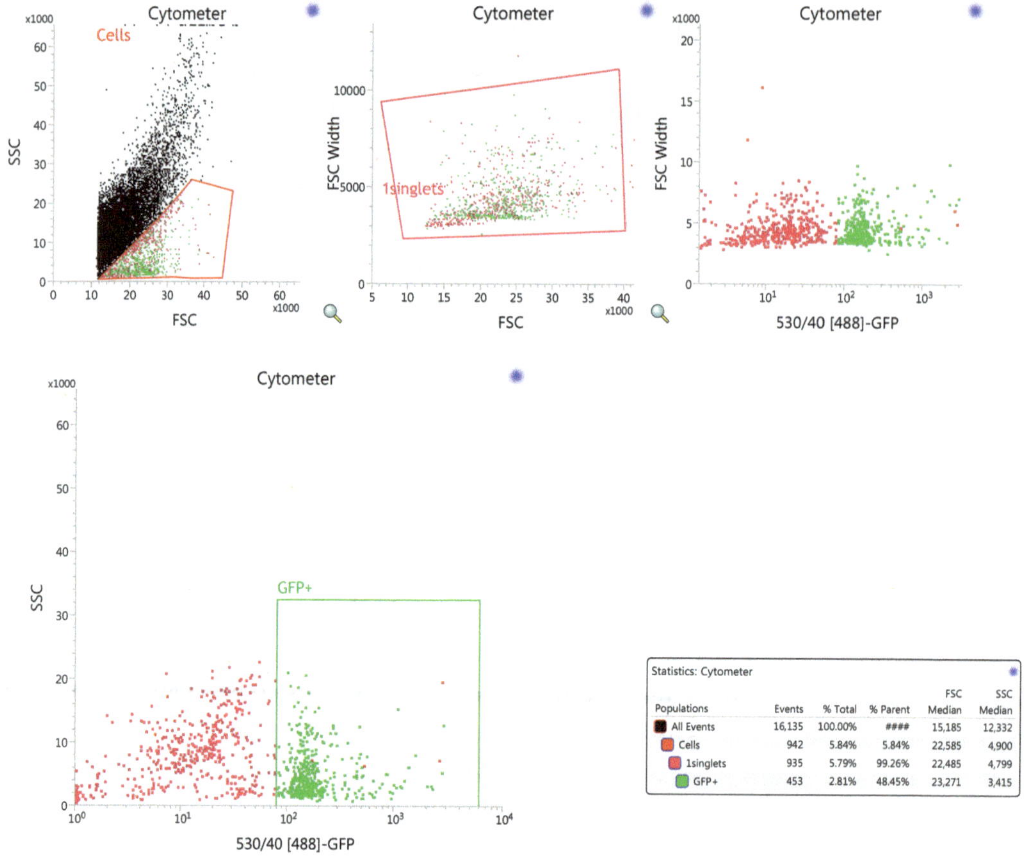

Fig. 2 FACS gating for exclusion of myelin debris and selection of GFP+ cells from Pdgfra-Cre RCE mice

3. Load 11 microliter of the cell suspension mix on a C1 Single-Cell AutoPrep IFC microfluidic chip designed for 10- to 17-μm cells, and process the chip on a Fluidigm C1 instrument using the "mRNA Seq: Cell Load (1772×/1773×)" script for 30 min at 4 °C.
4. Transfer the plate to an automated microscope (Nikon TE2000E), and acquire a bright-field image (20× magnification) for each capture site (Fig. 3) using μManager (http://micro-manager.org/). The expected time for sorting is approximately 15 min.
5. Manual inspection of each capture site is used to identify empty chambers, unhealthy cells, or chambers with duplets, triplets, or more cells (Fig. 3). Only chambers containing a single healthy-looking cell are further processed. In preliminary experiments, we had confirmed that cells judged unhealthy rarely yielded useable cDNA (Fig. 3).

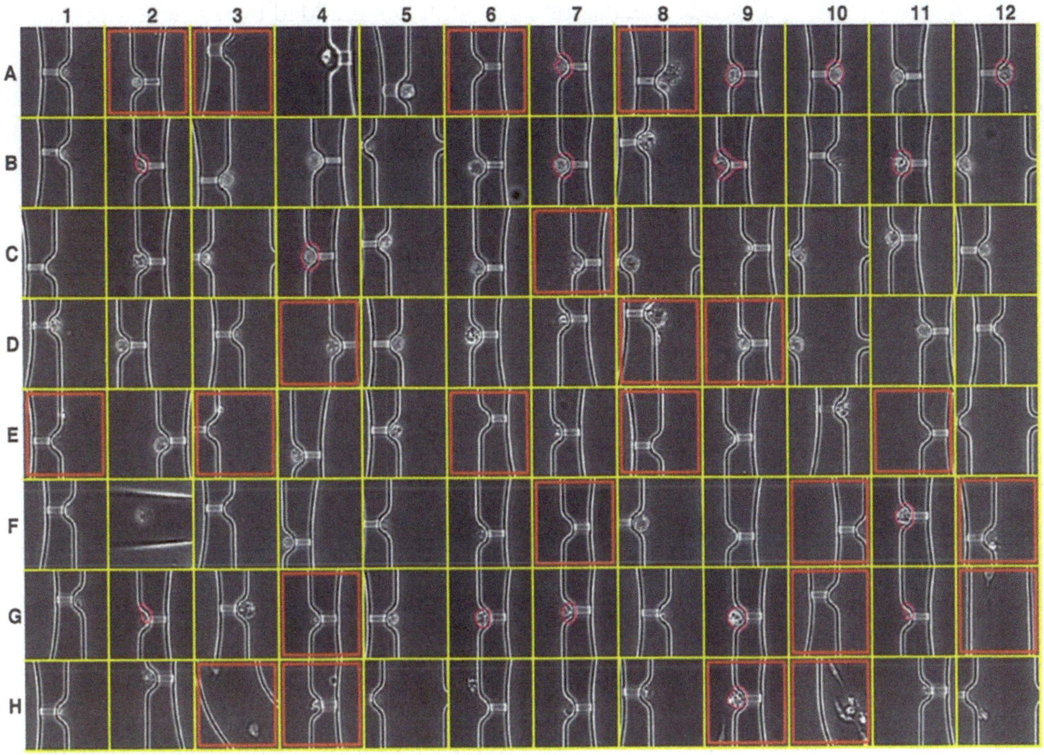

Fig. 3 Example of image analysis algorithm in MATLAB with cells excluded in red

6. After each capture experiment, the individual cell images are used to determine cell diameter by an automated image analysis algorithm in MATLAB.

3.5 Library Preparation

Current single-cell RNA-sequencing protocols vary in the way adapter and primers are designed, dictating the information that will be measured. In our study, we used STRT-sequencing, developed by the lab of Sten Linnarsson [28–30]. The initial first strand synthesis is performed with the use of a barcoded oligodT primer, with a template switching approach. A specific reverse transcriptase is used, which, when finalizing the first strand synthesis, will add a sequence of three cytosine nucleotides. A specially designed primer containing a unique molecular identifier (UMI) sequence is then added to the mix, to address the intrinsically noisy data of single-cell sequencing caused by PCR amplification bias. UMIs can then be used to identify the individual molecules that were present before the amplification steps. The UMI barcoded primer will anneal to three cytosines at the 3′ end of the first strand cDNA molecule, with three guanidines. Further elongation will incorporate the complementary sequence of the template switching oligonucleotide into the existing first strand of cDNA. Subsequently, a barcode specific for each single cell is added by tagmentation, using Tn5 transposase, after which library sequencing can be performed.

1. Return the plate to the lab and add 20 μL lysis buffer, reverse transcription mix, and PCR mix to the designated wells according to the manufacturer's instructions.
2. Place the plate in the Fluidigm C1 instrument and execute the "*mRNA Seq: RT + Amp (1772×/1773×)*" script. This step takes ~8.5 h and includes lysis, reverse transcription, and 21 cycle PCR.
3. When the run has finished, harvest the amplified cDNA in a total of 13 μL C1 Harvesting Reagent and assess the quality of cDNA on an Agilent BioAnalyzer. The typical yield is 1 ng/μL.
4. Prepare 96 different 10× transposome stocks, each with a different barcode sequence.
5. Mix six microliters of harvested cDNA with 5 μL tagmentation buffer, 11.5 μL nuclease-free water, and 2.5 μL 10× transposome stock. Incubate the mix for 5 min at 55 °C, then cool on ice.
6. Wash 100 μL Dynabeads MyOne Streptavidin C1 beads (Invitrogen) in 2× BWT, then resuspend in 2 mL 2× BWT.
7. Add 20 μL of beads to each well and incubate at room temperature for 15 min.
8. Pool all fractions, immobilize the beads, and remove the supernatant (thus removing all internal fragments and retaining only the 5′- and 3′-most fragments).
9. Resuspend the beads in 100 μL TNT, wash in 100 μL Qiagen Qiaquick PB, then wash twice in 100 μL TNT.
10. Resuspend the beads in 100 μL restriction mix, designed to cleave 3′ fragments carrying the PvuI recognition site. Incubate the mix for 1 h at 37 °C, then wash three times in TNT.
11. Finally, to elute DNA, resuspend beads in 30 μL ddH_2O and incubate for 10 min at 70 °C. Immediately bind beads to magnet and collect the supernatant.
12. To remove short fragments, use Ampure beads (Beckman Coulter) at 1.8× volume and elute in 30 μL.

3.6 Illumina High-Throughput Sequencing

1. Quantify the library molar concentration by qPCR using KAPA Library Quantification kit.
2. Estimate library fragment length using Bioanalyzer of a reamplified (12 cycles) library.
3. Perform sequencing on an Illumina HiSeq 2000 instrument using C1-P1-PCR-2 as the read 1 primer, and C1-TN5-U as the index read primer. Multiplex samples, with 96 cells run by

lane. The average read depth per cell is between 200 and 300 thousand reads.

4. Generate reads of 50 bp along with 8 bp index reads corresponding to the cell-specific barcode. Each read is expected to start with a 6 bp unique molecular identifier (UMI), followed by 3–5 guanines, followed by the 5′ end of the transcript.

3.7 Bioinformatic Analysis

In STRT sequencing, UMIs are located on the 3′ end of the cDNA molecule. The first sequence read from the original mRNA molecule will therefore be the 5′ UTR. This sequence provides sufficient information to be able to map the transcript to the genome. In addition, 5′ UTR sequencing allows discrimination of specific transcription start sites. Apart from providing unique molecule count information, UMIs also allow measurement of the amount of PCR duplication or bias. UMIs make possible to quantify the overall dynamic range of the library in terms of the rate of new discovered UMIs per read over time, which should plateau when all the UMIs have been sequenced in the sample. As such, in the downstream analysis, we analyze the number of unique molecules (UMIs) rather than reads. The median amount of UMIs per cell was around 7500 for the whole dataset, mature oligodendrocytes have a median count of 7500, and oligodendrocyte progenitor cells (OPCs) have a median count of 4000 UMIs per cell.

3.7.1 Quality Control

Many factors within the single-cell RNA-seq pipeline can introduce bias. Therefore, a quantification of cell quality and bias is paramount for successful downstream analysis. Several metrics can be used when spike-in data (such as ERCC spike-ins, *see* Subheading 3) is available for the dataset. This includes calculation of the capture efficiency, calculated by the total spike-in counts captured within a single sample compared to the total spike counts present. In STRT sequencing capture efficiencies vary between 10 and 30% up to 50%, dependent on the cell size and viability [28]. Other measures include the total counts, total number of genes expressed per cell, the percentage of mapped counts compared to spike-ins [31], and principal component analysis (PCA). In most single-cell datasets, there is a propensity for the first principal component of PCA to be correlated with number of expressed genes and total counts per cell. These correlations can sometimes be high ($R > 0.9$) and are often unwanted. The backSPIN clustering algorithm (*see* below) is less sensitive to these artifacts; however, normalization schemes might be useful for downstream differential analysis purposes and to quantify technical variation, cell-cycle variation, and biological variation (*see* **Note 3**).

3.7.2 Accounting for Technical Variability (Noise) in Single-Cell Data and Feature Selection

We mapped individual count reads to genes, resulting in a count expression matrix where genes are rows and individual cells are columns. The gene expression matrix is then fit to an estimation governed by $\log_2(\mathrm{CV}) = \log_2(m-a + k)$ by calculating the coefficient of variation (standard deviation divided by the mean) relative to the mean expression level (Fig. 4). Genes above this line have a greater chance of demonstrating biological variability [6, 28], and are therefore used for the first initial clustering with the backSPIN algorithm, described below.

3.7.3 The BackSPINV2 Algorithm

Many clustering methods for single-cell RNA-Seq data are currently available (*see* **Notes 4** and **5**). We used the backSPIN algorithm, which is modified from the sorting algorithm called "sorting points into neighborhoods" or SPIN [32]. The backSPIN algorithm sorts the distance matrix (often Euclidean distance) into a one-dimensional ordering for both genes and cells. Splitting of cells to generate clusters was performed based on the following parameters:

splitlev = 5; Nfeture1 = 500; Nfeture = 50; N_to_backspin = 150; N_to_cells = 200; mean_tresh = 0.1; fdr_th = 0.3; min_gr_cells = 5; min_gr_genes = 3;

BackSPINV2 uses the splitting algorithm as previously described [6]. Feature selection based on the coefficient of variance relative to the mean is performed after each split of the backSPIN algorithm after which subsequent splits are performed. The amount of splits is defined by the "splitlev" argument. Manual inspection of clusters can lead to a merger of subset of clusters, if

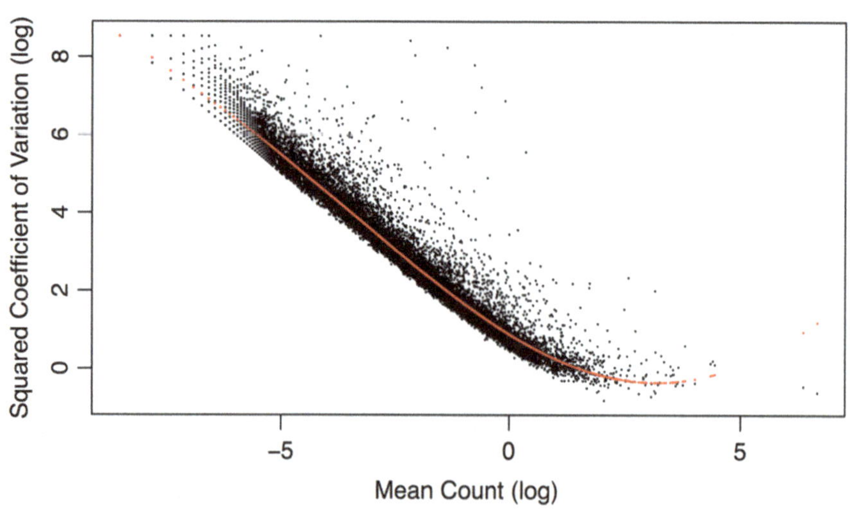

Fig. 4 Finding the relationship between gene expression and technical noise. Low mean counts have higher variation compared to high mean counts, support vector regression fit

inspection of the heatmap does not reveal obvious differences between the clusters [14] (which means the clusters were over split).

After this, an estimation of enriched genes for each cell cluster is performed based on the following criteria:

$$\mathbf{enrich}_{i,j}=\frac{1}{\langle k\in j\rangle}\sum_{k\in j}\mathbf{E}_{i,k}/\frac{1}{N}\sum_{k}\mathbf{E}_{i,k};\mathbf{posfrac}_{i,j}=\frac{1}{\langle k\in j\rangle}\sum_{k\in j}\mathbf{I}(\mathbf{E}_{i,k}>0)$$

This means that for the cluster *j*, and each gene *i* in cell *k*, the enrichment of the gene for the cluster is calculated compared to the whole dataset. Then, the second expression shows the fraction of cells that express the gene within the cluster.

Then, a ranking of genes is established by

$$S_{i,j}=\left(\mathbf{enrich}_{i,j}*\mathbf{postfrac}_{i,j}\right)^{\mathbf{power}}$$

Here genes are ranked based on a power value that sets the weight for the fraction of positive cells in the cluster. We used a power value of power = 0,0.5,1 was used to rank the genes. A cutoff value can then be used to rank the x most enriched genes.

3.7.4 Unsupervised Ordering of Cells in Pseudotime

Reconstruction of developmental time points is crucial when attempting to uncover differentiation dynamics. Single-cell RNA--seq data will provide a snapshot view of the cells contained within the sample. Often, information regarding the developmental trajectory of cells is lost. This is particularly true when analyzing samples collected at different time points during development or data obtained from differentiation studies. Cell ordering algorithms such as the one employed in Monocle [33] aim to reconstruct the lost differentiation trajectories by ordering the cells along an artificial differentiation axis called pseudotime. The principle of Monocle is based on an algorithm established to order microarray data [34]. Monocle v1.2 reduces dimensionality by means of independent component analysis (ICA), next a weighted graph is computed where cells are vertexes and edges represent the distance in ICA space. Then, a minimal spanning tree (MST) is computed on the graph, and cells differing from the MST are reassigned to the MST in a way that minimizes expression differences with neighboring cells to establish a stable ordering accounting for noise in the dataset. For full details *see* [33]. The oligodendrocyte lineage constitutes many intermediate states in the differentiation process as illustrated by single-cell RNA sequencing in our paper [14]. This broad differentiation spectrum allows us to reconstruct the putative differentiation axis with Monocle (Fig. 5).

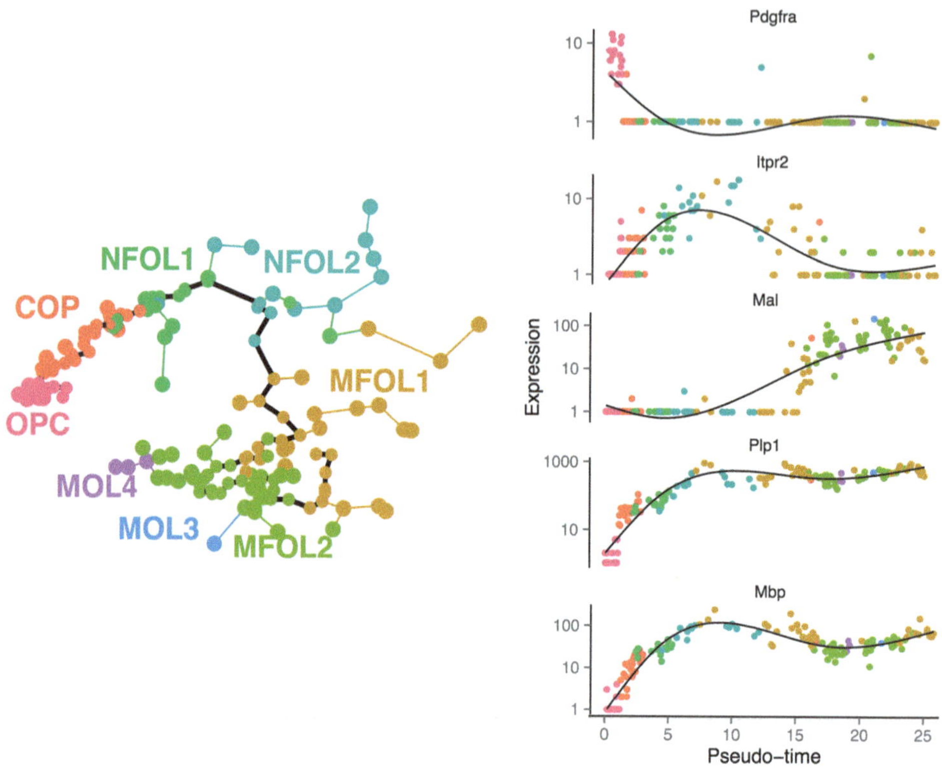

Fig. 5 Left: minimal spanning tree and pseudotime calculation using Monocle v1.2, right: expression as a function of pseudotime for several markers of oligodendrocyte development

4 Notes

1. Traditional low-throughput cell capture techniques compatible with single-cell RNA-seq are micromanipulation and laser-capture microdissection (LCM) [35]. Micromanipulation (using for instance a mouth pipette) allows for visual conformation of the cell morphology and quality, which reduces the likelihood of doublet libraries in downstream procedures. Micromanipulation has been used for single-cell sequencing in the context of low cell count experiments, such as analyzing preimplantation embryo, allowing capture of hundreds of cells when given enough time. However, low-throughput cell capture techniques are insufficient and time consuming when analyzing samples on the tissue level. Therefore, protocols involving enzymatic dissociation have been developed, an essential step in achieving high-throughput cell capture.
2. Oligodendrocyte lineage cells represent a quite large population of cells in the CNS. As such, they are a relatively easy cell group to be collected in an unbiased way (without selection).

However, the presence of myelin from P10 onward turns the analysis by single-cell RNA sequencing of any cell type in the juvenile/adult mouse brain into a challenging task. For capture of cells in C1 microfluidic device, a cell suspension as clean and deprived of debris (cellular or myelin derived) as possible is required. When analyzing cells other than the OL lineage, a thorough extraction of myelin debris can be done using Myelin removal beads (Miltenyi). However, the same approach is problematic when studying the OL lineage, since the likelihood of losing mature OLs during the process is high. Therefore, we found alternatives for this extra step of sample cleaning. One alternative is density gradients. In Zeisel et al. (2015) [6], when analyzing cell types diversity in the mouse juvenile cortex and hippocampus, an albumin-based gradient was used as part of the Papain Dissociation system from Worthington. Despite the successful collection of several cells along the oligodendrocyte lineage with this method, oligodendrocyte precursor cells were lost in the process of centrifugation, probably due to their small size (<7 μm) [36] and the low speed used in the gradient (70 × *g* as opposed to the standard 300 × *g* speed used for OPC primary culture preparation). Therefore, we further optimized the tissue dissociation protocol to avoid losing both the OPCs and the mature OLs (methods). For adult or juvenile CNS tissue derived from non-transgenic mice, an extra gradient step from the Adult brain dissociation kit (Miltenyi) was used. This gradient was able, as opposed to the Worthington's albumin gradient, to clean the myelin debris without losing either the OPCs or the mature stages.

3. Estimating sources of variance and normalizing data
   Estimating sources of variance in the dataset can be difficult, even when a principle component correlates with technical variance or other types of variance, such as the cell cycle. It is often not enough to simply remove the associated PCA component from the dataset. Methods have been developed to easily infer certain types of variances and even correct for it [9]. Moreover, normalization has been common practice when analyzing bulk RNA sequencing. However, normalization of single-cell data requires different approaches due to the intrinsically zero-inflated nature of the gene expression matrix, due to both technical and biological factors. A normalization method has been developed that utilizes a deconvolution approach, meaning normalization of pools of cells by summing the expression within the pools leading to fewer zeros. Then pools of cells are "deconvolved" and a size factor estimate is established for each individual cell [37].

4. Classical clustering approaches
   Hierarchical clustering
   A common divisive method of clustering, wherein groups of points are split according to certain criteria of which many variants exist. The most common criteria are single linkage, complete linkage, average linkage, or Wards. In general any clustering algorithm able to perform divisive clustering ($k = 2$) can be used. Two general approaches exist, bottom up (agglomerative clustering) and top down (divisive) [38]. The most common metric is Euclidean distance, and the algorithm produces a dendrogram as a result, visually depicting the hierarchy in the data.
   K-means clustering
   This very simple machine learning clustering approach requires a predetermined amount of clusters (k) as input. The algorithm then randomly assigns points to clusters and iteratively assigns each point to the cluster with the nearest mean, until all points are assigned [39].
   Partitioning Around Medoids (PAM) clustering
   Very similar to K-means clustering, PAM clustering assigns a point to a nearest median point representative of the cluster (medoid) [38].
   Louvain modularity
   Louvain modularity optimization is a frequently used method for clustering single cells by embedding them in a network. This is the standard clustering method for the Seurat 2.0 [40].
5. Alternative clustering approaches for single-cell RNA-Seq
   Affinity propagation
   This clustering approach is frequently used for single-cell data. The clustering algorithm does not require a predetermined number of clusters (k), instead, every point in the data set is a potential center of a cluster. The algorithm then attempts to identify likely centers (exemplars) by means of viewing each point as a node in a network, by which it then iteratively allows nodes to communicate through the edges in the network. The algorithm takes similarity measures as input, often negative squared error (Euclidean distance), and the resulting starting similarity $s(i, j)$ of point i and point j will result in a preference for points that are similar to each other. The algorithm then iteratively allows the nodes in the network to communicate through a local minima search using an energy function. The strength of the communication determines the current affinity a node has for choosing a different data point as exemplar, thus propagating affinity. For further details please *see* [41].

SNN-Cliq

This clustering algorithm attempts to find shared nearest neighbors using similarity, for example Euclidean distance, to generate a graph in which neighborhoods or highly interconnected subgraphs can be identified called cliques [42]. In short, a subgraph is defined by choosing a node in the graph and calculating all the nodes connected to it, then nodes are deleted from the graph based on the degree (amount of connections) within the subgraph, the deletion is governed by a cutoff value based on the size of the cluster and the degree within it. This is iteratively continued until the subgraph is below the cutoff value or contains three nodes, and regarded as a quasi-clique. Quasi-cliques are then merged and nodes are assigned generating clusters.

Pathway and gene set overdispersion analysis (PAGODA)

This clustering method is recently developed using an approach based on functional criteria as defined by gene ontology terms and related pathways. The intrinsically noisy data obtained from single-cell sequencing is reduced by evaluating overdispersion of groups of genes as defined using gene ontology or gene sets defined through other sources. First, the effective sequencing depth, drop-out rate, and amplification noise are estimated using a mixture-model, then variable genes are selected using an adjusted variance index which also dictates the contribution of each gene in downstream analysis. Then the first principal component is calculated for all gene sets that are provided to the algorithm, and overdispersion is defined as the amount of variance explained by the gene set above expectation (compared to other genesets). For more details please *see* [43].

Acknowledgments

We would like to thank Elisa Floriddia for proofreading and Amit Zeisel for comments. Work in G.C.-B.'s research group was supported by Swedish Research Council, European Union (FP7/Marie Curie Integration Grant EPIOPC, Horizon 2020 European Research Council Consolidator Grant EPIScOPE), European Committee for Treatment and Research in Multiple Sclerosis, Swedish Brain Foundation, Swedish Cancer Society, Ming Wai Lau Centre for Reparative Medicine, Petrus och Augusta Hedlunds Foundation and Karolinska Institutet.

References

1. Tang FC et al (2009) mRNA-Seq whole-transcriptome analysis of a single cell. Nat Methods 6:377–382. https://doi.org/10.1038/NMETH.1315
2. Pollen AA et al (2014) Low-coverage single-cell mRNA sequencing reveals cellular heterogeneity and activated signaling pathways in developing cerebral cortex. Nat Biotechnol 32:1053–1058. https://doi.org/10.1038/nbt.2967
3. Macosko EZ et al (2015) Highly parallel genome-wide expression profiling of individual cells using nanoliter droplets. Cell 161:1202–1214. https://doi.org/10.1016/j.cell.2015.05.002
4. Klein AM et al (2015) Droplet barcoding for single-cell transcriptomics applied to embryonic stem cells. Cell 161:1187–1201. https://doi.org/10.1016/j.cell.2015.04.044
5. Rosenberg AB et al (2018) Single-cell profiling of the developing mouse brain and spinal cord with split-pool barcoding. Science 360:176–182. https://doi.org/10.1126/science.aam8999
6. Zeisel A et al (2015) Brain structure. Cell types in the mouse cortex and hippocampus revealed by single-cell RNA-seq. Science 347:1138–1142. https://doi.org/10.1126/science.aaa1934
7. Patel AP et al (2014) Single-cell RNA-seq highlights intratumoral heterogeneity in primary glioblastoma. Science 344:1396–1401. https://doi.org/10.1126/science.1254257
8. Jaitin DA et al (2014) Massively parallel single-cell RNA-Seq for marker-free decomposition of tissues into cell types. Science 343:776–779. https://doi.org/10.1126/science.1247651
9. Buettner F et al (2015) Computational analysis of cell-to-cell heterogeneity in single-cell RNA-sequencing data reveals hidden subpopulations of cells. Nat Biotechnol 33:155–160. https://doi.org/10.1038/nbt.3102
10. Yan LY et al (2013) Single-cell RNA-Seq profiling of human preimplantation embryos and embryonic stem cells. Nat Struct Mol Biol 20:1131–1139. https://doi.org/10.1038/nsmb.2660
11. La Manno G et al (2016) Molecular diversity of midbrain development in mouse, human, and stem cells. Cell 167:566–580.e19. https://doi.org/10.1016/j.cell.2016.09.027
12. Telley L et al (2016) Sequential transcriptional waves direct the differentiation of newborn neurons in the mouse neocortex. Science 351:1443–1446. https://doi.org/10.1126/science.aad8361
13. Poulin JF, Tasic B, Hjerling-Leffler J, Trimarchi JM, Awatramani R (2016) Disentangling neural cell diversity using single-cell transcriptomics. Nat Neurosci 19:1131–1141. https://doi.org/10.1038/nn.4366
14. Marques S et al (2016) Oligodendrocyte heterogeneity in the mouse juvenile and adult central nervous system. Science 352:1326–1329. https://doi.org/10.1126/science.aaf6463
15. Marques S, Vanichkina D, van Bruggen D, Floriddia EM, Munguba H, Väremo L, Giacomello S, Falcão AM, Meijer M, Samudyata S, Codeluppi S, Björklund AK, Linnarsson S, Hjerling-Leffler J, Taft R.J, Castelo-Branco G. (2018) Transcriptional convergence of oligodendrocyte lineage progenitors during development. Dev Cell 46:504–517.e7
16. Lassmann H (2012) The birth of oligodendrocytes in the anatomical and neuropathological literature: the seminal contribution of Pio del Rio-Hortega. 1921. Clin Neuropathol 31:435–436. https://doi.org/10.5414/NP301002
17. Gill AS, Binder DK (2007) Wilder Penfield, Pio del Rio-Hortega, and the discovery of oligodendroglia. Neurosurgery 60:940–948. https://doi.org/10.1227/01.NEU.0000255448.97730.34 discussion 940–948
18. Vinet J et al (2010) Subclasses of oligodendrocytes populate the mouse hippocampus. Eur J Neurosci 31:425–438. https://doi.org/10.1111/j.1460-9568.2010.07082.x
19. Murtie JC, Macklin WB, Corfas G (2007) Morphometric analysis of oligodendrocytes in the adult mouse frontal cortex. J Neurosci Res 85:2080–2086. https://doi.org/10.1002/jnr.21339
20. Bjartmar C, Hildebrand C, Loinder K (1994) Morphological heterogeneity of rat oligodendrocytes: electron microscopic studies on serial sections. Glia 11:235–244. https://doi.org/10.1002/glia.440110304
21. Bakiri Y, Karadottir R, Cossell L, Attwell D (2011) Morphological and electrical properties of oligodendrocytes in the white matter of the corpus callosum and cerebellum. J Physiol 589:559–573. https://doi.org/10.1113/jphysiol.2010.201376
22. Anderson ES, Bjartmar C, Westermark G, Hildebrand C (1999) Molecular heterogeneity of oligodendrocytes in chicken white matter. Glia 27:15–21

23. Anderson ES, Bjartmar C, Hildebrand C (2000) Myelination of prospective large fibres in chicken ventral funiculus. J Neurocytol 29:755–764
24. Kessaris N et al (2006) Competing waves of oligodendrocytes in the forebrain and postnatal elimination of an embryonic lineage. Nat Neurosci 9:173–179. https://doi.org/10.1038/nn1620
25. Tripathi RB et al (2011) Dorsally and ventrally derived oligodendrocytes have similar electrical properties but myelinate preferred tracts. J Neurosci 31:6809–6819. https://doi.org/10.1523/JNEUROSCI.6474-10.2011
26. Klinghoffer RA, Hamilton TG, Hoch R, Soriano P (2002) An allelic series at the PDGF alpha R locus indicates unequal contributions of distinct signaling pathways during development. Dev Cell 2:103–113. https://doi.org/10.1016/S1534-5807(01)00103-4
27. Roesch K et al (2008) The transcriptome of retinal miller glial cells. J Comp Neurol 509:225–238. https://doi.org/10.1002/cne.21730
28. Islam S et al (2014) Quantitative single-cell RNA-seq with unique molecular identifiers. Nat Methods 11:163–166. https://doi.org/10.1038/nmeth.2772
29. Islam S et al (2011) Characterization of the single-cell transcriptional landscape by highly multiplex RNA-seq. Genome Res 21:1160–1167. https://doi.org/10.1101/gr.110882.110
30. Islam S et al (2012) Highly multiplexed and strand-specific single-cell RNA 5′ end sequencing. Nat Protoc 7:813–828. https://doi.org/10.1038/nprot.2012.022
31. Stegle O, Teichmann SA, Marioni JC (2015) Computational and analytical challenges in single-cell transcriptomics. Nat Rev Genet 16:133–145. https://doi.org/10.1038/nrg3833
32. Tsafrir D et al (2005) Sorting points into neighborhoods (SPIN): data analysis and visualization by ordering distance matrices. Bioinformatics 21:2301–2308. https://doi.org/10.1093/bioinformatics/bti329
33. Trapnell C et al (2014) The dynamics and regulators of cell fate decisions are revealed by pseudotemporal ordering of single cells. Nat Biotechnol 32:381–U251. https://doi.org/10.1038/nbt.2859
34. Magwene PM, Lizardi P, Kim J (2003) Reconstructing the temporal ordering of biological samples using microarray data. Bioinformatics 19:842–850. https://doi.org/10.1093/bioinformatics/btg081
35. Nichterwitz S et al (2016) Laser capture microscopy coupled with Smart-seq2 for precise spatial transcriptomic profiling. Nat Commun 7:12139. https://doi.org/10.1038/ncomms12139
36. Didar TF, Li K, Veres T, Tabrizian M (2013) Separation of rare oligodendrocyte progenitor cells from brain using a high-throughput multilayer thermoplastic-based microfluidic device. Biomaterials 34:5588–5593. https://doi.org/10.1016/j.biomaterials.2013.04.014
37. Lun AT, Bach K, Marioni JC (2016) Pooling across cells to normalize single-cell RNA sequencing data with many zero counts. Genome Biol 17:75. https://doi.org/10.1186/s13059-016-0947-7
38. Rousseeuw PJ, Kaufman L (1990) Finding groups in data: an introduction to cluster analysis. John Wiley & Sons, Hoboken, NJ
39. Hartigan JA, Wong MA (1979) Algorithm AS 136: A K-means clustering algorithm. J Roy Statist Soc Series C (Appl Statist) 28:100–108. https://doi.org/10.2307/2346830
40. Butler A, Hoffman P, Smibert P, Papalexi E, Satija R (2018) Integrating single-cell transcriptomic data across different conditions, technologies, and species. Nat Biotechnol 36:411–420. https://doi.org/10.1038/nbt.4096
41. Frey BJ, Dueck D (2007) Clustering by passing messages between data points. Science 315:972–976. https://doi.org/10.1126/science.1136800
42. Xu C, Su ZC (2015) Identification of cell types from single-cell transcriptomes using a novel clustering method. Bioinformatics 31:1974–1980. https://doi.org/10.1093/bioinformatics/btv088
43. Fan J et al (2016) Characterizing transcriptional heterogeneity through pathway and gene set overdispersion analysis. Nat Methods 13:241–244. https://doi.org/10.1038/nmeth.3734

Chapter 2

Isolation and Preparation of Cells from Focal Remyelinating Central Nervous System Lesions for RNA Sequencing

Claire L. Davies, Stefka Gyoneva, Anne Cotleur, Richard M. Ransohoff, and Veronique E. Miron

Abstract

Remyelination is the regenerative process whereby myelin sheaths are restored around axons following nervous system injury, allowing reinstatement of electrical impulse conduction, trophic/metabolic support, and axon health. Failure of remyelination in progressive multiple sclerosis is considered to contribute to axon loss, a correlate of clinical decline. Lack of approved pro-regenerative therapies for MS highlights the need to understand the cellular and molecular mechanisms underpinning successful remyelination. One approach is to conduct nonbiased gene expression analyses of cell types which regulate remyelination, such as microglia and monocyte-derived macrophages. Recent technological advances address the challenges of RNA sequencing of small tissue samples, thus allowing relatively small numbers of cells to be isolated from discrete lesions for analysis. Here, we present methods for FACS-based isolation of cells from focal remyelinating lesions of the adult mouse brain and subsequent RNA extraction for sequencing, using isolation of microglia/macrophages as an example.

Key words Microglia, Monocyte-derived macrophages, RNAseq, Demyelination, Remyelination, Stereotaxic, FACS, Cell isolation

1 Introduction

Myelin ensheaths axons in the nervous system to provide insulation for electrical impulse conduction and trophic/metabolic support. The importance of myelin is evident in the sensory, motor, and cognitive impairments observed in disorders in which myelin does not form properly or is damaged. An example of the latter is the autoimmune disease multiple sclerosis (MS), where in progressive phases the regeneration of myelin (termed remyelination) is often limited or fails altogether. Given the lack of approved therapies aimed at promoting remyelination, there is a pressing need to identify the cellular and molecular components of efficient regeneration.

David A. Lyons and Linde Kegel (eds.), *Oligodendrocytes: Methods and Protocols*, Methods in Molecular Biology, vol. 1936, https://doi.org/10.1007/978-1-4939-9072-6_2,

One approach to address this involves conducting gene expression analyses. Indeed, microarray analysis of MS cortical lesions, which are considered to remyelinate efficiently, identified prominent tissue repair/regeneration signatures [1]. Several experimental studies have also investigated gene expression signatures during remyelination. For instance, microarray analysis of whole remyelinating lesions identified new molecular players involved, such as retinoid-X-receptors [2] and the MHC-II/tumor necrosis alpha axis [3]. Additional pathways with potential roles in remyelination were also identified via more focused arrays, for miRNAs [4] or chemokines [5]. Moreover, microarrays have linked remyelination to inflammation, first, by identifying endothelin-2 as an inflammation-derived pro-remyelination factor [6] and second, by revealing regenerative functions of microglia in phagocytosis and oligodendrocyte progenitor cell recruitment and activation [7].

Although these studies revealed novel components of the remyelination process, microarrays detect expression of only a limited number of genes of known sequence and of relatively high expression, and can be subject to cross hybridization artifacts. In addition, gene expression is unlikely to be uniform in all cells within a whole lesion and signals from specific cell populations may often be masked. Thus, the ability to analyze gene expression in an unbiased manner from individual cell populations of a discrete brain region can contribute immensely to understanding the role of cell populations in remyelination. The advent of RNA sequencing now allows unbiased differential gene expression analyses as well as providing information on alternative splice variants, gene fusion, single nucleotide polymorphisms, post-translation modifications, and noncoding RNAs. Furthermore, RNA sequencing can be performed without a reference genome, allowing de novo sequencing of novel species or gene sequences. This technique has been successfully applied to differentiating gene signatures between microglia and monocyte-derived macrophages (MDMs) in a model of MS pathology, experimental autoimmune encephalomyelitis (EAE), at the onset, peak and recovery from disease [8, 9]. Although sequencing low numbers of cells or degraded samples is often challenging, recent advances in technology enable ultra-low input (10 pg–10 ng RNA, or 1–1000 cells) or single-cell sequencing, addressing some of the practical issues associated with next generation sequencing (NGS) [10]. Comparison of five RNA-seq methods for low-input samples revealed SMART (Takara Bio Inc., USA) and NuGEN (NuGEN Technologies Inc., USA) amplification of cDNA were preferable when samples were limited [10]. SMART RNA amplification has low levels of rRNA reads, low percentage of duplicated reads, high coverage of annotated exons, and good transcript coverage. Evaluation of low-input RNAseq protocols showed that, although the pre-amplification step can introduce some variance, it did not adversely affect the quality of data, and low-input protocols performed almost as well as high-input kits [11].

Here, we present methods to isolate microglia and MDMs from remyelinating lesions and subsequent preparation for RNA sequencing, given their established roles in promoting remyelination, via phagocytosis of myelin debris and secretion of growth factors [7, 12–20]. We describe induction of focal demyelination in the mouse brain, FACS-based isolation of small numbers of cells from the lesioned brain, and RNA extraction for downstream low-input RNAseq analysis. Although the techniques detailed herein focus on microglia and MDMs present in remyelinating lesions, the protocol can easily be adapted to assess gene expression in other cell types including oligodendrocytes present in the lesion or, indeed, in other disease models.

2 Materials

2.1 Equipment

1. Stereotaxic frame with holder for drill and Hamilton syringe.
2. Dental drill and drill bit (0.6 mm).
3. Hot bead sterilizer.
4. Inhalable anesthetic rig and/or injectable anesthetics.
5. Sterile surgical instruments (scissors, forceps, hemostat).
6. Franklin and Paxinos Mouse Brain Atlas (ISBN: 9780123910578).
7. Heat pads set to 37 °C—one for surgery, one for recovery.
8. Microinjector and pump.
9. Perfusion pump.
10. Dissection microscope.
11. Hamilton syringe.
12. 1 mL glass dounce.
13. Coronal brain matrix (1 mm).
14. Centrifuge (chilled, capable of holding falcon tubes and plates).
15. FACS machine—FACS Aria II (BD) or similar.
16. Agilent 2100 Bioanalyzer.
17. −80 °C freezer.
18. Thermal cyclers.
19. qPCR-enabled thermal cycler with SYBR Green capability.

2.2 Laboratory Materials

1. 30G needle for Lysolecithin (LPC) injections.
2. 27G needle for intraperitoneal injection.

3. Sterile surgical sutures.
4. Lachrilube eye ointment.
5. Sterile cotton swabs.
6. RNA-Zap.
7. 15 mL and 50 mL Falcon tubes.
8. 250 μm filters to fit 15 mL Falcon tubes.
9. 10 mL stripettes.
10. Non-treated 96 well plates.
11. FACS tubes with 30 μm filter caps.
12. Live/Dead marker for flow cytometry (Life Technologies).
13. Fc receptor (CD16/32) block (Biolegend).
14. Directly conjugated flow cytometry antibodies (*see* **Note 1**): CD11b clone M1/70, CD45 clone 30-F11, Ly6G clone 1A8, CD3 clone 17A2.
15. Compensation beads for single label controls.
16. 1.5 mL Eppendorf tubes.
17. RNase-free PCR tubes and plates.
18. Plate seals.
19. RNase-free reagent reservoirs (10 mL).
20. Agencourt AMPure XP beads.
21. Magnetic separation stands.

2.3 Kits, Buffers, and Solutions

1. PBS + 10% sodium citrate.
2. 1% LPC (L-α-lysophosphatidylcholine from egg yolk, Sigma).
3. Chlorhexidine antibacterial wound scrub.
4. Analgesia—([0.03 mg/mL] Buprenorphine and [0.05 mg/mL] Rimadyl).
5. 1× HBSS+HEPES.
6. 100% FBS.
7. 10% FBS (diluted with 1× HBSS+HEPES).
8. 100% isotonic Percoll (1 mL 10× HBSS, 9 mL Percoll).
9. 33% isotonic Percoll (10 mL 100% isotonic Percoll, 20 mL 1× HBSS+HEPES).
10. FACS buffer (481 mL HBSS, 5 g BSA, 2 mL 0.5 M EDTA, 12.5 mL 1 M HEPES, 4.5 mL 10% NaN_3).
11. Qiagen QIAshredder.
12. Qiagen AllPrep DNA/RNA/miRNA Universal Kit.
13. Bioanalyzer RNA 6000 Pico kit (Agilent).
14. Quant-iT RiboGreen RNA Assay kit (Life Technologies).

15. SMARTseq v4 Ultra Low Input RNA Kit for Sequencing (Takara).
16. Bioanalyzer High Sensitivity DNA kit (Agilent).
17. Quant-iT PicoGreen DNA Assay kit (Life Technologies).
18. Nextera XT DNA Library Preparation kit (Illumina).
19. Library Quantification Kit for NGS (KAPA Biosystems)—for Illumina platforms.

3 Methods

3.1 Focal Demyelination of Corpus Callosum by Stereotaxic Injection

Please *see* [21] for an overview and principles of stereotaxic surgery, using rats as an example.

The following steps of the protocol are carried out under sterile conditions, and in accordance with regulatory guidelines for your institute and country regarding animal work. Sterile technique must be used at all times, and instruments should be sterilized between animals.

Before starting surgery, determine coordinates of injection location using Franklin and Paxinos Mouse Brain Atlas. For example, injection coordinates for adult mouse corpus callosum are: anterior-posterior (AP) −1.2 mm, medial-lateral (ML) −0.5 mm, dorsal-ventral (DV) −1.4 mm (*see* **Note 2**). Adult mouse skull thickness is 300 μm, and must be accommodated for correct ventral coordinates.

Fill dead space in 30G needle with LPC and ensure no air bubbles are present. Attach needle to Hamilton syringe and draw up LPC to fill syringe. Attach syringe to microinjector and set aside.

LPC-induced focal demyelination has distinct stages of disease: oligodendrocyte progenitor cell (OPC) recruitment occurs at 3 days, OPC differentiation occurs at 10 days, and remyelination is complete at 21 days.

1. Place mouse in anesthetic box and induce anesthesia (e.g., 3% isoflurane in 0.2 mL O_2). Check depth of anesthesia by toe pinch reflex and monitor throughout surgical procedure. Ensure animal is unresponsive before proceeding. Breathing rate should also be monitored.
2. Move anesthetized mouse to stereotaxic frame, position correctly in nose cone, and reduce anesthesia to 2%. Secure mouse's head in frame using ear bars. Gently insert bars forward at 45° angle, then move backward until bar is in ear canal. The bar should be against bone, not soft tissue. The skull should be level and not move when touched gently with forceps.
3. Place a drop of Lachrilube on each eye to prevent drying out during surgery.

4. Administer analgesic mix by intraperitoneal injection (27G needle).
5. Using scissors, trim off fur between the ears. With a sterile swap, dab chlorhexidine onto the scalp to clean shaved area and remove any loose fur.
6. Lift scalp gently with forceps and make small incision with scissors. Lengthen the incision along the midline so that it stretches from level of ears to eyes (~1 cm).
7. Polish the skull gently with a clean swap to remove membrane until the skull is dry.
8. Identify and place drill tip on Bregma. Using the drill on max speed (*see* **Note 3**), drill a small mark on surface of the skull as a reference point (do not drill all the way through the skull). Take note of these coordinates as your reference point.
9. From the reference point, move the drill to injection location using AP and ML arms of the stereotaxic frame. Check by eye that drill is now in the correct position. Return to reference coordinates and repeat if incorrect.
10. Drill slowly through the skull to create a small burr hole for injection.
11. Attach the microinjector with loaded syringe onto the stereotaxic frame and line up the needle tip (with bevel facing away from midline) to the reference location (Bregma). With the needle tip touching the skull surface, determine new coordinates for reference point, including ventral coordinate.
12. Move needle to injection location using the AP and ML coordinates. Lower slowly to ensure that the needle will fit down the burr hole. If the needle does not match correctly with the hole, return to the reference point and redetermine coordinates for the injection location.
13. Lower needle down to DV coordinate, then continue to lower by 0.2 mm and return to DV coordinate to make a small pocket for LPC. Inject 2 μL LPC at a rate of 0.5 μL/min using the microinjector and automated pump. Leave needle in place after injection for a further 4 min to allow LPC to disperse through tissue and prevent reflux.
14. Slowly remove needle from brain and close wound with three sutures using the hemostat and forceps.
15. Remove ear bars and turn off isoflurane. Turn O_2 up to 0.4 mL/min and wait for the mouse to regain consciousness. Once the mouse awakens, place the mouse in a new cage with shredded bedding and mashed food pellets on the cage floor. Place the cage on a heat pad overnight. Monitor animals daily for adverse effects (*see* **Note 4**). Recover animals for time point relevant to experimental question.

3.2 Perfusion, Brain Extraction and Cell Isolation

Use RNA-Zap to clean instruments and surfaces of RNases prior to starting.

1. Anesthetize the mouse and ensure it is unresponsive before proceeding. Make a lateral incision across the abdomen, just below the rib cage. Cut diaphragm along the length of the ribcage and on either side of the ribcage make an incision up to the collarbone to expose the heart. Insert a 27G needle into left ventricle, snip the right atrium and perfuse with 20 mL of ice-cold PBS containing sodium citrate (10%).
2. Extract brain carefully by peeling off the skull and place in a 1 mm coronal brain matrix. Cut a 2 mm section at the lesion site (*see* **Note 5**). Cut brain section down the midline to separate lesioned and non-lesioned hemispheres, and collect into separate tubes containing 5 mL 1× HBSS+HEPES and place on ice.
3. Dissect out lesion site in corpus callosum (*see* **Note 2**) using dissecting microscope and collect white matter in 1 mL HBSS +HEPES.
4. Remaining gray matter can be kept to use as a control for RNA integrity following processing of tissue as few cells and low concentrations of RNA will be recovered from lesion site. Contralateral (non-lesioned) hemisphere white matter can be pooled and used as a "negative control" for sequencing.
5. Transfer white matter pieces to 2 mL glass dounce and manually homogenize (minimum 40 passes of dounce) until cell suspension is homogeneous (*see* **Note 6**).
6. Transfer cell suspension to a 15 mL falcon with a 250 μm filter on top. Rinse dounce with 1 mL HBSS+HEPES and add to falcon. Top up to 15 mL and discard filter.
7. Spin cell suspension at 600 × *g* for 5 min (with brake) at 4 °C.
8. Remove supernatant and resuspend cell pellet in 1 mL 100% FBS. Add 9 mL of 33% Percoll to cell suspension and mix well with stripette. Carefully overlay 1 mL 10% FBS on top of Percoll layer.
9. Spin for 15 min at 800 × *g* at 4 °C (lowest brake). Three layers will be visible after centrifugation with the density gradient: upper layer, myelin, and lower layer with cell pellet. Carefully remove the myelin layer using a 10 mL stripette and discard. Remove remaining supernatant without disturbing the pellet.
10. Resuspend cell pellet in 1 mL FACS buffer, mix well with pipette, and top up to 15 mL.
11. Spin for 10 min at 600 × *g* at 4 °C.
12. Resuspend cell pellet in 200 μL 1× HBSS (gray matter controls can be resuspended in larger volume to use for negative control). Protein-free buffer must be used as live/dead marker can bind to proteins.

3.3 Staining and Fluorescence-Activated Cell Sorting of Cells

1. Transfer 200 μL of cell sample to 96-well plate. Gray matter can be used as negative unstained control.
2. Centrifuge cells at 400 × *g* for 3 min (*see* **Note 7**).
3. Resuspend cell in 100 μL Live/Dead marker and incubate at room temperature for 15 min.
4. Centrifuge cells at 400 × *g* for 3 min.
5. Resuspend cells in 50 μL of Fc receptor block and incubate on ice for 10 min.
6. Without washing, add 50 μL 2× concentrated conjugated antibody cocktail to cells directly and incubate for 30 min on ice.
7. Add single labeled antibody controls to compensation beads as per manufacturer's instructions.
8. Centrifuge at 400 × *g* for 3 min.
9. Resuspend cells in 200 μL FACS buffer and transfer samples to FACS tubes.
10. Sort (*see* **Note 8**) samples using FACS machine with 100 μm nozzle (*see* **Note 9**) into FBS-coated Eppendorf tubes (*see* **Note 10**). Place sorted cells on ice until all samples are collected. *See* Table 1 for expected cell yield.
11. Once all samples are sorted, proceed with RNA extraction immediately.

3.4 RNA Extraction for RNAseq

1. Spin sorted cells at 800 × *g* for 5 min to remove FBS before proceeding with RNA isolation.
2. Homogenize cells using Qiagen QiaShredder kit as per manufacturer's instructions. Extract RNA using Qiagen AllPrep DNA/RNA/miRNA extraction kit as per manufacturer's instructions.
3. Quantify amount of recovered RNA with Quant-IT Ribo-Green RNA Assay kit as per manufacturer's instructions (*see* **Note 11**).

Table 1
Expected cell and RNA yield from focal lesion in corpus callosum

	# Cells	RNA concentration [pg/μL]
Microglia 3 days	900–2600	90–183
Microglia 10 days	800–8200	40–400
MDM 3 days	8–64	49–128
MDM 10 days	5–153	65–110

4. Analyze quantity (Table 1) and quality of RNA using Agilent 2100 Bioanalyzer with RNA 6000 Pico kit as per manufacturer's instructions (*see* **Note 12**).
5. Store extracted RNA at −80 °C until required for RNAseq.

3.5 Library Preparation

The preparation consists of two distinct steps: (1) conversion of polyadenylated RNA to cDNA and cDNA amplification and (2) fragmentation of DNA and addition of barcoding primers for sequencing.

1. Use the SMARTseq v4 Ultra Low Input RNA Kit to reverse-transcribe the RNA and amplify the cDNA (*see* **Note 12**). Follow the manufacturer's instructions through the "Purification of amplified cDNA using the Agencourt AMPure XP Kit."
2. Check the concentration of the amplified cDNA with the Quant-iT PicoGreen DNA Assay kit as per manufacturer's instructions (*see* **Note 13**).
3. Check efficiency and quality of amplification of the cDNA using Agilent 2100 Bioanalyzer with High Sensitivity DNA kit as per manufacturer's instructions (*see* **Note 14**).
4. For sequencing on Illumina platforms: use 250 ng of amplified cDNA as input for the Nextera XT DNA Library Preparation kit to fragment the DNA and add sequencing primers. Follow the manufacturer's protocol through the "PCR Clean-Up" step.
5. Validate the fragmentation using Agilent 2100 Bioanalyzer with High Sensitivity DNA kit as per manufacturer's instructions (*see* **Note 15**).
6. Quantify the concentration of the resulting libraries with the KAPA Library Quantification kit as per manufacturer's instructions.
7. Proceed with library pooling and sequencing.

4 Notes

1. Directly conjugated antibodies used in this experiment were CD11b-PeCy7 (eBioscience) [1/100], CD45-BV605 (Biolegend) [1/200], Ly6G-PerCp/Cy5.5 (Biolegend) [1/200], CD3-APC (Biolegend) [1/200].
2. Injection location and area of focal lesion to dissect (Adapted from Franklin and Paxinos Mouse Brain Atlas) are shown in Fig. 1.
3. Always use the drill on max speed. Do not start drill when it is touching the skull as it will slide on the skull surface. Instead, raise the drill slightly before turning on and then lower. Raise

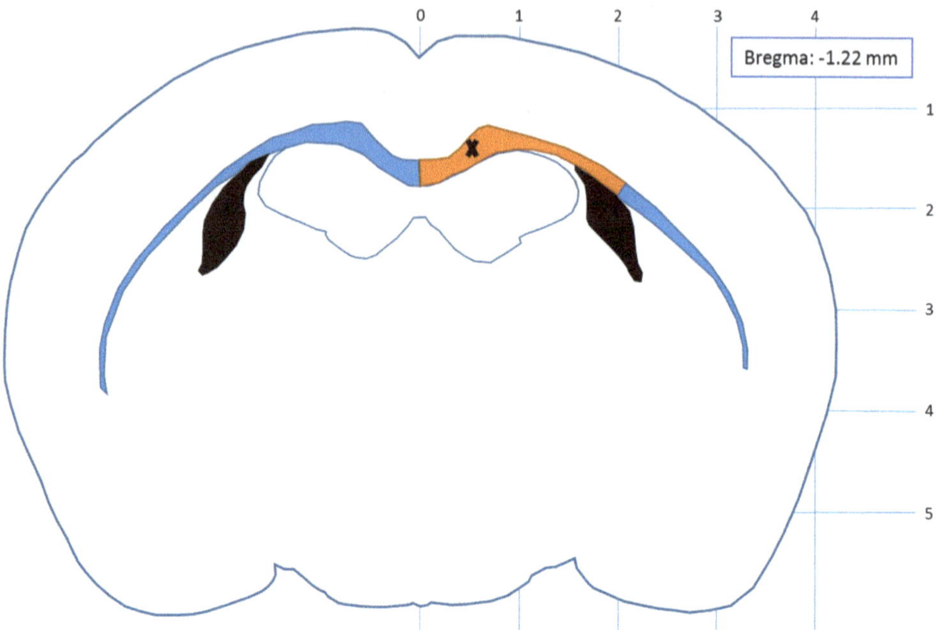

Fig. 1 Schematic of lesion site in corpus callosum and area to be dissected. Injection coordinates (marked by X) for adult mouse corpus callosum are: anterior-posterior (AP) −1.2 mm, medial-lateral (ML) −0.5 mm, dorsal-ventral (DV) −1.4 mm, from Bregma. The area to be dissected from the ipsilateral hemisphere is highlighted in orange. Adapted from Franklin and Paxinos Mouse Brain Atlas

and lower the drill while drilling slowly so as not to go too far and to reduce the drill causing heat damage to the skull/brain. When almost through the skull the hole will look a darker pink than skull surface. Depending on location, a small bleed may occur due to presence of blood vessels in area. Use a sterile swab to apply pressure to stop bleeding.

4. Adverse symptoms often present within 24 h after surgery and clear by 36–48 h, including ocular and nasal discharge, piloerection, social isolation, and head tilting. If symptoms are persistent or severe symptoms develop (circling), consult animal staff and/or veterinarian.
5. The lesion may be difficult to identify, particularly if the animal is recovered for a long time (e.g., 10–21 days). Cut the skull down the midline and peel back the skull on the contralateral hemisphere only. Locate the injection site on the surface of the ipsilateral skull (the small burr hole should be visible by eye). On the contralateral hemisphere mark the surface of the brain with a scalpel at the level of the injection site. Use this mark to determine where to cut the 2 mm section.
6. For cell isolation from a small amount of tissue, manual homogenization is preferred to enzymatic digestion. If tissue is not sufficiently homogenized, cells may not be separated

from the myelin causing low cell yields. Forty passes of the dounce gave consistent cell yields.

7. To remove the supernatant from 96-well plates efficiently, the plate can be turned upside down and flicked.
8. An example gating strategy used for microglia and macrophages isolation is shown in Fig. 2a. As it is difficult to set the gating strategy based on the number of cells obtained from the white matter lesion, stereotaxic injection of lipopolysaccharide (LPS) (1 μL [5 mg/mL], 24 h recovery) can be used as a positive control to set up gates around microglia and MDM. LPS will induce a robust recruitment of MDM and populations can be easily distinguished, as shown in Fig. 2b.
9. Nozzle size used for sorting cells will depend on cells of interest. A larger nozzle will have a lower pressure and will be gentler for cells, although sorting will take longer, potentially decreasing cell viability. Sorting cells from the lesioned corpus callosum takes ~10 min per sample; thus, a larger nozzle was used as the time taken to sort was not greatly increased.
10. FBS can be used as a buffer to limit damage to cells as they are sorted into Eppendorf tubes. To coat tubes, add 1 mL 100% FBS to Eppendorf tube and refrigerate at 4 °C for 1 h. Remove FBS prior to sort, leaving 50 μL as a cushion for cells. Alternatively, cells may be sorted directly into 250 μL lysis buffer containing 1% 2-mercaptoethanol and may increase RNA yield. Cells sorted directly into lysis buffer may be stored at −80 °C for a few days before RNA extraction.
11. Common lab methods for quantifying RNA concentration such as Nanodrop are not sensitive enough to accurately quantify RNA extracted from small tissue samples or low cell numbers. Using a fluorescent-based assay such as the Quant-iT RiboGreen assay with sensitivity as low as 1 ng/mL is preferable. It is possible to scale down the assay to a 96-well plate format. Simply divide all volumes by 10: final volume of each reaction will be 200 μL rather than 2 mL.
12. Extracting cells from a small region of the brain will result in a small population of cells and thus low concentrations of RNA. Often the amount of RNA obtained is insufficient to determine accurate RIN scores as the concentration is outwith the range required for the calculation. To overcome this issue, the remaining gray matter can be collected and processed alongside the white matter samples. If the RIN score for this sample is of good quality, it is fine to assume that white matter samples processed simultaneously will also have good RNA quality. Alternatively, the default settings on the Bioanalyzer can be adjusted to allow RIN calculation for samples with low

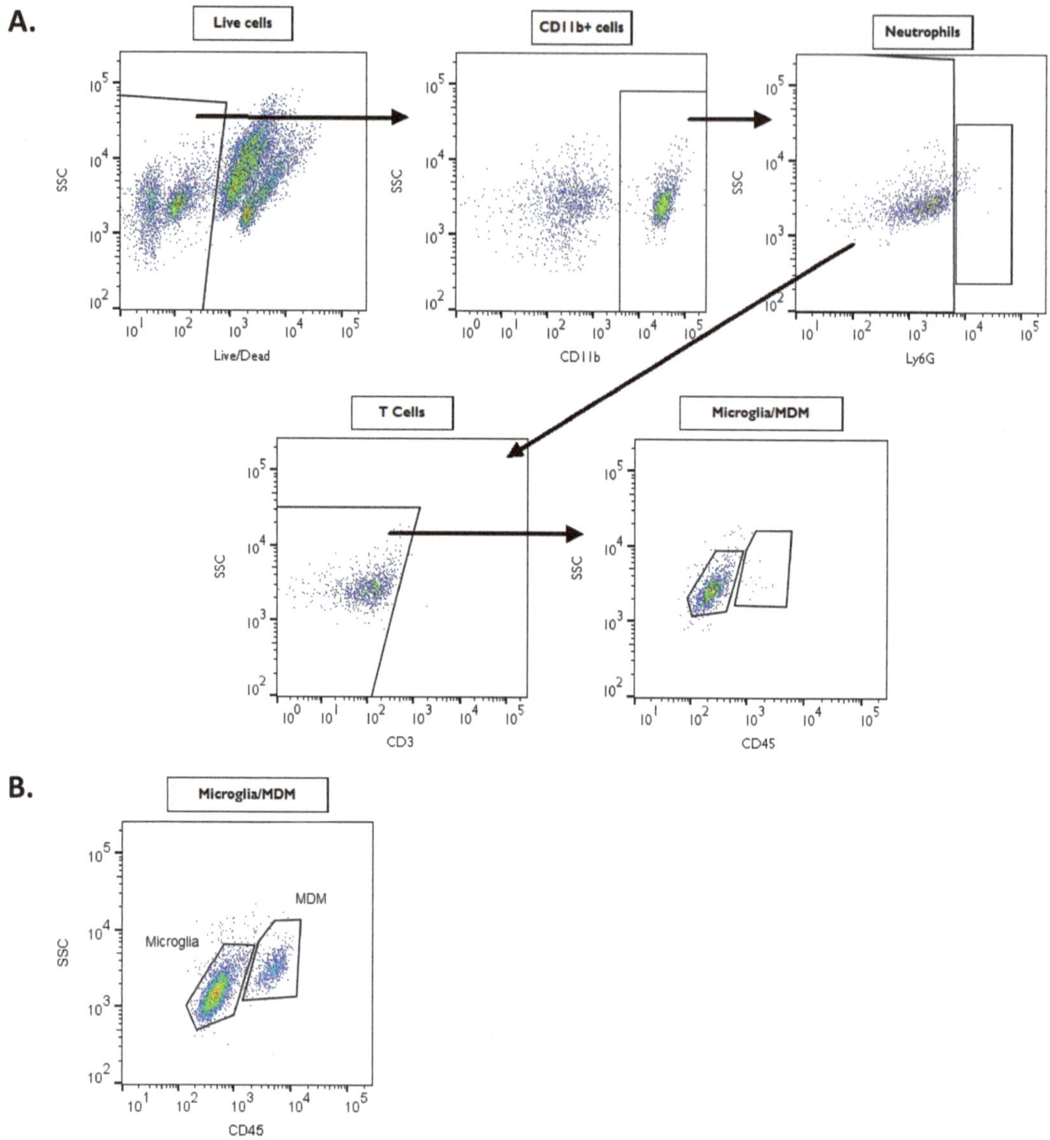

Fig. 2 Example gating strategy for isolation of microglia and MDM. (**a**) Dead cells were excluded using Live/Dead marker. Live cells were then gated on CD11b to exclude nonmyeloid cells. $CD11b^+$ cells were gated on Ly6G to exclude $Ly6G^+$ neutrophils and then onto CD3 to exclude $CD3^+$ T cells. Finally, the $CD11b^+$ $Ly6G^-$ $CD3^-$ population was gated onto CD45. Cells with $CD45^{lo}$ expression were classed as microglia while $CD45^{hi}$ cells were classed as MDM. (**b**) Identification of microglia and MDM isolated from LPS-injected brains

concentration. RIN scores above 8 are required for low RNA-input protocols for RNA-seq.

13. In our hands, as little as 1 ng of RNA can be used as starting material. The number of amplification cycles need to be empirically determined based on the starting material and thermal cycler. We found that ~12 cycles work well when using 1 ng

RNA while 9 cycles are sufficient if using 10 ng RNA as input. Include a positive control RNA sample, such as Mouse Brain Reference RNA, and a sample with no RNA input as negative control.

14. It is possible to scale down the assay to a 96-well plate format. Simply divide all volumes by 10: final volume of each reaction will be 200 μL rather than 2 mL.
15. The amplified cDNA should have DNA size distribution between 400 and 10,000 base pairs, which represents the typical length of mRNA in mammalian cells.
16. The fragmented and tagged DNA should have size distribution between 200 and 1000 base pairs. Titrating the amount of beads used can alter both the size distribution and amount of recovered DNA fragments. We recommend using 90 μL of AMPure XP beads manufacturer's instructions.

References

1. Fischer MT, Wimmer I et al (2013) Disease-specific molecular events in cortical multiple sclerosis lesions. Brain J Neurol 136 (Pt 6):1799–1815
2. Huang JK, Jarjour AA et al (2011) Retinoid X receptor gamma signaling accelerates CNS remyelination. Nat Neurosci 14(1):45–53
3. Arnett HA, Wang Y et al (2003) Functional genomic analysis of remyelination reveals importance of inflammation in oligodendrocyte regeneration. J Neurosci Off J Soc Neurosci 23(30):9824–9832
4. Kuypers NJ, Bankston AN et al (2016) Remyelinating oligodendrocyte precursor cell miRNAs from the Sfmbt2 cluster promote cell cycle arrest and differentiation. J Neurosci Off J Soc Neurosci 36(5):1698–1710
5. Zhao C, Li WW et al (2006) Differences in the early inflammatory responses to toxin-induced demyelination are associated with the age-related decline in CNS remyelination. Neurobiol Aging 27(9):1298–1307
6. Yuen TJ, Johnson KR et al (2013) Identification of endothelin 2 as an inflammatory factor that promotes central nervous system remyelination. Brain J Neurol 136 (Pt 4):1035–1047
7. Olah M, Amor S et al (2012) Identification of a microglia phenotype supportive of remyelination. Glia 60(2):306–321
8. Lewis ND, Hill JD et al (2014) RNA sequencing of microglia and monocyte-derived macrophages from mice with experimental autoimmune encephalomyelitis illustrates a changing phenotype with disease course. J Neuroimmunol 277(1–2):26–38
9. Yamasaki R, Lu H et al (2014) Differential roles of microglia and monocytes in the inflamed central nervous system. J Exp Med 211(8):1533–1549
10. Adiconis X, Borges-Rivera D et al (2013) Comparative analysis of RNA sequencing methods for degraded or low-input samples. Nat Methods 10(7):623–629
11. Combs PA, Eisen MB (2015) Low-cost, low--input RNA-seq protocols perform nearly as well as high-input protocols. PeerJ 3:e869
12. Ruckh JM, Zhao JW et al (2012) Rejuvenation of regeneration in the aging central nervous system. Cell Stem Cell 10(1):96–103
13. Miron VE, Boyd A et al (2013) M2 microglia and macrophages drive oligodendrocyte differentiation during CNS remyelination. Nat Neurosci 16(9):1211–1218
14. Kotter MR, Zhao C et al (2005) Macrophage-depletion induced impairment of experimental CNS remyelination is associated with a reduced oligodendrocyte progenitor cell response and altered growth factor expression. Neurobiol Dis 18(1):166–175
15. Voss EV, Skuljec J et al (2012) Characterisation of microglia during de- and remyelination: can they create a repair promoting environment? Neurobiol Dis 45(1):519–528
16. Li WW, Setzu A et al (2005) Minocycline-mediated inhibition of microglia activation impairs oligodendrocyte progenitor cell responses and remyelination in a non-immune

model of demyelination. J Neuroimmunol 158 (1–2):58–66

17. Lampron A, Larochelle A et al (2015) Inefficient clearance of myelin debris by microglia impairs remyelinating processes. J Exp Med 212(4):481–495
18. Neumann H, Kotter MR et al (2009) Debris clearance by microglia: an essential link between degeneration and regeneration. Brain J Neurol 132(Pt 2):288–295
19. Vogel DY, Vereyken EJ et al (2013) Macrophages in inflammatory multiple sclerosis lesions have an intermediate activation status. J Neuroinflammation 10:35
20. Peferoen LA, Vogel DY et al (2015) Activation status of human microglia is dependent on lesion formation stage and remyelination in multiple sclerosis. J Neuropathol Exp Neurol 74(1):48–63
21. JoVE Science Education Database (2016) Rodent stereotaxic surgery. Essentials of Neuroscience. https://doi.org/10.3791/5205

Chapter 3

Myelin: Methods for Purification and Proteome Analysis

Michelle S. Erwig, Dörte Hesse, Ramona B. Jung, Marina Uecker, Kathrin Kusch, Stefan Tenzer, Olaf Jahn, and Hauke B. Werner

Abstract

Molecular characterization of myelin is a prerequisite for understanding the normal structure of the axon/myelin-unit in the healthy nervous system and abnormalities in myelin-related disorders. However, reliable molecular profiles necessitate very pure myelin membranes, in particular when considering the power of highly sensitive "omics"-data acquisition methods. Here, we recapitulate the history and recent applications of myelin purification. We then provide our laboratory protocols for the biochemical isolation of a highly pure myelin-enriched fraction from mouse brains and for its proteomic analysis. We also supply methodological modifications when investigating posttranslational modifications, RNA, or myelin from peripheral nerves. Notably, technical advancements in solubilizing myelin are beneficial for gel-based and gel-free myelin proteome analyses. We conclude this article by exemplifying the exceptional power of label-free proteomics in the mass-spectrometric quantification of myelin proteins.

Key words Myelin, Oligodendrocyte, White matter, Density gradient ultracentrifugation, Mass spectrometry, Proteome/proteomics, Proteoform, Lipidome/lipidomics, Transcriptome/transcriptomics, Cyclic nucleotide phosphodiesterase (CNP), Proteolipid protein (PLP), Myelin basic protein (MBP), Cholesterol, Demyelination, Nerve conduction

1 Introduction

About 65 years ago, the emergence of novel lipid extraction methods enabled the first biochemical analyses of myelin. Notably, Marjorie Lees and Jordi Folch developed a protocol to extract lipids from nervous tissue using a chloroform-methanol mixture [1]. Compared to the previously used ethyl alcohol and ester, chloroform-methanol has allowed extracting a much wider range of lipid species, enhanced purity, and less impact on the chemical nature of the lipids. When the method was applied to rat brains, the lipid extracts also contained an abundant protein that like lipids was soluble in chloroform-methanol mixtures but insoluble in aqueous

David A. Lyons and Linde Kegel (eds.), *Oligodendrocytes: Methods and Protocols*, Methods in Molecular Biology, vol. 1936, https://doi.org/10.1007/978-1-4939-9072-6_3,

solutions [2]. Termed proteolipid protein (PLP), this protein is the most abundant constituent of central nervous system myelin [3–6] and associated with lipids [7–11]. Indeed, expression of PLP is required to enrich membrane cholesterol to the high levels normally found in central myelin [12].

More systematic molecular characterization of the entire myelin sheath required a method to enrich myelin from nervous tissue to a level of purity that was not achievable by chloroform-methanol extraction. It took over 15 years until such a protocol had evolved from classical approaches to fractionate entire brains by density centrifugation. Initially, myelin was found as an often-unwelcome contaminant of the nuclear fraction [13, 14]. However, such approaches led to the insight that myelin membranes have the highest lipid content and thus the lowest hydrated density of any membrane in the nervous tissue. In an early attempt specifically directed to exploit this characteristic for enriching myelin from nervous tissue, Saul Korey and his colleagues homogenized subcortical white matter in low-density (0.25 M) sucrose; before ultracentrifugation the homogenate was layered over sucrose of high and intermediate density (1.75 M and 0.5 M) [15]. Most myelin accumulated at the interface between 0.25 M and 0.5 M sucrose. Yet, the fraction also contained considerable amounts of cell debris and axonal fragments that were reduced only to a certain extent during a second ultracentrifugation, in which myelin floated up in a single layer of 1.0 M sucrose.

A subsequent modification of this protocol resulted in improved purity by using 0.8 M instead of 1.0 M sucrose in the second ("floating-up") ultracentrifugation and replacing buffered salt solutions by water in the washing steps [16]. The use of water, a hypotonic solution, represents an osmotic shock, which we now know contributes considerably to separating myelin from the axonal plasma membrane and other cellular debris. In a next stage of methodological evolution, brain tissue was homogenized in 0.32 M sucrose and layered over 0.66 M sucrose for ultracentrifugation; then two osmotic shocks were followed by a second ultracentrifugation using a similar "spin-down" gradient [17]. Importantly, the high purity of the resulting myelin fraction was carefully scrutinized. Indeed, the myelin fraction displayed a high lipid and proteolipid content, and when observed by electron microscopy contained a high proportion of intact-appearing, multilayered myelin sheaths. Equally important, the preparation appeared essentially free of nuclei, mitochondria, blood vessels, cell debris, endoplasmic reticulum, and axonal remnants.

Finally, William Norton and Shirley Poduslo combined the advantageous-appearing conditions of the earlier protocols. Rat brains homogenized in 0.32 M sucrose were layered on top of 0.85 M sucrose; most material sedimented during ultracentrifugation. Indeed, a discrete myelin fraction accumulated at the interface, which was subsequently subjected to two osmotic shocks [18]. It is noteworthy that this protocol is commonly cited as "*the*" method to isolate myelin from nervous tissue despite the fact that it lacks a second ultracentrifugation step, the omission of which is disadvantageous for the purity [19] and enrichment [20] of myelin membranes as we now know. The most relevant procedural adaptations since 1973 thus include the reintroduction of the second ultracentrifugation step, as well as replacing the previously used Dounce homogenizer by a high-shear disperser (Ultra-Turrax) [21] and the supplementation of solutions with protease inhibitors.

Together, by taking advantage of its low density, myelin can be highly enriched from homogenized nervous tissue using sucrose density gradient ultracentrifugation. Other cellular compartments and debris are almost quantitatively removed. Indeed, the method has enabled a fair yield of highly purified myelin across a number of laboratories despite moderate procedural differences. Considering the long-standing interest in myelin's composition of proteins [22–25] and lipids [26–28] it is not surprising that the myelin fraction has more recently been the subject of modern systematic approaches, thereby laying the foundations for myelin proteomics [4] and lipidomics [29]. Notably, a recent RNAseq-analysis of myelin purified from mouse brains at various ages has identified an unpredictably large transcript pool enriched in CNS myelin that is intimately interrelated to the myelin proteome but quite dissimilar to the transcriptomes of oligodendrocytes and brain regions [30]. The findings imply that the selective incorporation of transcripts into the myelin compartment enables both local adaptation and turnover of the myelin sheath. Yet, among the conceivable systematic approaches to myelin, proteome analysis has been most frequently used so far ([31] and references therein; [32–34]).

In the following, we provide our laboratory protocols for myelin purification from the mouse central nervous system, as well as modifications for the application to peripheral nerves or for RNA analysis. We then focus on techniques for gel-based and gel-free myelin proteome analysis. We conclude this article by exemplifying the exceptional power of label-free proteomics in the relative quantification of myelin proteins. We refer the reader to original articles for the related topic of proteome analyses of isolated oligodendrocytes [35–38], to earlier review articles for proteome analysis in the context of white matter disorder [39, 40] and to general overviews about the mechanisms and functions of myelination [41–44], which we do not cover extensively.

2 Materials

2.1 Biochemical Purification of Central Nervous System Myelin

2.1.1 Preparations (At Least One Day Ahead of the Experiment)

1. 500 mL ddH$_2$O.
2. 0.85 M sucrose: 29.09 g sucrose in 100 mL ddH$_2$O (sterile-filtered, cool to 4 °C).
3. 0.32 M sucrose: 10.95 g sucrose in 100 mL ddH$_2$O (sterile-filtered, cool to 4 °C).
4. Thermo Scientific WX+ Ultracentrifuge (or equivalent).
5. TH641 swing-out rotor (or equivalent).
6. 6 corresponding adaptors.

2.1.2 Preparations (On the Day of the Experiment)

1. Thermo Scientific WX+ Ultracentrifuge (or equivalent).
2. Protease inhibitors to 50 mL 0.32 M sucrose (dissolve at 4 °C, then keep on ice).
3. Large ice bucket.

2.1.3 Tissue Preparation

1. 6 Beckman centrifuge tubes, Ultra-clear, 14 × 89 mm.
2. IKA T10 basic Ultra-Turrax (or equivalent).
3. 70% EtOH.
4. ddH$_2$O.

2.1.4 First Sucrose Density Gradient

1. 6 Beckman centrifuge tubes, Ultra-clear, 14 × 89 mm.

2.1.5 Washing

1. 6 Beckman centrifuge tubes, Ultra-clear, 14 × 89 mm.
2. Forceps.
3. Pasteur pipette.

2.1.6 First Osmotic Shock

1. Paper tissue.

2.1.7 Second Osmotic Shock

1. Paper tissue.
2. Protease inhibitors to 50 mL 0.32 M sucrose (dissolve at 4 °C, then keep on ice).

2.1.8 Second Sucrose Density Gradient

1. 6 Beckman centrifuge tubes on ice.
2. Paper tissue.
3. 10× TBS: 1370 mM NaCl, 200 mM Tris/HCl, pH 7.4, 4 °C.

2.1.9 Washing

1. 6 Beckman centrifuge tubes on ice.
2. Pasteur pipette.
3. 10× TBS: 1370 mM NaCl, 200 mM Tris/HCl, pH 7.4, 4 °C.
4. −80 °C freezer.

2.2 1D-Gel Separation of Myelin for Proteome Analysis

2.2.1 1D-SDS-PAGE

1. Precast Bis-Tris 4–12% gradient gels (8 × 8 cm, 1 mm gel thickness, NuPAGE, Thermo Fisher Scientific) OR Tris-glycine 8–16% gradient gels (7 × 8 cm, 1 mm gel thickness, TG PRiME, Serva).
2. Running buffer containing 2-(*N*-Morpholino)ethanesulfonic acid OR 3-(*N*-morpholino)propanesulfonic acid (MOPS) buffer system (NuPAGE gels) OR Tris-glycine buffer system (TG PRiME gels).
3. 5–10 μg per lane of purified myelin.

2.2.2 Gel Staining

1. Gel fixative: 40% (v/v) ethanol, 10% (v/v) acetic acid.
2. ddH_2O.
3. Dye stock solution: 2% (w/v) orthophosphoric acid, 10% (w/v) ammonium sulfate, 0.1% (w/v) Coomassie Brilliant Blue (CBB) G-250. Completely dissolve 20 g orthophosphoric acid and 100 g ammonium sulfate in 800 mL ddH_2O; add 1 g of CBB G-250 while stirring; fill up to 1 L with ddH_2O. Prepare stock solution ≥24 h before staining; do not filter. Dye working solution: 80% (v/v) dye stock solution, 20% (v/v) methanol (prepare dye working solution just before use; add methanol while stirring of the stock solution).

2.3 2D-IEF/SDS-PAGE Separation of Myelin for Proteome Analysis

2.3.1 Delipidation / Precipitation/Solubilization

1. 150 μL of purified myelin (typical protein concentration ~4 μg/μL).
2. 2× 1.5 mL tube (preferably Eppendorf Safe-Lock).
3. vortex.
4. 600 μL methanol.
5. 300 μL chloroform.
6. 450 μL ddH_2O.
7. Centrifuge, e.g., Eppendorf 5415 with rotor F45-24-11 or equivalent.
8. Pipet.
9. Nitrogen stream.
10. Homogenization buffer: 30 mM Tris–HCl pH 9.0, 7 M urea, 2 M thiourea, 2% ASB-14.

2.3.2 Rehydration/ Isoelectric Focusing

1. IPG-strip (24 cm pH 3–12 NL, Serva).
2. IPGphor (GE Healthcare) or a Protean i12 (BioRad) IEF system.
3. 300 μg protein per IPG-strip.
4. Rehydration buffer I: 7 M urea, 2 M thiourea, 2% ASB-14, 1% ampholytes (Serva), 30 mM DTT.
5. Sonicator.

6. Shaker.
7. Rehydration buffer II: 7 M urea, 2 M thiourea, 2% ASB-14, 0.5% ampholytes (Serva), 1.2% DeStreak (GE Healthcare) or equivalent reagents based on organic disulfides.
8. Centrifuge (e.g., Eppendorf 5415 with rotor F45-24-11 or equivalent).
9. Mineral oil.
10. Water-wetted paper pads.

2.3.3 SDS-PAGE

10–15% acrylamide gradient gels (25.5 × 20 cm, gel thickness 0.65 mm).

2.4 Filter-Aided Sample Preparation (FASP) for Gel-Free Myelin Proteome Analysis

2.4.1 Procedure

Per sample:

1. 10 μg of purified myelin.
2. DTT: 1 M stock solution in water, store 200 μL aliquots at −20 °C.
3. Lysis buffer: 7 M urea, 2 M thiourea, 0.1 M Tris pH 8.5, store at −20 °C. Add 1% ASB-14 or 2% CHAPS and 10 mM DTT shortly before use.
4. Iodoacetamide: 450 mM stock solution in water, prepare shortly before use, dilute 1:10 in urea buffer w/o DTT.
5. Shaker at 37 °C.
6. Filter unit (e.g., Merck Millipore Microcon-30 kDa Centrifugal Filter or equivalent).
7. 100 μL 1% (v/v) formic acid.
8. Urea buffer: 8 M urea, 0.1 M Tris pH 8.5, store at −20 °C. Add 10 mM DTT just before use (do not add DTT when diluting iodoacetamide).
9. Centrifuge (11,000 × *g*, RT) or a vacuum manifold (Δ600 mbar, RT).
10. Sonicator.
11. 100 μL 45 mM iodoacetamide in urea buffer (prepare freshly just before use from 450 mM stock solution in water).
12. 3× 200, 300 and 450 μL 50 mM ammonium bicarbonate, 10% (v/v) acetonitrile.
13. 40 μL of trypsin working solution (0.01 μg/μL), 50 mM ammonium bicarbonate/10% acetonitrile.
14. 40 μL 50 mM ammonium bicarbonate in water.
15. 40 μL 1% (v/v) trifluoroacetic acid (TFA) in water.
16. 1.5 mL low protein-binding tubes (e.g., Eppendorf LoBind or equivalent).

17. 2 µL Yeast Enolase 1 digestion standard OR Hi3 EColi standard (Waters) (100 fmol/µL).
18. Glass autosampler vial for LC-MS analysis.

3 Methods

3.1 Biochemical Purification of Central Nervous System Myelin

3.1.1 Preparations (At Least One Day Ahead of the Experiment)

1. Cool 500 mL ddH_2O to 4 °C.
2. Prepare sucrose solutions, filter-sterilize them, and cool to 4 °C.
3. Cool rotor and adaptors to 4 °C.

3.1.2 Preparations (on the Day of the Experiment)

1. Cool centrifuge to 4 °C.
2. Add protease inhibitors to 50 mL 0.32 M sucrose (dissolve at 4 °C, then keep on ice).

3.1.3 Tissue Preparation

1. Cool 6 Beckman Centrifuge tubes on ice.
2. Freshly dissect brain or spinal cord of mice (*see* **Notes 1** and **2**).
3. Transfer half a brain or one spinal cord into 5 mL 0.32 M sucrose in one of the precooled Beckman centrifuge tubes.
4. Homogenize the tissue using IKA T10 basic Ultra-Turrax (Speed 3).
5. Clean the Ultra-Turrax thoroughly after each sample with 70% EtOH and ddH_2O (*see* **Note 3**).

3.1.4 First Sucrose Density Gradient

1. Place additional 6 Beckman centrifuge tubes on ice.
2. Transfer 6 mL of cold 0.85 M sucrose into each of the new tubes.
3. Carefully and slowly layer the homogenate (in 0.32 M sucrose) on top of the 0.85 M sucrose (Fig. 1) (*see* **Note 4**).
4. Fill one tube with cold 0.32 M sucrose up to 0.5 cm below the tube edge and tare the other tubes with 0.32 M sucrose exactly to the same weight (max. difference 0.01 g) (*see* **Note 5**).
5. Centrifuge at 75,000 × g for 30 min at 4 °C (rotor TH641 at 24,400 rpm, acceleration 7, deceleration 7); always wait for vacuum.
6. Crudely purified myelin accumulates at the 0.32/0.85 M sucrose interface as fluffy cloud (Fig. 1). The crude myelin fraction appears whitish if mice were transcardially perfused with physiological salt solution or orange-to-reddish if not.

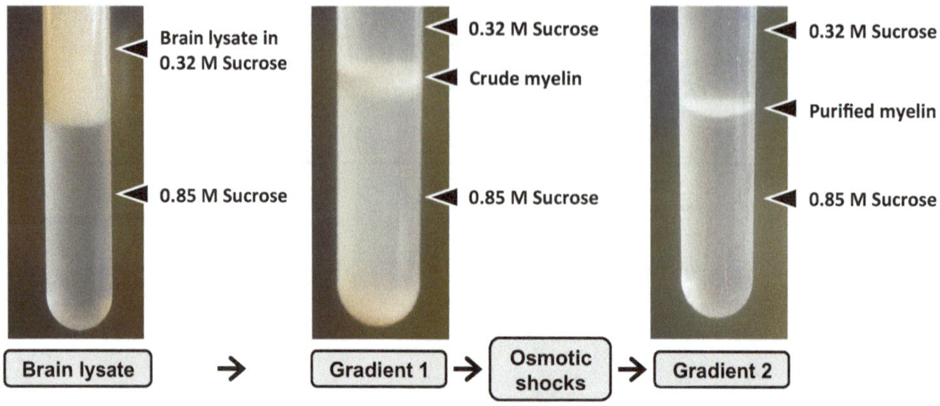

Fig. 1 To isolate CNS myelin, nervous tissue is first homogenized in 0.32 M sucrose (*see* **method 1**). The homogenate is layered over 0.85 M sucrose (image on the left) and subjected to sequential sucrose density gradient centrifugation, intermitted by osmotic shocks. During ultracentrifugation, the myelin-enriched light-weight membrane fraction accumulates at the interface

3.1.5 Washing

1. After each centrifugation step, place the adaptors on ice as vertically as possible.
2. Place 6 new Beckman centrifuge tubes on ice.
3. Carefully remove centrifuge tubes from the adaptor (forceps may help).
4. Collect the interface with a Pasteur pipette (be careful and try to collect as little of the 0.85 M sucrose as possible) and transfer it into the new tubes.
5. Add 1 mL of 4 °C cold ddH_2O and vortex.
6. Fill one tube with 4 °C cold ddH_2O up to 0.5 cm below the tube edge and tare the other tubes with ddH_2O.
7. Centrifuge at 75,000 × *g* for 15 min at 4 °C (rotor: TH641 at 24,400 rpm; acceleration 9, deceleration 9); always wait for vacuum.

3.1.6 First Osmotic Shock

1. Discard the supernatant and remove remaining fluid with paper tissue.
2. Add 1 mL cold ddH_2O to each tube and resuspend the pellet (vortex).
3. Fill tubes with cold ddH_2O up to 0.5 cm below tube edge and incubate 10 min on ice.
4. In the meantime tare the tubes with ddH_2O.
5. Centrifuge at 12,000 × *g* for 15 min at 4 °C (rotor TH641 at 9800 rpm; acceleration 9, deceleration 9); always wait for vacuum.

3.1.7 Second Osmotic Shock

1. Same procedure as for the first osmotic shock (*see* above).
2. During centrifugation, add protease inhibitors to 50 mL 0.32 M sucrose (dissolve at 4 °C, then keep on ice).

3.1.8 Second Sucrose Density Gradient

1. Place 6 new Beckman centrifuge tubes on ice.
2. Discard the supernatant and remove remaining fluid with paper tissue.
3. Add 1 mL cold 0.32 M sucrose to each tube and resuspend the pellet (vortex).
4. Add additional 4 mL cold 0.32 M sucrose.
5. Transfer 6 mL of cold 0.85 M sucrose into each of the new tubes.
6. Carefully and slowly layer the crude myelin (in 0.32 M sucrose) on top of the 0.85 M sucrose.
7. Fill one tube with cold 0.32 M sucrose up to 0.5 cm below tube edge and tare the other tubes with 0.32 M sucrose to the same weight (max. difference 0.01 g).
8. Centrifuge at 75,000 × *g* for 30 min at 4 °C (rotor TH641 at 24400 rpm, acceleration 7, deceleration 7); always wait for vacuum.
9. During centrifugation, dilute 10× TBS to 1× TBS and add protease inhibitors (dissolve at 4 °C, then keep on ice).
10. Purified myelin accumulates at the interface as a fluffy white cloud (Fig. 1).

3.1.9 Washing

1. Place 6 new Beckman Centrifuge tubes on ice.
2. Collect the interface with a Pasteur pipette and transfer it into the new tubes (on ice).
3. Fill one tube with cold ddH_2O up to 0.5 cm below tube edge and tare the other tubes with ddH_2O.
4. Centrifuge at 75,000 × *g* for 15 min at 4 °C (rotor TH641 at 24400 rpm; acceleration 9, deceleration 9); always wait for vacuum.
5. Discard the supernatant and remove remaining fluid with paper tissue.
6. Dissolve pellet in 200 μL cold 1× TBS (supplemented with protease inhibitor).
7. Transfer myelin (in 1× TBS) to a new 1.5 mL Eppendorf tube and store at −80 °C (*see* **Notes 6–9**).

3.2 1D-Gel Separation of Myelin for Proteome Analysis

One-dimensional separation of myelin and immunoblot analysis (Fig. 2) can serve to test for the enrichment of myelin proteins and the diminishment of contaminants during the biochemical purification of myelin. When aiming to establish myelin protein

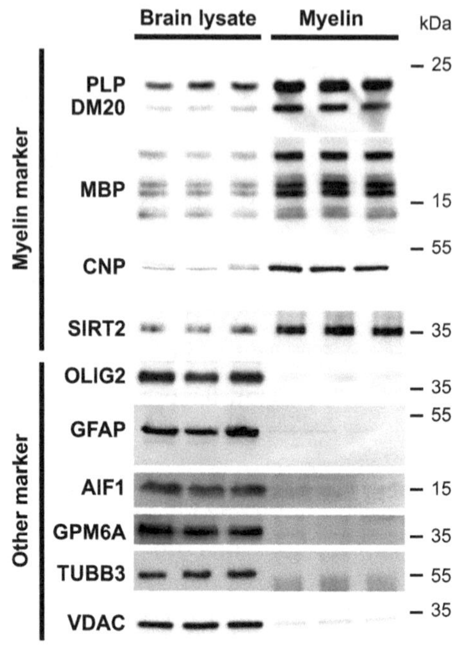

Fig. 2 Immunoblot illustrates that myelin proteins are enriched in purified myelin compared to brain lysate while markers for other cellular sources are diminished. The detected proteins localize to compact myelin (PLP/DM20, MBP), non-compact myelin (CNP, SIRT2), the oligodendroglial nuclear/cytoplasmic compartment (OLIG2), astrocytes (GFAP), microglia (AIF1/IBA1), the neuronal plasma membrane (GPM6A), axonal microtubules (TUBB3/TUJ1), and mitochondria (VDAC). The blot shows three biological replicates (male c57Bl6/N mice at postnatal day 75); equal amounts of brain lysate and myelin were loaded. *Figure modified from ref.* [30]

inventories, the use of various complementary proteomic approaches (e.g., with or without pre-fractionation of undigested proteins) will provide the most comprehensive, yet nonquantitative compilation of the protein constituents. One-dimensional (1D)-separation via SDS-PAGE (sodium dodecyl sulfate polyacrylamide gel electrophoresis) followed by in-gel tryptic digestion is possibly the simplest method to prepare myelin for the mass spectrometric identification of its proteins, including very hydrophobic transmembrane proteins (e.g., PLP, OSP/CLDN11) and proteins with extreme isoelectric points (e.g., MBP, MOBP). The comparatively low resolution provided by these gels is compensated by chromatographic separation of the proteolytic peptides prior to mass spectrometry (MS), an approach often referred to as "GeLC-MS."

3.2.1 1D-SDS-PAGE

For separation of purified myelin, we have mainly used precast Bis-Tris 4–12% gradient gels (8 × 8 cm, 1 mm gel thickness, NuPAGE, Thermo Fisher Scientific) according to the manufacturer's protocols. The running buffer containing 2-(*N*-Morpholino)

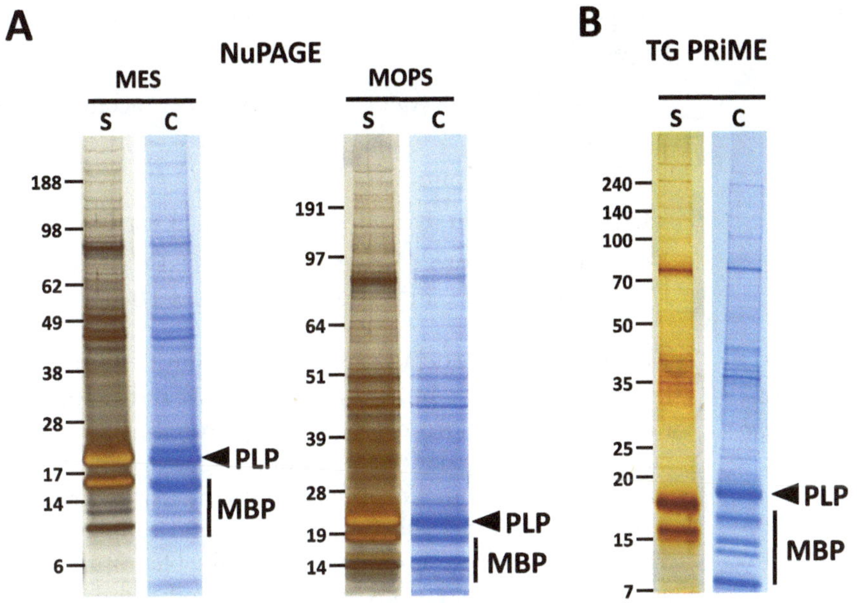

Fig. 3 Separation of purified CNS myelin in one-dimensional gels (*see* **method 2**). (**a**) Purified myelin was separated by SDS-PAGE using 4–12% Bis-Tris gels (precast, NuPAGE). Different buffer systems enhance separation in the low (MES buffer) or high (MOPS buffer) molecular weight range. Proteins were visualized by silver staining (S; load 0.5 μg per lane) or colloidal Coomassie (C; 5 μg loaded). *PLP* proteolipid protein, *MBP* myelin basic protein. Molecular weight standard is given in kDa. *Figure modified from ref.* [31] *with permission.* (**b**) More recently, we also use 8–16% Tris-glycine gels (precast, TG PRiME), yielding excellent separation of myelin

ethanesulfonic acid (MES) is recommended when aiming at good separation of myelin basic protein (MBP) and other small proteins, while the 3-(*N*-morpholino)propanesulfonic acid (MOPS) buffer system [45] enhances resolution in the molecular weight range above 25 kDa (Fig. 3a). More recently, we have also used Tris-glycine 8–16% gradient gels (7 × 8 cm, 1 mm gel thickness, TG PRiME, Serva), newly available precast gels related to the commonly used Laemmli system. These gels provide coverage of a broad molecular weight range and a remarkably high resolution for large proteins (Fig. 3b).

For subsequent mass spectrometric protein identification, typically 5–10 μg per lane of purified myelin is separated by 1D-SDS-PAGE. If a higher protein load is desired, it may be required to perform delipidation (*see* Subheadings 2.3.1 and 3.3.1) before loading, thereby preventing lipid-related distortion of separation particularly in the low molecular weight range (i.e., <20 kDa).

To deplete soluble and membrane-associated proteins and thereby enrich membrane-spanning proteins, the myelin fraction may be subjected to consecutive high-salt and high-pH washing/centrifugation cycles prior to loading on the gel as described previously [20].

3.2.2 Gel Staining

For gel staining, we recommend colloidal Coomassie according to [46].

1. Fix gel with 40% (v/v) ethanol, 10% (v/v) acetic acid for at least 1 h.
2. Wash 3 × 10 min with ddH_2O.
3. Stain the gel in dye working solution at least overnight.
4. Transfer gel into a new box with ddH_2O, wash until all Coomassie particles are removed (*see* **Notes 10–13**).

3.3 2D-IEF/SDS-PAGE Separation of Myelin for Proteome Analysis

Two-dimensional (2D) separation of purified myelin provides highly reproducible protein maps [20, 47] and enables the resolution of different "proteoforms" (i.e., protein isoforms, splice variants, and posttranslationally modified species) of myelin proteins. Indeed, the combination of isoelectric focusing (IEF) as the first dimension with SDS-PAGE as the second dimension (2D-IEF/SDS-PAGE; Fig. 4a) continues to be valuable for myelin proteome analysis despite limitations with respect to the underrepresentation of hydrophobic, transmembrane, and basic proteins. The separation of very basic proteins, or more generally the resolving power in the first dimension, can be improved by using longer IEF strips with wider pH gradients (Fig. 4b).

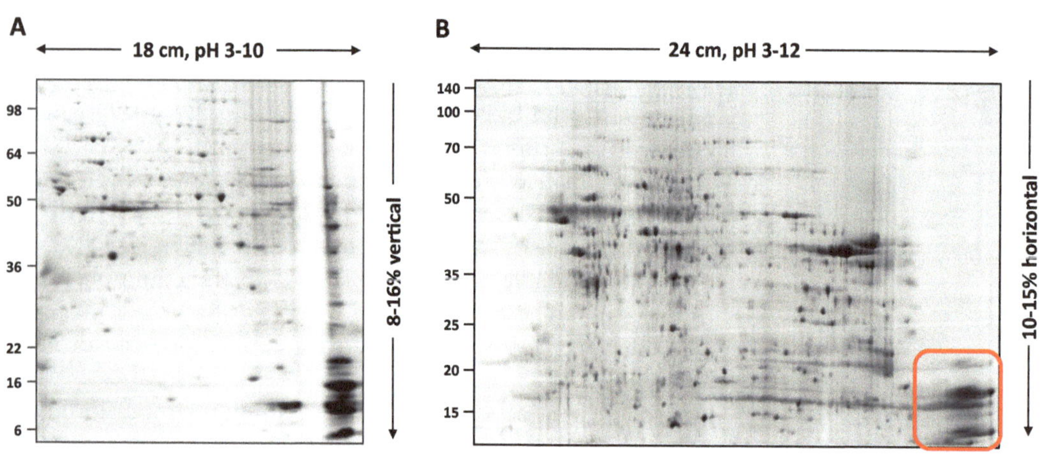

Fig. 4 Myelin proteome maps based on two-dimensional gels (*see* **method 3**). (**a**) Purified myelin was separated two-dimensionally using isoelectric focusing (IEF) in a nonlinear pH-gradient (pH 3–10, 18 cm gel) as the first and SDS-PAGE (self-cast 8–16% gel, vertically run) as the second dimension. Proteins were stained using colloidal Coomassie and identified by MALDI-TOF-MS. Molecular weight standard is given in kDa. *Figure modified from ref.* [20]. (**b**) More recently, we have changed the separation conditions to a pH 3–12 gradient (24 cm gel) as the first and horizontally run 10–15% precast gels as the second dimension, thereby gaining improved separation of basic proteins. We note that proteoforms of myelin basic protein (MBP, encircled in red) are still hardly resolvable due to their extreme basicity. Yet, moving MBP out of the more populated part of the gel allows the detection of previously masked spots

3.3.1 Delipidation/Precipitation/Solubilization

Prior to 2D-IEF/SDS-PAGE, purified myelin is first delipidated and precipitated with methanol/chloroform according to [48], as lipids and salts compromise the resolution during IEF. The solubility of hydrophobic and glycosylphosphatidylinositol (GPI)-anchored myelin proteins is greatly improved by the zwitterionic detergent amidosulfobetaine ASB-14 compared to the commonly used 3-[(3-cholamidopropyl)-dimethylammonio]-1-propanesulfonate (CHAPS) [49], thereby enhancing their resolution by 2D-IEF/SDS-PAGE.

1. Thaw myelin sample on ice (the precipitation procedure itself is performed at room temperature (RT)).
2. Transfer 150 μL of purified myelin (typical protein concentration ~4 μg/μL) into a 1.5 mL tube (preferably Eppendorf Safe-Lock) (*see* **Note 14**).
3. Add 600 μL methanol; vortex.
4. Add 300 μL chloroform; vortex.
5. Add 450 μL ddH_2O.
6. Centrifuge at 9000 × *g* for 1 min (proteins accumulate at the interface).
7. Discard upper phase by pipetting.
8. Add 450 μL methanol and vortex.
9. Centrifuge at 16,100 × *g* for 5 min to pellet precipitated proteins.
10. Discard supernatant by pipetting.
11. Dry pellet (use nitrogen stream; do not dry entirely).
12. Add 50 μL homogenization buffer and solubilize carefully (ultrasonic bath or freezing/thawing cycles help to solubilize the pellet; no vortexing to prevent foaming); prevent from heating up above 37 °C to avoid protein carbamylation (*see* **Note 15**).
13. Centrifuge at 10,000 × *g* for 2 min.
14. Transfer supernatant into a new 1.5 mL tube.
15. Determine protein concentration by 2D-Quant Kit (GE Healthcare) (*see* **Note 16**).

3.3.2 Rehydration/Isoelectric Focusing

Here, we describe a novel protocol for focusing myelin proteins on 24 cm immobilized pH-gradient (IPG) strips of nonlinear pH 3–12. For our previous protocol employing 18 cm IPG-strips of nonlinear pH 3–10 the reader is referred to a previous methods article [50].

1. Take the volume equivalent of 300 μg protein per IPG-strip and mix with the same volume rehydration buffer I.
2. Sonicate for 2 min, shake gently (700 rpm) for 15 min at RT.

3. Fill up to 450 μL with rehydration buffer II (*see* **Note 17**).
4. Sonicate for 2 min, shake gently (700 rpm) for 30 min at RT.
5. Centrifuge 10,000 × *g* for 1 min.
6. Transfer 450 μL of supernatant to focusing tray, apply IPG-strip facing the gel-side down. Let the strip re-swell for 15 min and cover slot with mineral oil.
7. After 12 h active rehydration at 50 V (70 μA per strip, 20 °C), apply water-wetted paper pads to both electrodes (the cathode pad should be soaking to compensate water flow from cathode to anode in highly basic pH-gradients).

Focus for about 40 kVh (125 μA per strip, 20 °C) by a ramped two-step protocol (0–8000 V /1 h, 8000 V/4 h) on an IPGphor (GE Healthcare) or a Protean i12 (BioRad) IEF system. The latter has the advantage that the program of choice is regulated for each strip individually.

3.3.3 SDS-PAGE

Here, we describe the use of precast gels polymerized on a non-fluorescent backing film, of which up to four can be run simultaneously in horizontal electrophoresis chambers (Serva HPE system). The major advantage of this high-performance electrophoresis system is the efficient cooling of the horizontal gels during the runs, which, together with other technical refinements, allows excellent reproducibility, spot resolution and detection sensitivity [51]. For our previous protocol employing self-cast 18 × 20 cm gels for the second dimension the reader is referred to a previous methods article [50].

We use 10–15% acrylamide gradient gels (25.5 × 20 cm, gel thickness 0.65 mm) according to the manufacturer's manual (Serva HPE system) and in the following only list our adaptations for the separation of myelin.

1. Increase the SDS content of the equilibration solution to 6%.
2. Use 6 mL cooling contact fluid (at least for the first generation HPE tower).
3. Run final step of electrophoresis 60 min longer (*see* **Notes 18** and **19**).

3.4 Filter-Aided Sample Preparation (FASP) for Gel-Free Myelin Proteome Analysis

The outstanding sensitivity of current mass spectrometers necessitates careful consideration of sample preparation. In particular, it is crucial that substances are efficiently removed (or at least diminished) that may interfere with chromatographic separation or electrospray ionization. Moreover, particle-free samples are required when direct injection setups without trapping columns are used to provide higher chromatographic separation [52].

In our previous studies [4, 34, 47] we have successfully subjected myelin to tryptic in-solution digest in the presence of the acid-labile surfactant RapiGest (Waters) (for protocols *see* refs. [34, 50]). Indeed, this method has facilitated sufficiently robust solubilization and unfolding of proteins. Yet, in our current view the particular requirements of myelin samples are met even better by filter-aided sample preparation (FASP) [53]. FASP allows the depletion of soluble low molecular weight compounds (including detergents) from a complex protein mixture by washing it with urea buffer in an ultrafiltration device, which is followed by buffer exchange, direct on-filter digestion, and recovery of the proteolytic peptides by filtration (for overview *see* Fig. 5). The ultrafiltration devices are designed as spin columns and typically

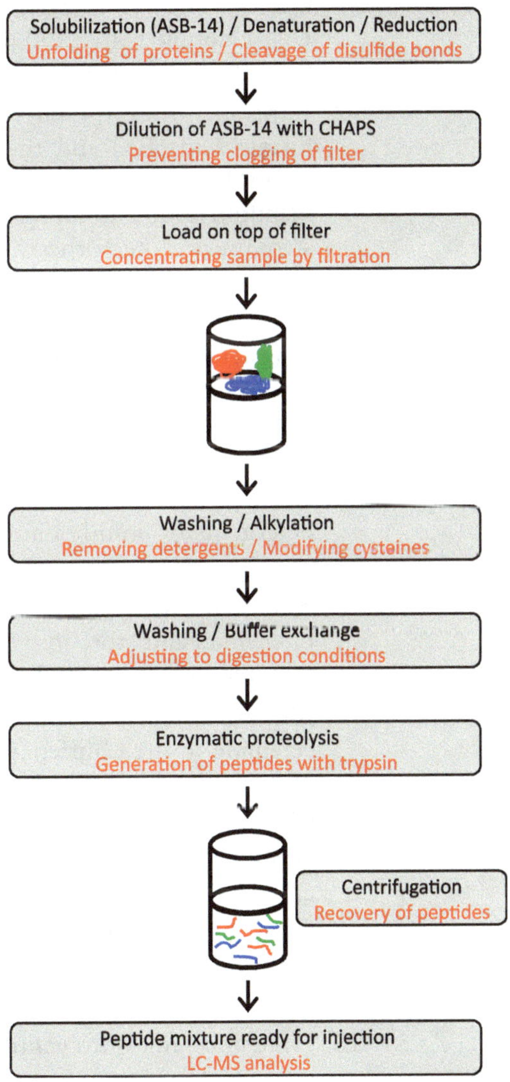

Fig. 5 Schematic overview of filter-aided sample preparation (FASP). For details *see* **method 4**

operated by centrifugation. However, reasoning that vacuum-driven removal of liquids can enable the automation of FASP we have implemented the procedure into a liquid handling workstation, the customized hardware and software components of which will be described elsewhere. The procedure itself is adapted from a previously reported manual protocol [52]. We consider FASP automation as important toward even higher reproducibility of sample preparation that is essential for quantitative label-free proteome profiling.

3.4.1 Procedure

1. Solubilize and denature 10 μg of purified myelin in 150 μL Lysis buffer containing 1% (w/v) ASB-14 and 10 mM DTT, sonicate for 5 min, incubate for 30 min at 37 °C while shaking.
2. Dilute sample with 1500 μL Lysis buffer containing 2% (w/v) CHAPS to reduce ASB-14 concentration to below 0.1% (w/v).
3. In parallel, rinse the filter unit first with 100 μL of 1% (v/v) formic acid (FA) and then with 200 μL 8 M urea buffer/10 mM DTT. A centrifuge (11,000 × *g*, RT) or a vacuum manifold (Δ600 mbar, RT) can be used in this and all following steps unless noted otherwise. In the following protocol, centrifugation/application of vacuum is referred to as "liquid removal."
4. Transfer 220 μL of myelin to the filter unit, remove buffer on top of the filter by liquid removal. Repeat until all volume is applied to the filter unit. Always sonicate the solubilized sample for 5 min before transferring an aliquot onto the filter.
5. Wash 2× with 200 μL urea buffer/10 mM DTT to deplete detergents by liquid removal.
6. Optional: Pause the procedure overnight. Cover the filter membrane with 100 μL urea buffer/10 mM DTT, cap the filter units and store in wet chamber at RT. Liquid removal is required before proceeding to the next step.
7. Alkylate cysteine residues by addition of 100 μL 45 mM iodoacetamide in urea buffer (prepare freshly just before use from 450 mM stock solution in water), incubate 20 min at RT in the dark, followed by liquid removal.
8. Wash 2× with 200 μL urea buffer/10 mM DTT to quench excessive iodoacetamide, followed by liquid removal.
9. Wash 3× with 200, 300, and 450 μL 50 mM ammonium bicarbonate/10% (v/v) acetonitrile for buffer exchange, followed by liquid removal.
10. Place filter unit onto centrifugation tube.
11. Wash 3× with 200 μL 50 mM ammonium bicarbonate/10% (v/v) acetonitrile, followed by liquid removal. Centrifuge 30 min at 10,000 × *g* (RT).

12. Place filter unit onto new centrifugation tube.
13. Add 40 μL of trypsin working solution (0.01 μg/μL) in 50 mM ammonium bicarbonate/10% acetonitrile to the filter, shake once gently (manually) and digest overnight at 37 °C in a wet chamber.
14. Centrifuge the filter for 30 min at 11,000 × *g* (RT) and collect the flow-through.
15. Add 40 μL 50 mM ammonium bicarbonate in water, centrifuge for 30 min at 11,000 × *g* (RT) and collect the flow-trough into the same tube.
16. Add 40 μL 1% (v/v) trifluoroacetic acid (TFA) in water, centrifuge for 30 min at 11,000 × *g* (RT) and collect the flow-trough into the same tube.
17. Combined flow-through is transfered to 1.5 mL low protein-binding tubes (e.g., Eppendorf LoBind or equivalent), document the volume.
18. Mix 2 μL Yeast Enolase 1 digestion standard or Hi3 EColi standard (Waters) (100 fmol/μL) with 20 μL of sample in a glass autosampler vial for LC-MS analysis (*see* **Notes 20–24**).

3.5 Gel-Free Myelin Proteome Analysis and Mass Spectrometric Quantification of a Myelin Protein Using CNP as an Example

For myelin proteome analysis, gel-free ("shotgun") approaches allow circumventing many of the limits of gel-based approaches, especially those of 2D IEF/SDS-PAGE (*see* above). The replacement of gel-based separation of entire proteins by their tryptic cleavage and subsequent separation of the resulting peptides via high resolution liquid chromatography (LC) before interfacing with mass spectrometry (LC-MS) goes at the expense of information about protein size and charge. This is a disadvantage when analyzing "proteoforms," i.e., protein isoforms, splice variants, or posttranslationally modified proteins. However, LC-MS is particularly well suited for quantitative and differential analyses of the myelin proteome with its many extraordinarily basic or hydrophobic proteins and its exceptional range of several orders of magnitude of protein abundance. Indeed, LC-MS enables the label-free quantification of proteins in complex samples such as myelin with the required dynamic range, yet it implements no bias against particular protein classes. Finally, only low amounts of starting material are required, about 100th compared to 2D-IEF/SDS-PAGE.

We note that the particular procedure of myelin proteome analysis will greatly depend on the locally available mass spectrometer and associated data acquisition and quantification method. For our investigations, we have decided for separation using nanoscale ultraperformance liquid chromatography (nanoUPLC) with quadrupole time of flight (QTOF)-mass spectrometers that allow a data-independent acquisition workflow with alternating low-energy and elevated-energy scans (e.g., QTOF-Premier or Synapt G2S, Waters). We refer the reader to previous articles for an overview about data-independent acquisition methods [54] or for a detailed protocol [52]. To determine the absolute abundance of myelin proteins, we utilized the data post-processing software ISO-Quant [55], which quantifies the validated mass spectra of the three most abundant peptides per protein according to the TOP3 method [56]. To compare the abundance of a myelin protein between two states (e.g., between mutant and control mice), its relative abundance was considered changed if significant according to two-tailed t-test and exceeding a threshold of 25%. This approach has enabled robust and reliable differential analysis of the entire myelin proteome in various mouse models of myelin-related pathology [34].

To exemplify the power of gel-free mass spectrometry for the differential quantification of myelin proteins, we have assessed the abundance of cyclic nucleotide phosphodiesterase (CNP) [57, 58], the third-most abundant CNS myelin protein [4]. Heterozygous *Cnp*$^{+/-}$ mice display a catatonia-depression syndrome upon aging [59]; yet their myelin appears morphologically normal. *Cnp*$^{+/-}$ mice provide a defined model in which the halved gene dosage should theoretically lead to a halved abundance of CNP in myelin. Indeed, when using the workflow outlined above, the abundance of CNP in myelin purified from *Cnp*$^{+/-}$ mice compared to wild-type littermates is 50.7% (±0.4%). The remarkable exactness of the quantification and the excellent sample-to-sample reproducibility (*see* Fig. 6) allows high confidence in unexpected findings yielded by differential myelin proteome analyses, notwithstanding that relevant-appearing results require validation by independent methods.

In conclusion, the method to purify myelin from nervous tissue has changed only moderately since the 1960s and 1970s, reflecting its efficiency and the excellent degree of myelin membrane enrichment that can be gained reliably and reproducibly. More recently, it was possible to combine myelin purification with modern "omics"-techniques to systematically determine molecular composition at the level of mRNA [30], lipids [29], and proteins. With respect to lipids, carbohydrates, and posttranslational modifications, their capabilities are not yet exploited to their full extent. However, myelin proteome analysis has proven a valuable resource for myelin research, as it has facilitated selecting proteins for functional investigations in mutant mice, revising the abundance of myelin proteins [4, 47], calculating molar ratios [60–62], analyzing models of myelin-related disorders [20, 39, 47, 63–65], discovering disease-causing genes [66], and identifying a novel molecular scaffold in the myelin sheath [34]. We believe that systematic approaches to myelin involving "omics"-techniques hold great promise for improving our understanding

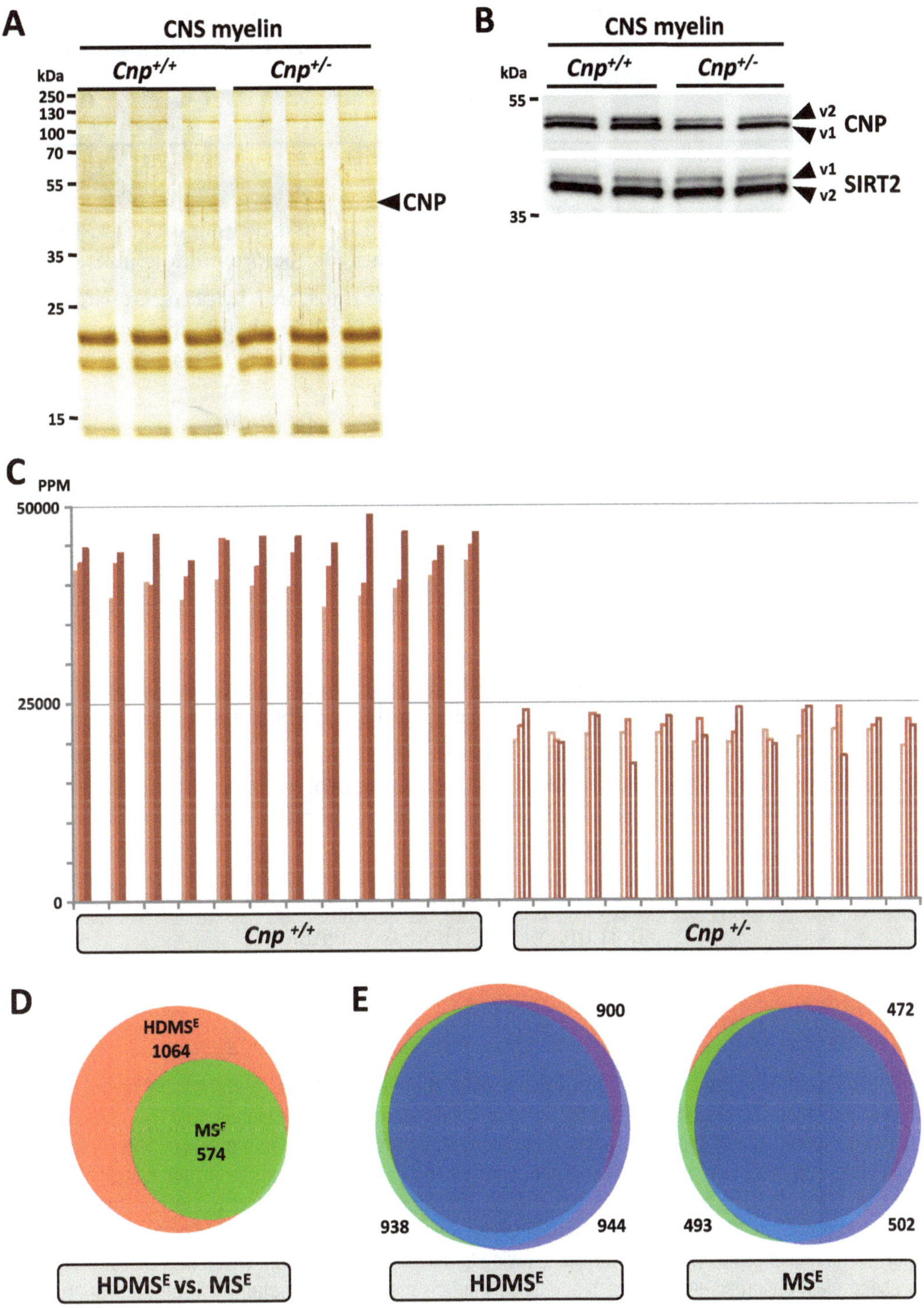

Fig. 6 Quantification of myelin proteins by label-free mass spectrometry using CNP as an example. (a, b) One-dimensional SDS-PAGE separation and silver staining (**a**) or immunoblotting (**b**) of myelin purified from mouse brains at P75 indicate that the abundance of CNP in myelin is reduced in heterozygous *Cnp*$^{+/-}$ mice compared to wild-type littermates. Sirtuin 2 (SIRT2) served as a loading control. *CNP* cyclic nucleotide

of the development, structure and function of the axon/myelin-unit in the context of the healthy and the pathological nervous system.

4 Notes

1. Optional: When dissecting the tissue earlier, store at −80 °C or keep frozen on dry ice until just before homogenization.
2. Optional: Animals may be transcardially perfused with physiological salt solution (HBSS, PBS), thereby clearing capillaries and reducing contamination of the myelin fraction with proteins derived from blood.
3. Optional: Store 400 μL of the homogenate for later use (−80 °C), e.g., for control by immunoblot analysis as exemplified in Fig. 2.
4. Do not add the homogenate drop by drop; best phase separation is achieved when the homogenate flows down slowly along the tube wall. The tube is placed vertically on ice; the pipette is tilted by 45°.
5. The tubes have to be weighed with adaptor and lid. Dry the centrifuge tubes before placing them into the adaptor. Dry the adaptors before placing them in the rotor.
6. From half a mouse brain, about 600 μg of purified myelin can be expected. For measuring protein concentrations, we recommend a Lowry assay (e.g., Biorad DC Protein Assay) or the precipitation-based assay kits 2-D Quant (GE Healthcare) or SingleQuant (Serva). Myelin should be well homogenized, e.g., using a Teflon pestle that fits exactly with 1.5 mL Eppendorf tubes (e.g., Thomas Scientific).

Fig. 6 (continued) phosphodiesterase. Arrowheads point at protein isoforms. (**c**) Differential myelin proteome analysis quantifies the abundance of CNP in myelin purified from heterozygous *Cnp*$^{+/-}$ mice as 50.7% (±0.4%) compared to wild-type littermates. The bar graph shows individual data points for three independent experiments with three biological replicates per genotype and four technical replicates per biological replicate. For details, *see* **method 5**. *PPM* parts per million. (**d**) Venn diagram comparing the number of proteins identified in myelin by MS^E (as used for quantification in Fig. 6c) with those identified by high-definition MS^E (HDMS^E). The higher proteome depth of HDMS^E owing to the ion mobility option, which may be considered as a gas phase electrophoresis-like device that equips modern mass spectrometers with an additional level of peptide separation. As this compresses the dynamic range of quantification, highly abundant proteins such as PLP, MBP, and CNP are more reliably quantified by MS^E (i.e., without ion mobility separation) at the cost of the number of identified proteins. (**e**) Venn diagram comparing the proteins identified in three biological replicates of myelin by HDMS^E (left image) or MS^E (right image). Note that both approaches allow excellent sample-to-sample reproducibility

7. Myelin samples to be compared in one experiment should be purified in parallel, i.e., in one rotor run.
8. Quality control may include electron microscopy [20] and immunoblot (Fig. 2).
9. Myelin purified according to this protocol can be used for protein and lipid analyses, RNA extraction, enzymatic assays, electron microscopy, and other applications.

   Optional: When posttranslational modifications are to be analyzed, the supplementation of solutions with respective inhibitors may be beneficial. For example, add phosphatase inhibitors (e.g., PhosStop cocktail tablets, Roche) or deacetylase inhibitors (10 mM Nicotinamide (Sigma-Aldrich), 0.5 μM Trichostatin A (Sigma-Aldrich)) to 0.32 M sucrose and 1× TBS.

   If aiming to isolate RNA from purified myelin [30], work under RNAse-free conditions is advantageous. Clean the bench, all materials, skull, and brain with RNAse AWAY (Sigma-Aldrich); use RNAse-free sucrose (e.g., Sigma-Aldrich) and DEPC-treated ddH_2O; after the last washing step dissolve myelin in QIAzol (Qiagen) instead of TBS; wear a mask and gloves at all times.

   Subfractions of low, medium, and high density can be isolated from purified myelin [67–71]. Note that it is speculative if these subfractions correspond to distinct subcompartments of myelin.

   Aiming to further purify the myelin fraction, purified myelin was subjected to two additional osmotic shocks and a third ultracentrifugation step using a "floating-up" protocol [67]. For the floating up gradient, myelin was resuspended in 0.83 M sucrose and overlayered with 0.3 M sucrose; during ultracentrifugation myelin accumulated at the interface. Note that in our experience this procedure does not yield better purity but results in considerable material loss and should thus be omitted.

   The purification of peripheral myelin requires different sucrose concentrations (0.27 M and 0.83 M sucrose) and only one gradient ultracentrifugation step [21, 47, 67]. A yield of up to 200 μg of purified myelin can be expected when pooling 30 sciatic nerves from 15 adult mice. Note that the achievable factor of enrichment is lower compared to central myelin.

   Myelin has been purified from other species ([72, 73] and references therein), including rats [74, 75], humans [29, 32, 76], clawed frogs [5], and zebrafish [77–79]. Note that sucrose concentrations may require adaptation if the density of myelin in a species differs from that in mice.

10. Only glass containers or plastic containers specified for gel staining should be used because low-quality plastic may emit substances that impede mass spectrometric identification.
11. Stained gels can be imaged using a near-infrared fluorescence scanner [80, 81] or a conventional flatbed scanner.
12. For mass spectrometric protein identification, manually cut out individual bands immediately after gel staining. Alternatively, excise entire gel lanes with a custom-made punching tool, resulting in 24 uniformly sized gel bands that can be subjected to automated in-gel digestion in microplate format [82]. Store gel bands or slices in adequate vials or microplates at −20 °C until further processed.
13. Wearing gloves and a hair net during these procedures minimizes the contamination of the samples with human keratins.
14. 150 μL is the maximal input volume for 1.5 mL tubes. If upscaling is desired, we recommend 5 mL Eppendorf Safe-Lock tubes and the corresponding fixed-angle rotor as the procedure appears to require the specific tube geometry for proper function
15. Optional: These buffer conditions also allow labeling with fluorescent dyes for quantitative comparison of proteomes by 2D difference gel electrophoresis (2D-DIGE). *See* refs. [83, 84] for protocols, as well as refs. [20, 47] for application to central and peripheral myelin, respectively.
16. For myelin delipidation/precipitation, expect protein loss of up to 50%.
17. This is a typical buffer composition for anodal cup-loading, which is commonly favored for focusing in highly basic gradients. Importantly, conventional cup-loading in our hands was incompatible with myelin, likely due to aggregation. However, the addition of DeStreak diminishes horizontal streaking also after rehydration loading, at least for separation of myelin proteins in broad pH-gradients of up to pH 12.
18. After gel staining (*see* Subheadings 2.2.2 and 3.2.2), cylindric gel spots are excised manually and subjected to automated in-gel digestion in microplate format, followed by analysis on a MALDI-TOF mass spectrometer [85].
19. As the improved coverage of membrane proteins comes at the cost of resolution, we currently see no major advantages in replacing IEF as the first dimension by size-dependent separation in presence of the cationic detergent 16-BAC (16-benzyl-dimethyl-*n*-hexadecylammonium chloride). For a detailed protocol of subjecting CNS myelin to 2D-16-BAC/SDS-PAGE we refer the reader to a previous methods article [50].

20. Pre-solubilization of myelin with ASB-14 under reducing conditions is crucial to increase accessibility of proteins for subsequent alkylation and enzymatic proteolysis. ASB-14 instead of CHAPS was chosen based on prior work [49] and our subsequent positive experience from 2D-IEF/SDS-PAGE (*see* Subheading 3.3). Although used in the original FASP protocol [53], we omit the anionic detergent SDS because potential remains of SDS (but not of the zwitterionic detergents ASB-14 and CHAPS) necessitate sample cleanup prior to injection to protect chromatography columns. Dilution with CHAPS buffer prevents clogging of the filter membrane.
21. Centrifugation is preferred when quantitative liquid removal is desired, as it cannot be entirely excluded that liquid remains within the membrane support in the bottom part of the filter unit after vacuum application.
22. Tryptic digestion in the presence of 10% acetonitrile facilitates the solubility of proteins after removal of urea and of nascent hydrophobic peptides.
23. Stepwise recovery of the tryptic peptides under different pH conditions is intended to increase peptide yield and leads to a sample that is ready to inject without further concentration or cleanup.
24. Centrifugation/vacuum application times may vary considerably depending on the protein input. A typical timing for FASP digestion of myelin is around 24 h with manual centrifugation being somewhat faster than automated vacuum-driven operation. However, considering that the latter is largely a walk-away process, its robustness and reproducibility clearly compensate the higher time need.

Acknowledgment

We thank K.-A. Nave for discussions and financial support made possible by an European Research Council (ERC) Advanced Grant.

References

1. Folch J, Ascoli I, Lees M, Meath JA, Le BN (1951) Preparation of lipide extracts from brain tissue. J Biol Chem 191(2):833–841
2. Folch J, Lees M (1951) Proteolipides, a new type of tissue lipoproteins; their isolation from brain. J Biol Chem 191(2):807–817
3. Lees MB (1998) A history of proteolipids: a personal memoir. Neurochem Res 23 (3):261–271
4. Jahn O, Tenzer S, Werner HB (2009) Myelin proteomics: molecular anatomy of an insulating sheath. Mol Neurobiol 40(1):55–72
5. Mobius W, Patzig J, Nave KA, Werner HB (2008) Phylogeny of proteolipid proteins: divergence, constraints, and the evolution of novel functions in myelination and neuroprotection. Neuron Glia Biol 4(2):111–127. https://doi.org/10.1017/S1740925X0900009X

6. de Monasterio-Schrader P, Patzig J, Mobius W, Barrette B, Wagner TL, Kusch K, Edgar JM, Brophy PJ, Werner HB (2013) Uncoupling of neuroinflammation from axonal degeneration in mice lacking the myelin protein tetraspanin-2. Glia 61(11):1832–1847. https://doi.org/10.1002/glia.22561
7. Brophy PJ, Horvath LI, Marsh D (1984) Stoichiometry and specificity of lipid-protein interaction with myelin proteolipid protein studied by spin-label electron spin resonance. Biochemistry 23(5):860–865
8. Yamaguchi Y, Ikenaka K, Niinobe M, Yamada H, Mikoshiba K (1996) Myelin proteolipid protein (PLP), but not DM-20, is an inositol hexakisphosphate-binding protein. J Biol Chem 271(44):27838–27846
9. Simons M, Kramer EM, Thiele C, Stoffel W, Trotter J (2000) Assembly of myelin by association of proteolipid protein with cholesterol- and galactosylceramide-rich membrane domains. J Cell Biol 151(1):143–154
10. Kramer-Albers EM, Gehrig-Burger K, Thiele C, Trotter J, Nave KA (2006) Perturbed interactions of mutant proteolipid protein/DM20 with cholesterol and lipid rafts in oligodendroglia: implications for dysmyelination in spastic paraplegia. J Neurosci 26 (45):11743–11752. https://doi.org/10.1523/JNEUROSCI.3581-06.2006
11. Ozgen H, Schrimpf W, Hendrix J, de Jonge JC, Lamb DC, Hoekstra D, Kahya N, Baron W (2014) The lateral membrane organization and dynamics of myelin proteins PLP and MBP are dictated by distinct galactolipids and the extracellular matrix. PLoS One 9(7): e101834. https://doi.org/10.1371/journal.pone.0101834
12. Werner HB, Kramer-Albers EM, Strenzke N, Saher G, Tenzer S, Ohno-Iwashita Y, De Monasterio-Schrader P, Mobius W, Moser T, Griffiths IR, Nave KA (2013) A critical role for the cholesterol-associated proteolipids PLP and M6B in myelination of the central nervous system. Glia 61(4):567–586. https://doi.org/10.1002/glia.22456
13. Mandel P, Borkowski T, Harth S, Mardell R (1961) Incorporation of 32P in ribonucleic acid of subcellular fractions of various regions of the rat central nervous system. J Neurochem 8:126–138
14. Eichberg J, Whittaker VP, Dawson RM (1964) Distribution of lipids in subcellular particles of Guinea-pig brain. Biochem J 92(1):91–100
15. Korey SR, Orchen M, Brotz M (1958) Studies of white matter. I. Chemical constitution and respiration of neuroglial and myelin enriched fractions of white matter. J Neuropathol Exp Neurol 17(3):430–438
16. August C, Davison AN, Maurice-Williams F (1961) Phospholipid metabolism in nervous tissue. 4. Incorporation of P32 into the lipids of subcellular fractions of the brain. Biochem J 81:8–12
17. Autilio LA, Norton WT, Terry RD (1964) The preparation and some properties of purified myelin from the central nervous system. J Neurochem 11:17–27
18. Norton WT, Poduslo SE (1973) Myelination in rat brain: method of myelin isolation. J Neurochem 21(4):749–757
19. Haley JE, Samuels FG, Ledeen RW (1981) Study of myelin purity in relation to axonal contaminants. Cell Mol Neurobiol 1 (2):175–187
20. Werner HB, Kuhlmann K, Shen S, Uecker M, Schardt A, Dimova K, Orfaniotou F, Dhaunchak A, Brinkmann BG, Mobius W, Guarente L, Casaccia-Bonnefil P, Jahn O, Nave KA (2007) Proteolipid protein is required for transport of sirtuin 2 into CNS myelin. J Neurosci 27(29):7717–7730. https://doi.org/10.1523/JNEUROSCI.1254-07.2007
21. Wiggins RC, Fuller GN (1981) Analysis of distribution of rat sciatic nerve protein among soluble, insoluble, and myelin subfraction. Neurochem Res 6(6):719–727
22. Gent WL, Gregson NA, Gammack DB, Raper JH (1964) The lipid-protein unit in myelin. Nature 204:553–555
23. Wolfgram F, Kotorii K (1968) The composition of the myelin proteins of the central nervous system. J Neurochem 15(11):1281–1290
24. Morell P, Greenfield S, Costantino-Ceccarini E, Wisniewski H (1972) Changes in the protein composition of mouse brain myelin during development. J Neurochem 19 (11):2545–2554
25. Ewald A, Kühne W (1874-1877) Ueber einen neuen Bestandtheil des Nervensystems. Verhandlungen des Naturhistorisch-Medizinischen Vereins zu Heidelberg 1:457–464
26. Norton WT, Poduslo SE (1973) Myelination in rat brain: changes in myelin composition during brain maturation. J Neurochem 21 (4):759–773
27. Finean JB (1953) Phospholipid-cholesterol complex in the structure of myelin. Experientia 9(1):17–19
28. Cuzner ML, Davison AN, Gregson NA (1965) The chemical composition of vertebrate myelin and microsomes. J Neurochem 12(6):469–481

29. Gopalakrishnan G, Awasthi A, Belkaid W, De Faria O Jr, Liazoghli D, Colman DR, Dhaunchak AS (2013) Lipidome and proteome map of myelin membranes. J Neurosci Res 91(3):321–334. https://doi.org/10.1002/jnr.23157
30. Thakurela S, Garding A, Jung RB, Muller C, Goebbels S, White R, Werner HB, Tiwari VK (2016) The transcriptome of mouse central nervous system myelin. Sci Rep 6:25828. https://doi.org/10.1038/srep25828
31. de Monasterio-Schrader P, Jahn O, Tenzer S, Wichert SP, Patzig J, Werner HB (2012) Systematic approaches to central nervous system myelin. Cell Mol Life Sci 69(17):2879–2894. https://doi.org/10.1007/s00018-012-0958-9
32. Manrique-Hoyos N, Jurgens T, Gronborg M, Kreutzfeldt M, Schedensack M, Kuhlmann T, Schrick C, Bruck W, Urlaub H, Simons M, Merkler D (2012) Late motor decline after accomplished remyelination: impact for progressive multiple sclerosis. Ann Neurol 71(2):227–244. https://doi.org/10.1002/ana.22681
33. Dagley LF, White CA, Liao Y, Shi W, Smyth GK, Orian JM, Emili A, Purcell AW (2014) Quantitative proteomic profiling reveals novel region-specific markers in the adult mouse brain. Proteomics 14(2–3):241–261. https://doi.org/10.1002/pmic.201300196
34. Patzig J, Erwig MS, Tenzer S, Kusch K, Dibaj P, Mobius W, Goebbels S, Schaeren-Wiemers N, Nave KA, Werner HB (2016) Septin/anillin filaments scaffold central nervous system myelin to accelerate nerve conduction. Elife 5. https://doi.org/10.7554/eLife.17119
35. Dumont D, Noben JP, Moreels M, Vanderlocht J, Hellings N, Vandenabeele F, Lambrichts I, Stinissen P, Robben J (2007) Characterization of mature rat oligodendrocytes: a proteomic approach. J Neurochem 102(2):562–576. https://doi.org/10.1111/j.1471-4159.2007.04575.x
36. Iwata K, Cafe-Mendes CC, Schmitt A, Steiner J, Manabe T, Matsuzaki H, Falkai P, Turck CW, Martins-de-Souza D (2013) The human oligodendrocyte proteome. Proteomics 13(23–24):3548–3553. https://doi.org/10.1002/pmic.201300201
37. Cassoli JS, Iwata K, Steiner J, Guest PC, Turck CW, Nascimento JM, Martins-de-Souza D (2016) Effect of MK-801 and clozapine on the proteome of cultured human oligodendrocytes. Front Cell Neurosci 10:52. https://doi.org/10.3389/fncel.2016.00052
38. Sharma K, Schmitt S, Bergner CG, Tyanova S, Kannaiyan N, Manrique-Hoyos N, Kongi K, Cantuti L, Hanisch UK, Philips MA, Rossner MJ, Mann M, Simons M (2015) Cell type- and brain region-resolved mouse brain proteome. Nat Neurosci 18(12):1819–1831. https://doi.org/10.1038/nn.4160
39. Werner HB, Jahn O (2010) Myelin matters: proteomic insights into white matter disorders. Expert Rev Proteomics 7(2):159–164
40. Farias AS, Pradella F, Schmitt A, Santos LM, Martins-de-Souza D (2014) Ten years of proteomics in multiple sclerosis. Proteomics 14(4–5):467–480. https://doi.org/10.1002/pmic.201300268
41. Sherman DL, Brophy PJ (2005) Mechanisms of axon ensheathment and myelin growth. Nat Rev Neurosci 6(9):683–690
42. Simons M, Lyons DA (2013) Axonal selection and myelin sheath generation in the central nervous system. Curr Opin Cell Biol 25(4):512–519. https://doi.org/10.1016/j.ceb.2013.04.007
43. Nave KA, Werner HB (2014) Myelination of the nervous system: mechanisms and functions. Annu Rev Cell Dev Biol 30:503–533. https://doi.org/10.1146/annurev-cellbio-100913-013101
44. Mobius W, Nave KA, Werner HB (2016) Electron microscopy of myelin: structure preservation by high-pressure freezing. Brain Res 1641(Pt A):92–100. https://doi.org/10.1016/j.brainres.2016.02.027
45. Good NE, Winget GD, Winter W, Connolly TN, Izawa S, Singh RM (1966) Hydrogen ion buffers for biological research. Biochemistry 5(2):467–477
46. Neuhoff V, Arold N, Taube D, Ehrhardt W (1988) Improved staining of proteins in polyacrylamide gels including isoelectric focusing gels with clear background at nanogram sensitivity using Coomassie Brilliant Blue G-250 and R-250. Electrophoresis 9(6):255–262. https://doi.org/10.1002/elps.1150090603
47. Patzig J, Jahn O, Tenzer S, Wichert SP, de Monasterio-Schrader P, Rosfa S, Kuharev J, Yan K, Bormuth I, Bremer J, Aguzzi A, Orfaniotou F, Hesse D, Schwab MH, Mobius W, Nave KA, Werner HB (2011) Quantitative and integrative proteome analysis of peripheral nerve myelin identifies novel myelin proteins and candidate neuropathy loci. J Neurosci 31(45):16369–16386. https://doi.org/10.1523/JNEUROSCI.4016-11.2011
48. Wessel D, Flugge UI (1984) A method for the quantitative recovery of protein in dilute

solution in the presence of detergents and lipids. Anal Biochem 138(1):141–143

49. Taylor CM, Pfeiffer SE (2003) Enhanced resolution of glycosylphosphatidylinositol-anchored and transmembrane proteins from the lipid-rich myelin membrane by two-dimensional gel electrophoresis. Proteomics 3(7):1303–1312. https://doi.org/10.1002/pmic.200300451
50. Jahn O, Tenzer S, Bartsch N, Patzig J, Werner HB (2013) Myelin proteome analysis: methods and implications for the myelin cytoskeleton. NeuroMethods 79:335–353
51. Moche M, Albrecht D, Maass S, Hecker M, Westermeier R, Buttner K (2013) The new horizon in 2D electrophoresis: new technology to increase resolution and sensitivity. Electrophoresis 34(11):1510–1518. https://doi.org/10.1002/elps.201200618
52. Distler U, Kuharev J, Navarro P, Tenzer S (2016) Label-free quantification in ion mobility-enhanced data-independent acquisition proteomics. Nat Protoc 11(4):795–812. https://doi.org/10.1038/nprot.2016.042
53. Wisniewski JR, Zougman A, Nagaraj N, Mann M (2009) Universal sample preparation method for proteome analysis. Nat Methods 6 (5):359–362
54. Distler U, Kuharev J, Tenzer S (2014) Biomedical applications of ion mobility-enhanced data-independent acquisition-based label-free quantitative proteomics. Expert Rev Proteomics 11 (6):675–684. https://doi.org/10.1586/14789450.2014.971114
55. Distler U, Kuharev J, Navarro P, Levin Y, Schild H, Tenzer S (2014) Drift time-specific collision energies enable deep-coverage data-independent acquisition proteomics. Nat Methods 11(2):167–170. https://doi.org/10.1038/nmeth.2767
56. Silva JC, Gorenstein MV, Li GZ, Vissers JP, Geromanos SJ (2006) Absolute quantification of proteins by LCMSE: a virtue of parallel MS acquisition. Mol Cell Proteomics 5 (1):144–156. https://doi.org/10.1074/mcp.M500230-MCP200
57. Lappe-Siefke C, Goebbels S, Gravel M, Nicksch E, Lee J, Braun PE, Griffiths IR, Nave KA (2003) Disruption of Cnp1 uncouples oligodendroglial functions in axonal support and myelination. Nat Genet 33(3):366–374. https://doi.org/10.1038/ng1095
58. Edgar JM, McLaughlin M, Werner HB, McCulloch MC, Barrie JA, Brown A, Faichney AB, Snaidero N, Nave KA, Griffiths IR (2009) Early ultrastructural defects of axons and axon-glia junctions in mice lacking expression of Cnp1. Glia 57(16):1815–1824. https://doi.org/10.1002/glia.20893
59. Hagemeyer N, Goebbels S, Papiol S, Kastner A, Hofer S, Begemann M, Gerwig UC, Boretius S, Wieser GL, Ronnenberg A, Gurvich A, Heckers SH, Frahm J, Nave KA, Ehrenreich H (2012) A myelin gene causative of a catatonia-depression syndrome upon aging. EMBO Mol Med 4(6):528–539. https://doi.org/10.1002/emmm.201200230
60. Sherman DL, Wu LM, Grove M, Gillespie CS, Brophy PJ (2012) Drp2 and periaxin form Cajal bands with dystroglycan but have distinct roles in Schwann cell growth. J Neurosci 32 (27):9419–9428. https://doi.org/10.1523/JNEUROSCI.1220-12.2012
61. Domenech-Estevez E, Baloui H, Meng X, Zhang Y, Deinhardt K, Dupree JL, Einheber S, Chrast R, Salzer JL (2016) Akt regulates axon wrapping and myelin sheath thickness in the PNS. J Neurosci 36 (16):4506–4521. https://doi.org/10.1523/JNEUROSCI.3521-15.2016
62. Yannakakis MP, Tzoupis H, Michailidou E, Mantzourani E, Simal C, Tselios T (2016) Molecular dynamics at the receptor level of immunodominant myelin oligodendrocyte glycoprotein 35-55 epitope implicated in multiple sclerosis. J Mol Graph Model 68:78–86. https://doi.org/10.1016/j.jmgm.2016.06.005
63. Derfuss T, Parikh K, Velhin S, Braun M, Mathey E, Krumbholz M, Kumpfel T, Moldenhauer A, Rader C, Sonderegger P, Pollmann W, Tiefenthaller C, Bauer J, Lassmann H, Wekerle H, Karagogeos D, Hohlfeld R, Linington C, Meinl E (2009) Contactin-2/TAG-1-directed autoimmunity is identified in multiple sclerosis patients and mediates gray matter pathology in animals. Proc Natl Acad Sci U S A 106 (20):8302–8307. https://doi.org/10.1073/pnas.0901496106
64. Fewou SN, Fernandes A, Stockdale K, Francone VP, Dupree JL, Rosenbluth J, Pfeiffer SE, Bansal R (2010) Myelin protein composition is altered in mice lacking either sulfated or both sulfated and non-sulfated galactolipids. J Neurochem 112(3):599–610. https://doi.org/10.1111/j.1471-4159.2009.06464.x
65. Martins-de-Souza D (2011) Proteomics as a tool for understanding schizophrenia. Clin Psychopharmacol Neurosci 9(3):95–101. https://doi.org/10.9758/cpn.2011.9.3.95
66. Soong BW, Huang YH, Tsai PC, Huang CC, Pan HC, Lu YC, Chien HJ, Liu TT, Chang MH, Lin KP, Tu PH, Kao LS, Lee YC (2013)

Exome sequencing identifies GNB4 mutations as a cause of dominant intermediate Charcot-Marie-Tooth disease. Am J Hum Genet 92 (3):422–430. https://doi.org/10.1016/j.ajhg.2013.01.014

67. Larocca JN, Norton WT (2007) Isolation of myelin. Curr Protoc Cell Biol Chapter 3: Unit3.25

68. Menon K, Rasband MN, Taylor CM, Brophy P, Bansal R, Pfeiffer SE (2003) The myelin-axolemmal complex: biochemical dissection and the role of galactosphingolipids. J Neurochem 87(4):995–1009

69. Cammer W, Bieler L, Fredman T, Norton WT (1977) Quantitation of myelin carbonic anhydrase-development and subfractionation of rat brain myelin and comparison with myelin from other species. Brain Res 138(1):17–28

70. Danks DM, Matthieu JM (1979) Hypotheses regarding myelination derived from comparisons of myelin subfractions. Life Sci 24 (16):1425–1440

71. Sheads LD, Eby MJ, Sampugna J (1977) Myelin subfractions isolated from mouse brain. Studies of normal mice during development, quaking mutants, and three brain regions. J Neurobiol 8(1):67–89. https://doi.org/10.1002/neu.480080106

72. Kirschner DA, Inouye H, Ganser AL, Mann V (1989) Myelin membrane structure and composition correlated: a phylogenetic study. J Neurochem 53(5):1599–1609

73. Waehneldt TV (1990) Phylogeny of myelin proteins. Ann N Y Acad Sci 605:15–28

74. Roth AD, Ivanova A, Colman DR (2006) New observations on the compact myelin proteome. Neuron Glia Biol 2(1):15–21. https://doi.org/10.1017/S1740925X06000068

75. Baer AS, Syed YA, Kang SU, Mitteregger D, Vig R, Ffrench-Constant C, Franklin RJ, Altmann F, Lubec G, Kotter MR (2009) Myelin-mediated inhibition of oligodendrocyte precursor differentiation can be overcome by pharmacological modulation of Fyn-RhoA and protein kinase C signalling. Brain 132 (Pt 2):465–481. https://doi.org/10.1093/brain/awn334

76. Ishii A, Dutta R, Wark GM, Hwang SI, Han DK, Trapp BD, Pfeiffer SE, Bansal R (2009) Human myelin proteome and comparative analysis with mouse myelin. Proc Natl Acad Sci U S A 106(34):14605–14610. https://doi.org/10.1073/pnas.0905936106

77. Nawaz S, Schweitzer J, Jahn O, Werner HB (2013) Molecular evolution of myelin basic protein, an abundant structural myelin component. Glia 61(8):1364–1377

78. Morris JK, Willard BB, Yin X, Jeserich G, Kinter M, Trapp BD (2004) The 36K protein of zebrafish CNS myelin is a short-chain dehydrogenase. Glia 45(4):378–391. https://doi.org/10.1002/glia.10338

79. Schaefer K, Brosamle C (2009) Zwilling-A and -B, two related myelin proteins of teleosts, which originate from a single bicistronic transcript. Mol Biol Evol 26(3):495–499. https://doi.org/10.1093/molbev/msn298

80. Luo S, Wehr NB, Levine RL (2006) Quantitation of protein on gels and blots by infrared fluorescence of Coomassie blue and Fast Green. Anal Biochem 350(2):233–238. https://doi.org/10.1016/j.ab.2005.10.048

81. Harris LR, Churchward MA, Butt RH, Coorssen JR (2007) Assessing detection methods for gel-based proteomic analyses. J Proteome Res 6(4):1418–1425. https://doi.org/10.1021/pr0700246

82. Schmidt C, Hesse D, Raabe M, Urlaub H, Jahn O (2013) An automated in-gel digestion/iTRAQ-labeling workflow for robust quantification of gel-separated proteins. Proteomics 13 (9):1417–1422. https://doi.org/10.1002/pmic.201200366

83. Viswanathan S, Unlu M, Minden JS (2006) Two-dimensional difference gel electrophoresis. Nat Protoc 1(3):1351–1358

84. Tannu NS, Hemby SE (2006) Two-dimensional fluorescence difference gel electrophoresis for comparative proteomics profiling. Nat Protoc 1(4):1732–1742. https://doi.org/10.1038/nprot.2006.256

85. Jahn O, Hesse D, Reinelt M, Kratzin HD (2006) Technical innovations for the automated identification of gel-separated proteins by MALDI-TOF mass spectrometry. Anal Bioanal Chem 386(1):92–103. https://doi.org/10.1007/s00216-006-0592-1

Chapter 4

In Vitro Generation and Electrophysiological Characterization of OPCs and Oligodendrocytes from Human Pluripotent Stem Cells

Dario Magnani, Siddharthan Chandran, David J. A. Wyllie, and Matthew R. Livesey

Abstract

The in vitro generation of defined cellular populations from induced human pluripotent stem cells (iPSCs) provides the opportunity to work routinely with human material and, importantly, allows examination of material derived from patients with clinically and genetically diagnosed disorders. In this regard, the ability to derive oligodendrocytes in vitro represents an important resource to examine human oligodendrocyte-lineage cell biology in normal and disease contexts. Oligodendrocytes undergo characteristic physiological maturation during differentiation in vitro, and patch-clamp electrophysiology allows a detailed examination of maturation state and, potentially, pathologically related variations of ion channel expression and regulation. Here, we detail our methodology to generate oligodendrocyte precursor cells and oligodendrocytes and characterize them electrophysiologically.

Key words Oligodendrocyte, Oligodendrocyte precursor cell, Culture, Human pluripotent stem cell, Electrophysiology, Patch-clamp

1 Introduction

The utility of induced human pluripotent stem cells (iPSCs) means it is possible to generate a variety of cell types with the genetic identity of the individuals from whom the iPSCs were derived. Several protocols, including ours [1], now exist that detail the in vitro generation of oligodendrocyte progenitor cells (OPCs) and oligodendrocytes from iPSCs [2–6], and direct conversion from patient fibroblasts [7]. For researchers, these present potentially large-scale platforms to explore human oligodendrocyte development, function, and pathology. In this respect, iPSC-derived oligodendrocytes obtained from amyotrophic lateral sclerosis [1], multiple sclerosis [2], and Pelizaeus-Merzbacher disease [4] patients have been examined. Furthermore, human iPSC-

David A. Lyons and Linde Kegel (eds.), *Oligodendrocytes: Methods and Protocols*, Methods in Molecular Biology, vol. 1936, https://doi.org/10.1007/978-1-4939-9072-6_4,

derived oligodendroglial cultures have been used to verify lead compounds identified from screening platforms designed to explore pharmacology that promotes oligodendrocyte differentiation [8, 9]. The capacity of iPSC-OPCs to differentiate into mature myelinating oligodendrocytes has been observed by the ability of iPSC-derived OPCs engrafted into the central nervous system of myelin-deficient mice to generate substantial axon myelination [2, 5, 10, 11]. This gives future scope for the clinical application of iPSC-derived OPCs and oligodendrocytes in regenerative approaches to demyelinating conditions [12].

It is notable that native OPCs receive synaptic-like connections from axons, and that the development and myelinating properties of oligodendrocytes in a variety of experimental models is modulated by neuronal activity [13–18]. The consequences of this signaling and functional modulation of oligodendroglia to modulate neuronal behavior, and vice versa, remain under-explored though are thought to be important in experience-dependent plasticity and learning [19]. Aberrant membrane ion channel regulation and function is also implicated in oligodendroglial pathologies [20]. This together with the fact that human–rodent species differences are present in the distribution and development of oligodendrocytes [21] indicates that it is of importance to better understand the membrane properties of human oligodendrocytes. We have been exploring these possibilities in our laboratory using a range of biochemical, immunohistochemical, and electrophysiological techniques.

Here we detail the methodology of our protocol to generate iPSC-derived oligodendrocytes and of our assessment of the functional properties of these cells using patch-clamp electrophysiology.

1.1 iPSC-Derived Oligodendrocyte Protocol

The majority of current protocols to obtain in vitro iPSC-derived oligodendrocytes are based on knowledge of the development of the vertebrate spinal cord. In the developing embryo oligodendrocytes originate from the ventral neural tube which is first patterned along the rostral-caudal axis, mainly by a gradient of retinoic acid (RA). Subsequently opposing gradients of SHH (ventral^{high}–dorsal^{low}) and BMPs/WNTs (dorsal^{high}–ventral^{low}) specify distinct progenitor domains along the dorso-ventral axes (for review *see* ref. [22]). Therefore, longer exposure to a higher concentration of SHH is associated with the formation of more ventral cell types including motor neurons (MNs) in the spinal cord, and several interneuron types. Spinal cord-derived OPCs originate within the OLIG2^{+} progenitor MN domain (pMN—which generate MNs) following the neurogenic phase [22].

Our protocol recapitulates in vitro spinal cord development exploiting these signaling pathways to generate ventral spinal cord-patterned OLIG2^{+} precursor cells, and it is currently comparable in time as well as in yields to another previously reported fast protocols, e.g., [2, 23]. Human iPSCs are differentiated into

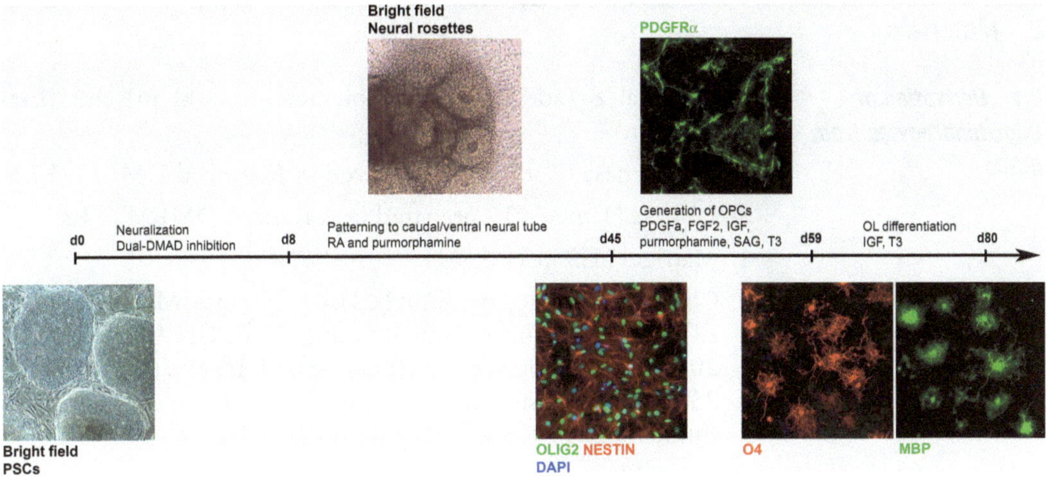

Fig. 1 Differentiation of oligodendrocytes from iPSCs. Pluripotent stem cells are first neuralized by dual-SMAD inhibition for 7 days. To achieve a high yield of OLIG2$^+$ cells neural progenitors are subsequently patterned to ventral spinal cord, with the application of retinoic acid (RA) and SHH agonist purmorphamine for a further 37 days. OLIG2$^+$ progenitors are finally expanded in the presence of PDGFα, FGF2, and other mitogens for up to 8 weeks. From day 59 oligodendrocyte differentiation from OPCs can be initiated by mitogen withdrawal

neuroectoderm by dual-SMAD inhibition [24]. Neurospheres subsequently are caudalized in suspension in the presence of RA and ventralized by induction of SHH signaling. Following patterning and induction of OLIG2$^+$ progenitor cells, addition of FGF2 and PDGFα over a relative extended period results into their conversion to PDGFRα$^+$-OPCs. Subsequent removal of these mitogenic agents induces the differentiation of O4$^+$/MBP$^+$-oligodendrocytes. An overview of our protocol design is given in Fig. 1.

1.2 Source of Human Pluripotent Stem Cells

iPSCs are typically derived from fibroblasts using viral transformation via the Yamanaka method [25, 26]. Numerous pluripotent stem cell lines, both embryonic stem cells and iPSCs, are available commercially or through numerous repositories. We assume that the operator has successfully sourced viable, healthy iPSCs, are experienced in standard cell culture practice and techniques, and has acquired the appropriate ethical approval from their host organization prior to receiving and using iPSCs.

For any iPSC study, there is potential for independent lines to exhibit variable or atypical properties, thus leading to potential erroneous conclusions (reviewed in ref. [27]). We therefore advocate the use of iPSC lines generated from several healthy and control subjects, particularly for disease-related studies. Recent advances in gene editing [27, 28] offers the ability to counter this issue if the gene of interest is known. In principle, such technology allows the correction or introduction of a mutant gene without affecting the rest of the genome, thus establishing causality between mutation and phenotype in an isogenic background.

2 Materials

2.1 Derivation of Oligodendrocytes from iPSCs

1. Essential 8 (add 10 mL supplement to 500 mL E8 basal medium).
2. Collagenase (2 mg/mL) prepared in Advanced DMEM/F12.
3. Dispase (1 mg/mL) prepared in Advanced DMEM/F12.
4. Matrigel (1:2 in DMEM).
5. Chemically defined medium (CDM): 250 mL IMDM, 250 mL F12, 2.5 g Bovine Serum Albumin, 5 mL CD Lipid (100×), 20 μL Monothioglycerol (100%), 350 μL Insulin (10 mg/mL), 250 μL Transferrin (30 mg/mL), 1 mL Antibiotic-Antimycotic (100×).
6. 6-well plates.
7. 15 mL falcon tubes.
8. DPBS (Ca^{2+}-, Mg^{2+}-free).
9. Liquid nitrogen.
10. 10% DMSO in Essential 8 medium.

2.2 Generation of Ventral Neural Progenitors from iPSCs

1. Neuralization medium (Phase I): 100 mL CDM (*see* Subheading 2.1, **item 5**), 200 μL *N*-acetyl cysteine (1.63 g in 20 mL of DPBS), 100 μL Activin Inhibitor (10 mg in 2.6 mL of DMSO), 100 μL LDN (100 μM).
2. Neuralization medium (Phase II): 100 mL CDM (*see* Subheading 2.1, **item 5**), 200 μL *N*-acetyl cysteine (1.63 g in 20 mL of DPBS), 25 μL FGF (20 μg/mL), 10 μL, Retinoic acid (100 μg in 9.9 mL of ethanol).
3. Neuralization medium (Phase III): 100 mL Adv DMEM/F12, 1 mL Antibiotic-Antimycotic (100×) 1 mL B27 (50×), 500 μL N_2 (100×) 500 μL Glutamax (100×), 25 μL FGF (20 μg/mL), 100 μL Purmorphamine (1 mM), 100 μL Retinoic acid (1 mM).
4. Neuralization medium Phase III-FGF: 100 mL Adv DMEM/F12, 1 mL Antibiotic-Antimycotic (100×), 1 mL B27 (50×), 500 μL N_2 (100×), 500 μL Glutamax (100×), 100 μL Purmorphamine (1 mM), 100 μL Retinoic acid (1 mM).
5. 90 mm dishes.
6. Orbital shaker set at 45 rpm.
7. Cell culture incubator at 37 °C.

2.3 Generation of OPCs and Oligodendrocytes from NPCs

1. Oligo Base: 500 mL Advance DMEM/F12, 5 mL Antibiotic-Antimycotic (100×), 5 mL, Glutamax (100×), 5 mL N_2 (100×), 5 mL B27 (50×), 500 μL Heparin (5 μg/mL).

2. Oligo-Pro media: 200 mL Oligo Base, 100 μL FGF (20 μg/mL), 200 μL PDGFα (20 μg/mL), 200 μL Purmorphamine (1 mM), 100 μL IGF (20 μg/mL), 300 μL T3 (60 μg/mL in Advance DMEM/F12), 200 μL SAG (1 mM).
3. Oligodendrocyte differentiation media: 200 mL Oligo Base, 100 μL IGF (20 μg/mL), 300 μL T3 (60 μg/mL), 2 mL Insulin-Transferrin-Selenium-X Supplement (100×).
4. Sterile razor blade.
5. Papain (Papain Dissociation System, Lorne Laboratories Ltd).
6. Matrigel (1:100), fibronectin (20 μg/mL), laminin (10 μg/mL) coated 13 mm coverslips (or plates).

2.4 Electrophysiological Materials

See **Note 1**.

2.4.1 Patch Pipette Recording Solution Materials

1. K-gluconate (155 mM).
2. $MgCl_2$ (2 mM).
3. Na-HEPES (10 mM).
4. Na-PiCreatine (10 mM).
5. Mg_2-ATP (2 mM).
6. Na_3-GTP (0.3 mM).
7. KOH (1 M).
8. 0.22 μm syringe filter.
9. High resistance, deionized water (18.3 MΩ).

2.4.2 Extracellular/Bath Recording Solution Materials

1. NaCl 152 mM.
2. KCl 2.8 mM.
3. HEPES (10 mM).
4. $CaCl_2$ (10 mM).
5. Glucose (10 mM).
6. NaOH (1 M).

3 Methods

3.1 Maintenance of iPSCs

1. Undifferentiated iPSCs are maintained on Matrigel-coated (1:60 dilution in cold Advance DMEM/F12) 6-well plates with 2 mL of Essential 8 medium and fed daily.
2. Once the cell monolayer reaches 80–95% confluence, cells are passaged by enzymatic digestion with incubation in an equal mixture of dispase (1 mg/mL) and collagenase (2 mg/mL)

solution prepared in Advanced DMEM/F12, for 20–30 min at 37 °C.

3. Detached stem cell colonies are then collected in 15 mL falcon tubes, washed with DPBS (Ca^{2+}-, Mg^{2+}-free; Life Technologies, 14190-094) and resuspended in Essential 8 medium at the required dilution and plated in fresh Matrigel-coated wells.
4. iPSCs can be banked and stored in liquid nitrogen by gentle freezing with (COOLCELL FTS30, VWR international) in 10% DMSO in Essential 8 medium.

3.2 Generation of Ventral Neural Progenitors from iPSCs

3.2.1 Neuralization (Phase I) (See **Note 2**)

1. Undifferentiated iPSCs are incubated with dispase/collagenase enzyme mix, detached, collected, and washed as previously described.
2. To allow embryonic bodies or neurospheres formation, cells are resuspended in 10 mL Phase I media, transferred into 90 mm dishes, and placed on an orbital shaker, set at 45 rpm in a cell culture incubator at 37 °C.
3. Neurospheres are maintained in Phase I media for 1 week with 10 mL media changes every 2–3 days. From this step onward neurospheres are cultured in suspension on an orbital shaker.

3.2.2 Phase II

1. On day 8, neurospheres are transferred to 10 mL Phase II media with withdrawal of SMAD inhibitors and addition of retinoid acid for caudal/spinal cord patterning of neural progenitors cells (NPCs).

3.2.3 Phase II–Phase III

2. On day 15 neurospheres are let adhere to cell culture plates, coated with 1:100 laminin and cultured in adhesion in Phase II media for 2–3 days.
3. At this stage, it is possible to select neurospheres with the use of a stereo microscope based on their morphology to further enrich for neural derivatives and/or to synchronize neuronal differentiation. These can subsequently be resuspended in Phase III media.
4. Differentiated neurospheres display a radial morphology with neural cell adhering to the substrate and forming neural processes.
5. Alternatively, neurospheres can be detached from the plates altogether with a cell scraper and resuspended in Phase III media containing SHH or SHH agonists for ventral patterning for an additional 7 days.

3.2.4 Phase III–Phase III-FGF

1. On day 29 neurospheres are transferred to Phase III-FGF media (*see* Subheading 2.2, **item 4**) and cultured for further 14 days. At this stage neurospheres will be highly enriched for $OLIG2^{+}$-progenitor cells.

3.3 Generation of OPCs and Oligodendrocytes from NPCs

3.3.1 OPC Maintenance

1. On day 45, neurospheres are transferred to oligodendrocyte proliferation media (Oligo-Pro media).
2. OPC spheres can be maintained long-term in Oligo-Pro media (up to 8 weeks).
3. If spheres reach between 5 and 10 mm they require mechanical chopping using a sterile razor blade. This may be required to be performed on a monthly basis (*see* **Note 3**).

3.3.2 Oligodendrocyte Differentiation from OPCs

1. Spheres are enriched for OPCs and ready to be plated down from 2 weeks in Oligo-Pro media.
2. "Oligo-spheres" containing OPCs are dissociated with papain and plated as a monolayer on Matrigel (1 in 100 dilution), fibronectin (20 μg/mL, F2006, Sigma-Aldrich), laminin (10 μg/mL, L2020, Sigma-Aldrich) coated 13 mm coverslips (or plates) at a density of 20,000–30,000 cells per 0.3 cm^2.
3. Oligo-spheres will be enriched for OPCs and ready to be plated down from 2 weeks in Oligo-Pro media.
4. OPCs are terminally differentiated in oligodendrocyte differentiation media by default upon withdrawal of cytokines and can be kept in culture in monolayer for up to 3 weeks.

3.4 Patch-Clamp Recording from iPSC-Derived Oligodendrocytes

Patch-clamp electrophysiology can be used to examine a range of cellular membrane activity from broad whole-cell to single-channel activity [29]. In combination with pharmacological tools it is possible to examine, in isolation, the properties of specific types of activity and/or ion channels of interest. Indeed, this method has been used to extensively characterize rodent oligodendrocyte-lineage cells [13, 30–33]. We will focus on whole-cell recording, the most common patch-clamp configuration to examine the properties of cells. While we include details regarding our specific setup, this review is written from the perspective that the reader has a patch-clamp electrophysiology rig setup and associated infrastructure to make a recording.

3.4.1 Live Staining of iPSC-Derived Oligodendrocytes

The above protocol generates heterogeneous cultures and identification of iPSC-oligodendrocytes is therefore required for patch-clamp electrophysiology. This can be achieved using a live immunolabeling approach.

1. Add 150 μL of O4 antibody (1:300, R&D Systems) to 100 μL of oligodendrocyte differentiation media that submerges the coverslips in 1 well of a 24-well plate.
2. Cells are incubated at 37 °C for 40 min.
3. Coverslips are then washed and incubated with secondary antibody (conjugated with GFP) diluted in media (1:1000) at 37 °C for a further 30–60 min, then assessed (*see* **Note 4**).

3.4.2 The Recording Bath

1. Transfer the 13 mm coverslip to the recording bath containing extracellular recording solution (*see* **Notes 5** and **6**).
2. Wash the coverslip with external recording solution to wash away any residual debris/media and observe the cells to see if the coverslip remains fully secure before committing to make a recording (*see* **Note** 7).

3.4.3 Identification of a Cell for Recording

Cells are first observed using differential interference contrast (DIC) and a high-magnification objective lens (*see* **Note 8**).

3.4.4 Patch Pipette Recording Solution Preparation

1. Prepare solutions using high resistance, deionized water (18.3 MΩ) and prepare with high-grade, research quality reagents.
2. The patch pipette electrode solution comprises (in mM): K-gluconate 155, $MgCl_2$ 2, Na-HEPES 10, Na-PiCreatine 10, Mg_2-ATP 2, and Na_3-GTP 0.3. The solution is brought to pH 7.3 using KOH (1 M).
3. The osmolarity of the solution is 300 mOsm and should be checked using an osmometer.
4. The solution is always passed through a 0.22 mm pore diameter filter before use to remove any debris that may block the patch electrode.
5. The intracellular solution may be stored in aliquots and kept as frozen stock for up to a month.

3.4.5 Extracellular Bath Solution Preparation

1. The extracellular electrode solution comprises (in mM): NaCl 152, KCl 2.8, HEPES 10, $CaCl_2$ 2, glucose 10, pH to 7.3 using NaOH (1 M); 320–330 mOsm.
2. Before placing cells into the recording bath, it is necessary to wash the perfusion system thoroughly with deionized water and then extracellular bath solution (*see* **Note 9**).

3.4.6 Manufacture of Patch-Clamp Electrodes

1. We fabricate our electrodes from thick-walled borosilicate glass (GC150F-7.5, Warner Instruments) using a specialist micropipette pulling system (P-1000, Sutter Instruments, USA).
2. The pulling system is empirically optimized according to manufacturer's guidelines to produce electrodes with a resistance of 3–6 MΩ and a taper of approximately 4–5 mm.
3. The electrode tip can be fire polished using a micro-forge (Narishige, Japan) to smooth the glass surface and to increase the resistance of the pipette.

3.5 Making a Recording

3.5.1 Filling the Electrode

1. Fill the electrode with pipette solution.
2. Attach the electrode to the electrode holder (*see* **Note 10**).

3.5.2 Electrode Positioning

1. The manipulators are used to position the tip of the electrode into the bath solution and approximately 10 μm above and 10 μm adjacent to the O4^{+}-oligodendrocyte cell body. This is performed while applying light positive pressure (~5 mm Hg) to the syringe and tubing connected to the electrode holder to remove any debris from the electrode tip.
2. Zero the pipette potential offset and apply a voltage test-pulse (5 mV, 10 ms) using the amplifier. The resistance of the electrode can be calculated from the current step observed, though modern electrophysiological data acquisition software can calculate this (*see* **Note 11**).
3. Liquid Junction Potential (LJP) can be estimated using PClamp electrophysiology software [34].
4. Membrane potential data can be corrected post hoc for LJP.

3.5.3 Cell-Attached Configuration

1. Under high power magnification use the fine gears of the manipulator and carefully position the electrode tip to be directly adjacent to the cell body.
2. A small reduction in the current step amplitude of the test pulse maybe observed as the electrode is brought close the cell membrane.
3. While positive pressure is maintained, a small indentation of the cell membrane may be observed due to the solution being ejected from the electrode tip.
4. Cessation of the positive pressure typically causes the membrane to seal to the electrode tip and the current step to effectively disappear. This signifies high-resistance seal formation. A seal of >2–3 GΩ is desirable.
5. On occasion a small amount of light negative pressure applied to the syringe connected to electrode holder/electrode is required to complete high-resistance seal formation.
6. Small "fast" decaying capacitive current transients may be observed which can be nulled using the fast capacitive cancellation on the amplifier.

3.5.4 Whole-Cell Configuration

1. Set the holding potential of the amplifier to the required holding potential. This invariably will be at a hyperpolarized potential (for example −70 mV) to prevent chronic depolarization of the cell of interest.

2. Using negative pressure carefully rupture the cell membrane at the tip of the electrode.
3. Successful rupture will be indicated by the simultaneous emergence of "slow" decaying capacitive transients in response to the test pulse. In the case of oligodendrocytes these can be potentially large given the large surface area of the cell membrane.
4. The holding potential is held at −70 mV (−84 mV including LJP) which is close to the reversal potential of K^+ ions.
5. The holding current should be no more than −150 pA at this holding potential.
6. The slow capacitive transients are canceled and the series resistance is determined using the amplifier. Series resistance should be no more than 25 MΩ and should drift no more than 15% during the course of an experiment. Series resistance compensation (>70%) is typically applied.
7. Removal of the test pulse allows a recording to go ahead.
8. Once successful whole-cell configuration has been obtained, then experimenter should proceed with recordings in their preferred recording configuration (voltage-clamp or current-clamp).
9. Various protocols can be employed to examine intrinsic membrane currents and activity by application of exogenous pharmacological compounds.
10. Recordings are typically low-pass filtered at 2 kHz and digitized at 10 kHz.

4 Notes

1. Prepare solutions using high resistance, deionized water (18.3 MΩ) and high-grade, research quality reagents.
2. The neuralization strategy for iPSCs is adapted from [24].
3. Chopping encourages the proliferation of progenitor cells and removal of postmitotic neurons.
4. The percentage of $O4^+$-iPSC-derived cells substantially overlaps with those that are MBP^+.
5. The bath must be suitably sized and positioned to accommodate the coverslip, perfusion inlets/outlets, bath Ag/AgCl electrode, and patch electrodes.
6. It is often suitable to secure the coverslip by gently pushing the coverslip against the glass at the bottom of the recording chamber. However, if this is not possible, then application of a small spot of petroleum jelly on the underside of the coverslip

works well. We employ a gravity-fed perfusion system and it is important to position the perfusion inlet and outlet points to ensure that the coverslip is super-fused with a laminar flow of solution, thus allowing reasonably efficient exchange of solution when various extracellular solutions are being employed. Faster solution exchange systems maybe required depending on experimental design.

7. It may be required to modify speed of perfusion or even examine the quality of the coverslip coating should cells detach while being super-fused.
8. Care should be taken when switching between microscope objectives as sudden movements or raising the bath solution to submerge the objective can dislodge cells or the coverslip. Switching on the excitation light source with the appropriate filter will reveal the distinctive morphology of live-stained $O4^+$-oligodendrocytes. The intensity of the DIC light source maybe required to be dimmed to observe the labeled cells. Only cells that do not appear punctate and exhibit uniform O4 labeling are examined electrophysiologically.
9. It may be required to perform experiments in the presence of selective pharmacological compounds and/or modify the composition of the recording solutions in order to examine specific activity or ion channels (*see* ref. [1], for example).
10. The degree of filling the electrode with solution is only required so that the Ag/AgCl wire is submerged and not to the end of the electrode. The latter is source of capacitance-based noise. It is important to periodically apply chloride to the wire to minimize offset errors generated by liquid-metal junction potentials.
11. A liquid junction potential (LJP) difference arising at the electrode tip due to the differences in the ionic activities of the respective electrode and extracellular recording solutions requires to be considered, particularly if the ionic compositions are not alike such as those described above (LJP offset = 14 mV). This is irrespective of the potential offset.

Acknowledgments

Funded by The Wellcome Trust (Grant 092742/Z/10/Z), MNDA, MRC, Euan MacDonald Centre, DBT-India, ISSF (WT/UoE), Royal Society of Edinburgh and Biogen/UoE Joint Discovery Research Collaboration. Authors state no conflict of interest.

References

1. Livesey MR, Magnani D, Cleary EM, Vasistha NA, James OT, Selvaraj BT, Burr K, Story D, Shaw CE, Kind PC, Hardingham GE, Wyllie DJ, Chandran S (2016) Maturation and electrophysiological properties of human pluripotent stem cell-derived oligodendrocytes. Stem Cells 34(4):1040–1053
2. Douvaras P, Wang J, Zimmer M, Hanchuk S, O'Bara MA, Sadiq S, Sim FJ, Goldman J, Fossati V (2014) Efficient generation of myelinating oligodendrocytes from primary progressive multiple sclerosis patients by induced pluripotent stem cells. Stem Cell Reports 3 (2):250–259
3. Stacpoole SR, Spitzer S, Bilican B, Compston A, Karadottir R, Chandran S, Franklin RJ (2013) High yields of oligodendrocyte lineage cells from human embryonic stem cells at physiological oxygen tensions for evaluation of translational biology. Stem Cell Reports 1(5):437–450
4. Numasawa-Kuroiwa Y, Okada Y, Shibata S, Kishi N, Akamatsu W, Shoji M, Nakanishi A, Oyama M, Osaka H, Inoue K, Takahashi K, Yamanaka S, Kosaki K, Takahashi T, Okano H (2014) Involvement of ER stress in dysmyelination of Pelizaeus-Merzbacher disease with PLP1 missense mutations shown by iPSC-derived oligodendrocytes. Stem Cell Reports 2(5):648–661
5. Wang S, Bates J, Li X, Schanz S, Chandler-Militello D, Levine C, Maherali N, Studer L, Hochedlinger K, Windrem M, Goldman SA (2013) Human iPSC-derived oligodendrocyte progenitor cells can myelinate and rescue a mouse model of congenital hypomyelination. Cell Stem Cell 12(2):252–264
6. Hu BY, Du ZW, Zhang SC (2009) Differentiation of human oligodendrocytes from pluripotent stem cells. Nat Protoc 4(11):1614–1622
7. Yang N, Zuchero JB, Ahlenius H, Marro S, Ng YH, Vierbuchen T, Hawkins JS, Geissler R, Barres BA, Wernig M (2013) Generation of oligodendroglial cells by direct lineage conversion. Nat Biotechnol 31(5):434–439
8. Najm FJ, Madhavan M, Zaremba A, Shick E, Karl RT, Factor DC, Miller TE, Nevin ZS, Kantor C, Sargent A, Quick KL, Schlatzer DM, Tang H, Papoian R, Brimacombe KR, Shen M, Boxer MB, Jadhav A, Robinson AP, Podojil JR, Miller SD, Miller RH, Tesar PJ (2015) Drug-based modulation of endogenous stem cells promotes functional remyelination in vivo. Nature 522(7555):216–220
9. Mei F, Mayoral SR, Nobuta H, Wang F, Desponts C, Lorrain DS, Xiao L, Green AJ, Rowitch D, Whistler J, Chan JR (2016) Identification of the kappa-opioid receptor as a therapeutic target for oligodendrocyte remyelination. J Neurosci 36(30):7925–7935
10. Marteyn A, Sarrazin N, Yan J, Bachelin C, Deboux C, Santin MD, Gressens P, Zujovic V, Baron-Van Evercooren A (2016) Modulation of the innate immune response by human neural precursors prevails over oligodendrocyte progenitor remyelination to rescue a severe model of Pelizaeus-Merzbacher disease. Stem Cells 34(4):984–996
11. Piao J, Major T, Auyeung G, Policarpio E, Menon J, Droms L, Gutin P, Uryu K, Tchieu J, Soulet D, Tabar V (2015) Human embryonic stem cell-derived oligodendrocyte progenitors remyelinate the brain and rescue behavioral deficits following radiation. Cell Stem Cell 16(2):198–210
12. Goldman SA, Kuypers NJ (2015) How to make an oligodendrocyte. Development 142 (23):3983–3995
13. Larson VA, Zhang Y, Bergles DE (2016) Electrophysiological properties of NG2(+) cells: matching physiological studies with gene expression profiles. Brain Res 1638 (Pt B):138–160
14. De Biase LM, Nishiyama A, Bergles DE (2010) Excitability and synaptic communication within the oligodendrocyte lineage. J Neurosci 30(10):3600–3611
15. Zonouzi M, Renzi M, Farrant M, Cull-Candy SG (2011) Bidirectional plasticity of calcium-permeable AMPA receptors in oligodendrocyte lineage cells. Nat Neurosci 14(11):1430–1438
16. Zonouzi M, Scafidi J, Li P, McEllin B, Edwards J, Dupree JL, Harvey L, Sun D, Hübner CA, Cull-Candy SG, Farrant M, Gallo V (2015) GABAergic regulation of cerebellar NG2 cell development is altered in perinatal white matter injury. Nat Neurosci 18 (5):674–682
17. Koudelka S, Voas MG, Almeida RG, Baraban M, Soetaert J, Meyer MP, Talbot WS, Lyons DA (2016) Individual neuronal subtypes exhibit diversity in CNS myelination mediated by synaptic vesicle release. Curr Biol 26(11):1447–1455
18. Mensch S, Baraban M, Almeida R, Czopka T, Ausborn J, El Manira A, Lyons DA (2015) Synaptic vesicle release regulates myelin sheath

number of individual oligodendrocytes in vivo. Nat Neurosci 18(5):628–630

19. Bergles DE, Richardson WD (2015) Oligodendrocyte development and plasticity. Cold Spring Harb Perspect Biol 8(2):a020453
20. Butts BD, Houde C, Mehmet H (2008) Maturation-dependent sensitivity of oligodendrocyte lineage cells to apoptosis: implications for normal development and disease. Cell Death Differ 15(7):1178–1186
21. Bradl M, Lassmann H (2010) Oligodendrocytes: biology and pathology. Acta Neuropathol 119:37–53
22. Richardson WD, Kessaris N, Pringle N (2006) Oligodendrocyte wars. Nat Rev Neurosci 7 (1):11–18
23. Douvaras P, Fossati V (2015) Generation and isolation of oligodendrocyte progenitor cells from human pluripotent stem cells. Nat Protoc 10(8):1143–1154
24. Chambers SM, Fasano CA, Papapetrou EP, Tomishima M, Sadelain M, Studer L (2009) Highly efficient neural conversion of human ES and iPS cells by dual inhibition of SMAD signaling. Nat Biotechnol 27(3):275–280
25. Yamanaka S (2008) Induction of pluripotent stem cells from mouse fibroblasts by four transcription factors. Cell Prolif 41(Suppl 1):51–56
26. Mostoslavsky G (2012) Concise review: the magic act of generating induced pluripotent stem cells: many rabbits in the hat. Stem Cells 30(1):28–32
27. Sandoe J, Eggan K (2013) Opportunities and challenges of pluripotent stem cell neurodegenerative disease models. Nat Neurosci 16(7):780–789
28. Hendriks WT, Warren CR, Cowan CA (2016) Genome editing in human pluripotent stem cells: approaches, pitfalls, and solutions. Cell Stem Cell 18(1):53–65
29. Sakmann B, Neher E (1995) Single-channel recording. Plenum Press, New York; London
30. Barres BA, Koroshetz WJ, Swartz KJ, Chun LL, Corey DP (1990) Ion channel expression by white matter glia: the O-2A glial progenitor cell. Neuron 4(4):507–524
31. Sontheimer H, Trotter J, Schachner M, Kettenmann H (1989) Channel expression correlates with differentiation stage during the development of oligodendrocytes from their precursor cells in culture. Neuron 2 (2):1135–1145
32. Káradóttir R, Attwell D (2007) Neurotransmitter receptors in the life and death of oligodendrocytes. Neuroscience 145 (4):1426–1438
33. Tripathi RB, Clarke LE, Burzomato V, Kessaris N, Anderson PN, Attwell D, Richardson WD (2011) Dorsally and ventrally derived oligodendrocytes have similar electrical properties but myelinate preferred tracts. J Neurosci 31(18):6809–6819
34. Barry PH (1994) JPCalc, a software package for calculating liquid junction potential corrections in patch-clamp, intracellular, epithelial and bilayer measurements and for correcting junction potential measurements. J Neurosci Methods 51(1):107–116

Chapter 5

Isolation and Culture of Oligodendrocytes

Marie-Theres Weil, Giselheid Schulz-Ëberlin, Chaitali Mukherjee, Wen Ping Kuo-Elsner, Isabelle Schäfer, Christina Müller, and Mikael Simons

Abstract

Primary cultures of brain-derived rodent cells are widely used to study molecular and cellular mechanisms in neurobiology. In this chapter, we describe methods of purifying and culturing oligodendroglial cells from mouse perinatal brains. In addition, we describe methods of coculturing the purified oligodendrocytes with neurons. When prepared and cultured according to these protocols, many essential aspects of the biology of oligodendrocytes, such as their proliferation, differentiation, and myelination, can be studied in culture.

Key words Oligodendrocyte primary culture, Coculture, Magnetic-activated cell sorting, Oligodendrocyte precursor cells, Oligodendrocyte differentiation

1 Introduction

The main function of oligodendrocytes is to myelinate axons within the central nervous system and once myelinated the long-term integrity of the axons depends on a functional myelin sheath [1]. The myelinating oligodendrocytes are formed from oligodendrocyte progenitor cells (OPCs) that are maintained lifelong [2–4]. OPCs proliferate in the brain at late fetal stages, and subsequently differentiate into mature oligodendrocytes (OLs) [5]. Within this lineage, distinct oligodendrocyte maturation states can be defined: proliferating OPCs with a bipolar morphology that express platelet-derived growth factor alpha receptor (PDGFαR). Premyelinating oligodendrocytes with multiple processes that contain galactosylceramide (O4 marker) on their surface. Mature, myelinating OLs are characterized by the expression of myelin-specific proteins, such as myelin basic protein [6–9].

Fundamental aspects of the biology of oligodendroglial cells, such as their proliferation, migration, differentiation, membrane or RNA trafficking, and myelination, can be studied in culture

David A. Lyons and Linde Kegel (eds.), *Oligodendrocytes: Methods and Protocols*, Methods in Molecular Biology, vol. 1936, https://doi.org/10.1007/978-1-4939-9072-6_5, © Springer Science+Business Media, LLC, part of Springer Nature 2019

[10–24]. Here, we present different protocols to isolate OPCs from mouse perinatal brains that can be applied to wild-type or transgenic animals to study their function in vitro.

The shake-off method was introduced by McCarthy and de Velis and relies on the adhesive properties of the glia cells [25]. The OPCs are grown in a mixed glia culture from newborn mice for 7–10 days and can be separated from astrocytes that are tightly adherent to the cell culture flask by differential shaking. The resulting purified OPCs can be differentiated into an OL-enriched population. This simple and inexpensive method also allows a high yield of OLs, and additionally allows researchers to isolate microglia and astrocytes with high purity [25, 26]. An alternative method is based on the isolation of OPCs by magnetic-activated cell sorting (MACS) via antibodies coupled to magnetic beads. After initial dissociation of the brain tissue, the cells can be purified by positive or negative selection within 1 day at a high purity. Depending on the question to be studied oligodendrocytes can be cocultured with neurons [27, 28].

2 Materials

1. Cell culture BME media: 500 mL BME-medium, 50 mL heat-inactivated horse serum (HS), 5 mL 5000 U/mL of penicillin/streptomycin, 5 mL 200 mM GlutaMAX. Do not use longer than 2 weeks. Store at 4 °C.
2. Coated T75 cell culture flasks: Incubate T75 cell culture flasks with 10 mL 0.01% poly-L-lysine solution for at least 1 h at 37 °C (*see* **Note 1**). Wash twice with PBS. Add 10 mL BME with 1% HS, 1% GlutaMAX, and 1% Pen/Strep to T75 flask and leave at 37 °C and 7.5% CO_2 overnight.
3. Super-SATO medium: 90 mL DMEM with 4.5 g/L glucose, 2 mL 50× B-27 supplement, 1 mL 200 mM GlutaMAX, 1 mL 100 mM sodium pyruvate, 1 mL heat-inactivated HS, 1 mL 5000 U/mL of penicillin/streptomycin, 10 μL 5 mM triiodothyronine, 13 μL 4 mM L-thyroxine.
4. Cell culture media for astrocyte culture: 88 mL DMEM with 4.5 g/L glucose, 10 mL heat-inactivated fetal calf serum, 1 mL 200 mM GlutaMAX, 1 mL 5000 U/mL of penicillin/streptomycin. Store at 4 °C for up to 2 weeks.
5. Neuronal media for cortical neurons: 10 mL 10× MEM, 2.9 mL 7.5% bicarbonate solution, 3 mL 20% glucose solution, 1 mL 200 mM GlutaMAX, 1 mL 100 mM sodium pyruvate, 1 mL 5000 U/mL of penicillin/streptomycin, 2 mL 50× B27 supplement. Add to 100 mL with distilled water and sterile filter. Store at 4 °C for up to 2 weeks. Neuronal media prevents astrocyte proliferation.

6. NMRI P0-2 mice.
7. C57BL6/J P3-P5 mice.
8. HBSS.
9. PBS pH 7.4.
10. 0.25% Trypsin-EDTA. Store in 10 mL aliquots at −20 °C.
11. DMEM with 4.5 g/L glucose.
12. 10× MEM.
13. 50× B-27 supplement. Store at −20 °C. Do not freeze-thaw.
14. 100 mM sodium pyruvate.
15. Horse serum (HS). Heat-inactivate at 56 °C for 30 min. Store 50 mL aliquots at −20 °C.
16. Fetal calf serum. Heat-inactivated at 56 °C for 30 min. Store 50 mL aliquots at −20 °C.
17. 0.5% poly-L-lysine (PLL) solution: Dissolve 1 g poly-L-lysine in 200 mL sterile PBS. Store in 10 mL aliquots at −20 °C.
18. 0.01% poly-L-lysine (PLL) solution: Dilute 10 mL 0.5% PLL solution in 500 mL sterile PBS. Store at 4 °C for up to 4 weeks.
19. Neural dissociation kit (P).
20. MACS gentle tissue dissociator.
21. μMACS separator.
22. MACS C-Tubes.
23. LS columns.
24. NeuroMACS medium.
25. MACS NeuroBrew21.
26. 200 mM L-Glutamine.
27. 4.2 mg/mL forskolin in DMSO. Store at −80 °C.
28. 10 μg/mL PDGF-aa in sterile Dulbecco's PBS plus 0.2% BSA. Store at −80 °C.
29. 1 μg/mL NT-3 in sterile Dulbecco's PBS plus 0.2% BSA. Store at −80 °C.
30. 10 μg/mL CNTF in sterile 10 mM sodium phosphate buffer.
31. 4 μg/mL T-3 in sterile 0.1 N NaOH and Dulbecco's PBS. Store at −20 °C.
32. Dissections tools (straight fine-tipped forceps, fine hippocampal scissors etc.).
33. 70 μm cell sieves.
34. 40 μm cell sieves.
35. Cell culture flasks.
36. Well dishes.

37. Petri dishes.
38. Glass pasteur pipettes.
39. Serological pipettes.

3 Methods

3.1 Primary Mouse Oligodendrocyte Cultures from Mixed Glial Culture of P0-2 Mice

3.1.1 Dissection of P0-2 Mouse Brains for Mixed Glial Cultures

1. One day prior to dissection prepare cell culture BME media and coat T75 cell culture flasks.
2. Dissection of the cerebral midbrain and cortex of the perinatal mouse brains are carried out under sterile conditions in a dissection hood. Prepare two petri dishes with pre-warmed HBSS, one to collect the brains and the other for the dissected and cleaned tissue.
3. Keep the newborn mice warm until dissection (*see* **Note 2**).
4. Sterilize dissection tools under the dissection hood upon spraying with ethanol and turning the ultraviolet light on for 1 h.
5. Pre-warm HBSS and 0.25% trypsin-EDTA to 37 °C.
6. Process five mice at once and process one step with all brains before moving to the next step. Cut the head close to the brain stem and rinse in pre-warmed HBSS.
7. Make an incision in the skin from the brain stem along the midline and a second transverse incision close to the olfactory bulbs to simultaneously flap over the skin to either side of the skull.
8. Then repeat the motion carefully to remove the skull from the brain. To remove the brain, cut at the midline with a fine straight hippocampal scissor and a transverse cut above the cerebellum. Remove the thin skull with forceps to either side simultaneously. The milky cartilage layer has to be removed as well to be able to excise the brain without damage. Remove the brain with a spoon and wash the blood away in a petri dish with pre-warmed HBSS.
9. Under a dissection microscope, remove the hindbrain with a two fine-tipped forceps. Cut the brains along the midline to separate the two hemispheres with the midbrain attached.
10. The meninges with the blood vessels are removed with fine-tipped forceps by pulling them in a lateral fashion from the olfactory bulb. The brains should now appear purely white. Process all brains to this step before proceeding.
11. Incubate brains for 10 min at 37 °C in a 3.5 cm petri dish with 5 mL 0.25% trypsin-EDTA in the incubator.
12. Remove brains from trypsin solution and put them into a 50 mL falcon with 5 mL supplemented BME media. The

brains are resuspended by pipetting up and down 20 times with a 5 mL serological pipette without introducing any bubbles. To further dissociate the tissue, triturate the cell solution 20 times through a glass Pasteur pipette that is narrowed by passing through the flame once. Repeat the procedure, by passing the cell solution through a glass Pasteur pipette that was narrowed to half the diameter (*see* **Note 3**).

13. Add a cell sieve on top of a 50 mL falcon. Pass the cell solution through the cell sieve.
14. Equally distribute the cell solution into four to five T75 cell culture flasks (*see* **Note 4**). Furthermore, add 5 mL supplemented BME media (from **step 1**) and 5 mL supplemented Super-SATO (*see* **step 6**) and store the flasks at 37 °C and 7.5% CO_2 (*see* **Note 5**). The trypsination of the perinatal cortices can be stopped with serum-containing BME media or the neurons can be washed using the neuron media (does not contain serum) twice to remove the trypsin.

3.1.2 Maintaining the Mixed Glia Culture

1. Perform a media change on day 3 or 4 after dissection by exchanging three-quarters of the media with fresh, supplemented BME media (*see* **Note 6**).
2. On day 5 or 6, when the astrocyte layer is almost complete and the first OPCs and microglia are visible, a microglia removal is performed, as follows in **steps 3** and **4**.
3. The cell culture flasks are firmly tapped twice to either side of the wall of the flask on a hard vertical surface. The wall opposite the flask neck is knocked on a horizontal surface. The flasks are then stored upright with the bottleneck facing vertically until four flasks are processed.
4. The media is aspirated. 10 mL of fresh, supplemented BME media is added to each flask and the flasks are then returned to the incubator.
5. On day 6–8, a stronger microglia removal is carried out.
6. The cell culture flasks are firmly tapped five times to either side of the wall of the flask on a hard vertical surface. The wall opposite the flask neck is knocked on a horizontal surface. The flasks are then stored upright with the bottleneck facing vertically until four flasks are processed.
7. The media is aspirated and 10 mL BME are used to wash the cell layer. After removal of the media used for washing, 10 mL of fresh, supplemented BME media is added to each flask and the flasks are then returned to the incubator.

3.1.3 Preparation of Coverslips for Culturing of OPCs

1. Prepare coverslips for OPC culturing a few days before the shake-off.
2. Add coverslips of desired diameter in a Schott bottle and incubate with concentrated HCl under the fume hood on a roller overnight. Take care to only fill the acid until the edge of the bottleneck when lying down vertically. Seal the lid with parafilm.
3. Perform several quick washes with tap water and continue until pH of the washing solution is pH 6–7. Neutralize wash fluid prior to discarding in the sink.
4. The coverslips are washed quickly with 70% ethanol to prevent them adhering together.
5. Separate coverslips on parafilm and leave to dry.
6. Bake coverslips in a glass petri dish at 200 °C for 6 h.

3.1.4 Preparation of Super-SATO Media for OPCs After Shake-off Procedure

1. One day prior to oligodendrocyte shake-off procedure prepare Super-SATO medium.
2. Add with DMEM up to 100 mL. Sterile-filter the media and store at 4 °C for 2 weeks.

3.1.5 Preparation of Cell Culture Dishes for OPCs

1. One day prior or on the day of the oligodendrocyte shake.
2. Add acid-treated coverslips of desired diameter into appropriate cell culture dish (if cells are used for biochemical analysis, the coverslips are not necessary as they can be lysed in the wells itself).
3. Add an appropriate volume to cover the bottom of the well with 0.01% poly-L-lysine solution and incubate for at least 1 h at 37 °C.
4. Wash twice with PBS.
5. Add an appropriate volume of Super-SATO-Media and equilibrate at 37 °C and 7.5% CO_2.

3.1.6 Shake-off Procedure of OPCs from Mixed Glia Culture

1. Depending on the microglia content of the culture, either a microglia removal is performed (*see* Subheading 3.1.2) or the following protocol is performed to remove oligodendrocytes. It is crucial that the astrocytic layer underneath is complete and not detaching as otherwise there is a large degree of astrocyte contamination in the cultures. The oligodendrocytes should be sitting on top of the astrocyte layer and appear as small round cells (*see* Fig. 1). If the shake-off is carried out too late, the microglia kill the OPCs.
2. Two flasks are shaken horizontally at a 45° angle for 30 times. If the majority of the oligodendrocytes are not removed as indicated by phase contrast microscopy, more shakes can be

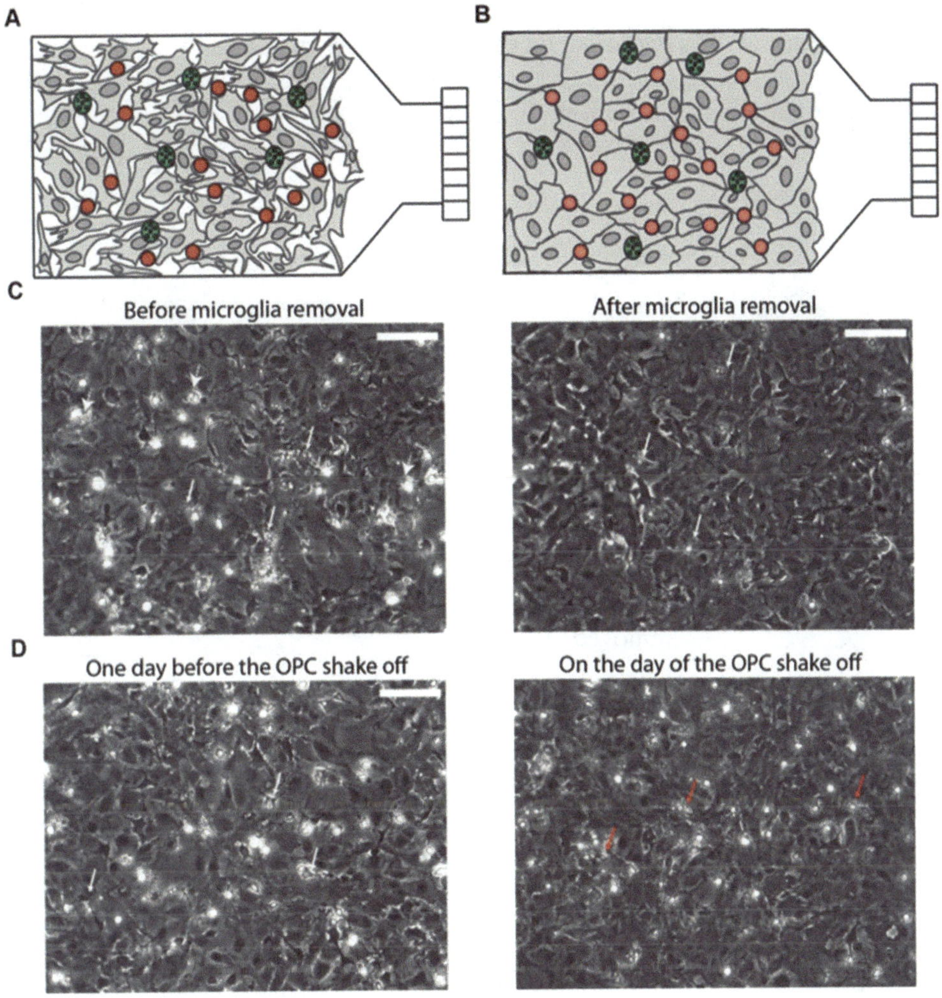

Fig. 1 Mixed glia culture at different stages during the cultivation. At day 3–4 after dissection, the astrocytic layer is not yet complete, but the first OPCs (red cells) can be seen in between the astrocytes (gray cells). The microglia (green, granular cells) are proliferating: thus, they should be removed before they outnumber and kill the OPCs (**a**). At later stages, the astrocytic layer completely covers the bottom of the flask, whereas the round OPCs are sitting on top of this layer (red cells with a sharp border). Microglia are usually still present at that stage and should be removed before the shake-off of the OPCs (**b**). Panel (**c**) shows representative phase contrast images of culture before and after the microglia removal. Microglia are the bright, bigger cells with irregular borders marked by the dashed arrow. Oligodendrocytes are small round cells with sharp borders indicated by the white arrow. Panel (**d**) displays representative images of culture 1 day prior and on the day of the OPC shake-off. The white arrows indicate oligodendrocytes that are not ready to be shaken off as they are still residing in between the astrocytes. The red arrows indicate oligodendrocytes that can easily be removed via the shake-off method as indicated by their round or drop-like shape and sharp borders and bright appearance

Table 1
Cultivation conditions of OPCs dependent on cell culture vessel and usage

Cell culture vessel	Absolute cell number	Coverslip size	Usage
6 cm petri dish	750,000–1 mio.	No coverslips	Biochemistry (isolation of RNA; proteins)
3 cm petri dish	125,000–500,000	No coverslips	Biochemistry
6 well dish	125,000–500,000	32 mm	Biochemistry
12 well dish	50–120,000	18 mm	Biochemistry, immunocytochemistry
24 well dish	20–40,000	12 mm	Immunocytochemistry
48 well dish	10–20,000	8 mm	Immunocytochemistry
96 well dish	~8000	No coverslips	Biochemistry

This table indicates the absolute number of OPCs dependent on the cell culture vessel with the coverslip size and subsequent usage options

performed. Nevertheless, not all oligodendrocytes can be removed.

3. The cell layer can be washed with 2.5 mL media and the cell suspension of the four flasks is pooled.
4. The flow-through is collected and subjected to centrifugation at 900 × *g* for 10 min. In the meantime, fresh media is added to the T75 cell culture flasks. These flasks can be used for another oligodendrocyte shake 5–7 days later.
5. The supernatant is carefully aspirated and the pellet is resuspended carefully without introducing any bubbles in 1 mL Super-SATO media with a 1 mL Eppendorf pipette. Later, up to 2 mL media can be added.
6. Optional: In order to increase the purity of the oligodendroglial culture, the cell suspension is allowed to settle on a 3.5 cm petri dish in the cell culture hood for 2 min.
7. The supernatant is removed and the cell number of the supernatant is counted.
8. An appropriate cell number is seeded on cell culture vessels (*see* Table 1 and **Note 7**) and incubated 1–5 days at 37 °C and 7.5% CO_2. For biochemical analysis and transfections, a higher cell density is used, whereas for immunocytochemistry a lower cell count is used to obtain single cells (*see* **Note 8**). We recommend to plate a portion of the cells for biochemical analysis also on coverslips to check for the purity of the culture by immunostaining for GFAP as astrocytic marker and Iba-1 as microglia marker.

3.2 Isolation and Culture of Oligodendrocytes from Pups Using MACS Technology

For this procedure, P7 are used for isolation of $O4^+$ OL, whereas P8-10 pups are used for $NG2^+$ OPCs for recovering the highest yield. Dissection of the brain and harvesting of desired tissue is carried out under sterile conditions in a dissection hood. Prepare two petri dishes with pre-warmed PBS: One for the dissected brain, the other for the tissue after the removal of meninges.

3.2.1 Preparation of Enzymes for Tissue Dissociation

1. Before sacrificing the pups, prepare the appropriate enzyme mix 1 and enzyme mix 2 as indicated in the table below using the reagents provided in the MACS neural dissociation kit (P).
2. Prepare the enzyme mix 1 in an MACS C-tube, taking care to not exceed a volume of 7.8 mL per C-tube. Pre-warm the enzyme mix 1 in the C-tubes to 37 °C.

Enzyme mix 1 (per 400 mg of brain tissue)		Enzyme mix 2 (per 400 mg of brain tissue)	
Enzyme P	Buffer X	Buffer Y	Enzyme A
50 μL	1900 μL	20 μL	10 μL

3.2.2 Dissection of P7-P10 Pups for Isolation of Oligodendrocytes by MACS

1. For sacrificing the pups, cut the head close to the brainstem blotting away excess blood onto clean tissue paper. Place the heads in a petri dish with the pre-warmed PBS.
2. Make an incision in the skin from the brainstem along the midline all the way until the olfactory bulb. Make two more incisions into the connective tissue attaching the skin to the ear. Using forceps, separate the skin from the skull such that the entire cranium is exposed.
3. Make two small incisions along the midline of the cranium and also a transverse incision along the plate above the hindbrain with Dumont fine forceps. Carefully lift the cranium and peel it away from the underlying brain without damaging the tissue until the entire brain is free of the cranium and can be extracted from the underlying skull.
4. Carefully separate the hindbrain (cerebellum and brainstem) from the forebrain with one clean incision using the fine forceps. Using a small spatula extract the forebrain and place it in the other petri dish containing fresh pre-warmed PBS.
5. Place the petri dish with the isolated forebrains under a dissection microscope and separate the two hemispheres along the midline.
6. Remove the meninges with the blood vessels using fine-tipped forceps by pulling them gently away from the brain tissue. The tissue should appear predominantly white.

7. Once the meninges are removed, collect the clean tissues into the falcon containing pre-warmed PBS.

Alternatively: the whole brain can be used without removing the meninges.

3.2.3 Dissociation of Tissue into Single Cell Suspension Using MACS

1. After briefly rinsing the brains in PBS, pool 3–4 brains for isolation of $NG2^+$ OPCs and up to 6 brains for isolation of $O4^+$ OL into one C-tube containing appropriate amounts of pre-warmed (37 °C) enzyme mix 1.
2. Dissociate the tissue in a coarse manner using fine forceps or 1 mL pipette tip to allow better enzymatic digestion of the tissue by the enzyme.
3. Following coarse dissociation, attach the C-tubes securely onto the Gentle MACS tissue dissociator and run the MACS program m_Brain01 twice per tube.
4. Incubate the C-tubes for 15 min at 37 °C with gentle agitation. In the meantime, pre-warm enzyme mix 2.
5. Attach the C-tubes to the MACS gentle tissue dissociator again and run program MACS program m_Brain02 twice per tube.
6. At this point, the tissue should be homogenously disrupted. Add the appropriate amount of enzyme mix 2 per C-tube as indicated in the table and incubate the tubes at 37 °C for 10 min with gentle agitation.
7. Attach the C-tubes once more onto the MACs gentle tissue dissociator and run the MACS program m_Brain03 twice per tube followed by 10 min incubation at 37 °C with gentle agitation. At this stage, the tissue should have been digested enough to reach a single cell suspension.
8. Centrifuge the C-tubes at 900 rpm and room temperature (RT) for 10 min and carefully aspirate the supernatant without disrupting the pellet.
9. Add 1 mL of pre-warmed DMEM and resuspend the pellet by gently pipetting up and down using a 1 mL pipette tip to achieve a homogenous single cell suspension. Add the remaining volume to 10 mL using the pre-warmed DMEM, pipetting up and down gently, thus maintaining a homogenous cell suspension.
10. In order to remove any undigested tissue or debris, pass the cell suspension through cell strainers. For $O4^+$ isolation, place a 70 μm cell strainer on a new 50 mL falcon and pass 5 mL of the cell suspension through it. Wash the strainer with 5 mL of fresh DMEM. Place a new cell strainer onto the falcon and do the same with the remaining 5 mL of the cell suspension, and washing again with 5 mL of fresh DMEM. Repeat this

procedure using a 40 μm cell strainer. For NG2⁺ OPC isolation, directly proceed with 40 μm cell strainers.

11. To assess the total amount of cells isolated, count the cells using the Neubauer chamber. On average, the protocol yields on average 9–10 × 10^6 cells per brain.

3.2.4 Isolation of Oligodendrocyte Precursors and/or Premature Oligodendrocytes Using MACS Antibody-Coupled Beads

At this stage, it is possible to isolate different cell populations from the cell suspension by either positive or negative selection. For that, antibodies coupled to magnetic beads (provided by Miltenyi) specific for cell surface markers expressed by each of cell populations are used. In this chapter, we will focus on isolating oligodendrocyte precursor cells and immature oligodendrocytes. Therefore, different markers and different protocols are used: To isolate OPCs, the cell surface proteoglycan NG2 is used as a marker for cells that have committed to the OPC fate. Newly differentiated pre-oligodendrocytes are isolated using O4 as a surface marker.

1. Centrifuge the cells in the 50 mL falcons at 900 rpm and RT for 10 min and carefully aspirate the supernatant without disrupting the pellet. For isolation of $O4^+$ OL, repeat centrifugation step upon resuspending the cells in 10 mL of pre-warmed DMEM. Add 90 μL of DMEM + 1% HS per 10^7 cells and 10 μL NG2 or O4 beads per 10^7 cells; mix gently with the cells (do not vortex).
2. For isolation of $NG2^+$ OPC, incubate the cell suspension at 37 °C for 3 h with gentle agitation. During this time, the cells start to re-express NG2 antigen on their surface (as sometimes the enzymatic digest can cleave NG2 from the cell surface). For isolation of $O4^+$ premature oligodendrocytes, one can directly proceed to isolation of the cells using the magnetic bead coupled antibodies.
3. Incubate the cells on ice for 15 min to facilitate binding of the antibodies to the specific cell populations, and then wash cells with 1–2 mL DMEM + 1% HS per 10^7 cells.
4. Centrifuge the tubes at 900 rpm and 4 °C for 10 min and carefully aspirate the supernatant without disrupting the pellet. Dissolve pellet in 3 mL DMEM + 1% HS by gently pipetting up and down using 1 mL pipette tip to attain a homogenous cell suspension.
5. Attach the LS column to the MACS magnet and activate the magnetic LS column with 3 mL of DMEM + 1% HS.
6. Add 3 mL of cell suspension to the column and collect the flow-through in a fresh falcon.
7. Wash the column three times with 3 mL DMEM + 1% HS and collect the flow-through after each wash. The flow-through contains the cells that are not bound to the antibody, whereas

the cells bound to the magnetically coupled antibody are retained in the column. Add new media onto the column only when all the liquid from the previous step has passed through the column.

8. Detach the LS column from the magnet, flush out the cells bound to the resin in the column with 5 mL of fresh DMEM + 1% HS using the plunger provided into a new falcon.
9. Count the number of cells in the flow-through, as well as the cells that were isolated from the magnetic LS column.

3.2.5 Plating of Isolated Oligodendrocytes and Maintaining the Culture

1. On average, it is possible to isolate approximately 1×10^6 $NG2^+$ cells or $O4^+$ per brain, respectively. Plate the cells at the desired density in DMEM + 1% HS initially onto plates previously processed and coated with poly-L-lysine as described in protocol 1 **step 5** in subheading 3.1.5 and **step 4–8** in subheading 3.1.6.
2. Two to 3 h after plating $NG2^+$ OPC, the cells are washed twice with pre-warmed PBS and the media is replaced with Neuro-MACs OPC cultivation media (100 mL NeuroMACS media, 2 mL MACS NeuroBrew21, 1 mL Pen/Strep, 1 mL L-Glutamine, 100 μL Forskolin, 100 μL CNTF, 100 μL PDGF, 100 μL NT-3). When plating $O4^+$ OL, replace the media with Neuro-MACS $O4^+$ OL cultivation media after 1 h.
3. After 24–48 h, the media is collected and centrifuged at 900 rpm for 5 min to remove dead cells and debris, while the cells are washed twice with pre-warmed DMEM. Without touching the bottom of the tube (debris), the conditioned OPC media is added back to the cells. Alternatively, half of the media can be replaced with fresh media.
4. The $NG2^+$ OPCs and $O4^+$ OLs are kept in culture at 37 °C for 3 or up to 5 days, respectively. For $NG2^+$ OPCs, the conditioned media is replaced on DIV4 with fresh OPC differentiation media (add 100 μL Forskolin, 100 μL CNTF, 1 mL T3 to 100 mL of the OPC cultivation conditioned media). For $O4^+$ OLs, replace half of the media with fresh $O4^+$ OL cultivation media on DIV3.

3.3 Coculture of Neurons and Oligodendrocytes

3.3.1 Preparation of Astrocytes

Prepare 4–5 days prior to dissection of embryos for cortical neurons

1. Use flasks containing an astrocytic layer on the day of the oligodendroglial shake-off. Remove the media and wash the cell layer with PBS.
2. Detach the cells with 1 mL 0.25% trypsin-EDTA and observe their detachment under the microscope. Usually, it takes around 2 min.
3. Add 8 mL supplemented DMEM media to stop trypsination.

4. Rinse the bottom of the cell culture flask ten times with media to breakdown the astrocyte layer.
5. Pellet the cells by centrifugation for 5 min at 900 rpm and RT.
6. Resuspend the pellet in 1 mL supplemented DMEM using a micropipette.
7. Perform a cell count and seed 60,000 cells per well in 1 mL supplemented DMEM media in a 12 well dish. The cell culture dishes do not need to be coated.
8. On day 3 (or 1 day prior to the preparation of cortical neurons), the media is exchanged to a 1:1 mixture of Super-SATO and neuron media.

3.3.2 Preparation of Cortical Neurons from Perinatal E14-E18 Mice

Pregnant NMRI mice are used for the preparation of cortical neurons. After killing the mother by cervical dislocation, the embryos are removed under sterile conditions in a dissection hood and hemispheres are used for preparation of cortical neurons. Perinatal E14-E18 mice can be used, but E16 mice are optimal (*see* **Note 9**). Prepare two petri dishes with pre-warmed HBSS, one for the heads of embryos and the other for the dissected and cleaned tissue.

1. Prepare neuronal media for cortical neurons (*see* **Note 10**).
2. Sterilize dissection tools under the dissection hood upon spraying with ethanol and turning the ultraviolet light on for 1 h.
3. Pre-warm HBSS and 0.25% trypsin-EDTA to 37 °C.
4. The mother is killed by cervical dislocation. The animal is sprayed with 70% ethanol, and then the abdominal cavity is cut open with a scissor under the dissection hood.
5. Remove the uterus sac containing the embryos by pulling on one end and cutting the other end with a scissor.
6. Hold the uterus sac above a petri dish with pre-warmed HBSS. Remove the embryos by pulling the uterus sac down and cutting the embryos out one by one using a fine scissor.
7. Cut their head immediately and transfer it to a dry petri dish. Three embryos usually give a high yield of cells.
8. Make an incision in the skin from the brain stem along the midline and a second transverse incision close to the olfactory bulbs to simultaneously flap over the skin to either side of the skull.
9. Then, repeat the motion carefully to remove the skull from the brain. To remove the brain, cut at the midline with a fine straight hippocampal scissor and a transverse cut above the cerebellum.

10. Remove the thin skull with forceps to either side simultaneously.
11. Remove the brain and transfer in a petri dish with pre-warmed HBSS.
12. Under a dissection microscope, remove the hindbrain with a two fine-tipped forceps.
13. Cut the brains along the midline and separate the hemispheres from the midbrain. Discard the midbrain.
14. The meninges from the hemispheres are removed with fine-tipped forceps by pulling them in a lateral fashion from the olfactory bulb. The cortices should now appear purely white. Process all brains till this step.
15. Incubate brains 10 min at 37 °C in a 3.5 cm petri dish with 5 mL 0.25% trypsin-EDTA in the incubator.
16. Remove cortices from trypsin solution and put them into a 50 mL falcon with 3 mL supplemented BME media to stop trypsination.
17. The brains are resuspended by pipetting up and down 20 times with a 5 mL serological pipette without introducing any bubbles.
18. To further dissociate the tissue, triturate the cell solution 20 times through a glass Pasteur pipette that is narrowed by passing through the flame once. Repeat the procedure, by passing the cell solution through a glass Pasteur pipette that was narrowed to half the diameter.
19. Add a 40 μm cell sieve on top of a 50 mL falcon. Pass the cell solution through the cell sieve.
20. The cell number is determined and ~250,000 cells are seeded per well on acid-treated, PLL-coated cell culture dishes for OPC coverslips in a 12 well dish in a 1:1 mixture of Super-SATO and neuronal media.
21. Incubate the cell culture plates at 37 °C and 7.5% CO_2.
22. More than 4 h after the dissection or the next day, the cortical neurons are transferred to cell culture plates with 4–5 days old astrocytes.
23. A sterilized metal ring is first placed on the confluent astrocytic layer, then the coverslip with the neurons is placed upside up onto the metal spacer.
24. Add 1–2 mL 1:1 mixture of Super-SATO and neuron media.
25. Incubate the cell culture plates at 37 °C and 7.5% CO_2.
26. Half of the media is replaced two to three times a week by fresh 1:1 mixture of Super-SATO and neuron media.
27. OPCs are shaken off from the mixed glia culture as described above.

28. On day 5–8 of the neuronal culture, 60,000 OPCs are seeded onto the cortical neurons. Incubate the cell culture plates at 37 °C and 7.5% CO_2. The coculture is usually maintained for additional 3 days in culture (*see* **Note 11**). For the cocultures, the astrocytic layer should be examined on a regular basis. If it contains too much microglia, it should be replaced (usually once a week) by fresh astrocytes as the microglia can affect the viability of the culture.

4 Notes

1. Do not leave PLL on the cell culture dishes longer than 24 h as it might cause increased cytotoxicity.
2. Quickly use the newborn mice for dissection. After prolonged separation from the mothers, the cultures do not turn out well.
3. Vigorous pipetting of the brain after trypsin digest can destroy the cells. This step is just to confer an initial separation of the digested tissue.
4. Never cultivate more than two brains of the newborn mice in one T75 cm^2 flask as the vast amount of astrocytes cannot adhere efficiently due to enhanced cell proliferation so that the dead cells detach again and leave holes in the astrocytic layer.
5. Super SATO is added to the mixed glia cultures after dissection to enhance the proliferation of young OPCs. Alternatively, 5 mL sterile-filtered conditioned media from cultures just before the oligodendrocyte shake-off can be used.
6. Do not exchange media before day 2 after dissection as it kills the cells.
7. After counting the OPCs, pipet the cell suspension up and down again before and also in between platting of cells to achieve a homogenous cell density in all wells.
8. Low density of OPCs ($<1 \times 10^4$ cells/mm^2) decreases their viability.
9. Protocol 2 can be used for rapid and pure cell isolation either to be cultured as mentioned above, or, alternatively, the isolated cells can directly be used for biochemical experiments. For this purpose, apply the following protocol:
   (a) Centrifuge the isolated cells at 900 rpm for 5 min and aspirate the supernatant without disturbing the pellet.
   (b) Wash the pellet twice with PBS, centrifuging the sample at 900 rpm for 5 min after each wash and aspirating the supernatant.

(c) The isolated cells can now either be used for biochemical experiments or further stained with other markers and used for flow cytometric measurements.

10. From the flow-through obtained after isolation of $NG2^+$ or $O4^+$ cells, one can further isolate neurons using PSA-NCAM beads, astrocytes using ACSA1 beads and microglia using CD11B beads by repeating **steps 4** and **5**.
11. The younger the perinatal mice for preparation of cortical neurons, the better the cultures are. E16 mice are a good compromise between the yield of cells and the quality of the cultures.

References

1. Nave K-A, Werner HB (2014) Myelination of the nervous system: mechanisms and functions. Annu Rev Cell Dev Biol 30:503–533. https://doi.org/10.1146/annurev-cellbio-100913-013101
2. Noble M, Wren D, Wolswijk G (1992) The O-2A(adult) progenitor cell: a glial stem cell of the adult central nervous system. Semin Cell Biol 3:413–422
3. Noble M, Wolswijk G (1992) Development and regeneration in the O-2A lineage: studies in vitro and in vivo. J Neuroimmunol 40:287–293
4. Wolswijk G, Noble M (1989) Identification of an adult-specific glial progenitor cell. Development 105:387–400
5. Marques S, Zeisel A, Codeluppi S et al (2016) Oligodendrocyte heterogeneity in the mouse juvenile and adult central nervous system. Science 352:1326–1329. https://doi.org/10.1126/science.aaf6463
6. Pfeiffer S, Warrington A, Bansal R (1993) The oligodendrocyte and its many cellular processes. Trends Cell Biol 3:191–197. https://doi.org/10.1016/0962-8924(93)90213-K
7. Bansal R, Pfeiffer SE (1992) Novel stage in the oligodendrocyte lineage defined by reactivity of progenitors with R-mAb prior to O1 anti-galactocerebroside. J Neurosci Res 32:309–316. https://doi.org/10.1002/jnr.490320303
8. Levine JM, Reynolds R, Fawcett JW (2001) The oligodendrocyte precursor cell in health and disease. Trends Neurosci 24:39–47. https://doi.org/10.1016/S0166-2236(00)01691-X
9. Baumann N, Pham-Dinh D (2001) Biology of oligodendrocyte and myelin in the mammalian central nervous system. Physiol Rev 81:871–927
10. Aggarwal S, Yurlova L, Snaidero N et al (2011) A size barrier limits protein diffusion at the cell surface to generate lipid-rich myelin-membrane sheets. Dev Cell 21:445–456. https://doi.org/10.1016/j.devcel.2011.08.001
11. Aggarwal S, Snaidero N, Pähler G et al (2013) Myelin membrane assembly is driven by a phase transition of myelin basic proteins into a cohesive protein meshwork. PLoS Biol 11: e1001577
12. Snaidero N, Möbius W, Czopka T et al (2014) Myelin membrane wrapping of CNS axons by PI(3,4,5)P3-dependent polarized growth at the inner tongue. Cell 156:277–290. https://doi.org/10.1016/j.cell.2013.11.044
13. Fitzner D, Schnaars M, van Rossum D et al (2011) Selective transfer of exosomes from oligodendrocytes to microglia by macropinocytosis. J Cell Sci 124:447–458. https://doi.org/10.1242/jcs.074088
14. Frühbeis C, Fröhlich D, Kuo WP et al (2013) Neurotransmitter-triggered transfer of exosomes mediates oligodendrocyte-neuron communication. PLoS Biol 11:e1001604. https://doi.org/10.1371/journal.pbio.1001604
15. White R, Gonsior C, Krämer-Albers E-M et al (2008) Activation of oligodendroglial Fyn kinase enhances translation of mRNAs transported in hnRNP A2-dependent RNA granules. J Cell Biol 181:579–586. https://doi.org/10.1083/jcb.200706164
16. Almeida RG, Czopka T, ffrench-Constant C, Lyons DA (2011) Individual axons regulate the myelinating potential of single oligodendrocytes in vivo. Development 138:4443–4450. https://doi.org/10.1242/dev.071001

17. Zuchero JB, Fu M-M, Sloan SA et al (2015) CNS myelin wrapping is driven by actin disassembly. Dev Cell 34:152–167. https://doi.org/10.1016/j.devcel.2015.06.011
18. Traka M, Wollmann RL, Cerda SR et al (2008) Nur7 is a nonsense mutation in the mouse aspartoacylase gene that causes spongy degeneration of the CNS. J Neurosci 28:11537–11549. https://doi.org/10.1523/JNEUROSCI.1490-08.2008
19. Lee S, Chong SYC, Tuck SJ et al (2013) A rapid and reproducible assay for modeling myelination by oligodendrocytes using engineered nanofibers. Nat Protoc 8:771–782
20. Lee S, Leach MK, Redmond SA et al (2012) A culture system to study oligodendrocyte myelination processes using engineered nanofibers. Nat Methods 9:917–922. https://doi.org/10.1038/nmeth.2105
21. Watkins TA, Emery B, Mulinyawe S, Barres BA (2008) Distinct stages of myelination regulated by gamma-secretase and astrocytes in a rapidly myelinating CNS coculture system. Neuron 60:555–569. https://doi.org/10.1016/j.neuron.2008.09.011
22. Seiberlich V, Bauer NG, Schwarz L et al (2015) Downregulation of the microtubule associated protein tau impairs process outgrowth and myelin basic protein mRNA transport in oligodendrocytes. Glia 63:1621–1635. https://doi.org/10.1002/glia.22832
23. Sharma K, Schmitt S, Bergner CG et al (2015) Cell type- and brain region-resolved mouse brain proteome. Nat Neurosci 18:1819–1831. https://doi.org/10.1038/nn.4160
24. Nawaz S, Sánchez P, Schmitt S et al (2015) Actin filament turnover drives leading edge growth during myelin sheath formation in the central nervous system. Dev Cell 34:139–151. https://doi.org/10.1016/j.devcel.2015.05.013
25. McCarthy KD, de Vellis J (1980) Preparation of separate astroglial and oligodendroglial cell cultures from rat cerebral tissue. J Cell Biol 85:890–902
26. Dubois-Dalcq M, Behar T, Hudson L, Lazzarini RA (1986) Emergence of three myelin proteins in oligodendrocytes cultured without neurons. J Cell Biol 102:384–392
27. Fitzner D, Schneider A, Kippert A et al (2006) Myelin basic protein-dependent plasma membrane reorganization in the formation of myelin. EMBO J 25:5037–5048. https://doi.org/10.1038/sj.emboj.7601376
28. Trajkovic K, Dhaunchak AS, Goncalves JT et al (2006) Neuron to glia signaling triggers myelin membrane exocytosis from endosomal storage sites. J Cell Biol 172:937–948. https://doi.org/10.1083/jcb.200509022

Chapter 6

A Neuron-Free Microfiber Assay to Assess Myelin Sheath Formation

Marie E. Bechler

Abstract

Important advances in our understanding of oligodendrocyte precursor cell biology and differentiation have stemmed from in vitro experiments using cultures of isolated primary oligodendrocyte precursor cells. To examine the process of myelination in the final stages of oligodendrocyte development, experimental systems have previously been limited to models utilizing neurons. Recent advances in three-dimensional culture systems, however, have opened the possibility to observe myelin sheath formation with only one cell type, the oligodendrocyte precursor cell. In this chapter, such a method is described for examining oligodendrocyte myelin sheath formation with isolated oligodendrocytes in the absence of neurons. This assay is ideal for gaining mechanistic insight into oligodendrocyte-specific regulation of myelin sheath formation. Oligodendrocyte heterogeneity can be readily assessed, determining whether different oligodendrocyte sources influence myelin sheath formation. As well, the direct impact of both physical and molecular cues on oligodendrocytes can be determined in this defined system. This assay extends the capability of two-dimensional oligodendrocyte cultures, permitting post-differentiation analysis of myelinating oligodendrocytes, the number of sheaths formed by individual oligodendrocytes, as well as the lengths of myelin sheaths formed.

Key words Myelination, Oligodendrocyte, Myelin sheath, Nanofiber, Microfiber

1 Introduction

Isolated oligodendrocyte precursor cell cultures have been invaluable to determine the effects of molecules, forces, and genetic modifications directly on oligodendrocyte survival, migration, proliferation, and differentiation. The mechanisms underlying oligodendrocyte biology can be identified in an isolated culture environment without the complexity of contribution from other cell types. The majority of isolated oligodendrocyte cultures have been conducted in traditional cell culture conditions, i.e., on flat surfaces. As oligodendrocyte precursor cells differentiate in these isolated cultures, myelin-specific protein expression is upregulated and the cells undergo dramatic changes in morphology [1–3]. The oligodendrocyte morphology becomes increasingly complex, with

David A. Lyons and Linde Kegel (eds.), *Oligodendrocytes: Methods and Protocols*, Methods in Molecular Biology, vol. 1936, https://doi.org/10.1007/978-1-4939-9072-6_6, © Springer Science+Business Media, LLC, part of Springer Nature 2019

the formation of highly branched processes that later elaborate flat membranes. Cellular specialization of non-compact cytoplasm-filled regions and compact membranes occurs, with the segregation of many of proteins and lipids akin to oligodendrocytes in vivo [4]. These morphological changes have often been treated as a proxy for oligodendrocyte maturity and myelin membrane capacity [5]. While this type of culture demonstrates changes in cellular morphologies, it precludes the examination of the three-dimensional wrapping that culminates in the generation of myelin sheaths. This is a particular shortcoming for in vitro assays with isolated oligodendrocytes, as forming the proper multilayered membrane architecture of the myelin sheath is critical for the regulation of axonal conduction and health [6–10].

For decades researchers developed and utilized cocultures of oligodendrocyte precursor cells with primary neurons to examine this three-dimensional process in vitro (refs. 11–14, for review *see* ref. 5). The coculture of oligodendrocytes with neurons provides the axo-glial interactions long believed to be essential for myelination. Due to mixed cell types present, a mixture of genetically modified cells must be used (i.e., wild-type neurons with modified oligodendrocytes or vice versa) to distinguish between neuron and oligodendrocyte-specific effects on myelination. While valuable for examining myelination, neuron-oligodendrocyte cocultures also have drawbacks: oligodendrocyte-specific effects from the addition of small molecules and recombinant proteins cannot be readily distinguished; there can be substantial variation in the neurons and myelination from preparation to preparation; as well, neurons are very sensitive to plating density or any alterations in culture conditions.

An alternative to cocultures is the development of a three-dimensional culture model that supports oligodendrocyte myelination. Accumulating evidence shows that oligodendrocytes can produce myelin-like membranes that surround cylindrical glass, carbon, and electrospun fiber substrates without the presence of neuronal instruction [15–18]. Advances in these techniques utilizing inert, electrospun polymers for culture with oligodendrocytes have now demonstrated the hallmark of myelin sheath formation: compact multi-lamellar membranes [19]. Likewise, myelin sheath lengths and responses to extrinsic signaling molecules have been shown to be comparable between fiber cultures and neurons [19]. We have found that this three-dimensional, neuron-free assay is less variable than neuronal cultures, making them more amenable to revealing oligodendrocyte-intrinsic mechanisms that dictate myelin sheath number and properties. Similarly, this culture model is better suited to the use of small molecules and recombinant proteins to assess oligodendrocyte-specific effects.

While isolated oligodendrocyte cultures with fibers have an immense capacity to inform the biology of myelin sheath regulation, like any experimental model, fiber cultures are not without limitations. Particularly the axo-glial interactions of the paranodes (reviewed in refs. 20, 21) are not present, limiting studies of complex axo-glial interactions found in vivo. Therefore, like any other experimental model, validating the findings gained from in vitro techniques in other models and/or in vivo will remain an essential approach.

In this chapter, such a neuron-free, three-dimensional fiber model is described. This chemically and physically defined oligodendrocyte culture model is easy to manipulate for pinpointing oligodendrocyte-specific mechanisms of myelin sheath regulation. This allows for comparisons of different sources of oligodendrocytes, genetic alterations, as well as the effects of small molecules, physical cues, or the presence of soluble and/or surface-coated recombinant proteins. This procedure utilizes commercially available electrospun fibers as a substrate for in vitro oligodendrocyte precursor cell myelination. Various sources of oligodendrocyte precursor cells can be used, but here is described a robust myelinating culture with rodent primary oligodendrocyte precursor cells using a chemically defined myelination medium [19], based upon previous myelination assays [22]. Immunofluorescence for myelin-specific proteins followed by confocal microscopy image acquisition permits the subsequent analysis of myelinating oligodendrocyte numbers, the length of myelin sheaths, as well as the number of sheaths made by individual oligodendrocytes.

2 Materials

2.1 Preparation of Fiber Scaffolds

Poly-L-Lactic Acid (PLLA) electrospun fibers: The following protocol is written for use with 2 μm aligned PLLA fibers in 12-well cell crown scaffolds from The Electrospinning Company. Other formats are possible with modification (*see* **Notes 1** and **2**).

1. Blunt forceps with grip, for handling scaffolds.
2. 12-well cell culture plate.
3. 70% Ethanol.
4. Sterile water.
5. Poly-D-Lysine (PDL), 2 mg/mL stock in H_2O.

2.2 Plating Cells

1. Cells: primary mouse or rat oligodendrocyte precursor cells (*see* **Note 3**).
2. Sterile water.
3. 21-gauge and 23-gauge needles.
4. 2 mL syringe.
5. Sterile conical tube.

6. Tabletop centrifuge.
7. Cell culture incubator.
8. 100× Modified SATO culture medium supplementation that supports oligodendrocyte differentiation [23]: 500 mL total volume made up in DMEM, high glucose, sodium pyruvate medium, 5.1 g BSA fraction V (170 mL of a 30 mg/mL solution in DMEM), 3.0 mg Progesterone (300 μL of a 10 mg/mL solution in ethanol), 805 mg Putrescine (5 mL of a 161 mg/mL solution in DMEM), 20 mg 3, 3′, 5-Triiodo-L-Thyronine Sodium Salt (T3) (400 μL of a 50 mg/mL in warm NaOH solution), 20 mg L-Thyroxine (T4) (400 μL of a 50 mg/mL in warm NaOH solution) (*see* **Note 4**). Store the modified SATO at −80 °C in 5 mL aliquots.
9. Myelin medium: 240 mL Neurobasal medium, 240 mL DMEM high glucose medium, 1× Pen/Strep, 5 mL ITS supplement, 5 mL modified SATO, 10 mL B-27(*see* **Note 5**), 500 μL *N*-acetyl cysteine (NAC) 1000× stock, to give 5 mg/mL, and 500 μL D-biotin 1000× stock, to give 10 μg/mL. Filter sterilize with a 0.22 μm filtration unit. After thawing components, the medium can be kept for 3–4 weeks when stored at 4 °C.

2.3 Fixing and Staining

1. Phosphate-buffered saline (PBS).
2. Fixative: 3.7% para-formaldehyde in phosphate-buffered saline (PFA).
3. Permeabilization buffer: 0.1% TritonX-100 in PBS.
4. Primary antibody: rat anti-myelin basic protein (MBP; Serotec).
5. Secondary antibody: Goat anti-rat Alexa 488 nm (Life Technologies).
6. Hoechst 33342 (10 mg/mL, or other suitable Nuclei stain).
7. 13 mm coverglass, No. 1.
8. Standard microscope slides.
9. Paper roll.
10. Appropriate sized box or cover to keep the scaffolds dark.
11. Blunt forceps with grip for handling scaffolds.
12. Pointed forceps for handling 13 mm coverglass.
13. Hard set mounting media, such as Fluoromount G (Southern Biotechnologies).
14. P1000 pipette tips.

2.4 Image Acquisition and Analysis

1. Confocal microscope with appropriate laser lines.
2. 40×/NA1.25 objective (or one that will achieve equivalent resolution).
3. Fiji ImageJ [24] or equivalent software.

3 Methods

3.1 Preparation of Fiber Scaffolds

In the following protocol the electrospun fibers are suspended across the bottom of a plastic insert, held in place only by a ring that snuggly fits around the base of the insert. The plastic insert fits tightly into 12-well cell culture plates, positioned by three "wings" on the sides of the scaffold (Fig. 1).

On the day prior to seeding cells (or several hours before), the fiber scaffolds should be soaked in ethanol to increase sterility and hydrophilicity, then washed and coated with poly-D-lysine (PDL).

1. Transfer an appropriate number of cell crown fiber scaffolds into a sterile 12-well cell culture plate, being careful to avoid touching the fibers or the ring at the base of the scaffold (*see* **Note 6**).
2. Add enough 70% ethanol to cover the entirety of each fiber scaffold for 10–20 min.
3. Aspirate the ethanol. Aspiration should be done at the edge of the well, on the outside of the scaffold to avoid disturbing the fibers. Immediately replace with sterile water to wash (*see* **Note 7**).

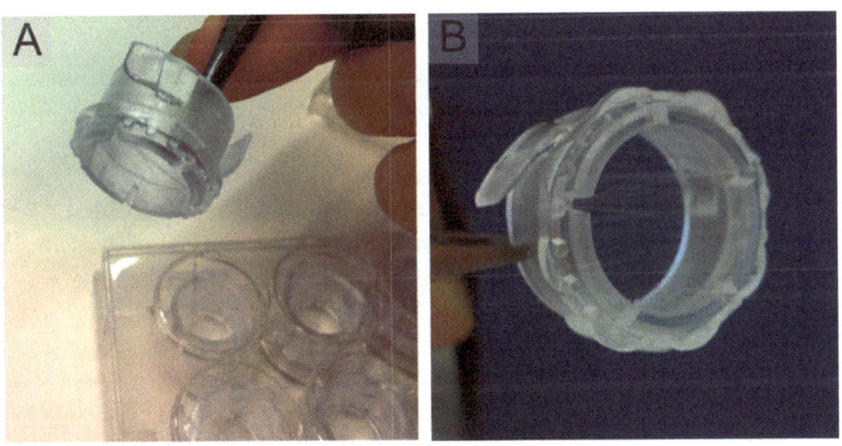

Fig. 1 The 12-well inserts holding electrospun microfibers. (**a**) An image showing the scaffolds containing the suspended microfibers. These scaffolds are positioned tightly into 12-well plates by "wings" at the top of the inserts. Scaffold inserts should be handled from the top to avoid disturbing the suspended fibers across the bottom. (**b**) View from the bottom of the scaffolds. Microfibers are suspended across the bottom of the inserts and held in place only by a ring around the base. Care should be taken to avoid touching the delicate fibers or the ring that secures them

4. Repeat aspirations and washes with sterile water two more times to remove residual ethanol. Immediately proceed to the next step (do not let fibers begin to dry).
5. Add enough PDL to cover fibers for each scaffold, approximately 0.75–1 mL (*see* **Note 8**).
6. Incubate for a minimum of 1 h to overnight in a cell culture incubator.

3.2 Plating Cells

1. Aspirate PDL and wash fibers two times with sterile water to remove residual PDL.
2. Rinse once with myelin medium to remove the residual water adhering to scaffolds.
3. Fill each well with 1.5–2 mL of pre-warmed (37 °C) myelin medium in preparation for the addition of cells.
4. Collect oligodendrocyte precursor cells, as prepared by the protocol in Chapter 5 (*see also* refs. 25, 26), by centrifugation 5 min 180 × *g*.
5. Triturate cells with 21-gauge needle followed by a 23-gauge needle, if necessary, to break apart into single cells (*see* **Note 9**).
6. Count cells using a hemocytometer and inspect for cell clumps.
7. Resuspend cells up to 150,000 cells/mL in myelin medium. Proceed immediately to plating to prevent cells from settling and clumping.
8. Add 0.2 mL (30,000) rat OPCs or 0.33 mL (50,000) mouse OPCs dropwise to the inside of the cell crown insert, above the fibers.
9. Place cells into a cell culture incubator at 37 °C, 7.5% CO_2, and 95% humidity.
10. Change media with fresh myelin medium every 2–3 days.

3.3 Fixation and Preparation for Microscopy

The timing for differentiation and myelin sheath formation should be determined for each oligodendrocyte precursor cell source. Typical rat or mouse OPCs isolated, as described in Chapter 5, will begin to express MBP in 3–7 days, maximally by day 7 in culture on fibers. Cells will begin to form myelin sheaths by day 7. Sheath formation and sheath lengths will reach a maximum by day 14 (Fig. 2). To analyze the timing of myelination, it is recommended to use times points of 3 and 7 days in culture. To analyze sheath number per individual oligodendrocyte and myelin sheath lengths, it is recommended to analyze cells at day 14 in culture.

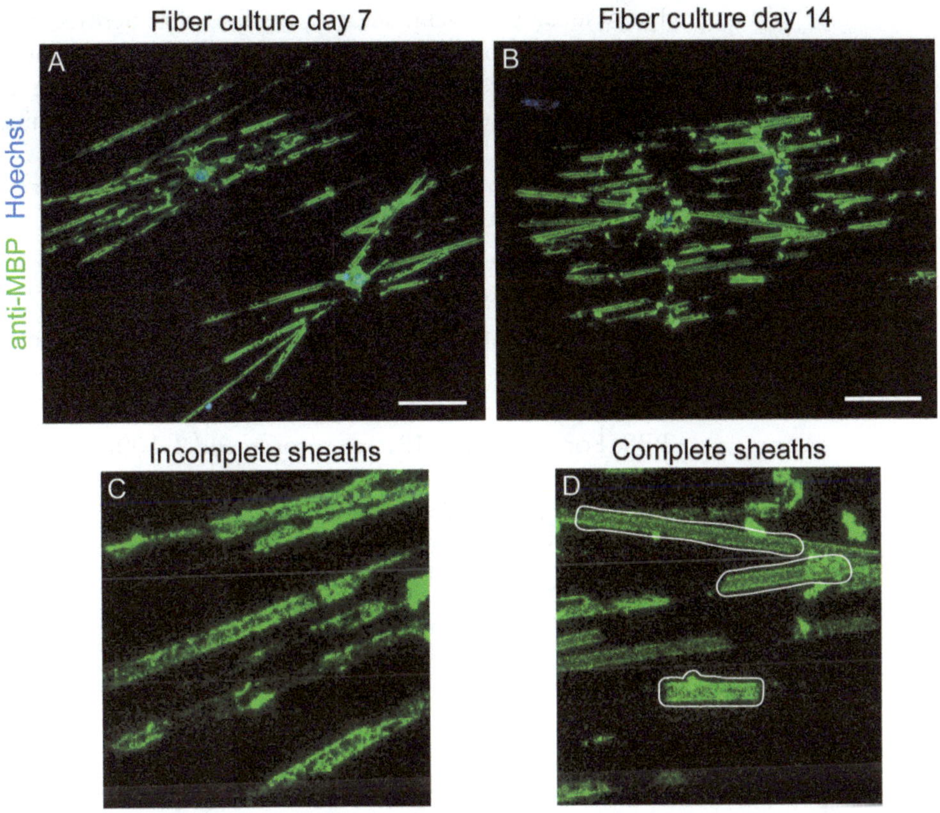

Fig. 2 Example confocal micrographs of oligodendrocytes ensheathing microfibers (**a**) Oligodendrocytes express myelin basic protein (MBP) by day 7 and begin to form myelin sheaths on the microfibers. Many sheaths by day 7 are not complete. (**b**) By 14 days in culture, most of the oligodendrocytes are forming sheaths, which completely surrounding fibers. Scale bars, 40 μm. (**c**) Examples of incomplete sheaths, magnified region from the image in A. (**d**) Examples of complete sheaths indicated by outlines, magnified section from the image in (**b**)

1. Gently aspirate medium and wash once with PBS.
2. Aspirate and replace PBS with 3.7% formaldehyde/PBS for 15 min.
3. Remove formaldehyde and wash 3 times with PBS for 5 min (*see* **Note 10**).
4. Permeabilize cells with 0.1% TritonX-100/PBS for 10 min.
5. Make primary antibody solution, 0.75 mL per scaffold, in PBS (blocking can be done if necessary, *see* **Note 11**). This volume is sufficient to completely cover the fibers as long as each scaffold is sitting flat on the bottom of the 12-well dish. Carefully push down each fiber scaffold to be sure that the scaffolds are sitting as low into the well as possible, gripping scaffolds only at the top to avoid disturbing the fibers.

For visualizing myelin sheaths, we find that Serotec rat anti-MBP diluted to 1:250 works well. This can be combined with other antibodies (not raised in rat) such as those to detect early oligodendrocyte precursor cells or astrocytes to determine differentiation and culture purity.

6. Fill several empty wells, if present, with PBS or water and seal the 12-well plate with Parafilm to prevent evaporation.
7. Place flat into 4 °C to incubate overnight with primary antibody (*see* **Notes 12** and **13**).
8. Wash three times with PBS for 5 min to remove unbound antibody.
9. Prepare secondary antibodies, 0.75 mL per scaffold, diluted in PBS. For detecting MBP we recommend 1:1000 diluted goat or donkey anti-rat Alexa 488 nm (or conjugated to another suitable fluorophore). Keep samples in the dark from here forward to prevent unwanted photobleaching.
10. Incubate scaffold with secondary antibodies for 1 h at room temperature.
11. Wash twice with PBS for 5 min to remove unbound antibodies.
12. Stain nuclei with Hoechst in PBS for 10 min.
13. Wash in PBS once for 5 min.
14. Prepare for mounting the fiber scaffolds onto glass microscope slides. Set out a sheet of paper roll with the appropriate number of glass microscope slides (2 fiber scaffolds per slide). Label the glass microscope slides with sample details. Also set out a small piece of paper roll or Kimwipe for blotting excess PBS and 13 mm coverslips for each scaffold. Have an appropriate sized box or cover to keep the scaffolds dark in between and after setting onto microscope slides.

    The following **steps 15–20** should be done for each scaffold one at a time. *See* Fig. 3 for details on handling the scaffolds during this process.

15. Add a drop (approx. 20 μL) of mounting medium to the slide for each scaffold.
16. Carefully remove fiber scaffold from the well and quickly blot the rim of the scaffold onto paper towel or a Kimwipe to remove excess PBS. Excessive residual PBS prevents proper hardening of the mounting medium. Avoid touching fibers or dislodging the ring on the bottom of the scaffold (*see* **Note 6**).
17. Place the fibers onto the drop of mounting medium. Working quickly, cover with two to three drops of additional mounting medium to completely cover the surface of the fibers. Immediately proceed to the next step.

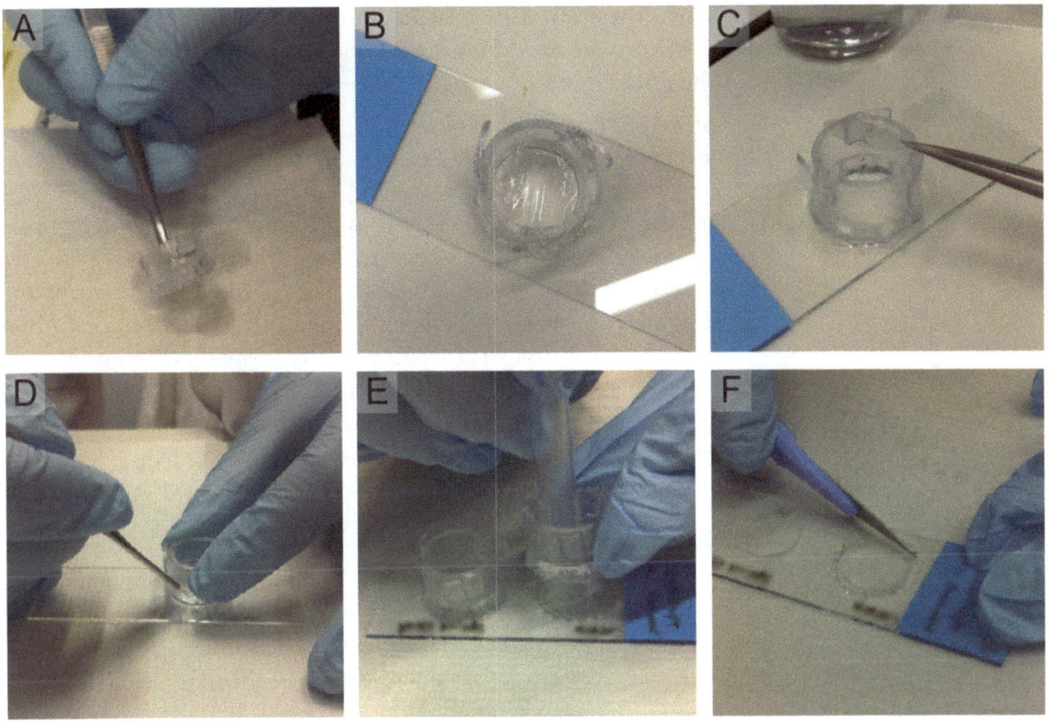

Fig. 3 Procedures for mounting fibers for confocal microscopy. The following images correspond to the Subheading 3.3: (**a**) **step 16** (**b**) **step 17** (**c**) **step 18** (**d**) **step 19** (**e**) **step 21** (**f**) **step 22**

18. Place a 13 mm coverslip inside of the fiber scaffold and press down gently to sandwich the fibers between the coverslip and the glass slide.
19. While holding the top of the scaffold, use the back of forceps to push down the ring surrounding the outside of the scaffold. Gently press the coverslip down again with the pointed end of forceps to make sure that no air bubbles are underneath and that the coverglass sits closely above the fibers (*see* **Note 14**). Carefully let go of the scaffold and cover the slide to keep dark until it sets (it will take 1–2 days to completely harden). Be extremely careful not to bump the scaffold at this stage and assure that the slide lies completely flat (*see* **Note 15**).
20. Let mounting medium set for 1–2 days in the dark.
21. To remove the plastic scaffolding gently hold the coverglass in place with the back of a P1000 pipette tip while lifting the plastic scaffold upwards. The scaffold and the ring should both come off from the slide. If there is a lot of resistance, do not apply excessive force (as this risks cracking the coverslip), instead run a scalpel along the inside between the coverglass and the scaffold.

22. Once the scaffolding has been removed, a scalpel can be used to trim away excess hardened mounting medium from around the coverglass.
23. Store slides in the dark at 4 °C short term or at −20 °C for longer-term storage.

3.4 Image Acquisition

Images should be of a sufficient resolution to accurately determine whether myelin sheaths are being formed around the fibers (*see* **Note 16**). Here we have outlined imaging recommendations for the downstream analysis of myelin sheath-forming oligodendrocytes, the number or sheaths formed by individual oligodendrocytes as well as myelin sheath lengths. Users should determine the appropriate settings to achieve sufficient resolution and to obtain sufficient numbers of cells and sheaths in images, as per confocal microscope system.

1. Bring slides to room temperature.
2. Set up imaging parameters: At least 1024 × 1024 without binning and 0.37 μm z-steps with a 40× objective are suggested to ensure that myelin sheaths (complete tubes) can be distinguished from cell membrane spreading along the fibers. For all images within an experiment, the same settings should be used.
3. Take a minimum of 10–15 random fields of view (296 μm × 296 μm area) per coverslip. In some cases it may be necessary to acquire adjacent fields of view to capture entire cells with all of their sheaths (*see* **Note 17**). These images can later be stitched using imaging software, such as Fiji ImageJ. More images may be necessary for analyzing sheaths formed by individual oligodendrocytes (*see* **Note 18**). Generally 100–200 sheaths and 15–20 individual cells should be imaged per coverslip to obtain large enough sample sizes for analysis (sample size calculations should be determined for the expected effect size and conditions of interest).

3.5 Analysis of Myelin Sheaths

Example images of cells with incomplete sheaths (such as at day 7) and complete sheaths (such as at day 14) are shown in Fig. 2. All conditions should be blinded to reduce bias prior to analysis. Original confocal z-series with metadata should be used to maintain scaling and accurate pixel information. Original z-series can be analyzed in Fiji ImageJ or suitable alternative. The maximum projections can be used to analyze images; however, it is recommended to keep the associated z-series open side-by-side for accurately determining which MBP positive membranes are complete tubes.

4 Notes

1. Other electrospun polymers may also be suitable substrates for myelination, such as polystyrene (as in refs. 18, 27). Suspended fibers are necessary for accurate analysis of myelination, as fibers spun onto flat surfaces result in oligodendrocytes producing a mixture of both flat membrane sheets on the bottom surface and myelin sheaths that surround the fibers. The effects on oligodendrocyte behavior and relative efficiency of myelination on other polymers or platforms should be determined before routine use.
2. The diameter has an impact on both the efficiency and length of myelin sheaths that oligodendrocytes produce [18, 19]. Fiber diameters above 0.5 μm and up to 4 μm work well; however, fiber diameters within an experiment should vary by less than a micron when sheath number and length comparisons are being made. Therefore, batch-to-batch variation in electrospun fiber diameters should be determined for each lot by scanning electron microscopy, or equivalent methods. Whenever possible, single batches/lots should be used for an entire experiment and biological replicates, to avoid unnecessary experimental variability.
3. Cells can be from other sources than what is noted. The key is to have relatively pure oligodendrocyte precursor cells, i.e., 95% pure, as the starting population to ensure cell-autonomous effects are analyzed. We have used both primary rat oligodendrocyte precursor cells prepared from mixed glial cultures by differential adhesion as well as immunopanned primary mouse oligodendrocyte precursor cells (*see* Chapter 5).
4. T3 and T4 do not dissolve readily in water or DMEM. Make in 0.1 N or, if necessary 1 N NaOH, and mix. Warm the tube in hot water for a few minutes until dissolved. Keep the tube warm (e.g., in a hot water bath) and add the solutions to DMEM before they cool.
5. Different sources and lots of B-27 can be extrememly variable and should be batch tested for use in this assay. To avoid variablility of commercial sources, a B-27 alternative can be prepared as described in Chen et al. [28]. Each batch should be tested to assure it supports myelination.
6. The ring at the base of the fiber scaffolds is the only support to keep the fibers in place. Therefore, take care not to bump and loosen this, as the fibers will detach from the scaffold.
7. Once wet, the fibers should remain wet at all times. When aspirating and exchanging liquid on the fibers do not leave

without liquid for extended periods of time as the fibers will splay apart due to surface tension/capillary action.

8. PDL-coated fibers can be additionally coated by protein adsorption by further incubation with the desired recombinant proteins. Additional protein coating should be verified (e.g., by antibody staining with appropriate antibody staining controls).
9. Cell density is critical to properly form sheaths. Cells that are not plated as single cells or are too dense will form clusters that do not myelinate properly. Similarly, proliferation of oligodendrocyte precursors on the fibers results in clumps of cells. Oligodendrocyte clumps adversely affect myelin sheath formation and preclude the ability to quantify myelin sheaths formed by individual oligodendrocytes. Cells at too low a density will result in a need for increased image acquisition for downstream analysis. We find primary rat oligodendrocytes are optimally plated at 226 cells/mm^2 and primary mouse oligodendrocytes at 376 cells/mm^2.
10. Ideally immunofluorescence would be done following fixation; however, fiber scaffolds can be kept at 4 °C for a week if kept in sufficient PBS and tightly sealed.
11. Blocking solutions can be used to reduce background of antibody staining. We have found that blocking was unnecessary with most lots of rat anti-MBP antibody. However, blocking may help reduce nonspecific binding/background with use of other primary antibodies. If desired, a 10% goat or donkey serum can be used for an hour prior to the addition of primary antibody and be used for diluting primary and secondary antibodies.
12. Alternatively, primary antibody incubation can be done at room temperature for 1–2 h. We find better anti-MBP signal with overnight incubation at 4 °C.
13. The volume of antibody solution is just enough to cover the fibers. Therefore, ensure that the plate sits flat when incubating with antibodies.
14. If the coverslip is not gently pressed down after loosening the ring along the base of the scaffold, there is a likelihood of introducing air bubbles that will increase the risk of cracking the coverslip when removing the scaffold. Pressing the coverslip gently also helps maintain a shorter working distance for imaging the fibers.
15. Once the ring along the base of the scaffold has been loosened, the fibers will be free to move. Therefore, it is important that the slides are not tilted or disturbed while the mounting medium sets.

16. The optical sections (z-steps) should allow for at least two to three images per fiber diameter to determine whether membrane is completely surrounding the fibers.
17. Highly elaborated cells or very long sheath lengths may not always be captured in a single 296 × 296 μm field of view. It is essential that adjacent fields are imaged and later stitched in these cases to avoid biasing the analysis of sheath lengths and the number of sheaths formed by individual cells.
18. Depending on cell density and distribution, more images may be required to obtain a suitable number for analyzing single cell behavior. In this case, it is important to have sufficient images to distinguish which sheaths belong to which oligodendrocyte. A culture that is too dense or has many small clusters of cells will prevent this analysis.

References

1. Gard AL, Pfeiffer SE (1989) Oligodendrocyte progenitors isolated directly from developing telencephalon at a specific phenotypic stage: myelinogenic potential in a defined environment. Development 106:119–132
2. Baumann N, Pham-Dinh D (2001) Biology of oligodendrocyte and myelin in the mammalian central nervous system. Physiol Rev 81:871–927
3. Pfeiffer SE, Warrington AE, Bansal R (1993) The oligodendrocyte and its many cellular processes. Trends Cell Biol 3:191–197
4. Aggarwal S, Yurlova L, Snaidero N et al (2011) A size barrier limits protein diffusion at the cell surface to generate lipid-rich myelin-membrane sheets. Dev Cell 21:445–456
5. Jarjour AA, Zhang H, Bauer N et al (2012) In vitro modeling of central nervous system myelination and remyelination. Glia 60:1–12
6. Nave K-A (2010) Myelination and support of axonal integrity by glia. Nature 468:244–252
7. Chernoff GF (1981) Shiverer: an autosomal recessive mutant mouse with myelin deficiency. J Hered 72:128
8. Klugmann M, Schwab MH, Pühlhofer A et al (1997) Assembly of CNS myelin in the absence of proteolipid protein. Neuron 18:59–70
9. Griffiths I, Klugmann M, Anderson T et al (1998) Axonal swellings and degeneration in mice lacking the major proteolipid of myelin. Science 280:1610–1613
10. Waxman SG (1980) Determinants of conduction velocity in myelinated nerve fibers. Muscle Nerve 3:141–150
11. Wood PM, Williams AK (1984) Oligodendrocyte proliferation and CNS myelination in cultures containing dissociated embryonic neuroglia and dorsal root ganglion neurons. Brain Res 314:225–241
12. Rosen CL, Bunge RP, Ard MD et al (1989) Type 1 astrocytes inhibit myelination by adult rat oligodendrocytes in vitro. J Neurosci 9:3371–3379
13. Chan JR, Watkins TA, Cosgaya JM et al (2004) NGF controls axonal receptivity to myelination by Schwann cells or oligodendrocytes. Neuron 43:183–191
14. Watkins TA, Emery B, Mulinyawe S et al (2008) Distinct stages of myelination regulated by γ-secretase and astrocytes in a rapidly myelinating CNS coculture system. Neuron 60:555–569
15. Althaus HH, Burgisser P, Klöppner S et al (1987) Oligodendrocytes ensheath carbon fibres and produce myelin in vitro. In: Althaus HH, Seifert W (eds) Glial-neuronal communication in development and regeneration. Springer Berlin Heidelberg, Berlin, Heidelberg, pp 779–798
16. Bullock PN, Rome LH (1990) Glass microfibers: a model system for study of early events in myelination. J Neurosci Res 27:383–393
17. Howe CL (2006) Coated glass and vicryl microfibers as artificial axons. Cells Tissues Organs 183:180–194
18. Lee S, Leach MK, Redmond SA et al (2012) A culture system to study oligodendrocyte myelination processes using engineered nanofibers. Nat Methods 9:917–922
19. Bechler ME, Byrne L, ffrench-Constant C (2015) CNS myelin sheath lengths are an intrinsic property of oligodendrocytes. Curr Biol 25:2411–2416

20. Zoupi L, Savvaki M, Karagogeos D (2011) Axons and myelinating glia: an intimate contact. IUBMB Life 63:730–735
21. Faivre-Sarrailh C, Devaux JJ (2013) Neuroglial interactions at the nodes of Ranvier: implication in health and diseases. Front Cell Neurosci 7:196
22. Wang Z, Colognato H, ffrench-Constant C (2007) Contrasting effects of mitogenic growth factors on myelination in neuron–oligodendrocyte co-cultures. Glia 55:537–545
23. Bottenstein JE, Sato GH (1979) Growth of a rat neuroblastoma cell line in serum-free supplemented medium. Proceed Natl Acad Sci U S A 76:514–517
24. Schindelin J, Arganda-Carreras I, Frise E et al (2012) Fiji: an open-source platform for biological-image analysis. Nat Methods 9:676–682
25. McCarthy KD, de Vellis J (1980) Preparation of separate astroglial and oligodendroglial cell cultures from rat cerebral tissue. J Cell Biol 85:890–902
26. Emery B, Dugas JC (2013) Purification of oligodendrocyte lineage cells from mouse cortices by immunopanning. Cold Spring Harb Protoc 2013:854–868
27. Lee S, Chong SYC, Tuck SJ et al (2013) A rapid and reproducible assay for modeling myelination by oligodendrocytes using engineered nanofibers. Nat Protoc 8:771–782
28. Chen Y, et al (2008) NS21: Re-defined and modified supplement B27 for neuronal cultures. J Neurosci Methods 171(2):239–247. https://doi.org/10.1016/j.jneumeth.2008.03.013

Chapter 7

Oligodendrocyte–Neuron Myelinating Coculture

Matthew Swire and Charles ffrench-Constant

Abstract

The preparation of oligodendrocytes and neurons independently in vitro has provided substantial insight into the biology of the process of myelin sheath formation. This chapter describes a myelination system of dorsal root ganglion neurons by independent isolation of oligodendrocyte progenitor cells from either rat or mouse cortex. This in vitro assay can be used to examine the molecular determinants of myelin sheath formation.

Key words Myelination, Oligodendrocyte, Myelin sheath, Oligodendrocyte progenitor, Dorsal root ganglion, Coculture, In vitro

1 Introduction

Oligodendrocyte biology has been investigated *in vitro* for over 60 years, initiating from studies of organotypic slice cultures [1] and central nervous system (CNS) cell aggregates [2, 3]. These initial investigations, although instrumental for the study of oligodendrocyte biology in vitro, offered little means of manipulation, particularly in a cell-specific manner.

Advancement in the methodology of oligodendrocyte lineage cell isolation in the 1980s [4] through the physical separation of CNS populations from mixed glial cultures permitted the study of these cells in relatively pure populations. In addition to this, a range of immunopanning [5, 6] magnetic-activated cell sorting and fluorescence-activated cell sorting [7, 8] techniques have since been developed for the isolation of rat oligodendrocyte progenitor cells (OPCs), using various surface antigens specific to the oligodendrocyte lineage. Due to their ease of use and robust nature, adapted versions of the mixed glial culture system still remain the gold standard for rat oligodendrocyte generation [4, 9, 10].

David A. Lyons and Linde Kegel (eds.), *Oligodendrocytes: Methods and Protocols*, Methods in Molecular Biology, vol. 1936, https://doi.org/10.1007/978-1-4939-9072-6_7,

Contrary to rat cells, mouse oligodendrocytes have proven more challenging to isolate. A mixed glial culture system remains relatively inefficient for mouse OPC isolation when compared to rat cells. To circumvent this issue, various methodologies have been established for the isolation of mouse OPCs including: fluorescence-activated cell sorting for the O4 antigen from monolayer cultures [11–14], however, the low yield from cell sorting makes it unsuitable for bulk isolation of OPCs. A more robust technology came from immunopanning for the platelet-derived growth factor receptor α (PDGFRα), present on OPCs at a stage permitting cell expansion in culture in response to OPC mitogens [15–17].

OPCs, isolated as above, can be grown in 2-dimensional systems allowing the investigation of range of oligodendrocyte lineage properties including progenitor cell proliferation, migration, survival, and differentiation. Furthermore, in the absence of axons, cultures of differentiated oligodendrocytes can produce "myelin-like" membranes, expressing a range of canonical myelin sheath proteins [18, 19]. How exactly the production of these myelin membranes equates to genuine in vivo myelination is poorly understood, emphasizing the need to manipulate oligodendrocyte biology in more representative, 3-dimensional assays of myelination.

Organotypic slice cultures and CNS aggregates permit the investigation of myelination in a 3-dimensional environment. However, these techniques offer little means of manipulation of the individual cell populations [20]. One technique allowing extensive and independent manipulation of the neuronal or glial mediators of myelination are neuronal coculture systems, where the neurons and oligodendrocytes are prepared independently. Initial reports of this system established neuronal populations from dorsal root ganglion (DRG) neurons that were dissected away from the spinal cord of embryonic day 19–21 rats, and maintained in the presence of nerve growth factor and antimitotic agents. Isolated sections of optic nerves from 1–2-week-old rats were used as a source of oligodendrocytes [21]. Overlaying optic nerves above DRG cultures led to the wrapping and formation of compact myelin around the DRG neurons. The past three decades have refined this protocol, exploiting the advances in the isolation of pure populations of OPCs [22], along with the generation of serum-free supplementation to support myelination [23–25]. The main advantage of the DRG coculture system involves the ease of manipulation of the distinct cellular populations used. Independent preparation, particularly of the oligodendrocyte lineage from a range of different sources, allows cellular manipulation prior to

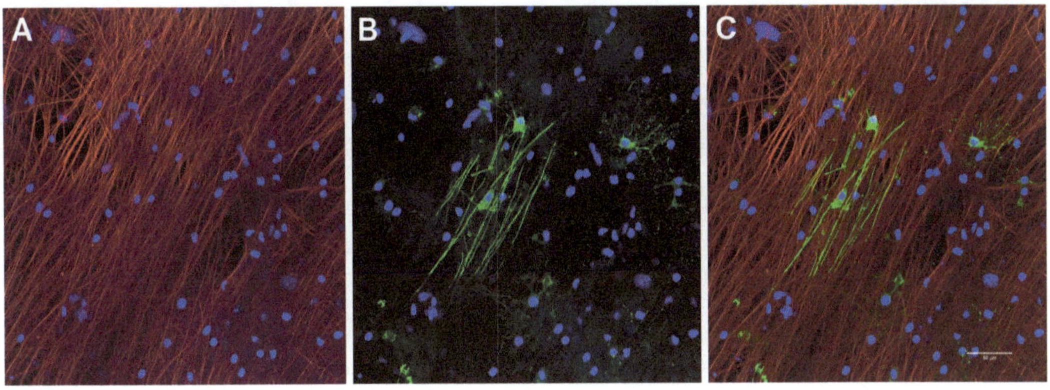

Fig. 1 (**a**) DRG neurons (red). (**b**) MBP expressing rat oligodendrocytes (green). (**c**) Overlay of MBP expressing oligodendrocytes myelinating DRG neurons, scale bar 50 μm

union. This system has been instrumental for a range of studies including the investigation of signals involved in myelin sheath formation [10, 26], the effect of mechanical stimulation on oligodendrocyte biology [27, 28], and the effect of neuronal activity on myelination [29, 30]. The major disadvantage of this system lies with the nature of the neurons themselves. With their neuronal cell bodies lying outside of the CNS, DRG neurons project axons connecting to both central and peripheral neurons. The composition of these neurons may therefore not represent that of a CNS axon, thus may not fully recapitulate CNS myelination. Nonetheless, oligodendrocytes have been identified to form myelin sheaths on DRG neurons analogous to those generated in vivo (*see* Fig. 1 and [2, 26, 31, 32]. In addition to DRG, CNS neuronal cultures can be established from retinal ganglion cells through immunopanning dissociated retinae for Thy1 [17]. While this system allows the manipulation of individual populations alike DRG cocultures, establishing a dense retinal ganglion cell network is highly technically challenging.

This chapter will outline techniques for the isolation and culture of: rat oligodendrocyte precursor cells using an adapted version of the mixed glial culture method presented by McCarthy and De Vellis [4, 9, 10], mouse oligodendrocyte progenitor cells through immunopanning for the platelet-derived growth factor receptor α (PDGFRα) modified from [15–17] and the culture of DRG neurons [22, 26, 33]. Combining these techniques allows the generation of multilayered myelin sheaths analogous to those seen in vivo [26, 31, 32], permitting the assessment of the impact that small molecules, recombinant proteins, and genetic manipulations have on the process of myelination.

2 Materials

2.1 Rat Oligodendrocytes—Mixed Glial Cultures

2.1.1 Preparation of Culture Medium

1. DMEM, high glucose, sodium pyruvate medium.
2. Penicillin/streptomycin mixture (Pen/Strep).
3. Fetal Calf Serum.

2.1.2 Preparation of Culture Flasks

1. Cell culture incubator.
2. Vented canted-neck 75 cm^2 tissue culture treated flasks.
3. Untreated tissue culture petri dishes.
4. Poly-D-lysine (PDL) (stock 2 mg/mL in sterile water).
5. Sterile water.

2.1.3 Dissection

1. Postnatal day 0–2 rats.
2. 70% ethanol.
3. Curved spring microdissection scissors.
4. Forceps.
5. Dissecting microscope.
6. Cold MEM.

2.1.4 Dissociation

1. Papain (30 U/mL).
2. DNase 1 Type IV (40 mg/mL).
3. L-Cysteine
4. MEM.
5. 0.2 μm filter unit.
6. 1 mL syringe.
7. Tabletop centrifuge.
8. Sterile conical tube.
9. P1000 pipette.

2.1.5 Detachment

1. Orbital shaker.
2. Parafilm.
3. 21-gauge needle.
4. Hemocytometer.
5. Untreated tissue culture petri dishes.
6. 50 mL sterile falcon tubes.

2.2 Mouse Oligodendrocytes—Immunopanning

2.2.1 Preparation of 100× Modified Mouse SATO

1. Progesterone (25 mg/mL in 100% EtOH).
2. Putrescine.
3. Apo-transferrin.
4. Sodium selenite (0.4 mg/mL in 0.1 N NaOH).
5. BSA.
6. DMEM (High glucose + Glutamax).

2.2.2 Preparation of Culture Medium

1. DMEM (high glucose + Glutamax).
2. B-27 supplement.
3. Penicillin/streptomycin mixture (Pen/Strep).
4. Modified mouse SATO.
5. *N*-Acetyl cysteine (NAC).
6. Insulin.
7. Forskolin.
8. Biotin.
9. Trace Elements B.
10. 0.22 μm filtration unit.
11. PDGF.
12. NT3.
13. CNTF.
14. 3,3′,5-Triiodo-L-Thyronine Sodium Salt (T3).

2.2.3 Preparation of Culture Flasks and Panning Dishes

1. Vented tissue culture treated flasks.
2. Poly-D-lysine (PDL) in sterile water (2 μg/mL).
3. Cell culture incubator.
4. PBS.
5. Untreated tissue culture petri dishes.
6. Parafilm.
7. Sterile water.
8. AffiniPure Goat Anti-Rat IgG (H + L) (Jackson ImmunoResearch—112-005-003) (2.4 mg/mL).
9. Anti-BSC1 Griffonia Simplicifolia Lectin (Vector Labs—L-1100) (2 mg/mL).
10. Rat Anti-Mouse CD140a (PDGFR-α) (BioLegend—135902) (0.5 mg/mL).
11. 0.2% BSA in PBS.

2.2.4 Dissection

1. Postnatal day 6–9 mice.
2. Curved spring microdissection scissors.

3. Fine forceps.
4. Dissecting microscope.
5. MEM.
6. 70% ethanol.

2.2.5 Dissociation and Panning

1. Neural Tissue Dissociation Kit (P) (MACS Miltenyi Biotec - 130-092-628).
2. MACSmix tube rotator (130-090-753).
3. Gentle MACS dissociator (130-093-235).
4. C tubes (130-093-237).
5. 70 μm cell strainers.
6. PDGF.
7. NT3.
8. Cell culture incubator.

2.2.6 Culture

1. PDGF.
2. NT3.

2.2.7 Detachment

1. 21-gauge needle.
2. Hemocytometer.
3. Phosphate-buffered saline (PBS).
4. TrypLE Express (Thermo Fisher).
5. Tabletop centrifuge.

2.2.8 Differentiation and Myelination

1. PDGF.
2. NT3.
3. CNTF.
4. 3,3′,5-Triiodo-L-Thyronine Sodium Salt (T3).

2.3 Rat Dorsal Root Ganglion Neuronal Culture

2.3.1 Preparation of Coverslips

1. PDL in sterile water (5 μg/mL).
2. Glass coverslips 13 mm.
3. Tissue culture 12-well plate.
4. Cell culture incubator.
5. Sterile water.
6. Tissue culture hood.
7. Matrigel growth factor reduced basement membrane matrix.

2.3.2 Dissection

1. Embryonic day 15 pregnant rat.
2. Dissection scissors.
3. 10 cm petri dish.

4. Cold MEM.
5. Dissecting microscope.
6. Fine forceps.
7. Bent/angled forceps.
8. Fine forceps (no. 5).

2.3.3 Dissociation

1. MEM.
2. Papain (30 U/mL).
3. DNase 1 Type IV (40 mg/mL).
4. L-Cysteine solution (24 mg/mL in MEM, made fresh).
5. MEM.
6. 0.2 μm filter unit
7. 1 mL syringe
8. Cell culture incubator.
9. DMEM, high glucose, sodium pyruvate medium.
10. Penicillin/streptomycin mixture (Pen/Strep).
11. Fetal calf serum.
12. Hemocytometer.
13. NGF (100 μg/mL).
14. FdU (100 mM).

2.4 Myelinating Coculture

2.4.1 Prepare 100× Rat Modified SATO

1. DMEM, high glucose, sodium pyruvate medium.
2. 30 mg/mL BSA fraction V solution in DMEM.
3. 10 mg/mL progesterone solution in ethanol.
4. 161 mg/mL putrescine solution in DMEM.
5. 3,3′,5-Triiodo-L-Thyronine Sodium Salt (T3).
6. 50 mg/mL L-Thyroxine (T4) in warm NaOH solution.

2.4.2 Prepare Myelination Media

1. DMEM, high glucose, sodium pyruvate medium.
2. Neurobasal medium.
3. Penicillin/streptomycin mixture (Pen/Strep).
4. 100× rat modified SATO.
5. ITS supplement.
6. B-27 supplement.
7. D-Biotin 1000× stock, 10 μg/mL.
8. *N*-Acetyl cysteine (NAC) 1000× stock, 5 mg/mL.
9. 0.22 μm filter for filter-sterilizing media.

2.4.3 Culture

1. Cell culture aspirator.

2.5 Fixation and Preparation for Microscopy

1. Phosphate-buffered saline (PBS).
2. 4% Formaldehyde in PBS
3. Forceps.
4. Staining box.
5. Normal goat serum.
6. Triton X-100.
7. Forceps.
8. Recommended antibodies: MBP (Rat) (Bio-Rad—MCA409), NFH (Chicken) (Biolegend—PCK-592P), CASPR (Rabbit) (Abcam—AB34151), Anti-rat Alexa 488 nm, Anti-rabbit Alexa 568 nm, Anti-chicken Alexa 647 nm.
9. Hoechst nuclei stain.
10. Hard set mounting media, such as Fluoromount G (Southern Biotechnologies).
11. Standard microscope slide.

3 Methods

3.1 Rat Oligodendrocytes—Mixed Glial Cultures

3.1.1 Preparation of Culture Medium

1. Mix DMEM, high glucose, sodium pyruvate medium with Penicillin/streptomycin mixture (Pen/Strep) 1% and Fetal Calf Serum 10%.

3.1.2 Preparation of Culture Flasks

1. Coat vented T75 flasks with PDL in sterile water (final concentration 2 μg/mL) at 37 °C for 1 h, or room temperature overnight. Allow 1.5 brains per flask.
2. Rinse coated flasks twice with sterile water and leave to dry.

3.1.3 Dissection

1. Cull postnatal day 0–2 rat pups by schedule one injection.
2. Spray down pups with 70% ethanol.
3. Use fine scissors to make a lateral incision through the base of the neck and one between the eyes.
4. Using scissors, cut through the skin connecting the two incisions. Peel back the skin exposing the skull.
5. Angling scissors away from the brain, cut through the skull using the two incisions made prior. Gently peel the skull away from the brain.
6. Ease the brain out into cold MEM.

7. Under microscope, use fine forceps to dissect the cortices away from hindbrain, midbrain, cerebellum, and olfactory bulbs.
8. Use fine forceps to pluck away the meninges covering the cortices.
9. Pool cortices in 1 mL MEM.
10. Mince cortices with fine scissors to a "smoothie-like" consistency.

3.1.4 Dissociation

1. Prepare papain digestion by mixing 1 mL MEM, 40 μL Papain (30 U/mL), 10 μL L-cysteine solution (24 mg/mL in MEM, made fresh), 10 μL DNaseI type IV (4 mg/mL).
2. Incubate digestion mix at 37 °C for 10 min.
3. Sterilize papain solution by passing through a 0.2 μm filter directly onto minced cortices.
4. Incubate at 37 °C for 1 h in incubator.
5. Dilute cells in 10 mL culture medium (3.1.1) to stop digestion.
6. Spin down for 5 min at 1000 rpm.
7. Carefully remove supernatant, leaving roughly 1 mL residual medium to prevent disturbing the cell pellet.
8. Triturate with a P1000 pipette and divide equally between PDL-coated flasks containing 10 mL culture medium, at a density or 1.5 brains per flask.
9. Place cells into a cell culture incubator at 37 °C, 7.5% CO_2, and 95% humidity.
10. Leave cells for 3 days before initial media change, then feed every 2–3 days.

3.1.5 Detachment

OPC detachment from mixed glial culture can be performed anywhere between 9 and 14 days in vitro.

1. Cover vented cap of flasks with parafilm and secure on an orbital shaker at 37 °C.
2. Shake at 240 rpm for 1 h to remove loosely attached microglia.
3. Aspirate microglia-containing medium and add 10 mL fresh culture medium.
4. Return to shaker and continue shaking for 16–18 h.
5. Remove medium and add to untreated tissue culture petri dishes, 1 flask per 10 cm dish, leave for 20–25 min in a tissue culture incubator.
6. Collect media from dishes, pool in 50 mL falcon tubes and spin cells for 5 min at 1000 rpm.
7. Resuspend cell pellet in desired media, triturate with a 21-gauge needle, and count cells using a hemocytometer.

8. Rat OPCs can now be added to coated coverslips, tissue culture plates, DRG neurons (*see* below), or electrospun microfibers (Chapter 6).

3.2 Mouse Oligodendrocytes—Immunopanning

3.2.1 Preparation of 100× Modified Mouse SATO

1. Mix DMEM (High glucose + Glutamax) 50 mL, Apo-transferrin 500 mg, BSA 500 mg, Putrescine 80 mg, Progesterone (25 mg/mL in 100% EtOH) 12.5 μL, Sodium selenite (0.4 mg/mL in 0.1 N NaOH) 500 μL.
2. Aliquot into 4 mL vials and store at −20 °C.

3.2.2 Preparation of Culture Medium

1. Mix 400 mL DMEM (High glucose + Glutamax), 4 mL Pen/-Strep, Modified mouse SATO, 8 mL B-27, 4 mL NAC (6 mg/mL), 400 μL Insulin (5 mg/mL), 400 μL Trace Elements B, and 40 μL Biotin (100 μg/mL).
2. Filter sterilize with a 0.22 μm filtration unit before adding 40 μL forskolin (20.5 mg/mL). Medium should be stored at 4 °C and used within 2 weeks.
3. Final culture media concentrations should be 100 μg/mL BSA, 100 μg/mL Apo-transferrin, 60 ng/mL progesterone, 16 μg/mL putrescine, 40 ng/mL sodium selenite, 60 μg/mL *N*-acetyl-cysteine, 10 ng/mL biotin, 5 μg/mL insulin, 2.05 μg/mL forskolin.
4. Add growth factors where required: PDGF (10 μg/mL), NT3 (5 μg/mL), CNTF (10 μg/mL), T3 (40 μg/mL). Use all at 1:1000. Final concentrations should be PDGF 10 ng/mL, NT3 5 ng/mL, CNTF 10 ng/mL, T3 40 ng/mL.

3.2.3 Preparation of Culture Flasks and Panning Dishes

1. Pre-coat vented flasks with PDL in sterile water (2 μg/mL) at 37 °C for 1 h, or room temperature overnight (*see* **Note 1**).
2. Prepare positive selection dishes, add 6 μL AffiniPure Goat Anti-Rat IgG (H + L) to 2 mL of PBS, and coat one untreated tissue culture dish (*see* **Note 2**). Wrap in Parafilm and leave overnight at 4 °C.
3. The following day, rinse coated dish twice with PBS. Add 3 μL primary antibody Rat Anti-Mouse CD140a (PDGFR-α) to 2 mL 0.2% BSA in PBS and coat an untreated petri dish at room temperature for 2 h.
4. Prepare two negative selection dishes per brain, add 2 μL Anti-BSC1 Griffonia Simplicifolia Lectin to 2 mL PBS in untreated dishes, leave at room temperature for 2 h.
5. Rinse coated flasks twice with sterile water and leave to dry.

3.2.4 Dissection

1. Cull postnatal day six to nine mouse pups by schedule one injection.
2. Spray down pups with 70% ethanol.
3. Use fine scissors to make a lateral incision through the base of the neck and one between the eyes.
4. Using scissors, cut through the skin connecting the two incisions. Peel back the skin exposing the skull.
5. Angling scissors away from the brain, cut through the skull using the two incisions made prior. Gently peel the skull away from the brain.
6. Ease the brain out into cold MEM.
7. Under microscope, use fine forceps to dissect the cortices away from hindbrain, midbrain, cerebellum, and olfactory bulbs.
8. Use fine forceps to pluck away the meninges covering the cortices.
9. Add brain to 1 mL MEM (*see* **Note 3**).

3.2.5 Dissociation and Panning

1. Follow MACS Neural Tissue Dissociation Kit protocol (*see* **Note 4**).
2. Wash negative selection plates twice with PBS.
3. Add brain solution to panning plates, gently swirl, and leave to sit at room temperature for 15 min.
4. Remove solution and repeat using the second negative selection plate.
5. After a further 15 min, wash positive selection dish twice with PBS, add brain solution to plate, gently swirl, and leave at room temperature for 45 min.
6. Aspirate solution, wash bound cells twice with preheated culture medium.
7. Add 2 mL medium containing freshly added PDGF and NT3 and scrape cells off plate.
8. Add scraped cells to medium containing PDGF and NT3 and transfer to culture flasks.
9. Transfer cells into a cell culture incubator at 37 °C, 7.5% CO_2, and 95% humidity.

3.2.6 Culture

1. Add fresh PDGF daily to flasks.
2. Every 2 days perform a half media change adding twice the amount of PDGF and NT3 in the fresh medium to equal 10 μg/mL PDGF and 5 μg/mL NT3 in the flask.

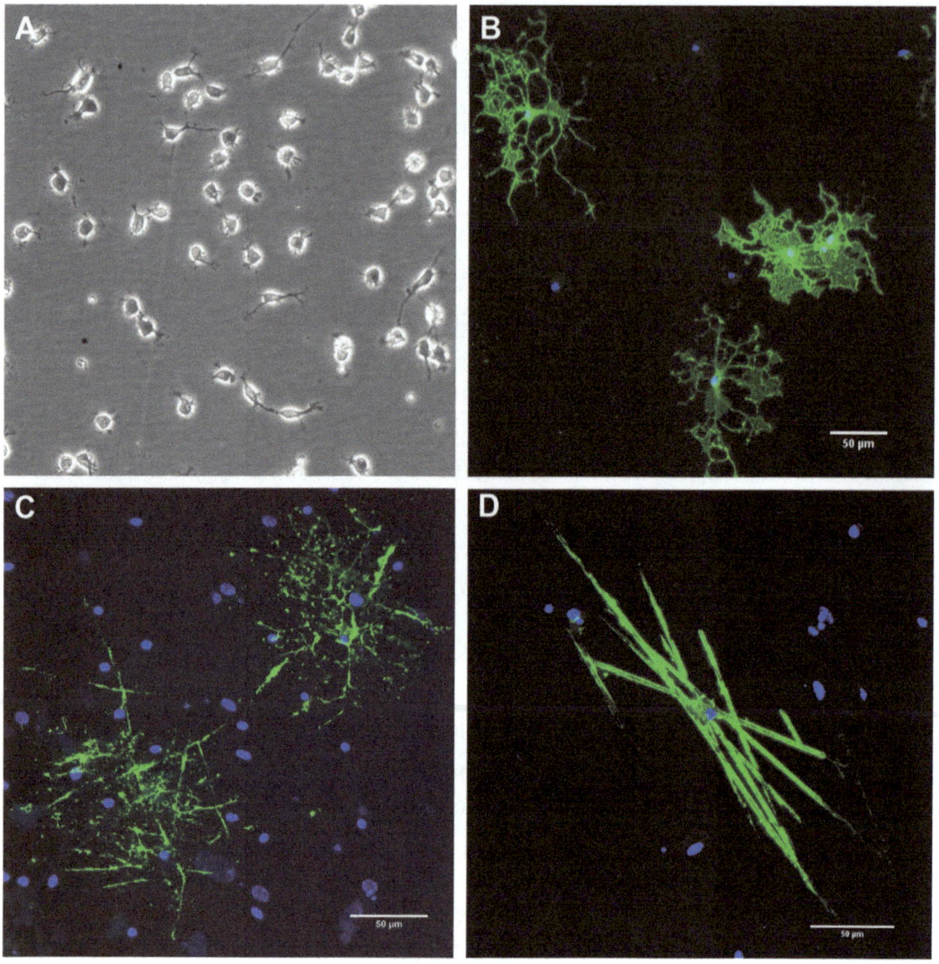

Fig. 2 (**a**) Bright-field image of mouse oligodendrocyte precursor cells proliferating in the presence of PDGF. (**b**) MBP (green) expressing mouse oligodendrocytes, 3 days in the absence of PDGF. (**c**) MBP (green) expressing oligodendrocytes beginning to myelinate DRG neurons. (**d**) MBP expressing oligodendrocyte myelinating electrospun microfibers

3.2.7 Detachment

After 6–10 DIV, cells will begin to reach confluency (Fig. 2a).

1. Wash cells twice with cold PBS, add enough TrypLE to cover the bottom of the flask, and return to the incubator for 5–10 min.
2. Rinse remaining cells off the flask using culture medium and collect the cell solution.
3. Centrifuge cell solution at 1000 rpm for 5 min.
4. Resuspend in fresh medium without disturbing cell pellet and spin again.
5. Triturate cells with 21-guage needle.
6. Count cells using a hemocytometer.

3.2.8 Differentiation and Myelination

1. OPCs can be further expanded in the presence of PDGF and NT3.
2. To analyze cell differentiation, withdraw PDGF from the medium and grow in the presence of NT3, CNTF, and T3. Within just 1 DIV OPCs will begin to differentiate and express myelin proteins such as MBP (Fig. 2b).
3. OPCs can also be added to DRG cocultures (*see* below Fig. 2c) or microfiber scaffolds (Fig. 2d) (Chapter 6).

3.3 Rat Dorsal Root Ganglion Neuronal Culture

3.3.1 Preparation of Coverslips

1. Add 300 μL PDL in sterile water (5 μg/mL) to each 13 mm coverslip in a tissue culture 12-well plate. Leave at 37 °C for 1 h, ensuring solution remains on coverslip.
2. Wash three times with sterile water and leave to dry in a tissue culture hood.
3. Dilute Matrigel 1:40 in ice-cold water, add 300 μL to each PDL-coated coverslip and incubate at 37 °C for 2–3 h.
4. Wash coverslips three times with sterile water and allow to dry before adding DRGs.

3.3.2 Dissection

1. Cull pregnant rat at embryonic day 15 by schedule one injection.
2. Use scissors to cut through skin and muscle of the abdomen.
3. Remove the uterus and transfer to a 10 cm petri dish.
4. Remove individual embryos from the uterus and transfer to a 10 cm petri dish containing cold MEM.
5. Under microscope, use fine forceps to decapitate the embryos.
6. Remove the organs exposing the vertebrae.
7. Using bent/angled forceps cut through the vertebrae.
8. Remove the spinal cord by pulling from the rostral end. If the spinal cord breaks, remove in pieces. Add spinal cord to a fresh petri dish of MEM.
9. Under microscope, use very fine forceps (no. 5) to pluck off individual ganglia from the spinal cord transferring to 15 mL of fresh MEM.

3.3.3 Dissociation

1. Prepare papain digestion mix: MEM 1 mL, Papain 40 μL (30 U/mL), L-cysteine solution (24 mg/mL in MEM, made fresh) 25 μL, and DNaseI type IV 25 μL (4 mg/mL).
2. Incubate at 37 °C for 10 min.
3. Spin DRGs at 1000 rpm for 5 min.
4. Remove excess media leaving a volume less than 2 mL.

5. Filter papain digestion mix through a 0.22 μm filter unit directly onto the DRGs and incubate at 37 °C 7.5% CO_2 for 1 h.
6. Add DMEM containing 10% FBS and 1% P/S to quench papain action.
7. Spin at 1000 rpm for 5 min.
8. Aspirate medium, resuspend in 1 mL fresh medium, and triturate five times with a 21-gauge needle.
9. Count DRG cell number using a hemocytometer.
10. Adjust media volume to 1×10^6 cells per mL, including NGF (1:1000-100 ng/mL).
11. Add a 150 μL droplet to the center of dried coverslips prepared earlier.
12. Transfer to incubator, taking care not to disturb the cell drops.
13. 3–24 h after seeding, add 2 mL media containing NGF (1:1000) and FdU (1:10,000-10 μM).
14. Change media every 2–3 days, alternating FdU addition for a total of 3 FdU incubations.

3.4 Myelinating Coculture

After 21 days in vitro, NGF can be withdrawn from the culture and oligodendrocyte precursor cells can be added to the neurons.

3.4.1 Prepare 100× Rat Modified SATO

1. Mix DMEM 500 mL, BSA fraction V 5.1 g (170 mL of a 30 mg/mL solution in DMEM), progesterone 3.0 mg (300 μL of a 10 mg/mL solution in ethanol), putrescine 805 mg (5 mL of a 161 mg/mL solution in DMEM), T3 20 mg (400 μL of a 50 mg/mL in warm NaOH solution), T4 20 mg (400 μL of a 50 mg/mL in warm NaOH solution).
2. Make 5 mL aliquots and store at −80 °C.

3.4.2 Prepare Myelination Media (See Note 5)

Mix DMEM (High glucose) 240 mL, Neurobasal medium 240 mL, Pen/Strep 5 mL, 100× Rat modified SATO 5 mL, B-27 10 mL, ITS supplement 5 mL, NAC (5 mg/mL) 0.5 mL, D-biotin (10 μg/mL) 0.5 mL.

3.4.3 Culture

1. Gently aspirate media and change DRGs into myelination media.
2. Add OPCs dropwise around the peripheries of the coverslips where the axons reside, away from the cell bodies in the center. For a 13 mm Coverslip add 50,000 rat or 75,000 mouse OPCs.
3. Change media every 2–3 days.

3.5 Fixation and Preparation for Microscopy

To analyze the timing of myelination, it is recommended to use times points between 7 and 21 days post OPC addition, with 14 days typically representing the greatest interface to observe positive and negative effects of any treatment or condition.

1. Gently aspirate medium from culture wells and wash once with PBS.
2. Aspirate and replace PBS with 4% formaldehyde in PBS for 15 min.
3. Remove formaldehyde and wash three times with PBS for 5 min.
4. Cultures can then be processed for immunofluorescence, in an antibody-specific manner. The following represents a standard protocol for analysis of myelination.
5. Prepare blocking buffer by mixing PBS, 10% Normal goat serum, and 0.2% Triton X-100.
6. Using forceps, carefully remove coverslips from tissue culture plates and place in a staining box. Add 200 μL blocking buffer per coverslip and leave at room temperature for 30 min.
7. Remove blocking buffer from coverslips by tilting.
8. Dilute primary antibodies in blocking buffer to MBP (Rat) 1:250 (Bio-Rad—MCA409), NFH (Chicken) 1:100,000 (Biolegend—PCK-592P), CASPR (Rabbit) 1:250 (Abcam—AB34151).
9. Add 200 μL primary antibody solution to each coverslip and leave at 4 °C overnight.
10. Wash coverslips 3× with PBS.
11. Dilute antibody species-specific secondary antibodies in blocking buffer, for example:

    Anti-rat Alexa 488 nm (1:1000), Anti-rabbit Alexa 568 nm (1:1000), Anti-chicken Alexa 647 nm (1:500).
12. Add 200 μL secondary antibody solution to each coverslip and leave at room temperature for 1–2 h, covered.
13. Wash coverslips 2× with PBS.
14. Stain nuclei with Hoechst in PBS for 5 min at room temperature.
15. Wash in PBS once for 5 min.
16. Place a drop (approx. 20 μL) of mounting medium on a standard microscope glass slide.
17. Carefully blot off excess PBS from the edge of the coverslips and place, cells down, on top of the drop.
18. Leave coverslips to set at room temperature overnight.
19. Store slides in the dark at 4 °C short term or at −20 °C for longer-term storage.

3.6 Analysis of Myelination

Oligodendrocytes can be scored based on morphology as: Contacting—processes touching axons, Extending—processes aligning but not fully surrounding axons, or Wrapping—full MBP positive tubes separated by two CASPR positive bands [2, 34]. Of the wrapping cells, internodes/myelin sheaths separated by CASPR bands can be counted and measured to assess myelin sheath number production and sheath length. Myelination can also be quantified through analysis of the co-localization of MBP to neurofilament area [10, 34, 35].

Images should be of a sufficient resolution to accurately determine whether myelin sheaths are being formed. All conditions should be blinded to reduce bias prior to analysis. Original z-series can be analyzed in Fiji ImageJ or a suitable alternative.

4 Notes

1. Flask sizes dependent upon experimental need. Within 6–10 DIV, isolated mouse OPCs will expand reaching confluency in T25 or T150 flasks.
2. If mice require genotyping, use one dish per brain. If pooling multiple mice of the same genotype, plate size can be scaled up.
3. If mice require genotyping, remove a small portion of the ear for subsequent DNA isolation. Label tubes and ear clips, keeping all brain separate.
4. Various protocols exist for brain dissociation utilizing various papain, accutase, ovomucoid combinations. We have found the MACS Neural Tissue Dissociation Kit protocol using the Gentle MACS dissociator to be the most consistent.
5. In addition, myelinating cultures can also be performed in the presence of serum-containing myelination media:
   (a) BME 480 mL
   (b) ITS 5 mL
   (c) Pen/Strep 5 mL
   (d) Glutamax 5 mL
   (e) Glucose (4.5% solution) 4.5 mL
   (f) Fetal Bovine Solution 2.5 mL

References

1. Hild W (1956) Myelin formation in central nervous system tissue cultures. Verh Anat Ges 53:315–317
2. Jarjour AA et al (2012) In vitro modeling of central nervous system myelination and remyelination. Glia 60(1):1–12
3. Matthieu JM et al (1978) Myelination in rat brain aggregating cell cultures. Neuroscience 3 (6):565–572
4. McCarthy KD, de Vellis J (1980) Preparation of separate astroglial and oligodendroglial cell

cultures from rat cerebral tissue. J Cell Biol 85 (3):890–902

5. Barres BA, Raff MC (1993) Proliferation of oligodendrocyte precursor cells depends on electrical activity in axons. Nature 361 (6409):258–260
6. Gard AL, Pfeiffer SE (1993) Glial cell mitogens bFGF and PDGF differentially regulate development of O4+GalC- oligodendrocyte progenitors. Dev Biol 159(2):618–630
7. Behar T et al (1988) Growth and differentiation properties of O-2A progenitors purified from rat cerebral hemispheres. J Neurosci Res 21(2–4):168–180
8. McMorris FA et al (1986) Insulin-like growth factor I/somatomedin C: a potent inducer of oligodendrocyte development. Proc Natl Acad Sci U S A 83(3):822–826
9. Milner R, Ffrench-Constant C (1994) A developmental analysis of oligodendroglial integrins in primary cells: changes in alpha v-associated beta subunits during differentiation. Development 120(12):3497–3506
10. Laursen LS, Chan CW, ffrench-Constant C (2009) An integrin-contactin complex regulates CNS myelination by differential Fyn phosphorylation. J Neurosci 29(29):9174–9185
11. Sontheimer H et al (1989) Channel expression correlates with differentiation stage during the development of oligodendrocytes from their precursor cells in culture. Neuron 2 (2):1135–1145
12. Trotter J, Schachner M (1989) Cells positive for the O4 surface antigen isolated by cell sorting are able to differentiate into astrocytes or oligodendrocytes. Brain Res Dev Brain Res 46 (1):115–122
13. Meier DH, Lagenaur C, Schachner M (1982) Immunoselection of oligodendrocytes by magnetic beads. I. Determination of antibody coupling parameters and cell binding conditions. J Neurosci Res 7(2):119–134
14. Campbell GL, Schanchner M, Sharrow SO (1977) Isolation of glial cell-enriched and depleted populations from mouse cerebellum by density gradient centrifugation and electronic cell sorting. Brain Res 127(1):69–86
15. Cahoy JD et al (2008) A transcriptome database for astrocytes, neurons, and oligodendrocytes: a new resource for understanding brain development and function. J Neurosci 28 (1):264–278
16. Rena Hesse A et al (2012) XIAP protects oligodendrocytes against cell death in vitro but has no functional role in toxic demyelination. Glia 60(2):271–280
17. Watkins TA et al (2008) Distinct stages of myelination regulated by gamma-secretase and astrocytes in a rapidly myelinating CNS coculture system. Neuron 60(4):555–569
18. Szuchet S, Polak PE, Yim SH (1986) Mature oligodendrocytes cultured in the absence of neurons recapitulate the ontogenic development of myelin membranes. Dev Neurosci 8 (4):208–221
19. Gard AL, Pfeiffer SE (1989) Oligodendrocyte progenitors isolated directly from developing telencephalon at a specific phenotypic stage: myelinogenic potential in a defined environment. Development 106(1):119–132
20. Fex Svenningsen A et al (2003) Rapid method for culturing embryonic neuron-glial cell cocultures. J Neurosci Res 72(5):565–573
21. Wood P, Okada E, Bunge R (1980) The use of networks of dissociated rat dorsal root ganglion neurons to induce myelination by oligodendrocytes in culture. Brain Res 196(1):247–252
22. Chan JR et al (2004) NGF controls axonal receptivity to myelination by Schwann cells or oligodendrocytes. Neuron 43(2):183–191
23. Bottenstein JE, Sato GH (1979) Growth of a rat neuroblastoma cell line in serum-free supplemented medium. Proc Natl Acad Sci U S A 76(1):514–517
24. Rosen CL et al (1989) Type 1 astrocytes inhibit myelination by adult rat oligodendrocytes in vitro. J Neurosci 9(10):3371–3379
25. Raff MC, Miller RH, Noble M (1983) A glial progenitor cell that develops in vitro into an astrocyte or an oligodendrocyte depending on culture medium. Nature 303(5916):390–396
26. Bechler ME, Byrne L, Ffrench-Constant C (2015) CNS myelin sheath lengths are an intrinsic property of oligodendrocytes. Curr Biol 25:2411–2416
27. Rosenberg SS et al (2008) The geometric and spatial constraints of the microenvironment induce oligodendrocyte differentiation. Proc Natl Acad Sci U S A 105(38):14662–14667
28. Hernandez M et al (2016) Mechanostimulation promotes nuclear and epigenetic changes in oligodendrocytes. J Neurosci 36 (3):806–813
29. Wake H, Lee PR, Fields RD (2011) Control of local protein synthesis and initial events in myelination by action potentials. Science 333 (6049):1647–1651
30. Wake H et al (2015) Nonsynaptic junctions on myelinating glia promote preferential myelination of electrically active axons. Nat Commun 6:7844
31. Murtie JC, Macklin WB, Corfas G (2007) Morphometric analysis of oligodendrocytes in

the adult mouse frontal cortex. J Neurosci Res 85(10):2080–2086

32. Câmara J et al (2009) Integrin-mediated axo-glial interactions initiate myelination in the central nervous system. J Cell Biol 185 (4):699–712
33. Wang Z, Colognato H, Ffrench-Constant C (2007) Contrasting effects of mitogenic growth factors on myelination in neuron-oligodendrocyte co-cultures. Glia 55 (5):537–545
34. Huang JK et al (2011) Retinoid X receptor gamma signaling accelerates CNS remyelination. Nat Neurosci 14(1):45–53
35. Jarjour AA et al (2015) The polarity protein scribble regulates myelination and remyelination in the central nervous system. PLoS Biol 13(3):e1002107

Chapter 8

Teasing of Ventral Spinal Cord White Matter Fibers for the Analysis of Central Nervous System Nodes of Ranvier

Andrew A. Jarjour and Diane L. Sherman

Abstract

In the central nervous system, the formation of nodes of Ranvier, the short, unmyelinated regions of the axon where voltage-gated sodium channels that mediate saltatory conduction in myelinated nerves are concentrated, is orchestrated by oligodendrocytes, the myelinating cells of the CNS. While transmission electron microscopy remains the gold standard for the study of how the nodal region is organized, this approach is both technically demanding and time-consuming. The availability of antibodies that can be used to label paranodal myelin and the underlying axonal domains that are formed as a result of myelination allows for the precise analysis of the nodal region. In this chapter, we describe the method used to prepare teased fiber preparations of CNS white matter. Teased fiber preparations facilitate the rapid, quantitative analysis of a large number of nodes of Ranvier per animal compared to conventional histological approaches.

Key words Oligodendrocyte, Myelin, Node of Ranvier, Paranode, Juxtaparanode, Internode, Septate junction, Voltage-gated sodium channel, Voltage-gated potassium channel, Caspr, Neurofascin, Ankyrin, Spectrin

1 Introduction

Myelin is the multi-lamellar extension of plasma membrane produced by oligodendrocytes in the central nervous system (CNS), which enables the rapid impulse conduction essential for CNS function in higher species including humans. It does this by clustering voltage-gated sodium channels at the short unmyelinated stretches of axon between myelin segments called nodes of Ranvier. The cytoplasm-containing spiral formed at the termini of each myelin segment forms a specialized adhesion with the axon, thus establishing a barrier that separates nodal sodium channels and juxtaparanodal potassium channels. Disruption of this barrier causes these channels to be mislocalized, and results in decreased conduction velocity. The oligodendroglial and axonal proteins involved in the formation of paranodal junctions and the

David A. Lyons and Linde Kegel (eds.), *Oligodendrocytes: Methods and Protocols*, Methods in Molecular Biology, vol. 1936, https://doi.org/10.1007/978-1-4939-9072-6_8,

establishment and maintenance of specialized axonal domains at and surrounding the node of Ranvier have been the object of intensive study over the past 15 years [reviewed by 1, 2–4].

The involvement of the CNS nodal region in neuropathology has also become clearer in recent years. Disruption of nodal and paranodal organization has been proposed to be an early event in the progression of multiple sclerosis, and the reassembly of nodes to be a critical early step in myelin repair and functional recovery [5, 6]. The finding that antibodies that recognize paranodal and nodal proteins are present in the cerebrospinal fluid of MS patients raises the possibility that the nodal region is the site where demyelination is initiated. More subtle changes in the morphology of the CNS node and/or paranodes have also been associated with aging [7], mild traumatic injury [8], spinal cord injury [9, 10], neonatal hypoxia [11], mild [12] or chronic [13] cerebral hyperperfusion, stroke [14], and noise-induced hearing loss [15]. While it is not clear whether pathological changes to nodal architecture are a proximal event or secondary to disrupted paranodal axo-glial adhesion, recent findings indicate that changes to nodal architecture may occur in healthy individuals as well. In neurons in the auditory brainstem involved in sound localization, axon diameter at the node of Ranvier can vary considerably along a single axon depending on the distance of a given node from the nerve terminal; the authors theorize that local differences in internode length and axon diameter contribute to the synchronization of impulses arriving at the giant calyx of Held presynaptic terminal by compensating for differences in axon length [16]. This finding raises the possibility that conduction velocity can be finely tuned in the CNS by altering the structural properties of both the myelin sheath and the node of Ranvier, and highlights the importance of precise analysis of paranodal and nodal architecture.

Electron microscopy remains the "gold standard" method used for analysis of the nodal region, as it can be used to observe the ultrastructure of the paranodal loops, their sites of adhesion to the axon, and the paranodal and nodal axon in the greatest possible detail. Conventional TEM is commonly used to image the node and paranodes, and is especially useful for the imaging of fine structures such as the transverse bands, electron-dense structures that represent specialized sites of paranodal axo-glial adhesion. More challenging, however, are analyses requiring the imaging of the node of Ranvier and both adjacent paranodes, as this requires that the tissue be sectioned in precisely the correct plane, and that the whole region of interest falls within a single square on an EM grid. Focused ion beam milling coupled to scanning electron microscopy (FIB/SEM) has been employed for the imaging and 3D reconstruction of relatively large volumes of tissue with TEM-like resolution, and has allowed for the visualization of how myelination proceeds, including the formation of paranodes, at a level of resolution not previously possible [17]. However, as this approach is both expensive and time-consuming, it would be impractical for the analysis of large numbers of nodes of Ranvier. Furthermore, electron microscopy is less useful for the measurement of internodal length.

The teasing of nervous tissue into individual nerve fibers on a flat substrate has long been in use by myelin biologists. Santiago Ramón y Cajal [18] credited the discovery of myelin to teased fiber studies carried out by Christian Ehrenberg in 1833 [19]. The modern use of this technique in the CNS dates back to pioneering studies carried out in the 1970s which used teased fiber preparations from cat spinal cord to establish the link between axon diameter and internode length [20], and how this relation is changed by demyelination and remyelination [21]. At the time, investigators were limited to using histological stains when using light microscopy to visualize their preparations, or using transmission electron microscopy (TEM) [22]. The principal advantage of this approach is that, once teased, fibers are arranged in the same plane, thus permitting the easy imaging and analysis of nodal and paranodal properties while eliminating distortions caused by variations between axon trajectories within white matter tracts, and circumventing the technical challenge of precisely orienting the tissue in the longitudinal plane prior to cryosectioning. Immunohistochemical labeling of teased fiber preparations for markers of internodal or paranodal myelin, or the nodal, paranodal, or juxtaparanodal axon is particularly useful for two types of analyses: (1) The measurement of the distance between heminodes or nodes of Ranvier, which can be used to quantify longitudinal myelin extension during development [23, 24], or internodal length (Fig. 1a, [25]); and 2) The visualization of the domain organization of myelinated axons (Fig. 1b, c; *see* Table 1 for a list of recommended commercially available antibodies), enabling the detection of fine structural alterations in the myelin sheath or underlying axon caused by changes in gene expression, age, or pathology to be detected using light microscopy. Super-resolution approaches such as STED allow for the visualization of these structures in even finer detail (Fig. 1d) [26], and will become more important to investigators as technology improves further, and as their use becomes more commonplace.

2 Materials

2.1 Teasing Nerves (See Note 1)

1. Stainless Steel slide rack.
2. Glass staining dishes.
3. 100% acetone.
4. 2% 3-triethoxysilylpropylamine (TESPA) in acetone.

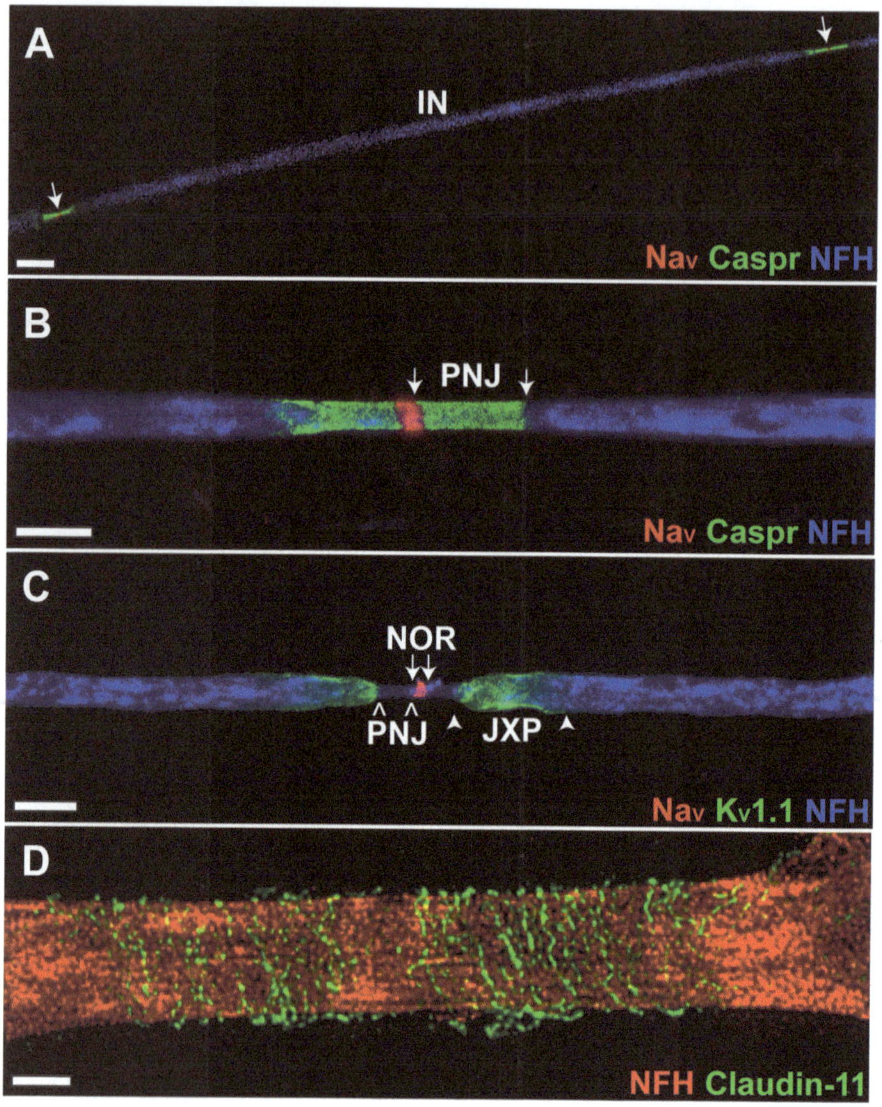

Fig. 1 Imaging and analysis of teased fiber preparations. (**a**) The lengths of myelin internodes can be determined by measuring the distance between adjacent nodes of Ranvier (IN, arrows). (**b**) The size of a paranodal domain can be determined by measuring the length of the band of axonal Caspr immunolabeling (PNJ, delimited by arrows) adjacent to the node of Ranvier (red). (**c**) The distance across the node of Ranvier can be determined by measuring the length of voltage-gated sodium channel (Nav) immunopositive band (NOR, arrows), the juxtaparanodal domain can be determined by measuring the length of the voltage-gated potassium channel (Kv1.1) immunopositive region (JXP, arrowheads), and the distance across the paranodal junction can be determined by measuring the distance between nodal and juxtaparanodal labeling (PNJ, circumflexes). (**d**) The paranodal "spiral" can be visualized in more detail when immunofluorescence is visualized using super-resolution imaging techniques such as gated STED (gSTED). (**a**) Scale bar = 20 μm. (**b**, **c**) Scale bar = 5 μm. (**d**) Scale bar = 2 μm

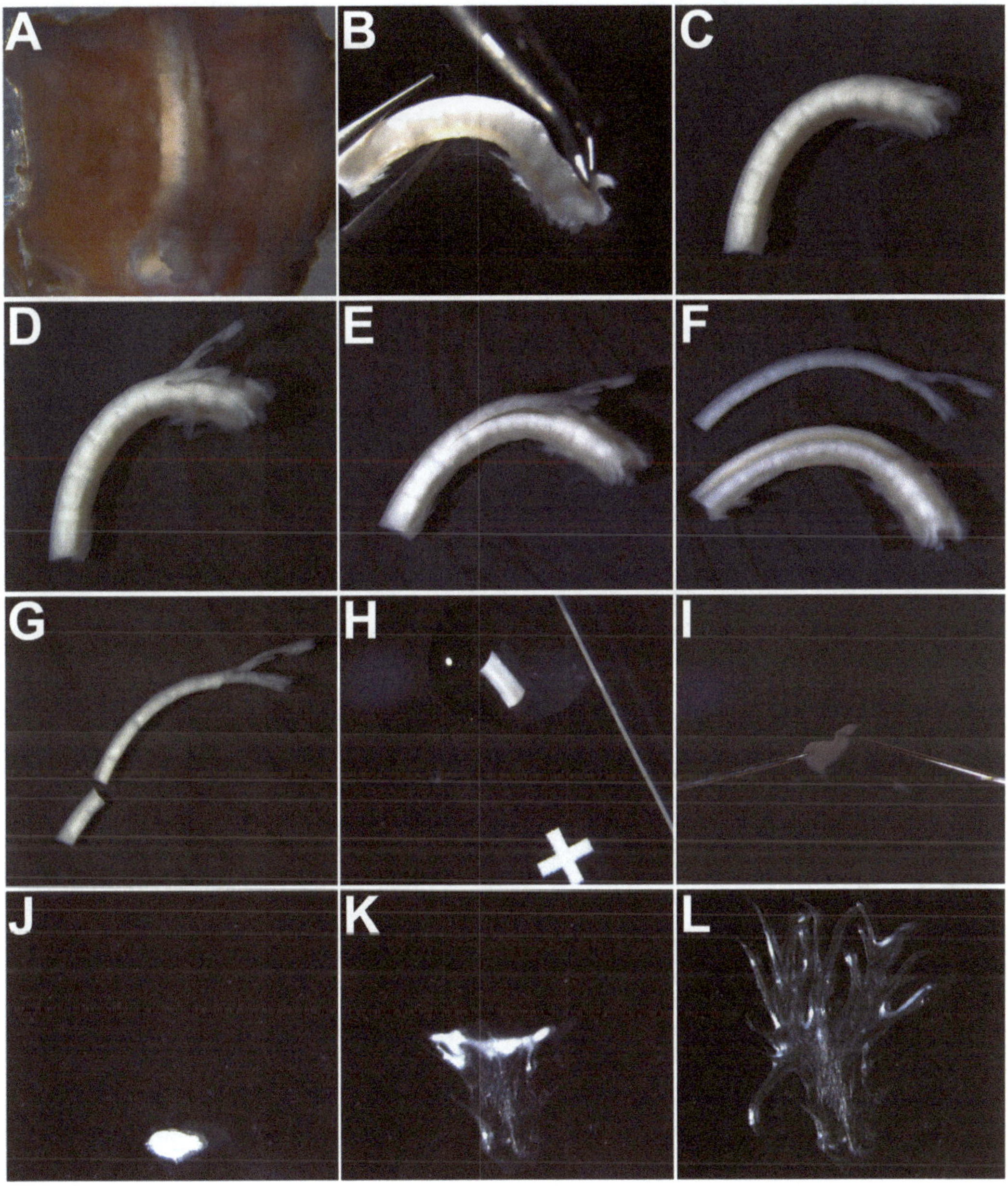

Fig. 2 Preparation of teased fibers from mouse ventral funiculus. (**a**) Exposed spinal cord following perfusion. (**b**) After dissecting out the spinal cord, the ventral white matter is gripped with #5/45 forceps. (**c–f**) A strip of ventral white matter is gradually pulled away until it is completely separated from the cord. (**g**) The strip of ventral white matter is further cut into smaller pieces. (**h**) A small piece of white matter is placed in a drop of PBS on the edge of a slide, outside the usable area. (**i**) White matter tissue is held using acupuncture needles. (**j–l**) Acupuncture needles are used to gently pull apart a piece of white matter tissue, spreading fibers into a thin layer on the slide

Table 1
Commercially available antibodies that have been used successfully to immunolabel CNS teased fiber preparations

Antigen	Location	Host species	Ordering information
Neurofilament heavy chain (NFH)	Axon	Chicken	Biolegend #822601
Myelin-associated glycoprotein (MAG)	Periaxonal myelin	Goat	Santa Cruz # sc-9544
Pan voltage-gated sodium channel (Na_V)	Node of Ranvier	Mouse	Sigma-Aldrich #S8809
βIV spectrin	Node of Ranvier	Mouse IgG2b	Antibodies Incorporated #75-377
βII spectrin	Myelinated regions of the axon	Mouse IgG1	BD Biosciences #612563
Caspr1	Paranodal axon	Rabbit	Abcam #ab34151
Caspr1	Paranodal axon	Mouse	Antibodies Incorporated #75-001
Pan neurofascin (NFC)	Paranodal myelin and Node of Ranvier	Mouse	Antibodies Incorporated #75-027
Ankyrin G	Paranodal myelin and Node of Ranvier	Mouse	Antibodies Incorporated #75-146
Voltage-gated potassium channel subunit 1.1 (K_V1.1)	Juxtaparanodal axon	Rabbit	Abcam #ab32433
Oligodendrocyte-specific protein (OSP)/Claudin-11	Non-compact mature myelin (paranodes, Schmidt-Lantermann incisures, inner and outer tongue)	Rabbit	Thermo Fisher Scientific #36-4500
Connexin 32 (Cx32)	Paranodal myelin	Mouse IgG2a	Thermo Fisher Scientific #35-8900

5. 4% Paraformaldehyde (PFA) in 0.1 M Sorenson's phosphate buffer with 0.5% NaCl, pH 7.4—Phosphate Buffered Saline (PBS) 1×, cold.
6. Petri plates (100 mm, though 30–60 mm plates are useful if tissue from each animal must be kept separate).
7. Treated slides or coverslips.
8. 5 mL bijou containers.
9. Straight-edge razor blades.
10. Fine iris scissors.

11. #5 forceps.
12. #5/45 forceps.
13. Rongeur.
14. Small spatula.
15. Acupuncture needles (25 mm long, 0.25 mm diameter).
16. Dissection microscope with light sources both above and below the stage (ideally with a substage mirror as it makes it much easier to see fibers during teasing).

2.2 Immunolabeling and Analysis

1. 1× Phosphate Buffered Saline (PBS).
2. Blocking solution: 3% normal donkey serum, 2% bovine serum albumin, 0.1% Triton X-100 in 1× PBS.
3. Orbital shaker.
4. Primary antibody of choice.
5. Secondary antibody of choice.
6. Mounting medium of choice.

3 Methods

3.1 Teasing Ventral Spinal cord

*If you are teasing onto pre-coated slides such as Superfrost Plus, proceed to **Step 5**.*

1. Place clean slides in a stainless steel slide rack.
2. Fill four staining dishes with the following solutions: 100% acetone, 2% TESPA in acetone, 100% acetone, 100% acetone.
3. Immerse slides in acetone for 1 min, then immediately into 2% TESPA in acetone for 2 min, followed by two quick rinses (10–20 s) in acetone.
4. Allow the slides to air dry.
5. Fix animal tissue by cardiac perfusion with 4% PFA in PBS.
6. Dissect out cervical spinal column. In adult animals, cut down each lateral edge of the dorsal spinal column and remove the bone using a Rongeur, exposing the spinal cord (Fig. 2a), and then transfer to bijou containers containing 4% PFA in 1× PBS.
7. Postfix for 30 min, rinse in 1× PBS, and store in 1× PBS at 4 °C until just before teasing. If immunolabeling for voltage-gated sodium channels be sure to not over-fix, as the anti-pan-Na_V antibody (*see* Table 1) is very sensitive to fixation.
8. In a petri plate containing cold PBS on ice, at most 48 h later (but preferably sooner), dissect the spinal cord out from the column, being careful to not damage the ventral white matter.

9. Remove the meninges. Use the new #5 forceps to hold the spinal cord at the lateral edges, while using the #5/45 forceps to gently peel the ventral funiculi away from the rest of the spinal cord (Fig. 2b–f). In all but the youngest animals, the tissue should peel away easily in thin strips (*see* Note **1**).
10. Transfer the strips of ventral funiculus to a fresh petri plate shallowly filled with cold 1× PBS. In very young (e.g., P4-P7) animals, peel off smaller pieces and proceed to **step 12**.
11. Using a razor blade, cut each of these strips into smaller pieces roughly 2 mm long (Fig. 2g), and use the razor blade or acupuncture needles to carefully subdivide these into pieces approximately 0.5 mm wide.
12. Using a small spatula, transfer one to three pieces of tissue to area of the slide that will not be used for teasing (we use the rightmost edge, to the right of the "+" marking on a Superfrost Plus slide, Fig. 2h). It should be in a small drop of PBS—just enough to keep it wet.
13. Gently lift the tissue out of the PBS using the 5/45 forceps, touch it to a dry area of slide edge to remove excess PBS, and place it on the usable area of the slide.
14. Using the needles to hold down the strip of tissue at each end, slowly drag the tissue along the slide whilst lightly pulling the needles away from each other (Fig. 2i) until individual fibers can be seen trailing behind under the dissection scope. If a "chunk" of tissue is left behind, it can be pulled again in another direction, repeating until the entire piece has been teased apart (Fig. 2j–l) (*see* **Note 2**).
15. Repeat until slide is sufficiently full. In our hands, one adult cervical spinal cord provides enough tissue for five slides (*see* **Note 3**).
16. Immediately after teasing, the slides are dried and stored at −20 °C. Slides stained immediately or can be stored long-term at −80 °C in slide boxes sealed within plastic freezer bags.

3.2 Immunolabeling and Analysis

1. Remove teased fiber preparations from the freezer immediately before use, and wash briefly with 1× PBS.
2. Incubate in blocking solution for 60 min at 4 °C on an orbital shaker.
3. Incubate with primary antibody diluted in blocking solution for at least 2 h at 4 °C on an orbital shaker (*see* **Notes 4** and **5**).
4. Wash 3 × 10 min in blocking solution at 4 °C on an orbital shaker.
5. Incubate with secondary antibody diluted in blocking solution for 60 min at 4 °C on an orbital shaker.

6. Wash 3 × 10 min in blocking solution at 4 °C on an orbital shaker.
7. Wash 2 × 10 min in 1× PBS at 4 °C on an orbital shaker.
8. Remove excess PBS, and mount slides using mounting medium of choice (*see* **Note 6**).
9. (*See* **Notes 7–9**).

4 Notes

1. While the ventral funiculus of the rodent spinal cord is most commonly used for teasing because of the large proportion of large-diameter axons, which are easiest to both tease and analyze, it should be noted that this approach can be used successfully with other species and regions of the CNS, such as zebrafish optic nerve [27].
2. It is important to have just the right amount of moisture on the tissue: If the tissue is too wet, fibers will not stick to the slide, while if it is too dry, it will become very sticky and fibers will not separate without applying a level of force that would be likely to damage them.
3. White matter from the rest of the cord can be used as well, but care should be taken to note what level of the cord the tissue is from, especially at earlier developmental time points when myelination in the caudal spinal cord may lag behind what is observed more rostrally.
4. If convenient, primary antibody incubation can be carried out overnight.
5. For a list of useful antibodies for the labeling of axons, myelin, and subdomains of the axon and myelin sheath, please *see* Table 1.
6. We use FluoroMount-G from EMS.
7. The protocol outlined here is an example and need not be rigorously followed; we have also obtained very clear immunolabeling using a solution of 5% fish skin gelatin and 0.2% Triton X-100 in PBS for the blocking and antibody incubation steps, while carrying out all steps at room temperature. It is also not essential to use an orbital shaker. We advise experimenters to adapt the protocol to suit the antibodies being used.
8. Examples of analyses that can be performed using teased spinal cord preparations are shown in Fig. 1. This includes measurements of internode length (Fig. 1a, distance IN between arrows), paranode length (Fig. 1b, distance PN between arrows), node of Ranvier length (Fig. 1c, distance NOR between arrows), and juxtaparanode length (Fig. 1c, distance

JXP between arrowheads). Because disruption to nodal or paranodal structure can result in changes to the localization of the markers for specialized axonal domains, it must be emphasized that these analyses should be used to complement, and not replace, analyses using electron microscopy.

9. Even when care is taken to tease fibers gently, some fibers will break, with breakages frequently occurring at nodes of Ranvier. Additionally, it is impossible to separate every fiber during teasing. Analysis of incompletely teased fibers (especially when measuring intermodal lengths) should be avoided as it is often difficult to discriminate between individual axons. Similarly, analysis of fibers at break points should be avoided, as the force applied is also likely to have stretched the axon and may distort the measurements obtained.

Acknowledgments

A.A.J. thanks the Wellcome Trust, Medical Research Council and UK Multiple Sclerosis Society for support (grants to Charles ffrench-Constant). D.L.S. thanks the Wellcome Trust (grant to Peter Brophy) and Medical Research council for support (grant to Peter Brophy and DLS). Many thanks to Peter Brophy and Charles ffrench-Constant for their helpful comments on this chapter.

References

1. Poliak S, Peles E (2003) The local differentiation of myelinated axons at nodes of Ranvier. Nat Rev Neurosci 4(12):968–980
2. Buttermore ED, Thaxton CL, Bhat MA (2013) Organization and maintenance of molecular domains in myelinated axons. J Neurosci Res 91(5):603–622. https://doi.org/10.1002/jnr.23197
3. Salzer JL (2003) Polarized domains of myelinated axons. Neuron 40(2):297–318
4. Susuki K, Otani Y, Rasband MN (2016) Submembranous cytoskeletons stabilize nodes of Ranvier. Exp Neurol 283(Pt B):446–451. https://doi.org/10.1016/j.expneurol.2015.11.012
5. Coman I, Aigrot MS, Seilhean D, Reynolds R, Girault JA, Zalc B, Lubetzki C (2006) Nodal, paranodal and juxtaparanodal axonal proteins during demyelination and remyelination in multiple sclerosis. Brain 129 (Pt 12):3186–3195
6. Howell OW, Palser A, Polito A, Melrose S, Zonta B, Scheiermann C, Vora AJ, Brophy PJ, Reynolds R (2006) Disruption of neurofascin localization reveals early changes preceding demyelination and remyelination in multiple sclerosis. Brain 129(Pt 12):3173–3185
7. Hinman JD, Peters A, Cabral H, Rosene DL, Hollander W, Rasband MN, Abraham CR (2006) Age-related molecular reorganization at the node of Ranvier. J Comp Neurol 495 (4):351–362. https://doi.org/10.1002/cne.20886
8. Volman V, Ng LJ (2014) Primary paranode demyelination modulates slowly developing axonal depolarization in a model of axonal injury. J Comput Neurosci 37(3):439–457. https://doi.org/10.1007/s10827-014-0515-7
9. Ouyang H, Sun W, Fu Y, Li J, Cheng JX, Nauman E, Shi R (2010) Compression induces acute demyelination and potassium channel exposure in spinal cord. J Neurotrauma 27

(6):1109–1120. https://doi.org/10.1089/neu.2010.1271

10. Sun W, Fu Y, Shi Y, Cheng JX, Cao P, Shi R (2012) Paranodal myelin damage after acute stretch in Guinea pig spinal cord. J Neurotrauma 29(3):611–619. https://doi.org/10.1089/neu.2011.2086
11. Ritter J, Schmitz T, Chew LJ, Buhrer C, Mobius W, Zonouzi M, Gallo V (2013) Neonatal hyperoxia exposure disrupts axon-oligodendrocyte integrity in the subcortical white matter. J Neurosci 33(21):8990–9002. https://doi.org/10.1523/JNEUROSCI.5528-12.2013
12. Reimer MM, McQueen J, Searcy L, Scullion G, Zonta B, Desmazieres A, Holland PR, Smith J, Gliddon C, Wood ER, Herzyk P, Brophy PJ, McCulloch J, Horsburgh K (2011) Rapid disruption of axon-glial integrity in response to mild cerebral hypoperfusion. J Neurosci 31(49):18185–18194. https://doi.org/10.1523/JNEUROSCI.4936-11.2011
13. Choi BR, Kim DH, Back DB, Kang CH, Moon WJ, Han JS, Choi DH, Kwon KJ, Shin CY, Kim BR, Lee J, Han SH, Kim HY (2016) Characterization of white matter injury in a rat model of chronic cerebral hypoperfusion. Stroke 47(2):542–547. https://doi.org/10.1161/STROKEAHA.115.011679
14. Hinman JD, Lee MD, Tung S, Vinters HV, Carmichael ST (2015) Molecular disorganization of axons adjacent to human lacunar infarcts. Brain 138(Pt 3):736–745. https://doi.org/10.1093/brain/awu398
15. Tagoe T, Barker M, Jones A, Allcock N, Hamann M (2014) Auditory nerve perinodal dysmyelination in noise-induced hearing loss. J Neurosci 34(7):2684–2688. https://doi.org/10.1523/JNEUROSCI.3977-13.2014
16. Ford MC, Alexandrova O, Cossell L, Stange-Marten A, Sinclair J, Kopp-Scheinpflug C, Pecka M, Attwell D, Grothe B (2015) Tuning of Ranvier node and internode properties in myelinated axons to adjust action potential timing. Nat Commun 6:8073. https://doi.org/10.1038/ncomms9073
17. Snaidero N, Mobius W, Czopka T, Hekking LH, Mathisen C, Verkleij D, Goebbels S, Edgar J, Merkler D, Lyons DA, Nave KA, Simons M (2014) Myelin membrane wrapping of CNS axons by PI(3,4,5)P3-dependent polarized growth at the inner tongue. Cell 156(1–2):277–290. https://doi.org/10.1016/j.cell.2013.11.044
18. Ramón y Cajal S (1909) Histologie du système nerveux de l'homme et des vertébrés. Consejo Superior de Investigaciones Cientificas, Madrid.
19. Rosenbluth J (1999) A brief history of myelinated nerve fibers: one hundred and fifty years of controversy. J Neurocytol 28(4–5):251–262
20. McDonald WI, Ohlrich GD (1971) Quantitative anatomical measurements on single isolated fibres from the cat spinal cord. J Anat 110(Pt 2):191–202
21. Gledhill RF, McDonald WI (1977) Morphological characteristics of central demyelination and remyelination: a single-fiber study. Ann Neurol 1(6):552–560. https://doi.org/10.1002/ana.410010607
22. Allt G, Gajree T (1975) Teased single C.N.S. fibres observed by light and electron microscopy. J Neurocytol 4(6):647–651
23. Court FA, Sherman DL, Pratt T, Garry EM, Ribchester RR, Cottrell DF, Fleetwood-Walker SM, Brophy PJ (2004) Restricted growth of Schwann cells lacking Cajal bands slows conduction in myelinated nerves. Nature 431(7005):191–195. https://doi.org/10.1038/nature02841
24. Zonta B, Tait S, Melrose S, Anderson H, Harroch S, Higginson J, Sherman DL, Brophy PJ (2008) Glial and neuronal isoforms of Neurofascin have distinct roles in the assembly of nodes of Ranvier in the central nervous system. J Cell Biol 181:1169–1177
25. Jarjour AA, Boyd A, Dow LE, Holloway RK, Goebbels S, Humbert PO, Williams A, ffrench-Constant C (2015) The polarity protein Scribble regulates myelination and remyelination in the central nervous system. PLoS Biol 13(3):e1002107. https://doi.org/10.1371/journal.pbio.1002107
26. D'Este E, Kamin D, Balzarotti F, Hell SW (2017) Ultrastructural anatomy of nodes of Ranvier in the peripheral nervous system as revealed by STED microscopy. Proc Natl Acad Sci U S A 114(2):E191–E199. https://doi.org/10.1073/pnas.1619553114
27. Munzel EJ, Becker CG, Becker T, Williams A (2014) Zebrafish regenerate full thickness optic nerve myelin after demyelination, but this fails with increasing age. Acta Neuropathol Commun 2:77. https://doi.org/10.1186/s40478-014-0077-y

Chapter 9

Whole-Cell Patch Clamp Recordings from Oligodendrocyte Lineage Cells in Brain Slices

Sylvia Agathou and Ragnhildur Thóra Káradóttir

Abstract

The whole-cell configuration of the patch-clamp technique is widely used to study electrically active cells and passive membrane properties, as well as the properties and pharmacology of ion channels, neurotransmitter receptors, and electrogenic transporters, in almost any cell type. In the brain, in addition to neurons, oligodendrocyte precursor cells (OPCs) that give rise to myelinating oligodendrocytes (OLs) are also excitable. Electrophysiological techniques provide the main tool for the thorough investigation of the electrogenic capacity of such cell types. Although there are many published protocols for whole-cell recordings, there are very few that touch upon the electrophysiological characteristics of oligodendrocyte lineage cells. Here we provide a detailed methodology for how to acquire and analyze whole-cell recordings from excitable cells, with a focus on oligodendrocyte lineage cells. We provide a protocol on how to successfully identify OPCs and OLs in brain slices, either with the use of transgenic animal models or through morphological and electrophysiological profiling. The method described can also be easily adopted for whole-cell recordings from oligodendrocyte lineage cells in vitro

Key words Whole-cell patch clamp, Oligodendrocyte precursor cells, Oligodendrocytes, Electrophysiology, Myelination

1 Introduction

The development of today's patch clamp technique is a result of the pioneering work of many scientists through the years. Seminal steps in the configuration of the technique were made between the late 70s and early 80s, when the high-resistance seal between the cell and a recording micro-pipette was developed [1, 2]. This provided the basis for the whole-cell configuration of the patch clamp technique—both voltage and current clamp mode [2, 3]—that has transformed modern cellular physiology and has allowed for real-time recordings of cellular activity.

The dynamics of excitable cell membranes were defined by Hodgkin and Huxley in 1952, as a result of ion movements across the cellular membrane. They showed that voltage-gated ion

David A. Lyons and Linde Kegel (eds.), *Oligodendrocytes: Methods and Protocols*, Methods in Molecular Biology, vol. 1936, https://doi.org/10.1007/978-1-4939-9072-6_9,

channels generate electrical excitation in neuronal cells [4]. Voltage difference across the membrane results from concentration differences of ions in the cytosol and the outside of the cell (a concentration gradient, described by the Nernst equation), and the cell membrane's selective permeability to ions (Na^+, K^+, Cl^-, Ca^{2+}). The membrane potential generated due to this difference is described by the Goldman-Hodgkin-Katz equation. Excitability is due to voltage-gated ion channels that abruptly allow specific ions to move across the cellular membrane altering the equilibrium potential across the cell membrane. Electrical charge is carried across the membrane through ion-specific channels (voltage- and ligand-gated channels) and transporters.

Changes in ion flow across a cell membrane and corresponding changes in membrane potential can be described by Ohm's law: $V = I \times R$, where I is current; V, voltage; and R, resistance. The cell membrane's resistance is relatively constant, and thus the main variability in resistance is due to the ion flow fluctuations through ion channels. The ease with which ions flow through the membrane channels can also be described by the reciprocal of resistance, conductance (G). Through this basic representation, we can predict with Ohm's law that if the voltage across the membrane is held constant, the subsequent current changes will correspond to changes in membrane resistance and thus changes in the opening of ion channels, ion channel composition, and/or ion channel properties ($I = V/R$). This concept provides the basis for the voltage-clamp technique during whole cell recordings that we describe in this chapter.

The cellular membrane acts both as a capacitor and as a resistor. The simplest representation of a cell membrane as a conventional RC-circuit contains a membrane capacitor and a membrane resistor, connected in parallel. Membrane capacitance (C_m, measured as farads (F)) results from the property of the phospholipid bilayer, as it both separates and stores charge. Thus, membrane capacitance is proportional to the cell's surface area, and normally, for most cells 1 cm^2 of cell membrane measures as 1 μF. Membrane resistance (R_m) is inversely related to the total area of the membrane and membrane leak conductance, which are mainly driven by potassium channels. These parameters are important to keep in mind during whole-cell recordings and for data analysis as set out below in Subheading 3.4.

The general principle of patch-clamp recordings is to establish a high-resistance seal between the cell membrane and the tip of a glass micro-electrode pipette. For the conventional whole-cell configuration of the patch-clamp technique this membrane patch is ruptured to establish a connection between the intracellular space of the cell and the internal solution of the pipette (Figs. 1 and 2). To obtain a high-resistance seal (giga-seal), the tip of the glass pipette is placed in close proximity to the cell membrane, and gentle

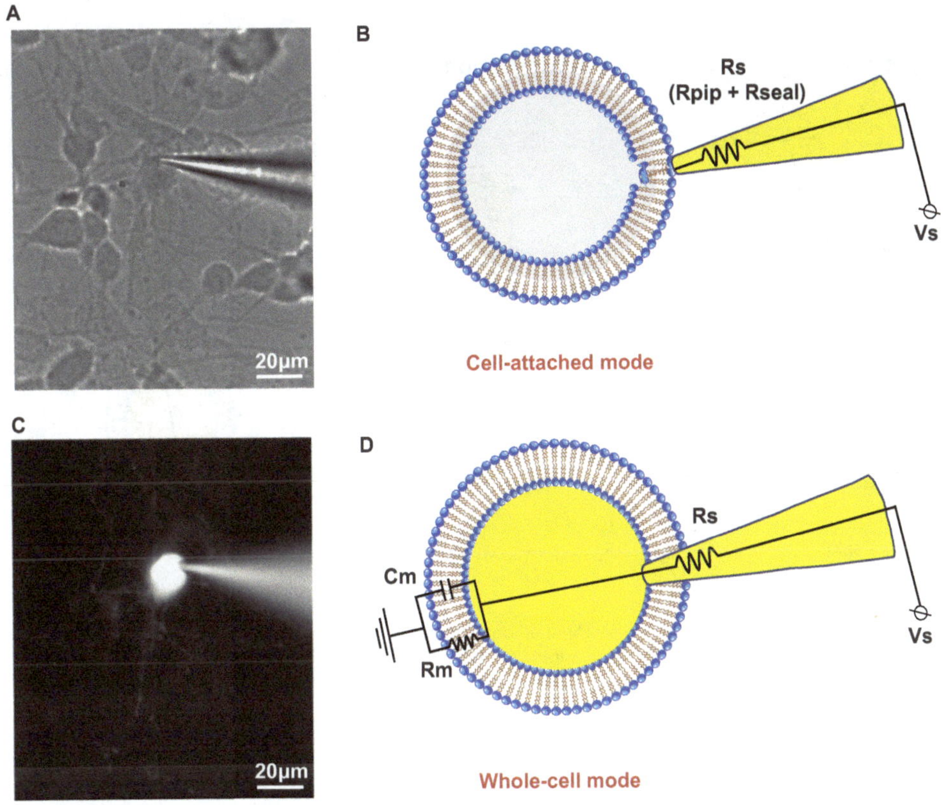

Fig. 1 Whole-cell patch clamp configuration. (**a**) A glass micro-electrode approaches the cellular membrane in order to establish a high-resistance giga-seal (>1 GΩ) between the cell membrane and the tip of the glass micro-electrode pipette, which is schematically represented in (**b**). Once the giga-seal is established, negative pressure is applied to the electrode in the form of a few brief suctions, so as to rupture the membrane and to establish the low-resistance connection with the intracellular space (**c**, **d**). The opening of the cell can also be visualized from the diffusion of the internal solution containing the Lucifer yellow dye (represented in yellow) inside the cell (**c**, **d**)

suction is applied (Figs. 1a, b and 2b). Once the giga-seal is established, a few brief suctions are applied to the electrode to rupture the membrane and to establish a low-resistance connection with the intracellular space (Figs. 1c, d and 2c). This allows the patch-clamp amplifier to monitor the pipette current through a voltage drop across a feedback resistor. Thus, in whole-cell recordings it is crucial to account for and compensate for the resistance between the cytosol and the pipette as it limits current resolution and the speed at which the cell membrane responds to the voltage clamp (*see* Subheading 3.4). The methodology described in this chapter for the acquisition of oligodendrocyte lineage cell electrophysiological properties is based on the abovementioned concepts of whole-cell patch clamp techniques.

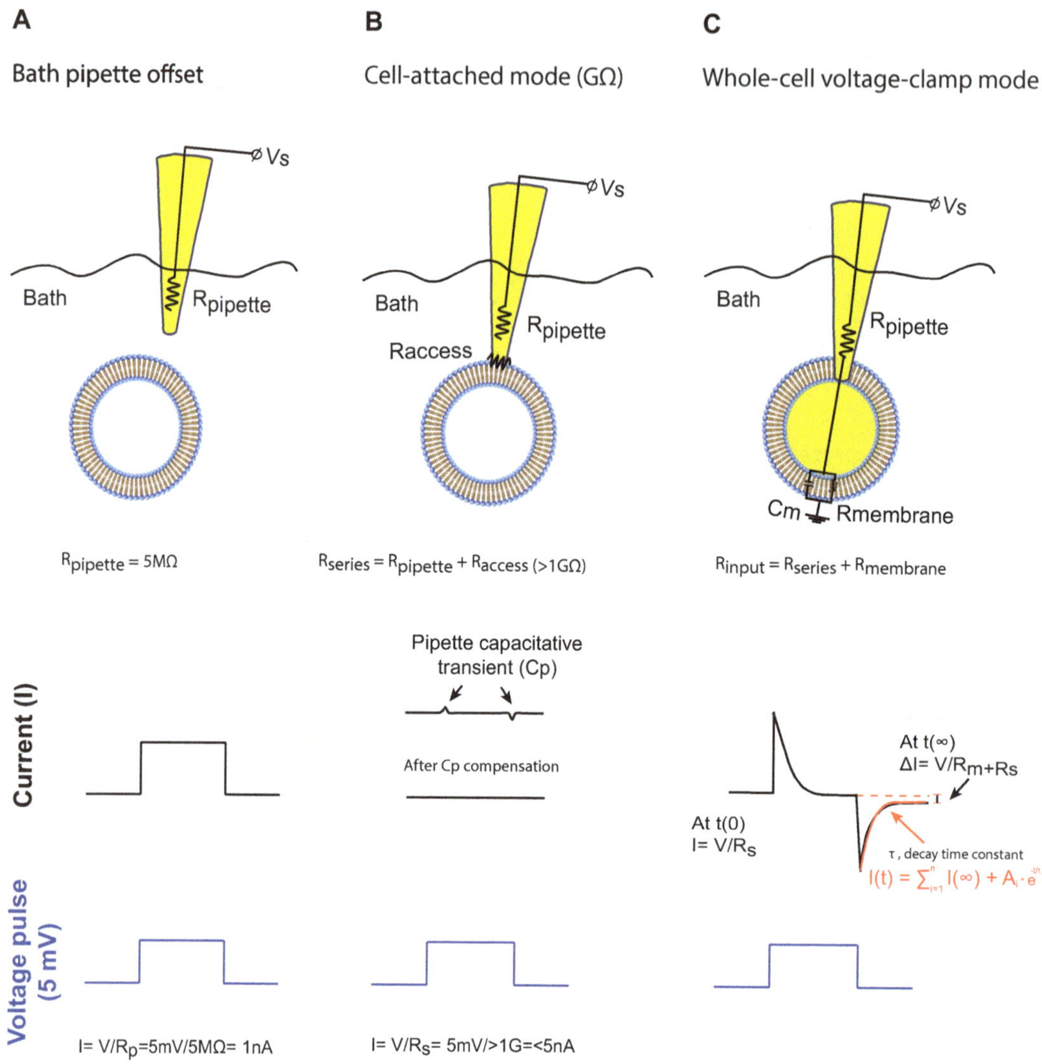

Fig. 2 Schematic representations of the patching approach and the recorded currents in voltage-clamp mode. Examples of recorded current in response to a small voltage step (5 mV) which is continuously monitored when the pipette is (**a**) in the bath, (**b**) attached to the cell surface with a giga-seal or (**c**) during whole-cell mode. During (**a**) the only resistance present is that of the pipette (R_p) whose value can be read off the amplifier. In (**b**) both the pipette resistance (R_p) and the access resistance (R_a) between the pipette and the cell surface are present. The main components of the R_s are the pipette resistance R_p, and access resistance Ra. In (**c**) R_s and the membrane resistance (R_m) exist. At the onset of the voltage step (at $t = 0$) the current only flows through the series resistor, thus R_s can be calculated using $I_{t=0}$ (using Ohm's law $V_s = I_{t=0} \times R$). At the steady state ($t = \infty$), the capacitor (cellular membrane) is fully charged and the current I_∞ equals the current flowing through the series resistor and the membrane resistor, which gives the value of $R_m + R_s$. To calculate membrane capacitance C_m for isopotential simple cells, the extrapolation of the decay time constant τ, from fitting the current response to a single exponential is needed (red line; *see* Subheadings 3.4.1 and 3.4.2 for how to calculate C_m for simple and complex cells, respectively)

The interest in studying the electrical properties of oligodendrocyte lineage cells and understanding neuron-oligodendroglia interaction has increased in recent years, partly because, like neurons, it is now clear that oligodendrocyte lineage cells express a plethora of ion channels as well as a variety of neurotransmitter receptors (such as glutamate and GABA receptors) [5–12]. Moreover, electrophysiological and electron microscopy (EM) studies have shown that oligodendrocyte precursor cells (OPCs) receive synaptic (both glutamatergic and GABAergic) inputs from unmyelinated axons, in both the gray [5, 6] and white matter [9, 13–15], and some OPCs are even capable of firing action potentials in response to depolarizing current pulses [6, 9]. However, the roles of synaptic input and neurotransmitter signaling, as well as the function of the variety of ion channels the cells express at different developmental stages are still unclear. However, the fact that new oligodendrocytes (OLs) are made throughout life [16] and are important for motor learning, [17] along with the fact that neuronal activity regulates OPC proliferation, differentiation, and myelination [15, 18, 19], makes it important to use electrophysiological methods to understand oligodendrocyte glia and myelin biology.

The theory and method of electrophysiological recordings has been a subject of many excellent reviews and protocols. These mostly refer to the applications of the technique on neuronal cells or cardiomyocytes or provide extensive details on set up, patch-clamp electronics and troubleshooting [20–26]. We recommend consulting these reviews for a more detailed technical understanding, methodology on setting up the patching equipment, troubleshooting, reducing electrical noise, or the analysis of electrical properties. We focus here on methods to perform whole-cell recordings from oligodendrocyte lineage cells and providing a morphological and electrophysiological profiling of OPCs and OLs.

2 Materials

2.1 Experimental Animals

The most convenient way to identify oligodendrocyte lineage cells for patch-clamp recordings is to have transgenic animals that express fluorescent proteins under a lineage specific promoter (such as NG2 or PDGFalphaR for OPCs [27], SOX10 or PLP for oligodendrocyte lineage [28], or MBP or CNP for differentiated OLs [12]). In addition to mice, now there is also a transgenic rat model, the Sox10-DsRed model, which is also suitable [28]. However, oligodendrocyte lineage cells can also be identified by their distinct cell body shape under IR-DIC and by their electrophysiological properties, and thus wild-type animals, or transgenic mouse/rat models not harboring lineage specific expression of recombinant proteins, can also be used [5, 7, 9, 15] (*see* below

Subheading 3.3; Figs. 5 and 6). All animal experiments must be performed in accordance with the relevant authorities' guidelines and regulations.

2.2 Buffer Recipes

2.2.1 Artificial Cerebrospinal Fluid (aCSF) Solutions for Slicing and External Bath Application

We prepare 10× stock solutions for making the slicing/resting and external (artificial cerebrospinal fluid) solutions, omitting glucose, $MgCl_2$, and $CaCl_2$, which are added fresh on the day of use. It is recommended not to use 10× stock solutions that are older than 1 month.

1. 10× Slicing and Resting solution recipe (Bicarbonate-based): 120 mM NaCl, 26 mM $NaHCO_3$, 1 mM NaH_2PO_4, and 2.5 mM KCl.
2. 10× External bath solution recipe (HEPES-based): 144 mM NaCl, 10 mM HEPES, 1 mM NaH_2PO_4, and 2.5 mM KCl.

   1× HEPES-based or 1× bicarbonate-based external solution and the 1× bicarbonate-based slicing solution are made fresh on the day by diluting the 10× stocks, and with the following additions.

   In the 1× bicarbonate slicing solution, we add 10 mM D-glucose, 2 mM $MgCl_2$, and 1 mM kynurenic acid to prevent excitotoxic cell death during slicing. Lastly, we add 2 mM $CaCl_2$ after at least 20 min of bubbling with 95% CO_2/5% O_2 to avoid its precipitation.

   On the day, in the 1× HEPES-based aCSF solution, we add 2.5 mM $CaCl_2$ and 10 mM glucose and set the pH to 7.35 with NaOH. For 1× bicarbonate external solution we add D-glucose, and then 2 mM $CaCl_2$ after at least 20 min of bubbling with 95% CO_2/5% O_2 to avoid its precipitation, and to equilibrate the buffer.

   Into the external solutions we add 1–2 mM of $MgCl_2$. Please note that Mg^{2+} blocks NMDA-Rs. If synaptic input from axons to OPCs needs to be recorded, we advise the lowering of $MgCl_2$ to 1 mM (closer to physiological Mg^2) in the external recording solution (*see* **Notes 1** and **2**).
3. 1× Slicing and resting aCSF (final composition): 120 mM NaCl, 2.5 mM KCl, 1 mM NaH_2PO_4, 26 mM $NaHCO_3$, 2 mM $CaCl_2$, 2 mM $MgCl_2$, 10 mM D-glucose, and 1 mM kynurenic acid (*see* **Note 3**).
4. 1× External HEPES-based aCSF solution (final composition): 144 mM NaCl, 2.5 mM KCl, 10 mM HEPES, 1 mM NaH_2PO_4, 2.5 mM $CaCl_2$, and 10 mM glucose, pH set to 7.35 with NaOH.
5. 1× External bicarbonate-based aCSF solution (final composition): 120 mM NaCl, 2.5 mM KCl, 1 mM NaH_2PO_4, 26 mM $NaHCO_3$, 2 mM $CaCl_2$, 2 mM $MgCl_2$, and 10 mM D-glucose, pH set by constantly bubbling with 95% CO_2/5% O_2.

2.2.2 Internal Solutions (Pipette Solutions)

1. *K^+/Cs^+-gluconate based*: 130 mM K^+ gluconate (or Cs^+ gluconate), 4 mM NaCl, 0.5 mM $CaCl_2$, 10 mM HEPES, 10 mM BAPTA, 4 mM MgATP, 0.5 mM Na_2GTP (2 mM K-Lucifer yellow; *see* below Subheading 2.2.3); pH adjusted to 7.3 with KOH or CsOH. Low chloride concentration E_{Cl} ~86 mV; junction potential likely to be close to −14 mV (depending on external aCSF used—needs to be measured prior to use). *See* **Note 4**.
2. *K^+/Cs^+ Chloride with BAPTA*: 130 mM CsCl or KCl, 4 mM NaCl, 0.5 mM $CaCl_2$, 10 mM HEPES, 10 mM BAPTA, 4 mM MgATP, 0.5 mM Na_2GTP, potassium Lucifer yellow 2 mM; pH set to 7.3 with CsOH or KOH. High chloride concentration E_{Cl} ~ 0 mV; junction potential likely to be close to −3 mV (depending on external aCSF ringer used—needs to be measured prior to use).
3. *K^+/Cs^+ Chloride EGTA* 130 mM KCl, 4 mM NaCl, 0.5 mM $CaCl_2$, 10 mM HEPES, 10 mM EGTA, 4 mM MgATP, 0.5 mM Na_2GTP, potassium Lucifer yellow 2 mM, pH set to 7.3 with KOH or CsOH. High chloride concentration E_{Cl} ~ 0 mV; junction potential likely to be close to −0 mV (depending on external aCSF used—needs to be measured prior to use).
4. *K^+/Cs^+ Chloride with EGTA and phosphocreatine*: K^+/Cs^+ Chloride: 126 mM CsCl (or KCl), 4 mM NaCl, 10 mM HEPES, 5 mM EGTA, 4 mM MgATP, 0.5 mM Na_2GTP, 12 mM phosphocreatine, 2 mM potassium Lucifer yellow; pH set to 7.3 with CsOH or KOH high chloride concentration E_{Cl} ~ 0 mV. In this internal, phosphocreatine is used along with ATP to give extra energy for maintaining cellular health and constant membrane potential. Junction potential likely to be close to −14 mV (depending on the external aCSF used—needs to be measured prior to use).

BAPTA has faster chelating kinetics than EGTA, and it is more calcium selective and less dependent on pH; for repeated glutamate agonist administrations we have found that BAPTA gives more stable recordings of oligodendrocyte lineage cells. Consideration as to which calcium chelator to use is particularly important when studying G-protein coupled receptors. Caesium (Cs^+) blocks potassium channels, thus providing a lower basal noise level and better signal to noise recordings (*see* **Note 5**).

Due to the difference in salt concentrations between the internal solution and the external aCSF solution, and because the reference electrode and the recording electrode are in different solutions a junction potential occurs (*see* **Note 6**). This junction potential needs to be measured for each internal/external combination and subsequent recordings should retrospectively be adjusted according to this potential difference.

Internal solutions should be aliquoted and stored at –20 °C in the appropriate volume needed for each experimental day. Once thawed, internal solution aliquots should not be refrozen or reused on another experimental day.

2.2.3 Dye-Filling for Internal Solutions

For morphological identification and post-hoc immunohistochemistry of the patched cells, appropriate labeling dyes can be added to the internal electrode solution. Available dyes include neurobiotin, Alexa dye, fluorescent-conjugated dextran, and Lucifer Yellow (LY). The Alexa dyes are costly but very bright, and thus can be used at a low concentration (0.2 mg/mL) or with the back-filling technique in order to just fill the tip of the electrode with the dye-filling solution and then back-fill the rest of the pipette with the same internal solution but lacking the dye [29]. Alternatively, neurobiotin (1.5 mg/mL) or LY (1 mg/mL) can be used. We mainly use LY for our recordings, as it is efficient and economical. We make our internal solution according to the above recipes with the final addition of 2 mM LY (1 mg/mL). Once LY is added we sonicate at RT the internal solution for a few seconds to dissolve the dye, and then pass it through a 0.22 μm filter.

2.3 Acute Brain Slicing Preparation

2.3.1 Instrumentation

1. Resting pre-chamber for brain slices (lab-made or commercially made e.g. Harvard Apparatus). A good incubation/resting chamber should allow the slice to rest undisturbed on a nylon mesh with constant exposure to oxygenated slicing solution, for the entire experimental duration (9–10 h). The chamber should also be sturdy, to prevent accidental slice floating and associated mechanical damage. Numerous resting chamber designs have been suggested by electrophysiologists [21–23]. We make our own chamber with the help of a nylon mesh, glued tubing with multiple holes to allow gas diffusion in the slicing buffer, and sand or any other heavy item in order to weight down the construction and avoid accidental tipping (Fig. 3b–d).
2. Vibratome (Leica or Camden).
3. Dissection tools: large and small pointed scissors, curved and straight forceps, large and small curved spatula.
4. Glass electrode puller.
5. Microforge (for visualizing pipettes (size, shape), tip polishing, etc.).

2.3.2 Consumables

1. 95% CO_2/5% O_2 gas cylinder.
2. Large polystyrene icebox.
3. Glassware: two small glass beakers (~50 mL) and a glass petri dish (~90 mm diameter).

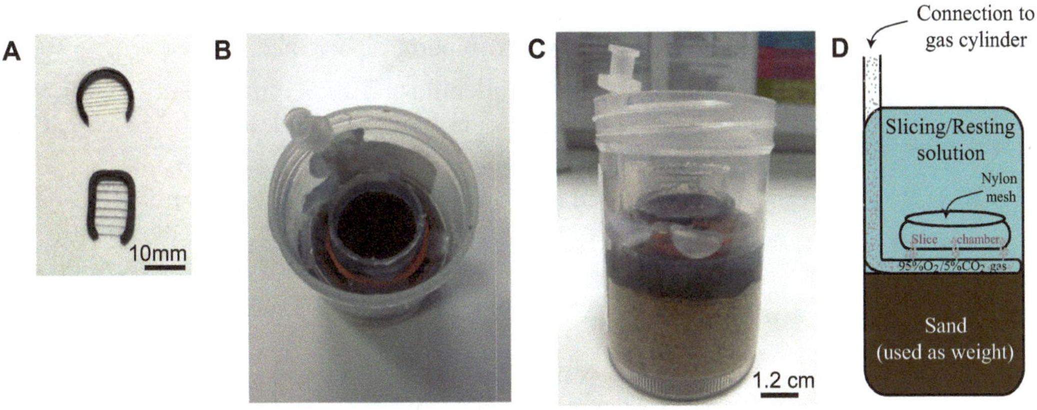

Fig. 3 Harp and slice resting chamber setup for electrophysiological recordings. (**a**) Harps, to use as slice anchors can be purchased ready-made or made in the lab with nylon mesh (can use strings from nylon stockings) and a U-shaped platinum wire. Top (**b**) and side (**c**) view of a lab-made slice resting chamber for tissue slices used for electrophysiological recordings. A good incubation/resting chamber should allow the slice to rest undisturbed on a nylon mesh with constant exposure to oxygenated slicing solution, for the entire experimental duration (8–10 h). The chamber should also be sturdy, to prevent accidental slice floating and associated mechanical damage. (**d**) Schematic representation of the resting slice chamber shown in **b**, **c** as a detailed visualization of all the components needed for a lab-made slice chamber. These include a nylon mesh slice compartment immersed in gas bubbling (95% CO_2/5% O_2) slicing solution, glued tubing with multiple holes to allow gas diffusion in the slicing/resting artificial cerebrospinal fluid solution and sand or any other heavy item to weigh down the construction and avoid accidental tipping

4. Cut-off glass Pasteur pipette with the thin end removed (lab-made or John Poulten Ltd.) with a suction rubber cap, useful for slice handling.
5. Agarose blocks: Fill almost 2/3 of a ~90 mm diameter petri dish with 5% agarose dissolved in Phosphate-buffered saline (PBS) and store at 4 °C.
6. Cyanoacrylate glue (i.e., superglue [commercial]).
7. Vibratome-compatible razor blades (Camden, or commercial sharp razor blades, e.g., Gillette [high-carbon steel blades sharper than stainless steel]).

2.4 Whole-Cell Patch Clamping Preparation

Once acute brain slice preparation is completed (*see* Subheading 3.1), the following materials and buffers should be prepared/acquired for successful whole-cell patch clamp recordings.

2.4.1 Consumables

1. Thick-walled borosilicate glass capillaries with a filament (outer diameter 1.5 mm, inner diameter 0.86 mm; type GC150F-10, Harvard Apparatus or Warner Instruments) work best for oligodendrocyte lineage cells, according to our experience. For some neurons, thin-walled (outer diameter 1.5 mm, inner diameter 1.17 mm; type GC150TF; Warner Instruments or Harvard Apparatus) electrodes are more suitable. However,

due to the thinner wall, sylgard, or dental wax, coating is needed to lower the pipette capacitance.

2. 100% O_2 gas cylinder for HEPES buffer solutions, or 95% CO_2/5% O_2 gas cylinder for bicarbonate-based aCSF solutions.
3. "Harp" used as a slice anchor (Harvard Apparatus or lab-made from platinum wire and nylon mesh, which can also be obtained from nylon stockings. Remove a thin string off the nylon stockings and use to string over the bent platinum wire; Fig. 3a).
4. Ag/AgCl "earth" wire pellet (Harvard Apparatus). The earth wire pellet needs to be well chlorinated at the start of an experiment—please check the condition of the earth wire pellet the day before commencing an experiment. If it is not well chlorinated (e.g., not displaying uniform deep grayish color), immerse it in bleach overnight or replace the pellet. Alternatively, place the silver wire (precleaned) in 1M HCl solution along with a platinum wire (acts as a cathode) and apply current for a few minutes until a good coating is achived.
5. Chlorinated silver wire for the preamplifier electrode. Please make sure, the day before commencing an experiment that the electrode sliver wire is chlorinated (clear dark gray coating without shiny breaks in the coating). If it is not well chlorinated, place it into bleach overnight. This is needed on regular intervals, depending on the usage.
6. Patching recording slice chamber (~1 mL volume, Scientifica, ALA Scientific, Harvard Apparatus, or lab-made).
7. Post-recording fixation solution: 4% Paraformaldehyde (PFA) in PBS. Keep at 25 °C in order to fix slices after electrophysiological recordings, for further post-hoc immunohistochemistry and cell identification.

2.4.2 Instrumentation

A schematic of the basic rig setup needed for electrophysiological recordings is shown in Fig. 4. More specifically the equipment required for a successful recording is listed below:

1. Pipette puller either horizontal or two-stage vertical.
2. Microforge.
3. Bunsen Burner.
4. Disposable scalpels for slice marking.
5. Upright fixed-stage infrared (IR) differential interference contrast (DIC) microscope with 4× and 40×/63× objectives. Note: For recording from cells within the white matter of older animals, 900 nm IR DIC optics are needed to visualize the cells, particularly in heavily myelinated tracts such as the cerebellar and optic nerve white matter.

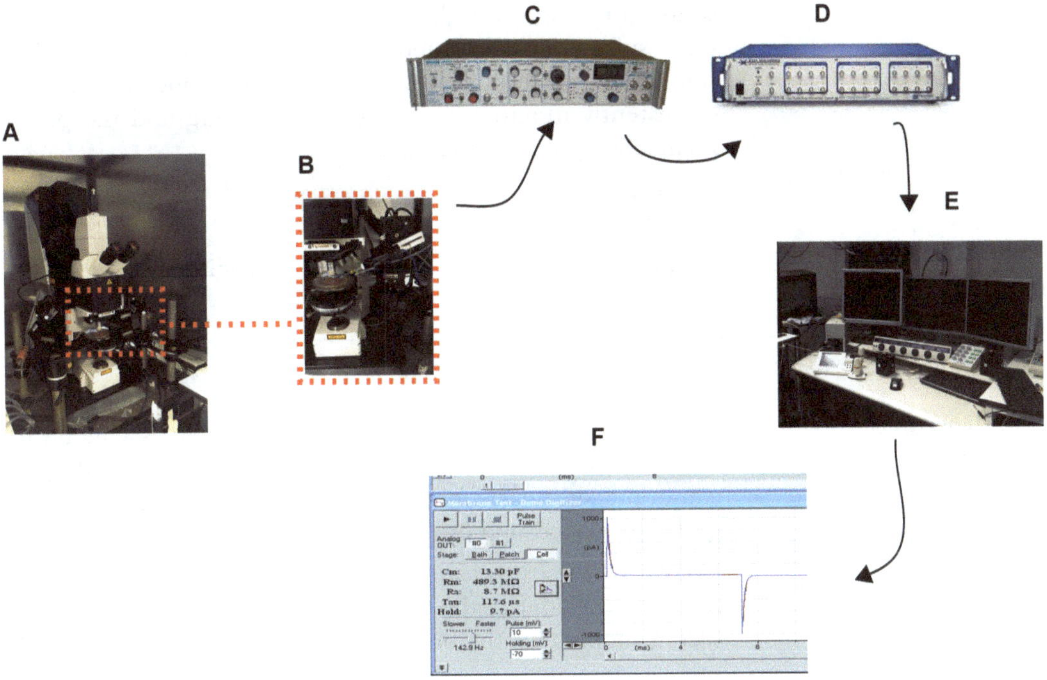

Fig. 4 Schematic representation of the basic components of the patch-clamp set-up. The experimental tissue or cells are mounted on the chamber of the upright light-microscope (**a**) and constantly perfused with an artificial cerebrospinal fluid solution (aCSF; *see* Subheading 2.2.1 for details). The microscope itself is mounted on an anti-vibration air table and surrounded by a Faraday cage, in order to avoid vibrations and electrical noise that will interfere with the recordings. The patch pipette with its internal recording electrode and the reference bath electrode are connected to the headstage (**b**), which is mounted on a micromanipulator (**b**). The recorded analog signal from the headstage is amplified (**c**) and passed to an analog to digital converter (**d**), where the signal is digitized and sent to a computer (**e**) for data analysis with specific electrophysiology software (**f**). Images adapted from Scientifica and Molecular Devices online resources

6. Anti-vibration air table.
7. Faraday cage.
8. Patch-clamp headstage, amplifier, and digitizer (Axon Instruments, Heka, Sutter, or Cambridge Electronic Design).
9. Halogen lamp and suitable fluorescence filters for the appropriate excitation and emission wavelengths of the fluorophore and reporter fluorescent protein in use.
10. Micromanipulators.
11. Peristaltic pump (multiple suppliers) or gravity-fed (lab-made) perfusion system.
12. Heating element for heating external solution to 35 °C with a thermometer in the recording bath, preferably with a feedback loop for constant temperature monitoring during recordings. Alternatively, for a lab-made heating setup, feed the external

solution tubing through resistors attached to a power supply. Once sufficient power is applied, the resistors will heat up. With the help of an in-bath thermometer, the temperature can be consistently monitored during the recording and the power manually regulated as required.

13. Microscope-attached monochromatic IR-CCD camera or similar.
14. Computer, screen monitors, and a TV monitor.

3 Method

3.1 Acute Brain Slicing

1. Prepare for acute brain slicing by oxygenating the slicing buffer with 95% CO_2/5% O_2 for at least 30 min on ice. Put the vibratome slicing chamber at −4 °C for 5 min, place all the glassware on ice, and prepare all the dissecting tools needed.
2. Evenly cut off a small square-shaped agarose block and superglue it to the edge of the vibratome-slicing platform to provide slicing support for the brain.
3. Remove the vibratome-slicing chamber from the freezer, place it on the vibratome according to the manufacturer's instructions, and fill it up with ice-cold oxygenated slicing buffer. Additionally, fill the exterior part of the chamber with ice to keep the temperature low while slicing.
4. Fill the resting slice pre-chamber with oxygenated slicing buffer, remove all bubbles adhering to the nylon mesh, using the cut-off Pasteur pipette. Ensure that the slice is oxygenated continuously until the end of the experiment.
5. Humanely and in accordance with the local animal welfare regulations sacrifice an animal. We normally use transgenic NG2-YFP knock-in mice [27] postnatally aged for 7–30 days but have used the same method to generate acute brain slices from embryonic day 13 to postnatal day 330, with slight modifications for embryonic slices [29, 30].
6. Separate the head from the body and quickly immerse it in a glass beaker containing ice-cold oxygenated slicing buffer.
7. Dissect the brain out by cutting with scissors between the eyes. Continue with lateral cuts close to the base of the skull, on both sides of the head from anterior to posterior.
8. Remove the excised skull carefully with forceps without damaging the brain.
9. Immediately place the brain in a glass beaker containing ice-cold oxygenated slicing buffer.

10. Remove meninges, and dissect the brain area of interest, giving a careful consideration to the angle of the axonal fiber tract orientation to ensure the axons are kept intact. Superglue the dissected part to the vibratome-slicing platform, adjacent to the agarose block for support and taking into consideration the correct brain orientation for the desired slices (i.e., sagittal, coronal, transverse) as well as the angle of the axonal tracts (consult a stereotaxic atlas for details).
11. Slice brain sections of 225–300 μm with the optimal vibration amplitude and frequency for your vibratome. Thicker slices are better for recording evoked synaptic inputs to OPCs.
12. Once each slice is cut off, transfer it using a cut and fire polished glass Pasteur pipette into a slice pre-chamber filled with oxygenated slicing buffer.
13. Allow the slices to rest for 1 h at room temperature (RT) prior to whole-cell patch clamp recordings (*see* **Note 7**).
14. Thaw an aliquot of PFA at RT (use caution and appropriate PPE when handling PFA) for post-recording fixation of slices/cells.

3.2 Whole-Cell (Patch-Clamp) Recording

1. Thaw an aliquot of the desired internal solution and keep it on ice and away from direct light during the experiment (*see* **Note 8**).
2. Fire-polish, with a Bunsen burner, both edges of the glass capillaries by holding them for a few seconds in the open flame. Or purchase glass pipettes that are already fire-polished.
3. Once they are fire-polished, use these capillaries to pull recording electrodes of the desired size/resistance. For whole-cell experiments, a large diameter pipette size (i.e., 2-4 MΩ) will provide a good access resistance but could damage or even be larger in diameter than a small cell soma (*see* **Note 9**). For small cells like OPCs, OLs or small neurons it is advisable to use pipette sizes of 5–7 MΩ, which provide a balance between a good access resistance and minimal cell damage.
4. Start up all the electronic equipment on the rig and oxygenate the external aCSF solution with 100% O_2 throughout the duration of the experiment if using HEPES-based external. Otherwise use 95% CO_2/5% O_2 gas for bicarbonate-based aCSF solutions; it is good practice to equilibrate the solution with the gas for few minutes before use. When recording at physiological temperatures, it is advised to use bicarbonate-based external.
5. Continuously perfuse the recording chamber on the rig, with oxygenated external aCSF solution with the help of a gravity-led perfusion system (or peristaltic pump).

6. Transfer a brain slice from the resting pre-chamber into the bath of the recording chamber on the rig with the help of a cut-off glass Pasteur pipette. To stabilize the slice inside the liquid, place a U-shaped harp (Fig. 3a) on top of the slice (*see* **Note 10**).
7. Visualize the slice first with a low-magnification water immersion objective (4×, 5× or 10×) to identify the desired patching region. Once in the right area, switch to a 40× (or 63×) water immersion objective (i.e., Olympus LumplanFI/IR; Leica HCX, NA = 0.80) with differential infrared interference contrast (IR-DIC) optics (higher wavelength infrared—900 nm instead of conventional 750 nm—IR-DIC optics allow for a better visualization of cells in heavily myelinated tracts).
8. By interchanging fluorescent and bright-field visualization, identify the cell of interest (*see* Subheading 3.3 on how to identify OPC/OLs; Figs. 5, 6 and 7) and note its location on the screen (Fig. 6c, d, h, I).
9. Once the desired cell for patch clamping has been identified, fill a glass electrode with internal solution (containing either neurobiotin, Alexa dye, fluorescent-conjugated dextran, or Lucifer Yellow for morphological identification and post-hoc immunohistochemistry; *see* Subheading 2.2.2; Figs. 5 and 6) and remove any bubbles by gently flicking it.
10. Place and tighten the electrode on the micromanipulator headstage (Fig. 4b) and lower it carefully up to one optical level higher than the cell to be patched.
11. Once the electrode is in the bath, start a protocol that delivers around every 30 ms, a ± 5 mV (or 10 mV) and 15 ms duration voltage pulse and display the current either on an oscilloscope or on a computer (using software such as Clampex, EPC, Igor, Spike, or Strathclyde). This protocol should be continuously on and monitored until the whole-cell clamp is established.
12. Make sure that the holding current is adjusted to 0, then adjust the pipette current offset to zero on the amplifier.
13. Start applying positive pressure by steadily but very gently blowing internal solution outward to avoid pipette clogging by residual debris. When close to the cell of interest, a slightly more positive pressure is applied to clean the cell surface before pipette attachment (we achieve the blowing and suctioning via a lab-made tube connected on one end to the micromanipulator and on the other end to a flow regulator and a 1 mL plastic syringe through which we apply pressure by mouth) (*see* **Note 11**).
14. The general principle of whole-cell recordings is to establish a high-resistance giga-seal (>1 GΩ) between the cell membrane

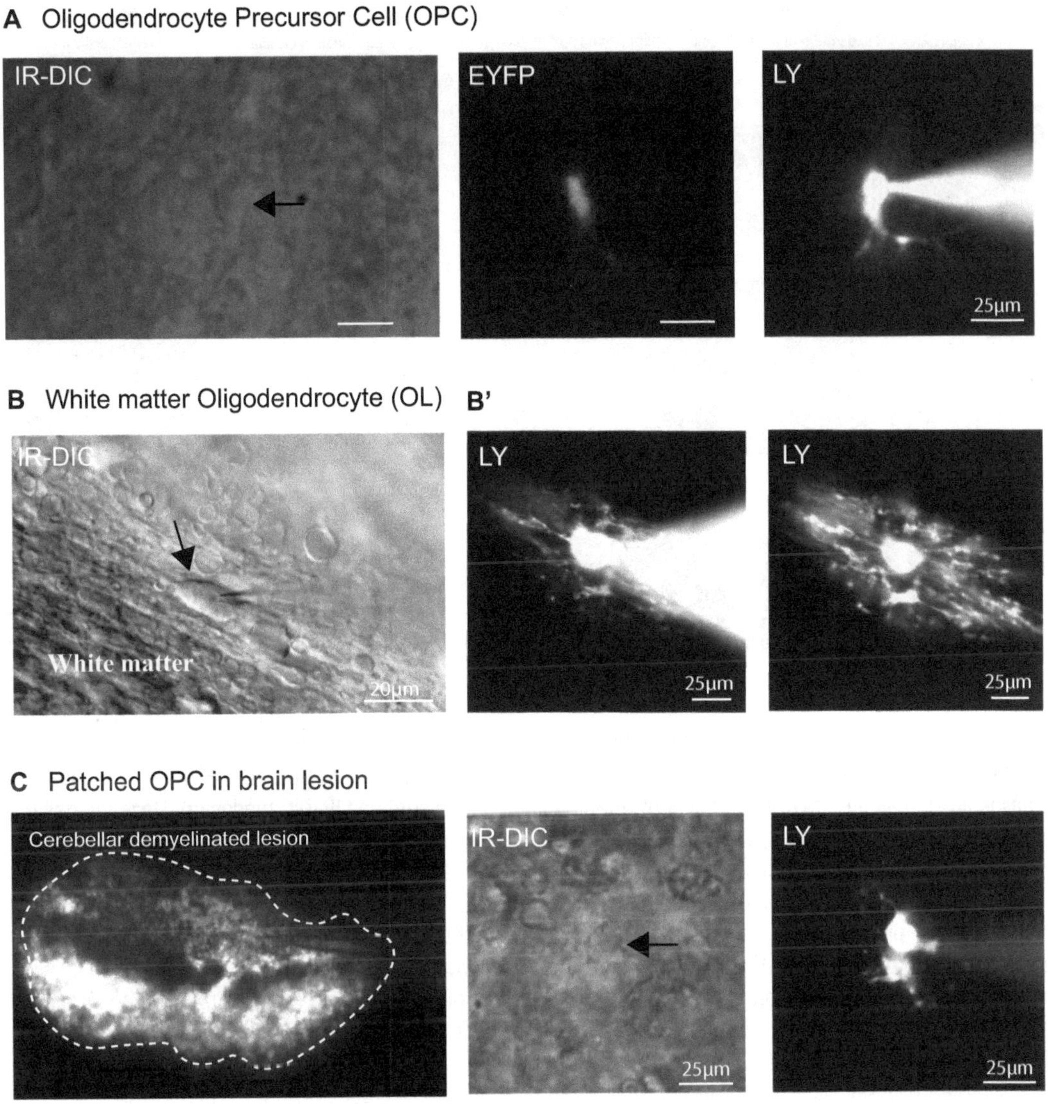

Fig. 5 Morphological properties of oligodendrocyte-lineage cells. Oligodendrocyte precursor cells (OPCs) have characteristic tear- and diamond-shaped somata as depicted in **a** (first panel, black arrow). They can easily be identified with the use of transgenic animal models, expressing fluorescent proteins under OPC-specific promoters. One example is the NG2-EYFP mouse model by Karram et al. (2008) [27]. With this model, OPCs can be identified live *ex vivo* by their expression of enhanced yellow fluorescent protein (EYFP), as shown in **a** (second panel). Once the cell is patched, Lucifer yellow (LY) from the internal solution fills the cell and provides a good morphological outline (**a**, third panel). (**b**) Mature white matter oligodendrocytes (OLs) tend to have larger somata with a more oval or eclipse shape and seem to lack visible processes, although processes are in fact present, as seen from a representative LY-filled OL in **b**′. Even without the use of transgenic reporters, OPCs can be identified (**c**, second panel) and patched (**c**, third panel) even in experimental demyelinating brain lesions, as depicted in **c**. The demyelinated lesion appears brighter than the myelinated area around, as myelin absorbs more light. Please note that the picture was taken with the electrode to note the location of the cell within the lesion and thus to aid identification for post-hoc immunohistochemistry (*see* Fig. 6)

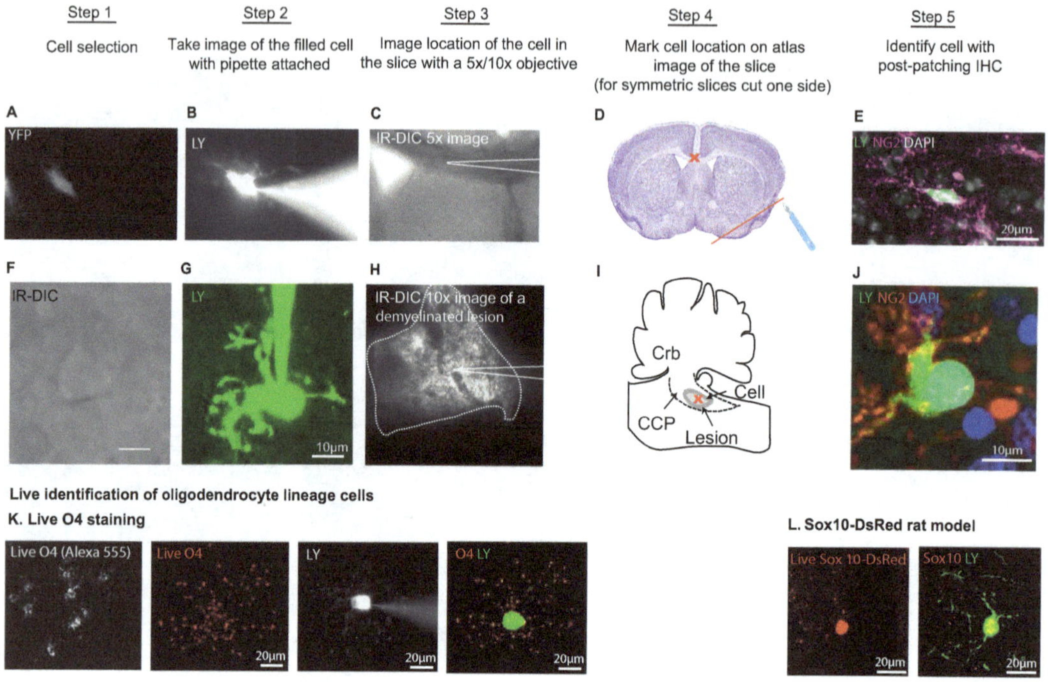

Fig. 6 Step guide for cell identification for patch clamp recordings. When ready to patch-clamp and when the appropriate tissue slice is mounted under the rig microscope, start by selecting a healthy looking cell (*see* Subheading 3.3), either with the help of the morphological cell properties in IR-DIC (BF; F) or if you are using a transgenic animal, (i.e., NG2-EYFP mice) by interchanging fluorescent and IR-DIC modes (**a**). Once the desired cell has been selected, fill a glass electrode with dye-labeled internal solution and proceed to the recording. Successful whole-cell recordings should fill the cell cytosol as depicted in (**b**), in a patch-clamped OPC. Capture a fluorescent image of the patched cell, focusing on the morphology of the cell, or draw a representative picture of the cell if the microscope is not fitted with a camera. (**c**) After the recording is over, capture a low-magnification IR-DIC image with the recording pipette still attached to the cell (**c**) so as to mark the location of the cell on the slice. Once removing the pipette from the cell, capture also a fluorescent image of the patched cell without the pipette, this will further facilitate cell identification for post hoc immunohistochemistry (IHC) (**e**). (**d**) For asymmetrical brain slices (as in **i**) mark the orientation of the slice. For symmetric brain slices cut one corner of the slice with a scalpel (red line). This will facilitate the identification of the slice orientation and cell positioning when preparing for imaging following IHC (**e**). The same procedure can be followed when not using transgenic animals by focusing on the shape of the OPC (**f**) in IR-DIC, and for identifying OPCs in induced demyelinated lesions with IR-DIC (**f–j**), where the location of the cell can be crucial for further analysis. Aside from the NG2-EYFP animal model (**a**) we have also successfully identified and patched OPCs in the Sox10-DsRed rat model, as depicted in **l**. If not using a transgenic animal model, in addition to identifying the shape of the cell soma of the OPC like in (**f**), OPCs and immature OLs can be visualized with the help of live ex vivo antibody labeling against the surface marker O4 (pictured) or NG2, prior to patching (**k**). Representative cells in **a** and **b** have been imaged with a 40× objective. Slice images in **c**, **k** (Live O4 panel) with a 5× objective and lesion slice image in **h** with a 10× objective

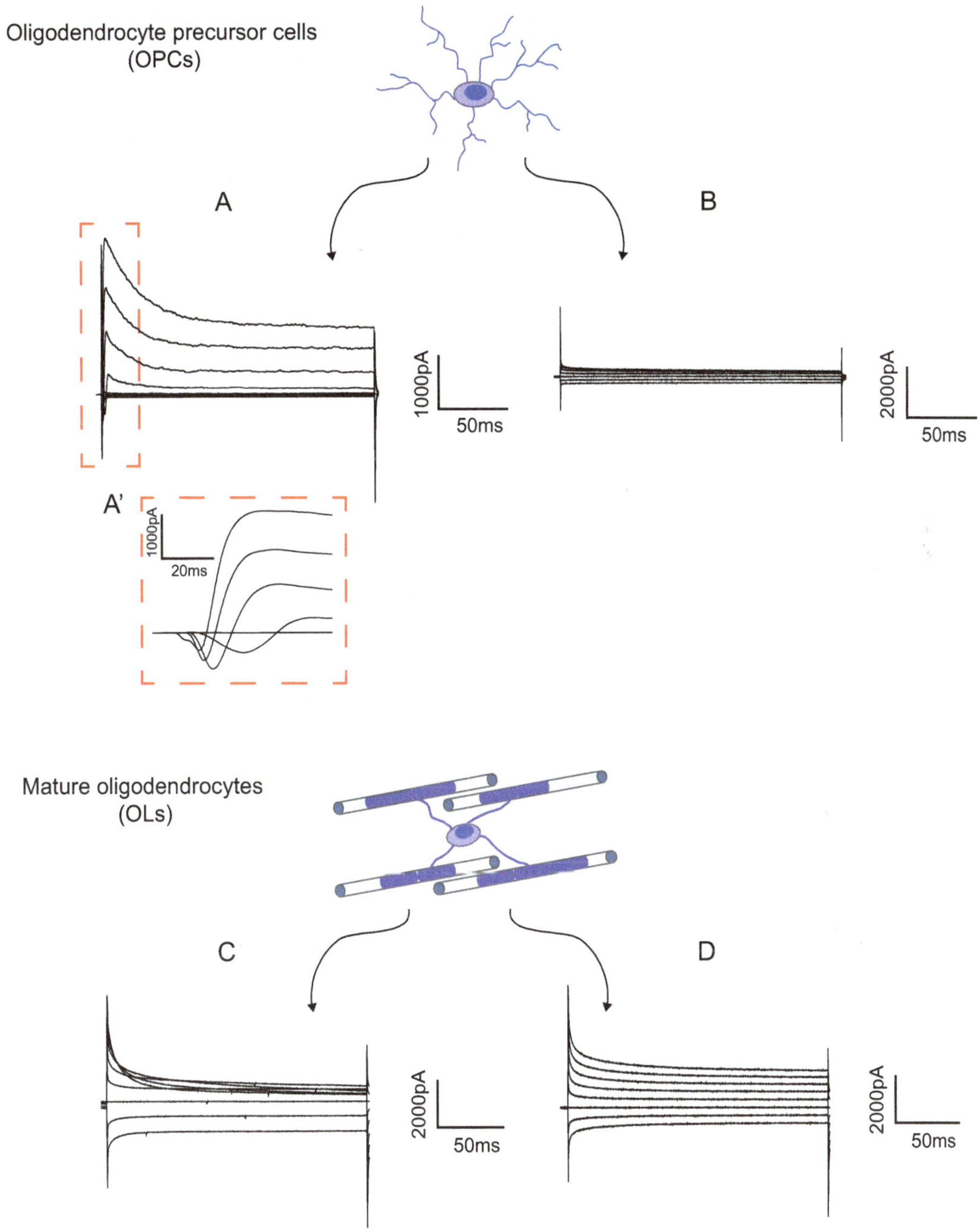

Fig. 7 Electrophysiological properties of oligodendrocyte-lineage cells. Voltage clamp, whole-cell recordings of oligodendroglia reveal changes in ion channel expression and membrane properties, as oligodendrocyte precursor cells (OPCs) mature into myelinating oligodendrocytes (OLs). The current response (including the capacitive transient) of the cells to voltage steps from holding potential (−70 mV) ranging from −120 to +40 mV in 20 mV increments, depict stereotypical responses of each "maturation" stage of oligodendrocyte-lineage cells. These stereotypic current–voltage (*I–V*) relationships can be used as a co-identifier with immunohistochemistry. Depending on the age and brain location of the cell, OPCs can express both

and the tip of the glass micro-electrode pipette and then to rupture this membrane patch in order to establish a low-resistance connection between the intracellular space of the cell and the internal solution of the pipette (Figs. 1 and 2). To obtain the giga-seal, place the tip of the patching pipette onto the cell membrane and switch from positive to negative pressure (suction) at the same time as changing the holding voltage on the amplifier to −70 mV (or − 60 mV, depending on junction potential) until the giga-seal is formed. The formation of the seal is marked by a striking reduction in current response to the applied continuous voltage pulse (*see* **step 11**). In some instances, the seal occurs rapidly; in other cases, it is essential to keep negative pressure (or sometimes to increase the negative pressure) until negligible current flows, or the resistance increases. This stage is crucial for a good recording. For OPCs, it is often sufficient to get "a good seal" by applying light suction upon touching the cell membrane. Some cells, however, need to have suction applied gently and steadily for some time. The level of suction needed is best judged by the change in current (change in resistance) evoked by the repeated voltage-pulse delivered (*see* **step 11**).

15. Once the giga-seal is formed and stabilized (could take 30–60 s), the pipette capacitance currents should be compensated (C_p fast and C_p slow compensation; Fig. 2b). For further information on all possible compensations and how to perform them, see the equivalent amplifier user manual (i.e., EPC 10, CED, or the Axon Instruments user manual).
16. Once the pipette capacitance is compensated and the recorded trace looks like Fig. 2b (current panel, after C_p compensation), apply stronger negative pressure in the form of sudden (not constant) suctioning bursts, to rupture the cell membrane and achieve a whole-cell connection with the pipette (Fig. 2c).

Fig. 7 (continued) voltage-gated sodium (Na_v) and potassium (K_v) channels, which generate a transient inward and a sustained outward current respectively, as demonstrated from the current response to the voltage steps applied of the patched cell (**a**). (**a′**) Magnification of the red dash-lined insert from **a**, for the clear depiction of Na_v and K_v expressed in the patched OPC. (**b**) Some OPCs however, have time-independent responses to voltage, indicating that these OPCs do not have Na_v or K_v channels, but have high input resistance. Such *I–V* relationships seem "ohmic" as the current changes almost linearly with each voltage step. All mature OLs lack Na_v and K_v currents, and have much lower input resistance and larger capacitance (evident by the decay of the capacitive current) compared to OPCs, as depicted in the I-V relationships of two representative OLs in (**c**) and (**d**). (**c**) Some mature OLs show a sag of outward current at positive potential whereas others are ohmic at every voltage step (**d**). Voltage protocol for all *I–Vs* consisted of eight 20 mV steps from −40 mV to +100 mV

17. Whole-cell connection is evident when a clear membrane capacitance current is visible in response to the continuously applied voltage pulse (*see* **step 11** and Fig. 2c current panel). If the membrane rupture is successful, the internal dye (LY) will fill the cell. This can be visualized with the appropriate fluorescent filters for LY.

    At this point, some extra negative pressure can be carefully applied (in short bursts) to decrease the series resistance of the recording (e.g., increase the peak current detected at the start of the voltage pulse), and thus improve quality of the recording. ***Caution***: If too much negative pressure is applied at this stage the seal can break and the recording becomes useless.
18. Once in whole-cell mode, series resistance (R_s) should be compensated (consult your amplifier manual). However, when recording from OLs, due to their high capacitance, it is nearly impossible to accurately compensate for R_s.
19. At this stage, it is good to apply varying voltage step protocols (stepping the membrane potential from −120 mV to +40 mV in 5, 10, or 20 mV steps) to obtain a "finger-print" of the cell recorded. OPCs, immature OLs and mature myelinating OLs have stereotypic voltage-current dependent properties, which can be used as an extra aid in cell identification (*see* Fig. 5).
20. Apply the desired stimulation and/or voltage protocols and/or agonists/antagonist applications according to your experimental needs.
21. Once all the desired data have been collected, capture both a fluorescent (LY) image of the cell patched and a bright-field image at lower magnification for identifying the location of the patched cell with the help of a microscope-attached camera (Fig. 6a–c, f–h). This will facilitate cell identification if further immunohistochemistry is required (Fig. 6e, j). For more details on post-hoc OPC immunohistochemistry *see* [29] (Fig. 6).
22. For post-hoc immunohistochemistry, mark the location of the cell in the brain slice at the end of the experiment (Fig. 6d, I). When working with a symmetric brain slice (i.e., coronal cortical slice) mark the orientation of the slice by cutting a corner of the slice with a scalpel (Fig. 6d), whereas when working with an asymmetrical brain slice, simply note the orientation of the slice (Fig. 6i). This will facilitate the identification of the slice orientation and the cell positioning when preparing the section for imaging [31].
23. Fix brain slices in 4% PFA overnight at 4 °C or for 1 h at RT on a shaking platform [31].

3.3 Identifying Oligodendrocyte Lineage Cells for Whole-Cell Recording

A crucial step for a successful whole-cell recording is the identification and selection of a healthy cell to patch. Selecting a good cell for patch clamping requires experience and differs depending on the brain region and cell type under examination. Generally, with IR-DIC optics, healthy cells are considered the ones with a smooth and clear cellular outline (Figs. 5a, b and 6f). Cells that look "shiny" with a perfect "hard" outline are very difficult to seal onto and are not good cells for patching. Dead or dying cells appear swollen and very round, and their nuclei can be clearly visualized.

Contrary to most neuronal types, OPCs and OLs are not as easy to visualize in a tissue slice and a combination of experience and expert tips are needed in order to identify them ex vivo. Without doubt, the use of transgenic models provides the easiest and quickest method for OPC lineage identification prior to patch-clamp recordings. We have routinely detected OPCs with the NG2-EYPF knock-in mouse model [27] (Figs. 5a and 6a–e), OPCs, pre-OLs/OLs with the Sox10-DsRed rat model, or Sox10-GFP mouse model [28] (Fig. 6l) and newly-made OLs with the PDGFRa/Tau/mGFP mouse model [17]. Nevertheless, if the experimental set-up does not allow the use of transgenic animals and/or fluorescently labeled proteins, we provide here a methodology on how to successfully identify oligodendrocyte-lineage cells without the use of transgenic models (Fig. 6f–k).

Morphological and electrophysiological OPC properties change depending on the developmental stage of the cell. The OPC cell somata appear small, diamond (or hexagon)-shaped (although in the corpus callosum OPC cell bodies can be narrow and ellipse-shaped), and without many processes, when visualized with IR-DIC optics (Fig. 5a and 6f). On the contrary, as they mature to OLs, they tend to have larger and rounder somata (or ellipse-shaped but fuller than OPCs) and can be easily identified in white matter tracts (Fig. 5b). OL cell somata often align a few in a row, a particularly evident sight in the optic nerve and corpus callosum. In addition to learning to identify cell soma shapes on the IR-DIC image, it is possible to use lineage specific antibodies that bind to membrane-bound proteins. We have successfully used antibodies against both NG2 [9] and O4 in slices to identify OPCs and immature OLs (Fig. 6k).

Mature and pre-myelinating OLs do not express voltage-gated sodium or potassium channels, and as they reach their final maturity stage, their electrophysiological properties resemble those of ohmic cells (Fig. 7c, d). On the contrary, OPC precursors have a fairly time-dependent response to voltage-steps and might express voltage-gated sodium and potassium channels (Fig. 7a, b). It is worth noting that microglial cells might be mistaken for OPCs when patching due their similar cell size. However, their electrophysiological properties differ, and if a microglial cell has been accidentally patched, it can be identified from the lack of any voltage-gated channels, extremely high membrane resistance, ohmic cell profile, and the appearance of an outward current after 200 ms of keeping the membrane voltage at positive membrane potential, above +40 mV.

3.4 Post-Patching Analysis: How to Measure Membrane Capacitance and Resistance (C_m, R_m), Series Resistance (R_s), Pipette Capacitance and Resistance (C_p, R_p)

Measuring R_m and C_m gives a good indication of the membrane's non-voltage gated channel profile, which determines the leak conductance and gives an estimation of the cell surface area and passive membrane properties.

3.4.1 Capacitive Transient Analysis

By analyzing the current response to a voltage step, it is possible to measure the pipette series resistance, cell membrane resistance, and cell capacitance. Series resistance (R_s) is the sum of all resistances between the input from the patch clamp amplifier and the cellular membrane. Its main components are pipette resistance (R_p) and access resistance (Raccess) localized between the pipette tip and the interior of the cell (Fig. 2b). In a simple RC-circuit, R_s is connected is series with R_m and C_m (Figs. 2c and 8b). Both R_s and R_m resistances are assumed to be constant during voltage pulses and can thus be described by Ohm's law ($V - IR$). Thus, it is important to monitor R_s during the recording, and if it changes more than 10–20%, the recordings should be discarded (*see* **Note 12**).

At the onset of the voltage step ($t = 0$), the voltage across the series resistor equals the applied voltage step (V_s), because the membrane capacitor has not yet reached its full charge (Q) yet. There is a current $I(t = 0)$ flowing through the resistors, but not through the capacitor (Figs. 2c and 8).

Consequently, we can calculate the R_s following Ohm's law:

$$R_s = \frac{V_s}{I(t = 0)} \tag{1}$$

After the application of the voltage step, the capacitor begins to charge until it reaches a steady state ($t = \infty$) in which no current flows through it. The R_m can then be calculated as the difference between the total resistance, obtained from the change in the steady state current $I(t = \infty)$ in response to the voltage step and the series resistance, R_s:

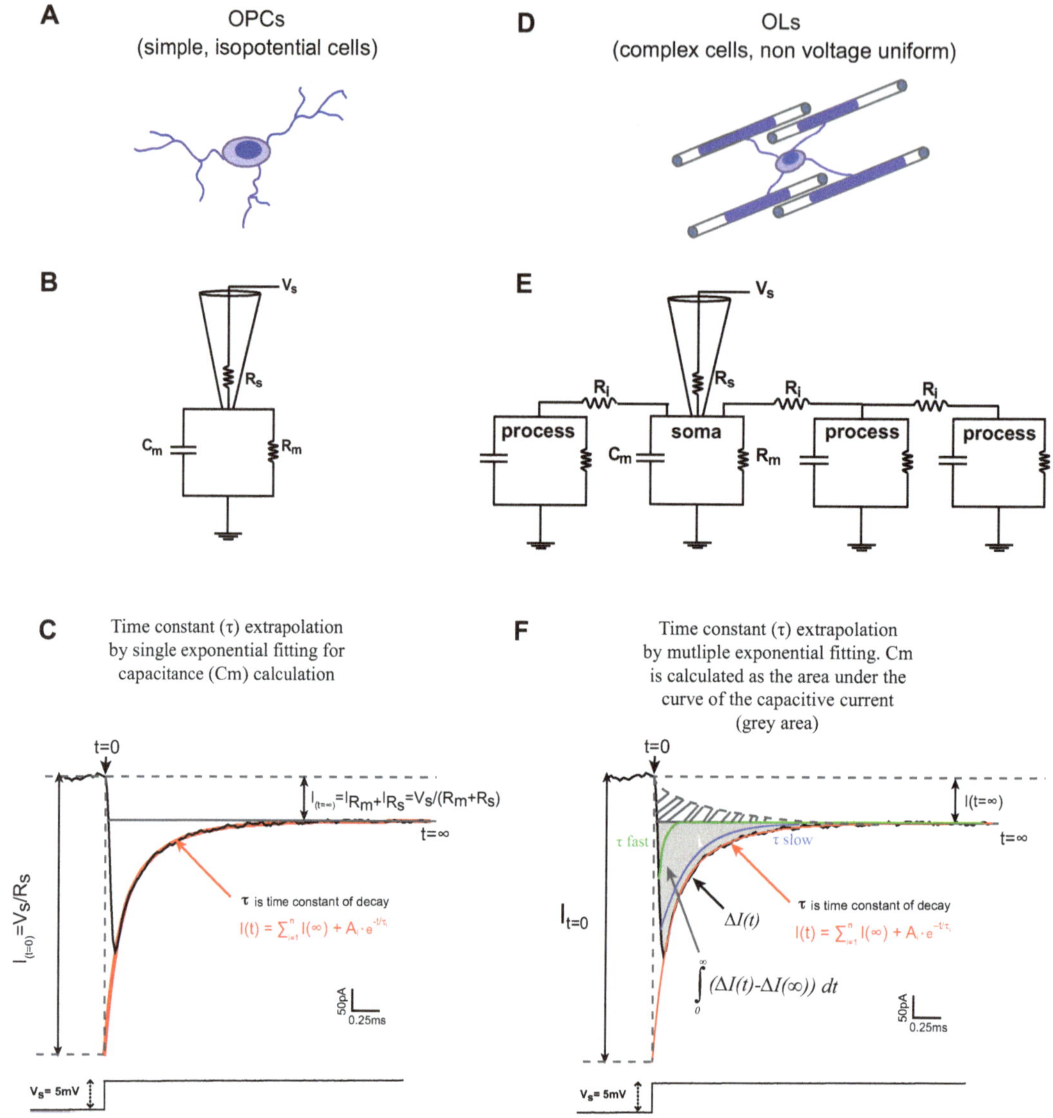

Fig. 8 Measurements of cell capacitance of whole-cell patched OPCs and OLs. OPCs are spatially compact cells with few processes (**a**) and when whole-cell patch-clamped they can be well represented by the simple RC-circuit depicted in (**b**), where the series resistor (R_s) is connected in series with the membrane resistor (R_m) and the membrane capacitor (C_m). (**c**) Current response of a patched OPC evoked by a 5 mV voltage pulse. Since OPCs are spatially compact, their capacitive transient can be fitted with a single exponential for the extrapolation of the decay time constant (τ; red line) and the eventual calculation of the cell capacitance (C_m; *see* Subheading 3.4.1). C_m calculations for more morphologically complex cells, like OLs (**d**) is difficult as they are not voltage uniform. OLs have long processes and their theoretical RC-circuit representation is depicted in (**e**). R_i is the intracellular resistance along the cell processes. This circuit would generate a multiexponential capacitive transient. (**f**) Representative OL current response to a 5 mV pulse. The capacitive transient response is multiexponential, and here it has been fitted with the sum of two exponentials here (a fast component: τ fast, green line and a slow component: τ slow, blue line). To measure the R_s, the current at the time of the pulse here, is extrapolated after fitting τ with the sum of two exponentials. Once R_s and τ is calculated, an estimation of C_m for OLs can be acquired by calculating the capacitive transient (gray area together with gray-dashed region; *see* Subheading 3.4.2)

$$R_{\mathrm{m}} = \frac{V_{\mathrm{s}}}{I(t=\infty)} - R_{\mathrm{s}} \tag{2}$$

Substituting R_s from eq. (2) gives:

$$R_{\mathrm{m}} = V_{\mathrm{s}} \frac{I(t=0) - I(t=\infty)}{I(t=0) \cdot I(t=\infty)} \tag{3}$$

According to the above equations, calculating Rs and Rm correctly requires the value of the current at $t = 0$ ($I(t = 0)$). Although the value of the current can be read off the recording, the voltage step applied does not instantly reach its full amplitude in practice, and therefore the peak of the current is not yet reached. The smoothing of the voltage step occurs due to low-pass filtering in the circuit, e.g., from the pipette. The glass pipette functions as a thin insulator separating two conductors, the bath solution and the solution in the pipette, and therefore behaves as a capacitor. The capacitor of the pipette needs to be charged before current flows through it, hence the delay in the actual voltage step.

Since the precise value of the current at $t = 0$ cannot be read directly from the current trace, the behavior of a circuit without filtering needs to be modeled using information about the current at later time points ($t = \infty$), when the filtering effect is insignificant.

However, the current recording can be fitted to this exponential equation to find the $I(t = 0)$, and hence R_s and R_m [32].

$$I(t) = \left(\frac{V_{\mathrm{s}}}{R_{\mathrm{m}} + R_{\mathrm{s}}}\right) \cdot \left(1 + \frac{R_{\mathrm{m}} \cdot \mathrm{e}^{t/\tau}}{R_{\mathrm{s}}}\right) \tag{4}$$

where t is the time after onset of the voltage step and τ the decay constant of the current.

The decay constant of the current (τ) is described by the following equation:

$$\tau = C_{\mathrm{m}} \times \frac{R_{\mathrm{s}} \times R_{\mathrm{m}}}{R_{\mathrm{s}} + R_{\mathrm{m}}} \tag{5}$$

Rearranging eq. (5) to gives

$$C_{\mathrm{m}} = \tau \times \frac{R_{\mathrm{s}} + R_{\mathrm{m}}}{R_{\mathrm{s}} \times R_{\mathrm{m}}} \tag{6}$$

Substituting R_s from eq. (1) and R_m from eq. (2), we can calculate the membrane capacitance (C_m) from the following equation:

$$C_{\mathrm{m}} = \tau \times \frac{I(t=0)^2}{V_{\mathrm{s}} \times (I(t=0) - I(t=\infty))} \tag{7}$$

The time constant τ, can be obtained by fitting the current transient with a single exponential:

$$I(t) = \sum_{i=1}^{n} I(\infty) + A_i \times e^{-t/\tau_i} \quad (8)$$

We use either the ClampFit software developed by Molecular Devices or MatLab for this analysis. An example of a fitted capacitance transient is shown in Fig. 8c. R_s, R_m, and C_m can be calculated in MS-Excel or MatLab using lab written scripts based on the above-described equations. The R_m reflects the ion channel composition of the cellular membrane whereas C_m reflects the surface area as well as the thickness of the membrane. Thus, measuring the R_m and C_m provides us with important information about the ion channel's composition and cell size, respectively.

This analysis assumes that the cell is isopotential, and predicts a single exponential decay of the capacity current (Fig. 8c). This is the case for electrically compact cells, so it is a reasonable assumption for OPCs, with their small somata and relatively small and few processes (Fig. 8a, b). For cells with more complex morphology, however, the resistance of the cell processes, or axon and dendrites in the case of neurons, makes the cell non-uniform in voltage, and the capacity current decays as the sum of two or more exponentials (the sum of multiple exponentials; Fig. 8d–f). However, the moment the voltage step is applied, the cell capacitance is uncharged, so all the applied voltage step appears across the pipette series resistance, and R_s can be calculated from eq. (1). Even when fitted with multiple exponentials, R_m can similarly be estimated. The non-spherical shape of OLs, with thin and long processes that wrap a number of times around an axon (or neurons, with somata and large complex branching dendritic trees), may cause a lack of uniformity of the voltage-clamp. A degree of non-uniformity in neurons has been reported, since the voltage control at most distal dendritic sites in CA1 pyramidal neurons is extremely poor [33]. An even greater problem might be expected to occur in the long processes and the number of wraps of the cell membrane of mature myelinating OLs, than is predicted for morphologically complex neurons. The voltage non-uniformity in morphologically complex cells can also introduce a problem when measuring currents where distal currents are filtered, smoothed, and reduced in size before they reach the recording site in the cell soma. Thus, to calculate the membrane capacitance of complex cells such as myelinating OLs, there are two options: either (a) only fit the 10–30% or 50% of the decaying peak of the capacitive current transient with a single exponential or (b) calculate the area of the current transient peak of the capacitive transient (Fig. 8).

3.4.2 Measuring Membrane Capacitance of Complex Cells

For complex cells (Fig. 8d–f) which have a multicomponent decay, and thus a sum of exponentials with different time constants and a number of membrane capacitors in series, it is less straight forward to estimate the membrane capacitance since the exact resistance in series with each capacitor is not known. In this case, C_m can be perhaps better estimated from the current trace in response to a voltage step as follows.

The membrane capacitance is given by the following equation:

$$C_{\mathrm{m}} = \frac{\Delta Q}{\Delta V_{\mathrm{m}}(\infty)} \tag{9}$$

where ΔQ is the gain in charge, accumulated on the membrane capacitor during the initial (capacitive) current transient, and $\Delta V_{\mathrm{m}}(\infty)$ is the voltage difference across the capacitor at the new steady state.

ΔQ can be calculated by integrating the capacitive current over time:

$$\Delta Q = \int_0^{\infty} I_{\mathrm{C}}(t)\mathrm{d}t \tag{10}$$

Since we only measure the total current $\Delta I(t)$, we now express $I_{\mathrm{C}}(t)$ in terms of $\Delta I(t)$. The capacitive current $I_{\mathrm{C}}(t)$ is the difference between the total $\Delta I(t)$ and the resistive $\Delta I_{R\mathrm{m}}(t)$ current:

$$\begin{aligned} I_{\mathrm{C}}(t) &= \Delta I(t) - \Delta I_{R_{\mathrm{m}}}(t) = \Delta I(t) - \frac{\Delta V - \Delta V_{R_{\mathrm{s}}}(t)}{R_{\mathrm{m}}} \\ &= \Delta I(t) - \frac{\Delta V - \Delta I(t)R_{\mathrm{s}}}{R_{\mathrm{m}}} \end{aligned} \tag{11}$$

This can be rearranged the following way:

$$I_{\mathrm{C}}(t) = \left(\frac{R_{\mathrm{m}} + R_{\mathrm{s}}}{R_{\mathrm{m}}}\right)(\Delta I(t) - \Delta I(\infty)) \tag{12}$$

Combining eqs. (10) and (12) gives

$$\Delta Q = \left(\frac{R_{\mathrm{m}} + R_{\mathrm{s}}}{R_{\mathrm{m}}}\right)\int_0^{\infty}(\Delta I(t) - \Delta I(\infty))\mathrm{d}t \tag{13}$$

The integral part of eq. (13) can be calculated using either Clampex software (Molecular Devices) or MatLab, by determining the area under the curve of the capacitive current graph with the baseline set to be the current at the new steady state $\Delta I(\infty)$, as illustrated in Fig. 8f.

To calculate C_m, according to eq. (9), $V_m(\infty)$ is needed. As at $t = \infty$, the capacitive current is equal to zero. Therefore, the current through the membrane resistor is the same as the current through the series resistor:

$$\Delta V_{\mathrm{m}}(\infty) = \frac{R_{\mathrm{m}}}{R_{\mathrm{m}} + R_{\mathrm{s}}}\Delta V \tag{14}$$

Combining eqs. (9), (13), and (14) gives a way of measuring the membrane capacitance:

$$C_m = \left(\frac{R_m + R_s}{R_m}\right)^2 \int_0^\infty (\Delta I(t) - \Delta I(\infty))dt \tag{15}$$

Equation (15) provides an estimation of the membrane capacitance of complex cells when the current is described as a sum of exponentials. The presence of extra resistors in series with additional capacitors makes their charging slower and does not account for the calculated C_m, making it slightly underestimated.

4 Notes

1. For the activation of NMDA-Rs in OPCs, the external aCSF solution should be supplemented with 0.1 mM glycine and 0.005 mM strychnine to block glycine receptors. Mg^{2+} should be omitted in your solution as it blocks NMDA-Rs.
2. To detect spontaneous synaptic inputs, lower Mg^{2+} in your aCSF solution to 1mM for increased neuronal firing (*high magnesium is good for long stable recordings*). Synaptic input can also be induced "chemically" by applying hypertonic solutions or ions/drugs that increase vesicular release.
3. Kynurenic acid is added to the slicing buffer to block glutamate receptor-mediated cell damage during the slicing process.
4. For internal solutions, methanesulfonate is sometimes used instead of gluconate and is completely interchangeable with gluconate in our solution recipes above. There are pros and cons for both ions. We opt for gluconate as most "side-effects" of gluconate are mitigated by adding BAPTA into the internal solution.
5. Various internal solutions can be used depending on the experimental question. For example, if blocking of voltage-gated K^+ channels is needed, that can be achieved with a Cs^+-gluconate internal (*see* Subheading 2.2.2). Cs^+ also reduces general noise levels in whole-cell recordings.
6. Specific NMDA (60 μM, Tocris) and Kainate (30 μM, Tocris) receptor agonists can be applied to the extracellular solution for the activation of the respective ionotropic glutamatergic receptors. NBQX (25 μM, Tocris) can be used to block AMPA-Rs and Kainate-Rs. Voltage-gated sodium channels (Na_v) can be blocked by the pore blocker tetrodotoxin (1 μM, TTX, Tocris). Voltage-gated potassium channels (K_v) can be blocked with the nonspecific pore blocker Barium ($BaCl_2$, 200 μM). All agonists and antagonists should be dissolved in the same aCSF external solution as is used for the whole-cell patch clamp recordings.
7. Slices can be used up to 9–10 h after slicing, provided they are kept in oxygenated slicing buffer at RT.

8. Keep the internal solution on ice and protected from the light, while patching to ensure the stability of temperature sensitive components like ATP and GTP.
9. For larger cell somata (i.e., pyramidal neurons), patching pipettes of 2.5–5 MΩ provide a good balance between access resistance and cell integrity during whole-cell patch clamping.
10. Make sure that you thoroughly remove all bubbles inside the resting pre-chamber before adding the brain slices, as bubbles will cause the slices to float and fold increasing tissue damage.
11. While approaching the desired cell with the glass electrode, the blowing of the internal solution has to be done with a steady flow. Fluctuations in this flow could result in the appearance of debris inside the patching electrode and render it unusable. In that case, discard the debris-filled electrode and refill a new one with internal solution.
    Suction techniques differ for each cell type. For OPCs it is often sufficient to get "a good seal" by applying light suction upon touching the cell membrane. However, other cells need to have suction applied gently and steadily upon approaching their membrane. The level of suction needed is best judged by the change in current (change in resistance) evoked by a repeated 5 mV voltage-pulse (50 ms duration) and by trial and error.
12. The membrane capacitance and resistance set both the temporal resolution of the whole-cell recording and the accuracy of the voltage-clamp. As membrane capacitance cannot be altered but R_s can, it becomes essential to have R_s as low as possible to allow for good temporal resolution and strong control of the voltage-clamp. Particularly when large currents are involved, there is a greater risk of losing the control of the clamp (as the voltage error generated by high currents and resistance is according to Ohms law: $V_{error} = IR_s$).

References

1. Neher E, Sakmann B (1976) Single-channel currents recorded from membrane of denervated frog muscle fibres. Nature 260 (5554):799–802
2. Hamill OP, Marty A, Neher E, Sakmann B, Sigworth FJ (1981) Improved patch-clamp techniques for high-resolution current recording from cells and cell-free membrane patches. Pflugers Arch 391(2):85–100
3. Sakmann B, Neher E (1984) Patch clamp techniques for studying ionic channels in excitable membranes. Annu Rev Physiol 46(1):455–472
4. Hodgkin AL, Huxley AF (1952) Currents carried by sodium and potassium ions through the membrane of the giant axon of *Loligo*. J Physiol 116(4):449–472
5. Bergles DE, Roberts JD, Somogyi P, Jahr CE (2000) Glutamatergic synapses on oligodendrocyte precursor cells in the hippocampus. Nature 405(6783):187–191
6. Chittajallu R, Aguirre A, Gallo V (2004) NG2-positive cells in the mouse white and grey matter display distinct physiological properties. J Physiol 561(Pt 1):109–122
7. Káradóttir R, Cavelier P, Bergersen LH, Attwell D (2005) NMDA receptors are expressed in oligodendrocytes and activated in ischaemia. Nature 438(7071):1162–1166

8. Bakiri Y, Burzomato V, Frugier G, Hamilton NB, Káradóttir R, Attwell D (2009) Glutamatergic signaling in the brain's white matter. Neuroscience 158(1):266–274
9. Káradóttir R, Hamilton NB, Bakiri Y, Attwell D (2008) Spiking and nonspiking classes of oligodendrocyte precursor glia in CNS white matter. Nat Neurosci 11(4):450–456
10. Spitzer S, Volbracht K, Lundgaard I, Káradóttir RT (2016) Glutamate signalling: a multifaceted modulator of oligodendrocyte lineage cells in health and disease. Neuropharmacology 110:574–585
11. Káradóttir R, Attwell D (2007) Neurotransmitter receptors in the life and death of oligodendrocytes. Neuroscience 145 (4):1426–1438
12. Marinelli C, Bertalot T, Zusso M, Skaper SD, Giusti P (2016) Systematic review of pharmacological properties of the oligodendrocyte lineage. Front Cell Neurosci 10:27
13. Kukley M, Capetillo-Zarate E, Dietrich D (2007) Vesicular glutamate release from axons in white matter. Nat Neurosci 10(3):311–320
14. Ziskin JL, Nishiyama A, Rubio M, Fukaya M, Bergles DE (2007) Vesicular release of glutamate from unmyelinated axons in white matter. Nat Neurosci 10(3):321–330
15. Gautier HOB, Evans KA, Volbracht K, James R, Sitnikov S, Lundgaard I et al (2015) Neuronal activity regulates remyelination via glutamate signalling to oligodendrocyte progenitors. Nat Commun 6:8518
16. Young KM, Psachoulia K, Tripathi RB, Dunn S-J, Cossell L, Attwell D et al (2013) Oligodendrocyte dynamics in the healthy adult CNS: evidence for myelin remodeling. Neuron 77 (5):873–885
17. McKenzie IA, Ohayon D, Li H, Paes de Faria J, Emery B, Tohyama K et al (2014) Motor skill learning requires active central myelination. Science 346(6207):318–322
18. Lundgaard I, Luzhynskaya A, Stockley JH, Wang Z, Evans KA, Swire M et al (2013) Neuregulin and BDNF induce a switch to NMDA receptor-dependent myelination by oligodendrocytes. PLoS Biol 11(12):e1001743
19. Gibson EM, Purger D, Mount CW, Goldstein AK, Lin GL, Wood LS et al (2014) Neuronal activity promotes oligodendrogenesis and adaptive myelination in the mammalian brain. Science 344(6183):1252304
20. Verkhratsky A, Parpura V et al (2014) Methods Mol Biol 1183:1–19
21. Sontheimer H (1995) Whole-cell patch-clamp recordings, in: Patch-clamp applications and protocols. Humana Press, New Jersey, pp. 37–74. https://doi.org/10.1385/0-89603-311-2:37
22. Gibb JA, Edwards FA (1994). Patch clamp recording from cells in sliced tissues. In: Ogden D (Second Eds.) Plymouth workshop. Microelectrode techniques: the plymouth workshop handbook. Company of Biologists, pp. 255–274
23. Moyer JR, Brown TH (1966) Patch clamping chapter 5. Dent Tech 35:135–193
24. Molleman A (2003) Patch clamping: an introductory guide to patch clamp electrophysiology, vol 186. Wiley, Chichester
25. Kornreich BG (2007) The patch clamp technique: principles and technical considerations. J Vet Cardiol 9(1):25–37
26. Sakmann B, Neher E (1995) Single-channel recording. Springer, New York, p 700
27. Karram K, Goebbels S, Schwab M, Jennisen K, Seifert G, Steinhäuser C et al (2008) NG2-expressing cells in the nervous system revealed by the NG2-EYFP-knock in mouse. Genesis 46(12):743–757
28. Chen Y, Spitzer S, Agathou S, Karadottir RT, Smith A (2017) Gene editing in rat embryonic stem cells to produce in vitro models and in vivo reporters. Stem Cell Reports 9:1262–1274
29. Blanco S, Dietmann S, Flores JV, Hussain S, Kutter C, Humphreys P et al (2014) Aberrant methylation of tRNAs links cellular stress to neuro-developmental disorders. EMBO J 33 (18):1–20
30. Spitzer S, Sitnikov S, Kamen Y, Evans KA, Kronenberg-Versteeg D, Dietmann S, de Faria O, Agathou S, Káradóttir R (2018) Oligodendrocyte progenitor cells become regionally diverse and heterogeneous with age, Neuron. (in press)
31. Káradóttir R, Attwell D (2006) Combining patch-clamping of cells in brain slices with immunocytochemical labeling to define cell type and developmental stage. Nat Protoc 1 (4):1977–1986
32. Tessier-Lavigne M, Attwell D, Mobbs P, Wilson M (1988) Membrane currents in retinal bipolar cells of the axolotl. J Gen Physiol 91 (1):49–72
33. Major G (1993) Solutions for transients in arbitraruly branching cables: III voltage clamp problems. Biophys J 65(1):469–491

Chapter 10

Ex Vivo Slice Cultures to Study Myelination, Demyelination, and Remyelination in Mouse Brain and Spinal Cord

Sowmya Sekizar and Anna Williams

Abstract

In vitro culture systems have been invaluable in understanding the cell biology of oligodendrocytes; the monoculture of primary oligodendroglia has helped characterize different stages of oligodendrocyte maturation in the absence of neurons. However, oligodendrocyte monocultures do not model the interaction of oligodendrocytes with neurons where they form myelin wraps. To circumvent this problem, coculture systems were developed; oligodendrocytes and neurons are cultured together, facilitating the study of myelin wraps and the interaction between the two cell types. However, this coculture system also has limitations, as other cells are not present and it does not represent the three-dimensional multicellular structure seen in vivo. Some of these limitations are resolved by using ex vivo slice cultures to serve as a three-dimensional culture system that is more similar to in vivo and can be used to study myelination, demyelination, and remyelination, over extended periods of time. Slice cultures are economical compared to in vivo studies and live imaging using them is less challenging. The focus of this chapter is to describe how to culture brain and spinal cord slices of mice and use them to study myelination, demyelination, and remyelination.

Key words Ex vivo, Slice culture, Spinal cord, Brain, Myelination, Oligodendrocytes

1 Introduction

To design effective therapies for central nervous system myelin repair, we need to understand how an oligodendrocyte produces myelin during development and during remyelination after damage to myelin. In vitro monoculture of oligodendroglia has been very successful in understanding the biology of oligodendrocytes and allowed characterization of the different stages of oligodendrocyte precursor cell (OPC) differentiation [1]. The expression of antigen markers during the stages of differentiation in vitro matches that found in oligodendrocytes during development in vivo. Primary oligodendrocytes cultured in two dimensions, in the absence of axons, produce myelin sheets that spread out around the cell, and the presence and complexity of these sheets can be used to classify

David A. Lyons and Linde Kegel (eds.), *Oligodendrocytes: Methods and Protocols*, Methods in Molecular Biology, vol. 1936, https://doi.org/10.1007/978-1-4939-9072-6_10,

the extent of maturation of oligodendrocyte in the absence of true myelin wraps [2]. Cell cultures can be manipulated by transduction of cells with viruses or adding shRNA, and changes in morphology can be analyzed to study the effect of the gene/pathway manipulated. These cultures are useful to study the intrinsic properties of oligodendrocytes without the influence of neurons and other glia. Even though myelin sheets are formed in two-dimensional monocultures, they do not truly represent myelination due to the absence of neurons. Coculture systems such as dissociated central nervous system cultures, oligodendrocyte-dorsal root ganglia cultures [3–6], and aggregate cultures have been able to overcome the limitation of oligodendrocyte monocultures [7–9]. These cultures are suitable for studying myelin wrap formation and interaction of oligodendrocytes with neurons.

While the abovementioned culture systems permit the study of intrinsic properties of oligodendrocytes and cell–cell interactions, they may not represent the three-dimensional structure and architecture of cells present in vivo [10, 11]. To circumvent this, without using in vivo models, slice cultures have been developed as a suitable alternative. Slice cultures consist of 300–400 μm tissue sections, cultured on coverslips or membrane inserts at a liquid–air interface. Slice cultures retain the three-dimensional architecture and cell–cell interactions that are seen in vivo, and all the cell types except blood cells can be studied. This system was initially developed for electrophysiology studies, and initial attempts to culture mammalian central nervous system and observe myelin was made in 1956 [12], using refractive patterns on light microscopy. However, these cultures could not be maintained for long periods, so to overcome this problem, in 1958, cerebellum from kitten and rat were cultured on collagen drops, and the presence of myelin was examined with luxol fast blue and sudan black histological stains [13]. In the early 1960s, initial wrapping of axons with myelin and compaction of myelin sheaths was observed in slices of rat cerebellum by electron microscopy [14–16]. Spinal cord cultures survived longer when longitudinal slices were made and grown on cellulose-coated fluoroplastic dishes [17]. Longer-term survival of slices with better health improved with the ability to make 200–500 μm slices using a tissue chopper and growing them on membrane inserts at an air–liquid interface. Currently slices are cultured on 0.4 μm porous membrane made of polytetrafluoroethylene (PTFE), where spinal cord and cerebellum slices from mice taken at P0–P2 can be robustly maintained for 4 weeks [18, 19]. At present, in our laboratory, we are able to maintain spinal cord cultures for 2–-3 months. PTFE membranes are permeable and inert, allowing molecules or therapeutics added to the medium to permeate the slice. Lysophosphatidyl choline (LPC), a chemical detergent, added to the medium, globally demyelinates slices, which then remyelinate in 14 days [18, 20]. Drugs or molecules can be added to the medium to change the rate of myelination or remyelination, allowing screening of potential pro-myelinating or pro-remyelinating compounds [18, 21–23].

Slice cultures can also be genetically manipulated by adding lentiviruses expressing reporters under the control of ubiquitous or cell specific promoters, and/or RNAi to knock down genes. These can either be added directly to slices or added to cells e.g., oligodendrocyte precursor cells (OPCs), in monoculture and then the cells transplanted by pipetting onto the surface of the slices, where they enter the tissue and integrate. As these cultures are long, lentiviral transduction is necessary to allow manipulation of expression over adequate time. Transgenic mice can also be used as a source of tissue in slice cultures to investigate downstream biology. This is especially useful if the transgene causes early death of the animal through other effects (e.g., cardiovascular) as it allows investigation of the myelin phenotype.

Transgenic mice expressing reporters in the central nervous system can be live imaged in vivo, but this is challenging, expensive, and difficult in terms of imaging white matter, which is deep in the brain. However, live imaging of slices from such brains is more economical and relatively less difficult. Brain and spinal cord slices can be imaged using an inverted microscope visualized through the bottom of the plate and the membrane insert. An oil or water immersion objective can be used, and a water pump can be used to keep water continuously present on the objective. This is the most convenient method of imaging as the slice is still in contact with the medium and there is less evaporation as the 6-well plate with the slice can be closed and the empty wells can be filled with water to prevent it from drying out. Brain slices from NG2cre: EGFP mice, which express EGFP in NG2-positive cells, have been live imaged using an inverted microscope with a 10× objective, and this was sufficient to see cell migration and divisions [24]. To visualize myelin sheaths and processes a higher magnification (at least 25×) and large working distance objective is needed (which are usually very expensive and necessarily available in all laboratories). To overcome the working distance problem, the insert can be cut from its surrounding plastic ring and be placed on a cover slip with the slice surface touching the coverslip and imaged using an inverted microscope. This helps in gaining a couple of millimeters of working distance as the plastic bottom of the dish and height of the feet of the insert are not present. To ensure health of the slice, the system requires perfusion (with oxygenated artificial cerebrospinal fluid) and once imaged the cut membrane is placed on a new insert with the slice facing up [25]. An upright microscope can be used to overcome the working distance problem, using a dry objective to image at low magnification, but for higher magnification, either water or oil immersion objectives are needed. It is possible to add small volume of culture medium onto slices and image using a water immersion objective for a short duration, but if the slice has to be imaged repeatedly for long time intervals then adding medium on the slice can cause the slice to lift off the membrane.

Although slice cultures have many advantages, they also have some disadvantages, which includes variability among slices and within a slice. If cutting through cerebellum, then slices along its sagittal axis will have differing amounts of gray and white matter, due to the cerebellum not being uniform. The same applies to sagittal sections of the spinal cord, where some slices will have more white matter than others. Chopping brain or spinal cord slices damages the tissue at these surfaces, which can affect both axons and myelin, and this variability must be taken into account while studying the influence of therapeutics that induce or accelerate myelination and a sufficiently large sample size used. Furthermore, slice cultures are easier to maintain from young animals (P0–P7), and those from adult animals (more than P7) cannot currently be maintained for more than a week. We know that remyelination is more efficient in younger animals than older ones, and therefore this can make improvements in remyelination difficult to detect. The rate of remyelination must be used instead, detecting more efficient remyelination over time rather than measuring the extent of remyelination. This should be acknowledged when screening pro-remyelination therapies as these are being sought mostly for translation into therapies for adult human patients. However, there is evidence from several different molecular pathways that an alteration in myelination or remyelination found in slice cultures is recapitulated when tested in vivo in rodents [26, 27]. Therefore, this slice culture system serves as a good way of identifying pathways which are relevant to in vivo situations, but is more easily manipulable and visualized, is less expensive, and does not use live animals.

2 Materials

2.1 Preparation of Slice Medium

1. Minimum essential medium (MEM).
2. Earl's balanced salt solution (EBSS).
3. Heat inactivated Horse serum.
4. Penicillin/streptomycin.
5. Glucose.
6. GlutaMax 100×.
7. Amphotericin B (125 μg/mL).
8. 0.22 μm filtration unit
9. 6-well dishes

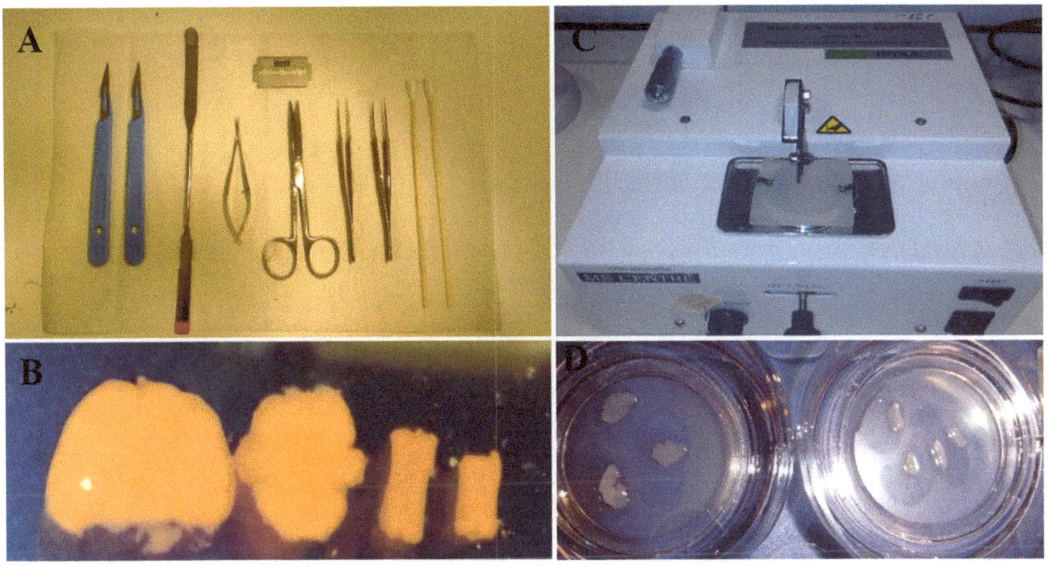

Fig. 1 Dissection instruments and tissue chopper. (**a**) An image showing the dissection and slice transfer instruments. The instruments should be sterilized by dipping in 70% ethanol for 15 min and then allowed to dry. (**b**) Cortical hemispheres, cerebellum (with a part of the midbrain and brainstem) and spinal cord dissected. Blood vessels and meninges are removed with fine forceps, but care should be taken to avoid damage to the tissue. (**c**) Tissue chopper with silicone pad. (**d**) Spinal cord and cerebellum slices (300 μm) placed on millicell® inserts soon after the slices are made

10. millicell® inserts.
11. Sterile forceps.
12. Tissue culture laminar flow hood.
13. 37 °C incubator.

2.2 Preparation of Slices (See Fig. 1)

1. P0–P2 mice (*see* **Note 1**).
2. Tissue chopper (McIlwain) (*see* **Note 2**).
3. Fine Forceps to remove blood vessels.
4. Small scissors.
5. Flat spatula.
6. Lebovitz-15 (L15) medium 30 mL (Life Technologies).
7. Pipette 200 to remove excess medium or moisten tissue.
8. 10 mL of slice medium in a petri dish
9. 120 mm petri dish
10. Scalpel blade with handle to separate slices, pipette tips can be used.
11. 6-well cell culture plate.
12. 0.4 μm millicell®-CM low height culture plate inserts (Millipore).

13. Razor blades.
14. Silicone pads (J-flex rubber products) (*see* **Note 3**).
15. 70% ethanol.
16. Cotton swabs.
17. Cell culture incubator 37 °C, 5% CO_2.

2.3 Adding Cells to Slices

1. Microglia or Oligodendrocyte progenitors from rat or mouse (*see* Chapters 2 and 5).
2. Tabletop centrifuge.

2.4 Demyelination

1. Lysophosphatidylcholine (LPC).
2. Phosphate-buffered saline (PBS).
3. Rat anti-Myelin Basic Protein (MBP; 1/300;AbD Serotec-MCA409S).
4. Chick anti-Neurofilament (NF;1/10000; Biolegend–822601).

2.5 Fixation and Preparation for Microscopy

1. Phosphate-buffered saline (PBS).
2. Fixative: 4% paraformaldehyde in phosphate-buffered saline (PFA).
3. Saturation solution: 0.5% TritonX-100 in PBS, 2% BSA, and 3% heat inactivated horse serum.
4. Primary antibody: rat anti-Myelin Basic Protein (MBP; 1/300; AbD Serotec-MCA409S).
5. Primary antibody: chick anti-Neurofilament (NF;1/10000; Biolegend-822601).
6. Secondary antibody: goat anti-rat Alexa 488 nm (Life Technologies).
7. Secondary antibody: goat anti-chick Alexa 594 nm (Life Technologies).
8. Hoechst 33342 (10 mg/mL, or other suitable Nuclei stain).
9. 25 × 50 mm glass cover slip #1.
10. Standard microscope slides.
11. Blunt forceps with grip to transfer membrane with slice.
12. Hard set mounting media, such as Fluoromount G (Southern Biotechnologies).

2.6 Image Acquisition and Analysis

1. Confocal microscope with appropriate lasers.
2. 10×, 25× and 40× objective (to achieve sufficient subcellular resolution).
3. Image analysis software such as Fiji.

3 Methods

3.1 Preparation of Slice Medium

1. Combine 50% (100 mL) MEM, 25% (50 mL) EBSS, 1% (2 mL) Pen/Strep, 1× (2 mL) Glutamax, 2.6 mg/mL (2.89 mL) of 45% glucose, 2.5 μg/mL (1 mL) Amphotericin B and 50% (100 mL) horse serum. Filter sterilize with a 0.22 μm filtration unit. The medium can be stored up to 3–4 weeks when stored at 4 °C.
2. Prepare a 6-well dish with millicell® inserts; use sterile forceps to transfer the inserts under a tissue culture laminar flow hood. Add 1 mL of slice medium below the insert and keep the plates at 37 °C.

3.2 Preparation of Slices

Mice pups are decapitated with sharp scissors, the head and trunk is placed in 15 mL cold L-15 medium in a petri dish that is kept on ice.

3.2.1 Spinal Cord Dissection

1. Cut off the head with scissors (Fig. 1).
2. Use small scissors to cut the skin and fat parallel to the spine on the back of the mice to expose the vertebral column.
3. At the cut end, you will see the white spinal cord enclosed in the vertebral column.
4. Taking care not to cut the spinal cord, use the small scissors and cut the bone laterally on both sides, to open up the vertebral column, and carefully ease off the posterior aspect of the column, revealing the white shiny spinal cord lying inside.
5. Run forceps carefully down each side of the spinal cord, to detach it from spinal roots.
6. Cut transversely across the spinal cord about 5–10 mm from the cut end and carefully lift the cervical cord out whole using blunt forceps or small scissors into 15 mL of L-15 medium in a petri dish on ice.
7. Carefully remove the meninges and blood vessels using fine forceps.

3.2.2 Cerebellum Dissection

1. Remove the head and cut the skin open in the center of the head along the midline in a sagittal direction, to expose the skull.
2. Carefully place small scissor blade points in the foramen magnum and cut into the bone to enlarge the opening.
3. With the head placed on its under surface in a petri dish containing medium, take the blunt side of the scissors and press over the junction between the cerebral hemispheres and the cerebellum, visible through the skull surface.

4. The brainstem and cerebellum will then extrude through the enlarged foramen magnum into the medium. Practice will ensure that the brain is not damaged, though with older mice, the brain may need to be dissected out.
5. Make sure there is brain stem attached to the cerebellum (*see* **Note 4**).
6. Try and remove some meninges and large blood vessels using fine forceps.

3.2.3 Cerebral Hemisphere Dissection

1. Cut the skin open in the center of the head along the midline in a sagittal direction, to expose the skull.
2. Use small scissors to carefully cut the skull along the midline, between the cerebral hemispheres.
3. Peel back the skull to expose the brain.
4. Cut the optic nerves and olfactory bulbs at the front of the brain (unless wanted) to release the brain anteriorly.
5. Cut between the hemispheres and the cerebellum posteriorly, to release it, and now the cerebral hemispheres can be scooped out using a small spatula or blunt forceps.
6. Remove some meninges and blood vessels using fine forceps.
7. The whole hemisphere can be taken if the corpus callosum is required or it can be sliced to study deep white matter, hippocampus, or gray matter as required.

3.2.4 Cutting of Slices of Brain, Cerebellum, or Spinal Cord

1. Slices are cut between 200 and 300 μm thickness, with the force and the speed of the blade adjusted according to the machine used. *See* **Note 5**.
2. An autoclaved silicone pad is placed on the metal holder, and a cotton swab is used to wipe the pad and the blade with ethanol.
3. After the ethanol evaporates, the brain or spinal cord is transferred onto the silicone pad using a spatula. Excess medium is removed using a pipette, and care should be taken to avoid damaging the tissue (Fig. 1).
4. Spinal cord is placed parallel to the blade on the tissue chopper and longitudinal slices are cut. In case of the brain (whole or hemisphere) coronal sections or sagittal slices can be cut. The cerebellum is cut into sagittal sections by placing it parallel to the blade with the ventral surface facing up (*see* **Notes 5** and **6**).
5. When chopped, slices are removed with small amount of slice medium on a spatula; the blunt edge of a scalpel blade can be used to gently push the slices onto the spatula.
6. The slices are gently transferred to a petri dish with slice medium and separated into individual slices using the blunt edge of scalpel blades, needles, or pipette tips.

7. The separated slices that are undamaged are taken gently with the spatula and placed onto inserts in a 6-well dish. In the case of spinal cord slices, we place four to five slices obtainable from one pup on a single insert, and in the case of cortical hemispheres or cerebellum, we place three slices per well.
8. Medium is changed every 2 days by aspirating it and carefully replacing 1 mL of medium at the bottom of the well.
9. Slices thin down to 150–200 μm in about 48 h. Healthy slices are transparent and have flattened out, while unhealthy slices can be identified if they have black or brown patches or holes on slices and debris around the slice.

3.3 Adding Cells to Slices

1. Cells (microglia/OPCs/other cells such as T/B lymphocytes) that are to be added onto slices should be spun down using a benchtop centrifuge at 5000 rpm (1957 × *g*) for 5 min and resuspended into a small volume (e.g., 100 μL) of slice medium.
2. For cerebellum and cortex, we use 100,000 OPC per slice in 5 μL of medium (*see* **Note 7**).
3. For spinal cord slices we use 10,000 OPC per slice in 1–2 μL of medium (*see* **Note 7**).
4. The cells normally fall to the edges of slice, so care should be taken to keep the volume to a minimum to avoid this.
5. If OPCs are added, then myelination from these cells can be seen 7–14 days after.

3.4 Demyelination

Slice cultures can be used to study demyelination and remyelination. The chemical demyelinating agent Lysophosphatidyl choline (LPC, Lysolecithin) is used to induce demyelination. The cerebellum is well myelinated at P10, the cervical spinal cord at P15, and cortex at P30 [28], LPC can be added to the medium after these time points to achieve demyelination. LPC from egg yolk and bovine origin are used (*see* **Note 8**).

1. LPC is dissolved in PBS, and a stock of 100 mg/mL is made and frozen. LPC is best not stored for more than a year. Aliquots are thawed at room temperature or heated up to 40 °C to dissolve before using.
2. For demyelination, LPC is diluted in slice medium to a concentration of 0.5 mg/mL and 1 mL is added to the well.
3. After 18–20 h, the medium is changed to fresh normal medium and the experiment continued until the desired stop time.
4. To ensure demyelination has occurred, test slices are visualized at day 2 after adding LPC, then fixed and analyzed using antibodies to MBP, looking for destruction of myelin sheaths

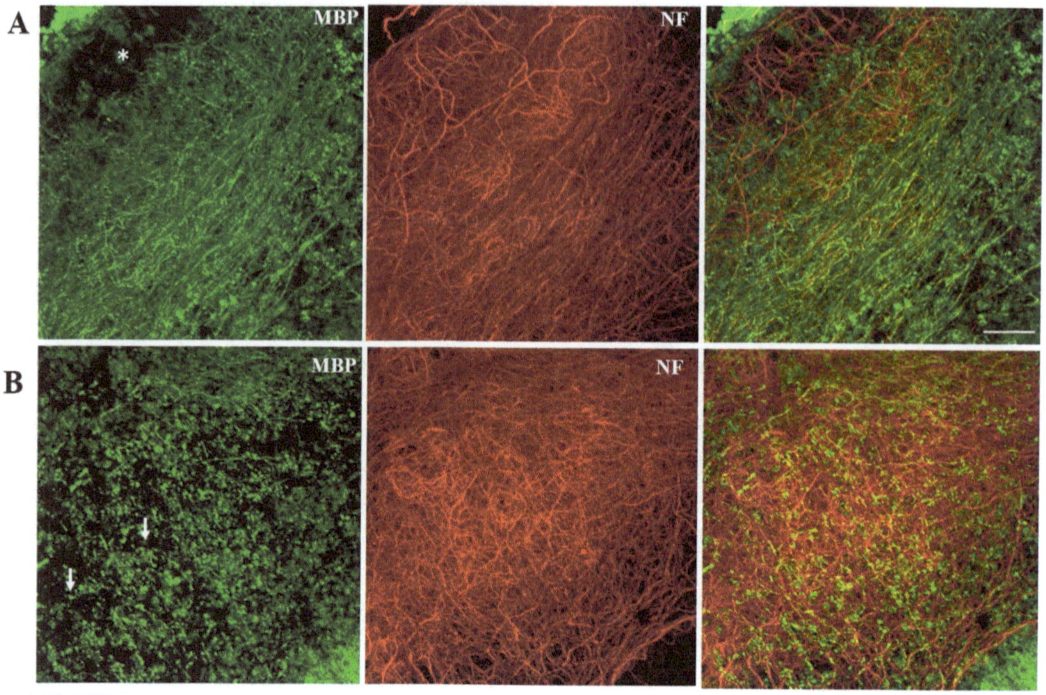

Fig. 2 Confocal images of myelinated spinal cord slices and spinal cord after demyelination. (**a**) The spinal cord is well myelinated at postnatal day P15. The image shows control spinal cord (not demyelinated) immunostained with antibodies against myelin basic protein (MBP) and neurofilament heavy chain (NF). Myelin (green) wraps around axons (red). (**b**) Chemical demyelination by 0.5 mg/mL lysophosphatidyl choline (LPC) for 18 h and immunostained against myelin basic protein (MBP) and neurofilament heavy chain (NF). Myelin debris (green dots) is seen around axons (red). *indicate the non-myelinated edges of control spinal cord slice. White arrow indicates myelin debris. Scale bar: 100 μm

and myelin debris and anti-Neurofilament H antibody to check the health of the axons (Fig. 2).

3.5 Fixation and Preparation for Microscopy

The slices can be fixed at different desired time points (*see* **Note 9**).

1. Gently aspirate medium and wash once with PBS.
2. Aspirate and replace PBS with 1 mL 4% paraformaldehyde in PBS adding it over and under membrane insert and leave for 1 h at room temperature.
3. Cut the membrane so that the plastic insert is removed, being careful to avoid the tissue.
4. Remove the paraformaldehyde and wash three times with PBS agitating gently for 5 min on a rocker.
5. Add 1 mL saturation solution (*see* Subheading 2.4, **item 3**) to the tissue to reduce nonspecific antibody binding for 1 h on a shaker.

6. Dilute primary antibodies in saturation solution, using 0.5 mL per membrane and incubate the plate at 4 °C overnight on a rocker set at a gentle speed (*see* **Note 10**).
7. Wash with 1 mL saturation solution for 10 min.
8. Wash three times for 1 h with 1 mL of saturation solution on a rocker to remove unbound antibody.
9. Prepare secondary antibodies, 0.5 mL per membrane, diluted in saturation solution. Incubate at 4 °C overnight on a rocker, keeping samples in the dark from here forward to prevent unwanted photobleaching of fluorescent signal.
10. Wash twice with 1 mL PBS for 10 min to remove unbound antibodies.
11. Wash three times for 1 h with 1 mL of saturation solution on a rocker to remove unbound antibody.
12. Stain nuclei with Hoechst in PBS (1:10,000) for 10 min.
13. Wash in 1 mL PBS once for 5 min.
14. During the last rinse, mounting preparation should be started. Label the glass microscope slide with details of the experiment and place them on a black surface or paper.
15. Remove the membrane from the 6 well plate using blunt forceps and place it on a drop of mounting medium on the slide with the slice side up. Place another drop of mounting medium on top of the slide and carefully place a coverslip over this, avoiding trapping bubbles.
16. Let the mounting medium dry, and after a day the edges of the cover slip can be sealed with nail varnish to avoid accumulation of bubbles over time as medium evaporates.
17. Store slides in the dark at 4 °C.

3.6 Image Acquisition and Analysis

Image acquisition of fixed slices is possible from an inverted or upright microscope. The microscope should be able to take images of tissue of 200 μm thickness, with a stack of images, which generally requires a confocal microscope or similar optical sectioning microscope. To visualize myelin sheaths and cells we use a Z step size of 0.5 μm. Users should determine the appropriate settings to achieve sufficient resolution and to obtain sufficient numbers of cells and sheaths in images, as per each individual confocal microscope system. Image can be analyzed on software such as Fiji [29].

A z-stack projection of the confocal stack can be used to represent an image after immunolabeling and qualitative analysis, but for quantitative analysis of overlap of two stains, e.g., myelin basic protein and neurofilament, individual focal planes must be analyzed to avoid overlap generated by the maximum projection, rather than true overlap.

To quantify myelin, the area of immunofluorescence staining for MBP can be measured [22, 23, 30], but this measurement includes the cell body of oligodendrocytes, and myelin debris as well as true myelin sheaths. To avoid this, our lab developed a macro to measure myelin associated with axons [18]. To use this macro, the tissue is immunolabeled to stain MBP in green and the axonal protein NFH in red. Where these labels overlap (green and red) in each confocal plane represents the myelin on axons, and helps us avoid the myelin present in cell bodies. A threshold was defined to avoid counting very small areas of overlap (usually background staining). To determine how much myelin there is per axon, the number of co-localized pixels was divided by the red pixels counted in the neurofilament photograph, giving the "myelination index." This measure can then be used to compare the efficacy of drugs added to slices that have potential to increase myelination/remyelination. Another method used to quantify changes in myelination and measure more compact myelin is by looking at the focal expression pattern of axonal adhesion protein Caspr (Contactin-associated protein) at the paranodal regions and excluded from nodes in the presence of compact myelin [27, 31].

4 Notes

1. Slice cultures are carried out at different developmental stages of mouse (or rat) based on the experiment. For electrophysiology studies, adult mouse and rat slices are often used and only are maintained for up to a week, in different culture medium with oxygenation by bubbling oxygen though the medium when the slices are cut. Cultures from neonatal mouse and rats can be kept for longer periods and do not need oxygenation in this manner.

   Cervical spinal cord begins to myelinate early, so slices taken between P0 and P2 survive better compared to slices taken at later time points. Spinal cords taken from older rodents are myelinated, and cutting these induces myelin damage which is very inhibitory to growth [32] and these slices slowly degrade. Cortical hemispheres myelinate much later and can therefore be taken later, e.g., at P7, and can survive for a month. Rat brain can be cultured from P0 to P10 on membrane inserts, and it has been shown that rat brains taken between P8 and P10 can survive for several months [33].
2. Alternatively, brain and spinal cord can be sliced on a vibrating microtome, in this case it is advisable to embed the brain in 4% low melting point agarose in PBS (temperature should be not more than 37 °C) to hold the tissue together and maintain the architecture better. Allow the agarose to solidify around the

brain on ice and then this block is stuck to the tissue holder using superglue. At medium speed, the brain can be cut into 200–300 μm thick slices into cold L-15 medium. The slices are transferred into cold slice medium in a 6-well dish kept on ice. The agarose surrounding the tissue is gently removed using forceps and the slice then placed on inserts. We prefer to remove the agarose to give space for the tissue to flatten and grow; if the agarose is kept on then there is tissue debris on the insert surrounding the slice.

3. Brain and spinal cord are placed onto autoclaved silicone pads to avoid the razor blade striking the metal surface, blunting the blade and moving the brain. The silicone pads act as a cushion and prevent the brain from moving, so that there is no need to use glue to fix the brain. This is useful especially in the case of spinal cord as the tissue is very small in size and glue cannot be used as it damages the tissue. Any firm material that can be sterilized could replace the silicone pad.
4. Cerebellum slice cultures are healthy and survive for long periods if the brainstem is maintained. Having a portion of the midbrain and brainstem helps in separating the slices as they can be used as points where the tissue can be held down with a pipette tip or scalpel while the slices are separated without damaging the cerebellum.
5. With our McIlwaine chopper (MTC/2E), the force is at 85° angle from the minimum position and the speed is at 15° angle from the minimum position (there are no numbers on the dial).
6. Spinal cords are known to survive longer as longitudinal slices than transverse slices [17]. The cerebral hemispheres can be cut into sagittal or coronal slices depending on the purpose of the experiment, for instance if the corpus callosum is being studied then the whole brain is cut coronally. If deep brain structures are required, then one hemisphere is sufficient and can be cut coronally. For successful long-term survival of cerebella and if myelin is to be studied, then sagittal sections are optimal.
7. The number of cells added to the slice depends upon the purpose of the experiment. If OPC (either rat/mouse/human) are added to slices, this mentioned number of cells is more than sufficient to see myelination. In most of the cases, the cells accumulate at the edges of the slice and myelinate axons there. To avoid this, the volume of cells should be kept low so the liquid does not form a pool around the sides of the slices. This number may vary for other cell types. Rat OPCs will myelinate mouse slices and vice versa. Human OPCs can also be added to rodent slices, and survive and form mature oligodendrocytes but as myelination is slow, it is difficult to see convincing compact myelin in the time frame possible.

8. LPC is bought from Sigma-Aldrich in most of the laboratories that use this for demyelination across the world. LPC is dissolved in PBS or saline at 40 °C (maximum). If it does not dissolve, the tube can be vortexed and be left for longer period of time at 40 °C. The LPC type widely used is the egg yolk alpha LPC, but there are laboratories that use the bovine origin LPC. To be certain of the concentration and duration of LPC needed for demyelination, it is advisable to try different concentrations of LPC and decide on a concentration that is suitable for the slices (especially as there is sometimes interbatch variation of LPC).
9. It takes about 2 days after placing the slice on the insert for it to thin down, to about 150–200 μm. It is better to wait for at least 24–48 h before fixing the slice for immunolabeling after demyelination, to allow the LPC effect to occur and for myelin debris to be partly cleared by microglia.
10. The primary antibody can be left at 4 °C overnight if this is more convenient. If the secondary antibody is left on for more than 48 h then there is an increased risk of high background staining.

References

1. Baumann N, Pham-Dinh D (2001) Biology of oligodendrocyte and myelin in the mammalian central nervous system. Physiol Rev 81:871–927
2. Szuchet S, Polak PE, Yim SH (1986) Mature oligodendrocytes cultured in the absence of neurons recapitulate the ontogenic development of myelin membranes. Dev Neurosci 8:208–221
3. Wood P, Okada E, Bunge R (1980) The use of networks of dissociated rat dorsal root ganglion neurons to induce myelination by oligodencrocytes in culture. Brain Res 196:247–252
4. Wood PM, Bunge RP (1986b) Myelination of cultured dorsal root ganglion neurons by oligodendrocytes obtained from adult rats. J Neurol Sci 74:153–169
5. Chan JR et al (2004) NGF controls axonal receptivity to myelination by Schwann cells or oligodendrocytes. Neuron 43:183–191
6. Huang JK, Jarjour AA et al (2011) Retinoid X receptor gamma signalling accelerates CNS remyelination. Nat Neurosci 14:45–53
7. Bornstein MB (1972) Immunopathology of the central and peripheral nervous systems. Verh Dtsch Ges Inn Med 78:777–790
8. Yavin E, Yavin Z (1974) Attachment and culture of dissociated cells from rat embryo cerebral hemispheres on polylysine-coated surface. J Cell Biol 62:540–546
9. Jarjour AA, Zhang H, Bauer N, ffrench-constant C, Williams A (2012) In vitro modelling of central nervous system myelination and remyelination. Glia 60:1–12
10. Avossa D, Rosato-Siri MD, Mazzarol F, Ballerini L (2003) Spinal circuits formation: a study of developmentally regulated markers in organotypic cultures of emb1ryonic mouse spinal cord. Neuroscience 122:391–405
11. Schnadelbach O, Ozen I, Blaschuk OW, Gour BJ, Meyer RL, Fawcett JW (2001) N-cadherin is involved in axon-oligodendrocyte contact and myelination. Mol Cell Neurosci 17:1084–1093
12. Hild W (1956) Myelin formation in central nervous system tissue cultures. Verh Anat Ges 53:315–317
13. Bornstein MB, Murray MR (1958) Serial observations on patterns of growth, myelin formation, maintenance and degeneration in cultures of newborn rat and kitten cerebellum. J Biophys Biochem Cytol 4:499–504
14. Field EJ, Hughes D, Raine CS (1969) Electron microscopic observations on the development of myelin in cultures of neonatal rat cerebellum. J Neurol Sci 8:49–60

15. Kim SU (1971) Electron microscope study of mouse cerebellum in tissue culture. Exp Neurol 33:30–44
16. Ross LL, Bornstein MB, Lehrer GM (1962) Electron microscopic observation of rat and mouse cerebellum in tissue culture. J Cell Biol 14:19–30
17. Bunge RP, Wood P (1973) Studies on the transplantation of spinal cord tissue in the rat. I. The development of a culture system for hemisections of embryonic spinal cord. Brain Res 57:261–276
18. Zhang H, Jarjour AA, Boyd A, Williams A (2011) Central nervous system remyelination in culture—a tool for multiple sclerosis research. Exp Neurol 230:138–148
19. Jarjour AA, Bull SJ, Almasieh M, Rajasekharan S, Baker KA, Mui J, Antel JP, Di Polo A, Kennedy TE (2008) Maintenance of axo-oligodendroglial paranodal junctions requires DCC, netrin-1. J Neurosci 28:11003–11014
20. Birgbauer E, Rao TS, Webb M (2004) Lysolecithin induced demyelination in vitro in a cerebellar slice culture system. J Neurosci Res 78:157–166
21. Huang JK, Jarjour AA, Nait-Oumesmar B, Kerninon C, Williams A, Krezel W, Kagechika H, Bauer J, Zhao C, Baron-Van Evercooren A, Chambon P, ffrench-Constant C, Franklin RJM (2011) Retinoid X receptor gamma signaling accelerates CNS remyelination. Nat Neurosci 14(1):45–53
22. Mi S, Miller RH, Tang W, Lee X, Hu B, Wu W, Zhang Y, Shields CB, Zhang Y, Miklasz S, Shea D, Mason J, Franklin RJ, Ji B, Shao Z, Chédotal A, Bernard F, Roulois A, Xu J, Jung V, Pepinsky B (2009) Promotion of central nervous system remyelination by induced differentiation of oligodendrocyte precursor cells. Ann Neurol 65:304–315
23. Miron VE, Ludwin SK, Darlington PJ, Jarjour AA, Soliven B, Kennedy TE, Antel JP (2010) Fingolimod (FTY720) enhances remyelination following demyelination of organotypic cerebellar slices. Am J Pathol 176:2682–2694
24. Hill RA, Medved J, Patel KD, Nishiyama A (2014) Organotypic slice cultures to study oligodendrocyte dynamics and myelination. J Vis Exp 90:e51835. https://doi.org/10.3791/51835
25. Seidl AH, Rubel EW (2010) A simple method for multiday imaging of slice cultures. Microsc Res Tech 73(1):37–44
26. Miron VE, Boyd A, Zhao JW, Yuen TJ, Ruckh JM, Shadrach JL, van Wijngaarden P, Wagers AJ, Williams A, Franklin RJ, ffrench-Constant C (2013) M2 microglia and macrophages drive oligodendrocyte differentiation during CNS remyelination. Nat Neurosci 16:1211–1218
27. Yuen TJ, Johnson KR, Miron VE, Zhao C, Quandt J, Harrisingh MC, Swire M, Williams A, McFarland HF, Franklin RJ, ffrench-Constant C (2013) Identification of endothelin 2 as an inflammatory factor that promotes central nervous system remyelination. Brain 136:1035–1047
28. Foran DR, Peterson AC (1992) Myelin acquisition in the central nervous system of the mouse revealed by an MBP-LacZ transgene. J Neurosci 12(12):4890–4897
29. Schindelin J, Arganda-Carreras I, Frise E, Kaynig V, Longair M, Pietzsch T, Preibisch S, Rueden C, Saalfeld S, Schmid B, Tinevez JY, White DJ, Hartenstein V, Eliceiri K, Tomancak P, Cardona A (2012) Fiji: an open-source platform for biological-image analysis. Nat Methods 9:676–682
30. Ghoumari AM, Ibanez C, El-Etr M, Leclerc P, Eychenne B, O'Malley BW, Baulieu EE, Schumacher M (2003) Progesterone and its metabolites increase myelin basic protein expression in organotypic slice cultures of rat cerebellum. J Neurochem 86:848–859
31. Eisenbach M, Kartvelishvily E, Eshed-Eisenbach Y, Watkins T, Sorensen A, Thomson C, Ranscht B, Barnett SC, Brophy P, Peles E (2009) Differential clustering of Caspr by oligodendrocytes and Schwann cells. J Neurosci Res 87:3492–3501
32. Neumann H, Kotter MR, Franklin RJ (2009) Debris clearance by microglia: an essential link between degeneration and regeneration. Brain 132(Pt 2):288–295
33. Marksteiner J, Humpel C (2008) Beta-amyloid expression, release and extracellular deposition in aged rat brain slices. Mol Psychiatry 13:939–952

Chapter 11

Forward Genetic Screen Using Zebrafish to Identify New Genes Involved in Myelination

Linde Kegel, Maria Rubio, Rafael G. Almeida, Silvia Benito, Anna Klingseisen, and David A. Lyons

Abstract

Zebrafish are now well established as the preeminent vertebrate model with which to carry out gene discovery/forward genetic screens to identify the molecular genetic basis of biological processes. Gene discovery screens in zebrafish have already provided novel insight into mechanisms of glial cell development and function. The vast majority of genetic screens in zebrafish are based around a three generation screen that starts with the random induction of mutations in adult males using the chemical mutagen ENU. Here we outline the methods that underlie this type of screen, detailing each step, from ENU mutagenesis, through the breeding schemes required to recover homozygous mutant animals in subsequent generations, the screening procedure itself, with a focus on the analysis of myelinating glia, and the subsequent confirmation of mutant phenotypes.

Key words Zebrafish, ENU, Forward genetics screen, Myelin, Oligodendrocyte, Whole genome sequencing

1 Introduction

1.1 Forward Genetic Screens in Zebrafish

Gene discovery screens represent an unbiased approach to elucidating the molecular basis of biological processes of interest. Typically, mutations are introduced at random in adult animals and these mutations are bred to homozygosity in subsequent generations, at which point animals are screened for phenotypes of interest. Following the identification of mutant animals with phenotypes of interest the molecular basis of the causative mutations must be established, to link phenotype and genotype. Large-scale gene discovery screens were first carried out in invertebrate models. Indeed, initial screens in *Drosophila melanogaster* provided fundamental insights into the molecular mechanisms of embryonic patterning and cell fate specification without always knowing the causative mutations. Such was the fundamental importance of this work that it culminated in the award of the 1995 Nobel Prize in

David A. Lyons and Linde Kegel (eds.), *Oligodendrocytes: Methods and Protocols*, Methods in Molecular Biology, vol. 1936, https://doi.org/10.1007/978-1-4939-9072-6_11,

Physiology or Medicine to Ed Lewis, Christiane Nüsslein Volhard, and Eric Wieschaus *"for their discoveries concerning the genetic control of early embryonic development."*

Although zebrafish had been shown to be amenable to mutagenesis by multiple methods from the early 1980s, it wasn't until two large-scale screens were carried out, by Dr. Nüsslein Volhard in Tubingen, Germany, and her former PhD student, Dr. Wolfgang Driever, as a group leader in Harvard, USA, that zebrafish were firmly established as the preeminent vertebrate model with which to carry out gene discovery. The results of the screens in these laboratories resulted in 37 papers being published as a special issue of the journal *Development* in 1996 [1], with individual papers reporting the identification of mutant lines that disrupted specific biological processes from early embryonic patterning through to axonal pathfinding.

Following these screens and the large proliferation of laboratories established by investigators working on various biological processes the community expanded greatly. It is essential also to note the parallel pioneering work of groups at the University of Oregon that established much of the knowledge base for the use of zebrafish as a model.

In the following 20 years numerous further gene discovery screens have been carried out using zebrafish, which have provided fundamental insights into multiple biological processes. Phenotypic assays have become ever more sophisticated, employing transgenic reporters of cell types and tissues of interest, or behaviors as diverse as optic reflexes and sleep. Gene discovery screens in zebrafish have also provided novel insights into the molecular basis of myelinating glial cell development. For example, a completely novel g-protein coupled receptor essential to myelination by Schwann cells, gpr126, was identified from a gene discovery screen in zebrafish [2], with its function in mammalian myelination subsequently confirmed by analysis of knockout mice [3]. Numerous insights into oligodendrocyte biology have also derived from genetic screens, e.g., the demonstration that the kinesin motor protein kif1b is essential to the transport of mRNA encoding key myelin proteins form the oligodendrocyte cell body to myelinating processes [4].

Despite significant progress in our understanding of myelinating oligodendrocyte development from studies in zebrafish and mammals, there remain many gaps in our knowledge, and of myelination in particular. Therefore, there still remains a pressing need to carry out further unbiased approaches to complete our understanding, such that we may be able to consider rational approaches to the treatment of human diseases of myelin and oligodendrocytes, from multiple sclerosis (MS) to genetically determined leukodystrophies [5]. Zebrafish represent a powerful model for the study of oligodendrocyte biology and we believe that further screens will provide novel insights into molecular mechanisms [6, 7].

The protocols we outline here are broadly applicable to ENU-based mutagenesis screens that aim to elucidate the genetic basis of any biological process using zebrafish.

1.2 ENU Mutagenesis and Breeding Scheme to Generate Families of Zebrafish to Screen

As detailed in the protocol below, the first step in the genetic screen is the induction of mutations in the sperm stem cells of adult males. The mostly widely used approach in zebrafish is chemical mutagenesis using *N*-ethyl-*N*-nitrosourea (ENU), which can induce numerous mutations at random in spermatogonia [8]. Adult male zebrafish are exposed to the compound by simply adding the ENU to the water as described in Subheading 3.1.4. ENU is an alkylating agent that results in the transfer of an ethyl group to DNA bases, giving rise mostly to AT to TA transversions and AT to GC transitions. Small deletions and other more complex genome rearrangements can be observed too, although less frequently [9].

After introducing mutations in adult males, the efficiency of mutagenesis must be assessed, and this is carried out by crossing mutagenized males to females that carry a known mutation that results in an easy-to-screen phenotype, e.g., those in genes such as *sox10* or *mitfa* that are required for pigment formation [10, 11]. The process by which mutagenesis efficiency is tested is described in Subheading 3.1.6. Following the assessment of mutagenesis efficiency, well-mutagenized males are selected as founders for the genetic screen. These males are then crossed with individual females to generate the first generation (F1) animals for the screen. *See* Fig. 1 for overview of screen breeding scheme. The number of F1 animals generated will determine the scale and scope of the genetic screen. This is because each F1 animal is used to generate individual F2 families of fish. It is the progeny of these F2 families that are typically screened, because at this point homozygous mutations can be recovered in the F3 generation (Fig. 1). Typically, F3 animals are screened during embryonic or early larval stages. The number of F2 families that can be screened will be determined by the number of personnel involved, aquarium capacity, and budget.

1.3 Introducing Transgenic Reporters into Screening Pipeline

The small size, optical transparency, relative simplicity, and rapid development of zebrafish embryos allow direct observation of biological events as they occur in living animals. Over the past years tools have been developed to visualize myelination at high resolution in live zebrafish. For example, our lab generated the Tg (mbp:EGFP-CAAX) line, in which mbp-expressing myelinating cells specifically express membrane-tethered GFP (Fig. 2) [12]. In a recent genetic screen we used this reporter to study the effect of mutations that affect myelination. The transgenic reporter was introduced into the screen by crossing with F1 mutagenized animals as outlined in Fig. 1.

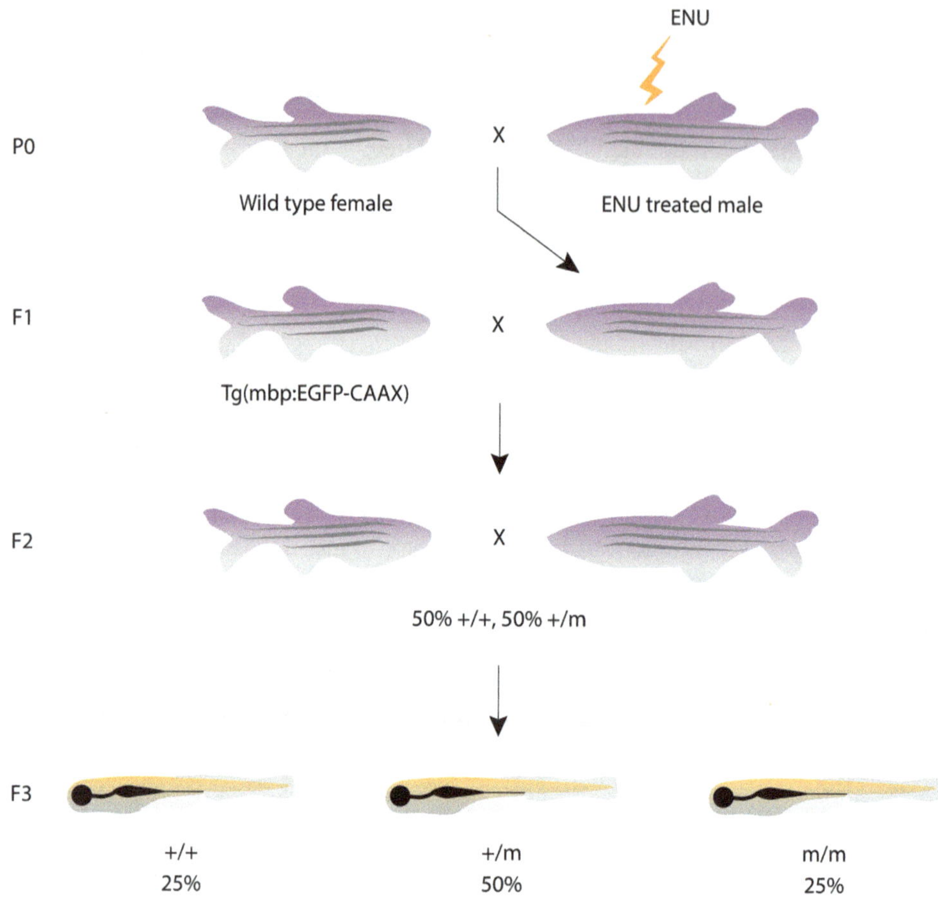

Fig. 1 Breeding strategy after ENU treatment. Wild-type (+/+) males in the parental (P) generation are treated with ENU and outcrossed to generate F1 generation fish, which have a uniquely mutagenized genome. F1 zebrafish are crossed with Tg(mbp:eGFP-CAAX) (Fig. 2). F2 family members are incrossed and transparent F3 embryos are screened for myelin phenotypes

1.4 Screening the F3 Generation

To identify disruption to myelinating glial cell development in the progeny of F2 families we used the Tg(mbp:EGFP-CAAX) reporter and examined animals at 5 days post fertilization, by which time a significant number of axons in the zebrafish central nervous system (particularly hindbrain and spinal cord) are myelinated (Fig. 2). This reporter also allows identification of mutants with disruption to peripheral nerve myelination, given the early myelination of the posterior lateral line (pLL) nerve, a prominent sensory nerve in developing zebrafish (Fig. 2). As outlined in Subheading 3.2.2, transgenic fish are mounted in a lateral view orientation for consistency and observed under a fluorescent microscope at high magnification to identify phenotypes of interest. Given that phenotypes are typically manifest upon recovery of homozygous mutations, 25% of offspring from an affected pair should exhibit specific

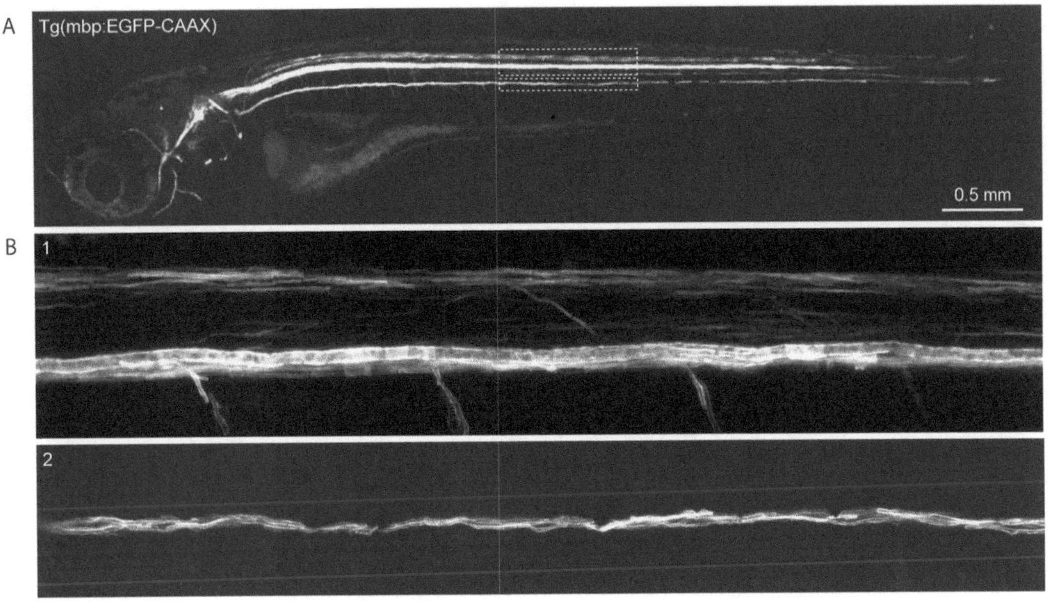

Fig. 2 Transgenic zebrafish allow visualization of myelin in vivo. (**a**) A transgenic mbp:EGFP-CAAX zebrafish larva at 5 days post fertilization (dpf) which expresses membrane-tethered green fluorescent protein (EGFP-CAAX) under the control of myelin basic protein (mbp) promoter, which is active in all myelinating glia in the CNS and PNS. Scale bar: 0.5 mm. (**b**) Higher magnification view of the myelinated spinal cord (1) and the posterior lateral line (pLL) (2) as indicated by the boxed rectangles

effects. Similarly, due to the nature of the breeding scheme, F2 fish will include carriers and noncarriers, on average ¼ of the breeding pairs within an affected family should allow for recovery of homozygous mutants. We aim to screen 6 pair matings per family, which provides an 87.5% chance of finding a single pair exhibiting any phenotypes present in that family.

1.5 Identifying the Mutation that Causes the Phenotype of Interest

As the screening process is underway and mutants are confirmed, one should identify the specific gene disruption that results in the phenotype. To confirm mutant phenotypes, F2 individuals that generate F3 progeny exhibiting phenotypes are outcrossed with a genetically diverse strain (Fig. 3). The F4 progeny of these newly generated F3 animals are then assessed to confirm penetrance of phenotype. The outcrossing step also has the advantage of introducing genetic diversity into the family, which facilitates genetic mapping and identification of causative mutations. Following confirmation of phenotypes, the DNA of homozygous F4 mutants can be collected for genetic mapping and candidate gene identification. Until recently the predominant method was to carry out PCR-driven recombination-based genetic mapping to identify regions of the genome linked with the phenotypic mutants [13, 14]. Because recombination between the causative mutation

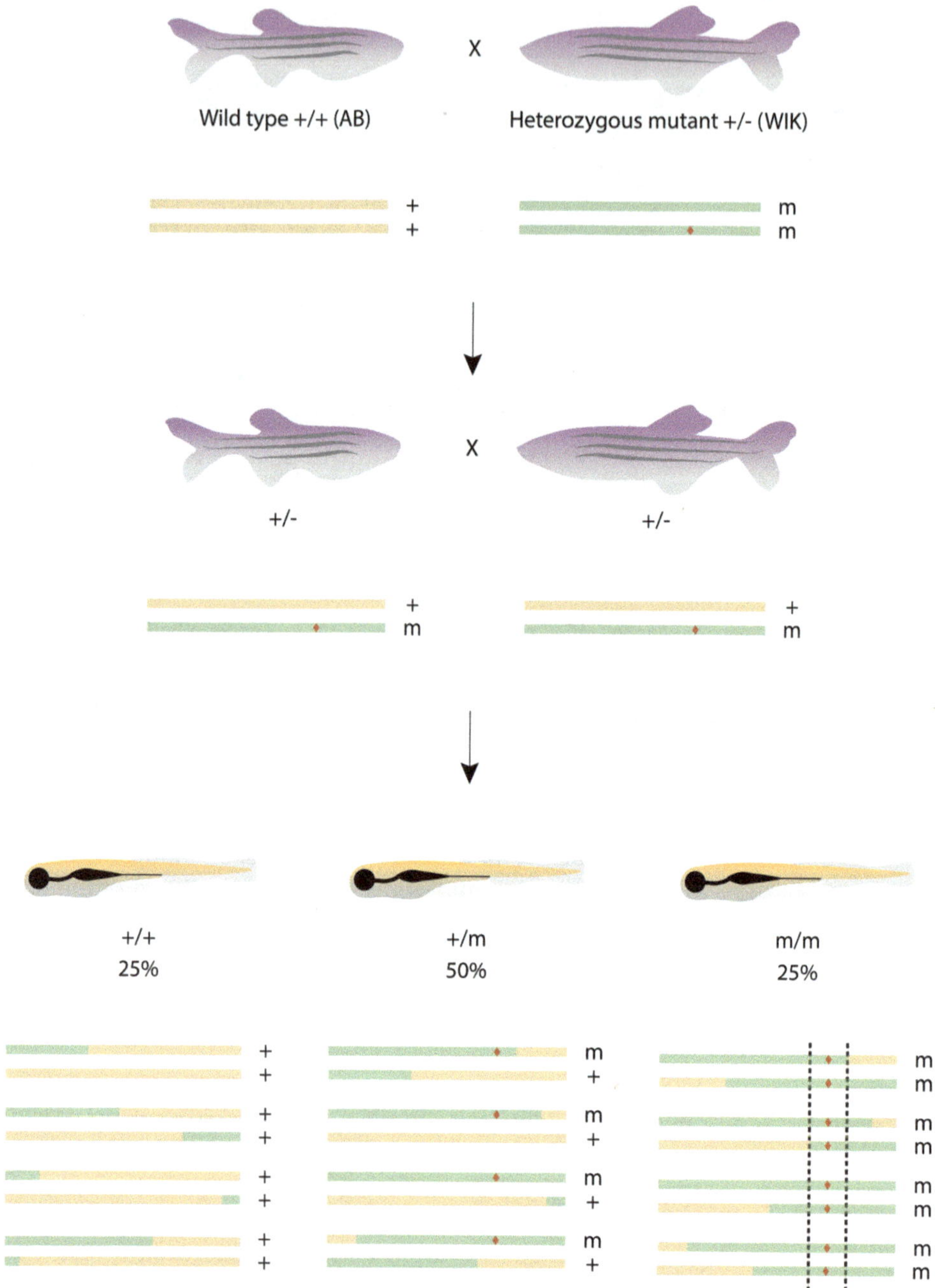

Fig. 3 Identifying candidate regions for the mutation that causes the phenotype of interest. Mutation carriers are outcrossed with wild type (WT) (+/+) from a different strain. F1 offspring are incrossed and F2 screened for the myelin phenotype. DNA of multiple mutant embryos is pulled and whole genome sequenced. The region in the genome that is equal between all mutants (dotted lines) will be identified as the region where the mutation resides

and nearby genetic markers will be rare, unlinked markers are typically randomly segregated, whereas linked markers co-segregate. With the availability of panels of genetic markers mapped to specific regions of the genome, one can define the candidate region in which the causative mutation resides. It is now becoming increasingly more common to take a next-generation sequencing approach to genetic mapping and candidate lesion identification. The reduced cost of sequencing, coupled with powerful bioinformatic approaches that allow in silico mapping of sequence reads to a template genome, allows the laborious process of PCR-based mapping to be circumvented (reviewed in [15, 16]). In addition to allowing one to map the location of phenotype-linked regions of the genome, the sequence-based approach also provides lists of candidate causative mutations that can be assessed for their likelihood to cause the phenotypes observed in a screen. In Subheading 3.3.4 we give a short example of the methods we have recently used to identify the causative mutations associated with phenotypes of mutants isolated in the recent forward genetic screen that was carried out in our lab.

1.6 Testing Candidate Mutations for Causation of Phenotype

After candidate mutations that may cause a specific phenotype (e.g., disruption to mbp:EGFP-CAAX) have been identified, these mutations need to be assessed for direct causality, for which there are several approaches:

1. PCR-restriction enzyme based genotyping assays that differentiate between the putative wild type and mutant sequence; all phenotypic mutants should carry the candidate mutation, while all phenotypically normal zebrafish embryos should have a wild type or heterozygous genotype.
2. Morpholino antisense oligonucleotide knockdown to phenocopy the mutant of interest [17].
3. Rescue of the mutant phenotype with expression of the wild-type version of the gene, typically by injecting synthetic mRNA.
4. For mutation verification examples of myelin studies in zebrafish we refer to [2, 4].

Ultimately creating additional mutant alleles, e.g., using Crispr/Cas gene targeting methods [18], in the same gene can be carried out to phenocopy the original mutant and to provide further important information on the relationship of gene/protein structure and function.

2 Materials

2.1 ENU Treatment

2.1.1 Making ENU

1. 1 g ENU (*N*-Ethyl-*N*-nitrosourea) ($C_3H_6N_3O_2$)
2. Fume hood.
3. Protective covers for hood surface.
4. Lab coat.
5. 2 boxes of nitrile long gloves
6. Squeezy bottle with 1 M NaOH.
7. Designated sharps container (opening should fit the ENU bottle).
8. 2 bottles of 1 L 1 M NaOH.
9. 1 large 5 L beaker for NaOH.
10. 1 Bottle of 100 mL 10 mM acetic acid.
11. 6× 25 mL syringe.
12. 2× 10 mL syringe.
13. 6× 18–22 gauge needles.
14. Disposable tissues.
15. Magnetic stirring plate.
16. Magnetic stir bar.
17. 4× 50 mL Falcon tubes.
18. Paint can-type container.

2.1.2 Spectrophotometry to Assess Concentration of ENU

1. Protective covers for hood.
2. Box of nitrile gloves.
3. Lab coat.
4. Squeezy bottle with 1 M NaOH.
5. Squeezy bottle with H_2O.
6. 1 Bottle of 1 L 1 M NaOH.
7. 1 large beaker for NaOH.
8. 10 mM Sodium phosphate buffer (Na_2HPO_4/NaH_2PO_4), pH 6.
9. UV quartz cuvette.
10. Squeezy bottle with H_2O.
11. Disposable tissues.
12. Micropipettes with relevant tips (p1000 and p20).

2.1.3 ENU Treatment

1. Wild-type male zebrafish.
2. Freshly made or thawed stock of ENU (usually 85.5 mM).

3. Tricaine (MS-222) (Ethyl 3-aminobenzoate methanesulfonate) A 25× stock solution contains 400 mg tricaine powder in 98 mL MilliQ water with 2 mL 1 M Tris–HCl (pH 9.0).
4. 1 L 2 M NaOH.
5. 1 L 1 M NaOH.
6. Fish water-buffered to 10 mM Sodium Phosphate (Na_2HPO_4/NaH_2PO_4), pH 6.
7. Stock of 0.5 M Sodium Phosphate.
8. 6× breeding cages.
9. 2× breeding inserts.
10. 10 L sized fish tank.
11. Nets for fish.
12. 2 Spill trays for breeding boxes.
13. Colored tape and pens.
14. Double-bagged biohazard bags.

2.1.4 ENU Treatment Efficiency Test

1. Nacre or sox10 colorless adult female zebrafish.
2. ENU-treated male zebrafish.
3. Light microscope.
4. Tricaine (MS-222) (Ethyl 3-aminobenzoate methanesulfonate) A 25× stock solution contains 400 mg tricaine powder in 98 mL MilliQ water with 2 mL 1 M Tris–HCl (pH 9.0).

2.2 Breeding

1. Adult zebrafish of interest.
2. Fish nets.
3. Wash bottle for E3 embryo medium.
4. Breeding tanks for adult zebrafish.
5. Tea strainers.
6. 10 cm petri dishes
7. 50× E3 stock: 250 mM NaCl, 8.5 mM KCl, 16.5 mM $CaCl_2$, 16.5 mM $MgSO_4$, ddH_2O.
8. Plastic pipettes.

2.3 Screening for Myelin Phenotypes

1. F3 zebrafish embryos.
2. Dissection microscope.
3. Tricaine (MS-222) (Ethyl 3-aminobenzoate methanesulfonate) A 25× stock solution contains 400 mg tricaine powder in 98 mL MilliQ water with 2 mL 1 M.
4. 1.3% low melting point agarose.
5. Hot plate (to keep agarose molten during mounting).
6. Sharp forceps.

7. 22 × 50 mm coverslips.
8. Microscope slides.
9. 150 mm length glass pasteur pipettes with teat.
10. 25 mL syringe filled with high vacuum grease.
11. Fluorescent microscope with 10×, 20× and 40× high NA lens.
12. Marker pen.

2.4 Collecting Zebrafish for Whole Genome Sequencing

1. 1.5 mL microcentrifuge collection tubes.
2. Dissection microscope.
3. Liquid nitrogen.
4. 2 Microsurgical blades, each attached to a plastic pipette.

2.5 DNA Isolation for Whole Genome Sequencing

1. Qiagen Puregene Core Kit A.
2. 55 °C water bath.
3. 70% Ethanol in freezer.
4. Clean bag of 15 mL Polypropylene Conical Tubes.
5. Clean bag of 1.5 mL centrifuge tubes.
6. 37 °C shaking incubator.
7. Table centrifuge with 15 mL tube holders cooled at 4 °C.
8. Isopropanol.
9. Freezer at −20 °C.
10. Clean piece of absorbent paper.
11. 65 °C heat block
12. Shaker that can hold 1.5 mL tubes/nutator mixer.
13. Nanodrop.
14. 1% agarose gel.
15. Gel electrophoresis system.

3 Methods

3.1 Mutant Generation: ENU Treatment and Generating Families for Screening

ENU is highly toxic and one should discuss plans to prepare ENU stock solutions and perform ENU treatments of adult males with local health and safety advisors.

3.1.1 Fume Hood Preparation for ENU Stock Preparation and Subsequent Treatments

1. Clear out all nonessential items from fume hood and line the hood with appropriate disposable lining.
2. Wear lab coat and long gloves throughout entire procedure. Use double gloving procedure. Keep a spare open box of gloves

in the hood in case of any contact with ENU with outside glove.

3. Prepare 1 L of 1 M NaOH and keep in a 5 L beaker in the hood to place disposable items in throughout procedure. NaOH rapidly oxidizes ENU and breaks it down to a nontoxic form.
4. Have a second bottle and a squeeze bottle of 1 M NaOH in the hood conveniently located in the case of any spillage.
5. Get appropriate syringes and needles prepared.
6. Get all buffers and plastic ware into the hood. Get magnetic stirring plate into appropriate position and have stirrer bar ready. Also make sure an appropriate sharps waste container is in the hood.

3.1.2 Stock Solution Preparation

The ENU stock is made up in 10 mM Acetic acid. It is easiest to dissolve when the ENU stock reaches a concentration of around 80–90 mM. Use 100 mL of 10 mM acetic acid to make the stock, assuming a 1 g dry weight stock solution (if weight is exactly 1 g that is 85.5 mM). The powder stocks can vary in weight, hence the need to confirm ENU concentration by spectrophotometry.

Never attempt to weigh out ENU powder. It is far too dangerous.

1. Using a 25 mL syringe with an 18–22 gauge needle add 10 mL of 10 mM acetic acid through the rubber stopper of the ENU container.
2. Gently mix the contents so that the powder is fully immersed in the liquid and unlikely to disperse through the air. Once powder is immersed, remove the syringe and needle. Dip the needle in the 1 M NaOH solution for 10–20 s in order to inactivate any ENU that may have contacted the tip. Place the syringe and needle into appropriate sharps waste container. Do not attempt to remove needle from syringe.
3. First place a new 18–22 gauge needle (without syringe) through the rubber stopper in the ENU vial. This will allow air in bottle to be displaced when adding the remainder of the liquid. Do not remove this needle in the next step.
4. Add the remaining 90 mL of Acetic acid to the vial with 25 mL syringes and 18–22 gauge needles. When adding liquid try to run it down sides of bottle to wash any residual powder down to the bottom of the bottle.
5. Remove syringes and two needles. Dip needles in 1 M NaOH and place needles and syringes in sharps container.
6. Once the ENU powder is fully immersed in the liquid, then very carefully open the rubber stopper and add a magnetic stir bar (ENU-dedicated). Replace the stopper. Stir at room

temperature for about 1 h by which time the ENU should be in solution.

7. Transfer the stock volume equally between three 50 mL Falcon tubes. This is the point where spillages are most likely. Do the transfer over the 1 M NaOH container, so that any major spill would go straight into inactivating solution.
8. Add 1 M NaOH to the initial original ENU vial and leave for at least 1 h to inactivate. Dip the vial in the 1 M NaOH. Dispose the vial in the glass waste. Add anything that came into contact with ENU to the 1 M NaOH beaker to be inactivated (*see* **Note 1**).
9. Once all items have been decontaminated in 1 M NaOH, add water to the 5 L beaker to dilute the NaOH. Continue to dilute the 1 M NaOH while washing the solution down the sink. Once NaOH is sufficiently diluted wash the items in the beaker with water, and dispose of or retain for future use. Wipe down any exposed surfaces with 1 M NaOH to decontaminate, using the squeeze bottle.
10. Prepare samples from each of the three vials for spectrophotometric determination of the ENU concentration (*see* below Subheading 3.1.3) and store the remainder at −20C in a secondary container, e.g., the can that the ENU initially came in. Get a paint can-type container if it did not come in one. Label all tubes and containers accordingly with the actual concentration once measured (*see* **Note 2**). During ENU treatment period, we kept a lockable freezer dedicated to ENU.

3.1.3 Measuring ENU Stock Concentration

The best wavelength to measure the absorbance of ENU is at 238 nm, where ENU has a well-defined extinction coefficient of 5830 M^{-1} cm^{-1}.

1 g of pure ENU in 100 mL 10 mM acetic acid should be almost exactly 85.5 mM, but the original amount of ENU in purchased vial can vary, so this will typically change vial to vial, requiring calibration.

1. Dilute the stocks of the 1 g/100 mL 10 mM acetic acid 1:500 in 10 mM Sodium Phosphate Buffer (pH 6). For accuracy use 100 μL of the ENU stock from the 50 mL tubes. Do this in the hood.
2. Add 10 mM Sodium Phosphate Buffer (pH 6) to clean the cuvette and as a standard blank. Take a blank reading at 238 nm.
3. Measure the concentration of each ENU stock at 238 nm (*see* **Note 3**).

4. Measure the absorbance (A) at 238 nm. For a 1 cm cuvette: Concentration of the stock solution (M) = $A_{238}/11.66$ because:

$$c_{\mathrm{ENU}}(M) = A_{238} \times \text{length of light path(cm)}/\text{dilution factor} \times \text{extinction coefficient}$$

$$c_{\mathrm{ENU}}(M) = A_{238} \times 1/0.002 \times 5830$$

5. Add 1 M NaOH to the remaining diluted solution and leave it in the beaker with 1 M NaOH for at least 1 h to inactivate.
6. Note precise concentrations of ENU stocks on stock vials ahead of treatments.

3.1.4 ENU Treatment of Adult Male Zebrafish

During the ENU treatment procedures the adult zebrafish will be very sensitive. Hence one should work in a completely silent work environment and preferably keep the treatment tanks in the dark. Prior to treatment of adult males with ENU, the adults should be crossed to ensure high fecundity and absence of any unexpected phenotypes in progeny, e.g., by test crossing with known mutants in pigment genes (*see* **Note 4**).

Until recently the 3 mM concentration of ENU that has been shown to lead to efficient mutagenesis also caused significant lethality when fish were left to recover, as fish went into shock. It has since been found, however, that by keeping the fish lightly sedated for a short time during the ENU treatment and for an extended period after the procedure almost completely prevents lethality of 3 mM treatment.

Use double gloves throughout the procedure and have extra gloves conveniently located. A lot of gloves will be used throughout the day. Have double-bagged biohazard bags at the side of the hood to dispose of any gloves that are exchanged routinely. Any gloves that may touch ENU throughout the day should be dipped in the 1 M NaOH.

ENU needs buffering beyond that provided by normal fish water.

This is done by using 10 mM (final concentration) phosphate buffered fish water at pH 6.5. The acetic acid in the ENU stock will not significantly affect the final pH.

Use a stock of 0.5 M phosphate buffer at pH 6.0, which when diluted (1:50) in the fish water is about pH 6.5 (the pH of phosphate buffers depends on its concentration) (*see* **Note 5**).

All following procedures should be carried out in the fume hood:

1. Thaw ENU stocks 2–3 h prior to use, if not using freshly made.
2. Have 1 L of 2 M NaOH in a 5 L beaker and a second 1 L bottle of 1 M NaOH ready in fume cupboard.

3. Calculate volumes of stock ENU required to have a final concentration of 3 mM ENU in 600 mL of incubation solution in 10 mM phosphate buffered fishwater. You will need 12 mL 0.5 M phosphate buffer per tank (600 mL final including the ENU), *n* mL of ENU stock and 600–12-n mL of fish water).
4. Add fish to equilibration tanks with 12 mL 0.5 M phosphate buffer and 100 minus 12 minus *n* mL of fish water (*see* **Note 6**). *n* = volume of ENU stock required to generate 3 mM final concentration of ENU.
5. Let the fish adjust to this new environment for 1 h.
6. Gently pour the *n* mL of stock ENU gently to the equilibration tank to bring to the final concentration of 3 mM.
7. While waiting, make up the recovery tank solutions of fish water plus 15 mg/L tricaine. This low dose of tricaine sedates the fish, but does not anesthetize them. The typical stock Tricaine solution of 400 mg/100 mL (4000 mg/L), i.e., a dilution factor of 267. Recovery tanks can be 750 mL. If making up two recovery tanks per ENU tank and have two ENU tanks add 11.25 mL Tricaine stock to a final volume of 3 L fish water.
8. After 55 min of the fish being in ENU, add tricaine to those tanks to a FINAL concentration of 15 mg/L. Given the dilution factor of 267, for a 600 mL tank this means adding 2.25 mL of stock tricaine directly to each ENU tank.
9. After 5 min transfer the fish very gently to the first recovery tanks. Leave for at least 1 h (*see* **Note 7**).
10. After the fish have been in recovery tanks for 1 h, transfer to the second recovery tank and leave for another several hours.
11. After several hours of monitoring, add the fish to a 10 L tank with 10 mg/L tricaine. Cover with a box to keep in dark and transfer back to fishroom (*see* **Note 8**). NO feeding until the next day.
12. Transfer the fish back to the system the next morning and carefully monitor them twice a day throughout the procedure.

3.1.5 Cleanup After ENU Treatment

Avoid cleanup until the fish have been removed from the hood, to avoid disturbance.

1. Add the ENU solution to an equal volume of 2 M NaOH.
2. Dip the ENU tanks and breeding inserts into 1 M NaOH for 10–20 s and wash down with further 1 M NaOH followed by water.
3. Dispose of the 2 M NaOH (plus ENU solution) and replace with new 2 M NaOH. Add the recovery tank water to this, and

similarly dip the recovery tanks into 1 M NaOH and wash down.

4. Dip anything else that might have contacted ENU into the 1 M NaOH. Obviously any gloves that certainly touched ENU at any point should be placed in NaOH immediately.
5. Wash down the surfaces of the hood with 1 M NaOH and then generously with water.
6. Place any (de)contaminated gloves in double-bagged biohazard waste.

The whole ENU treatment procedure (*see* Subheadings 3.1.4 and 3.1.5) should be repeated twice more with a weekly interval, such that males ultimately have had 3 one-hourly treatments with ENU.

3.1.6 Selection of ENU-Treated Adult Male Fish by Determination of Mutation Efficiency

High mutagenic dosages can yield more mutants in a single screen but will also increase the mutation load in the individual genomes. This can later complicate the identification of the causal mutation due to closely linked mutations that do not contribute to the mutant phenotype.

1. One week after the final mutagenesis treatment, set up the males to breed with fecund females. There should be reduced male fertility as judged by numerous unfertilized eggs and also obviously disrupted embryos due to mutational load.
2. One week later set up the males to breed again. They should appear less disrupted than the previous week.
3. One week later, keep each male separate and give them a unique ID number (*see* **Note 9**).
4. Start the test of mutagenesis efficiency by crossing each male at weekly intervals to females with known mutations in pigment loci, for example, golden or sox10 colorless. (Methods for breeding and collecting eggs are described in Subheading 3.2.1.) Aim to test at least 1000 embryos per mutagenized male, 2000 if female is a heterozygous carrier, e.g., of colorless.
5. For the rest of the procedure only use clutches from males that give morphologically normal, non-mosaic clutches, where the known pigment loci have been hit at a frequency of >1/1000.

3.1.7 Generating Families

1. Breed each mutagenized parent male (P0) with wild-type females of choice. The embryos obtained from such matings are F1 individuals, each of which will have a uniquely mutagenized genome containing multiple potential function disrupting mutations and many additional silent mutations (Fig. 1) (*see* **Note 10**). Each egg laid will be fertilized by one sperm cell with its unique set of mutations. Hence, every F1

fish will also have its unique set of mutations and will for every mutation be heterozygous (+/m) (*see* **Note 11**).

2. Depending on scope and scale of screen, one can raise up to 500 F1 offspring from each mutagenized P0 male to adulthood. Keep each F1 family derived from each individual P0 male separately.
3. Depending on the nature of the screen F1 adults can either be interbred to generate F2 families, outcrossed with wild types, or outcrossed with specific transgenic reporter lines of interest (Figs. 1 and 2). Individual F2 animals will be either wild type (+/+) or heterozygous carriers (+/m) for any one specific mutation. Depending on whether transgenic reporters were heterozygous or homozygous, individual F2 fish will again either be 50% wild type and 50% heterozygous carriers of the transgene, or 100% heterozygous carriers of the transgene respectively (*see* **Notes 12** and **13**).

F2 families are kept separate and as many families as can be accommodated and processed during the screen are generated. Care must be taken with respect to the timing of F2 family generation with the time of screening F3 progeny in mind. Depending on the throughput of the F3 screen and the number of families per week that can be screened one may want to stagger the generation of F2 families accordingly.

3.2 Screening for Myelin Phenotypes

3.2.1 Generating F3 Clutches

1. To generate F3 embryos, cross individual pairs of F2 fish from the same family. Both can be positive for the transgene or one positive and one negative. Do this the afternoon prior to the day that is planned for egg collection. Try to screen family by family. Ten to twenty breeding tanks are filled with water from the aquaculture system per family, in order to try to ensure collection of 6 clutches per family. Using the fish net, adult fish are transferred to the breeding tanks, each tank containing one male and one female in the inner container (*see* **Note 14**).
2. Label the tanks and put a lid over them to prevent fish from jumping out (*see* **Notes 15** and **16**).
3. Next day, eggs are laid and they will fall through a grid at the bottom of the inner container and collect in the outer container, where the adults cannot eat them. Examine each breeding tank for eggs collected at the bottom following a successful mating.
4. Put F2 pairs that generated eggs separated into individual tanks and label accordingly so that pairs that give rise to interesting phenotypes can be identified at the end of the screening protocol (*see* **Note 17**).

5. Remove the inner container, and drain the water from one breeding tank through a tea strainer. Eggs will collect in the strainer.
6. Invert the strainer over a Petri dish and rinse the strainer with E3 embryo medium to flush the eggs into the Petri dish. Keep all clutches in separate petri dishes and number them according to the clutch number.
7. Remove unfertilized eggs, which look cloudy white, using a plastic pipette. Do not keep more than 50 embryos in a 50 mL petri dish, as it is known that overcrowded dishes can lead to developmental abnormalities. Exchange medium daily.

3.2.2 Screening for Myelin Phenotypes of Interest

The day of screening depends on the developmental stage of interest.

1. For each clutch anesthetize the embryos using 1× tricaine.
2. Assess embryos for any disruption to morphological development and general health.
3. Sort the embryos for fluorescence. Only keep the embryos that express the transgene.
4. Mount the embryos positive for the transgene in 1.3% agarose drops on coverslips (Fig. 4). Mount at least 20 embryos total for each clutch.
5. Examine the embryos with a microscope of choice that is useful for the phenotype and transgene of interest (*see* **Note 18**). An example we found in our screen using Tg(mbp:eGFP-caax) line is given in Fig. 5.
6. Test at least six pairs of one F2 family to reduce risk that phenotypes are missed (*see* **Note 19**).

3.3 Identifying Causative Mutations

The ultimate goal is to identify the causative mutation for the observed phenotype in which myelin is affected.

In order to map a mutation on the zebrafish genome, outcrossing of F2 generation carriers with another polymorphic strain is required [5]. For example, when AB wild-type zebrafish have been used to create mutants, the polymorphic strain for outcrossing could be WIK, TL, or TUE (Fig. 3). Genomic DNA is isolated from a pool of 100–150 mutants, which is used for whole genome sequencing. Bioinformatic analysis of the sequenced data will create a list of candidate genes.

3.3.1 Outcrossing of F2 Carrier Fish

1. Heterozygous F2 carriers are crossed with a polymorphic wild-type strain (Fig. 3) (*see* **Note 20**).
2. The next generation is incrossed to generate homozygous mutant embryos and their siblings (Fig. 3).

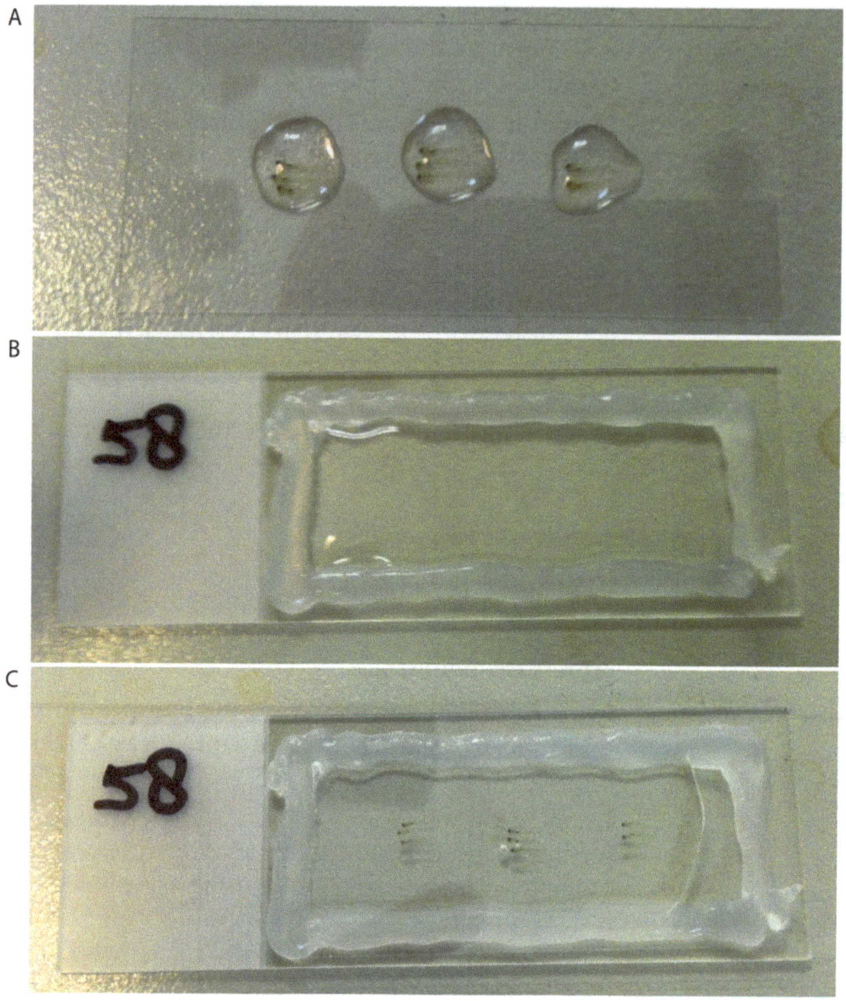

Fig. 4 Mounting embryos in agarose drops on coverslips. (**a**) Two to five embryos are in one drop of 1.3% agarose on a coverslip. (**b**) Make a square of grease filled with 1× Tricaine in embryo medium on a microscope slide marked with the family number. (**c**) When the agarose with the embryos on the coverslip has dried, put the coverslip on the microscope slide with the agarose drops facing the 1× Tricaine embryo medium

3. These incrosses are screened to identify the phenotype of interest according to Subheading 3.2.2.

3.3.2 DNA Extraction for Whole Genome Sequencing

Embryos of the same phenotype (mutant versus wild type) are pooled and collected in 1.5 mL tubes (*see* **Notes 21** and **22**).

1. Cut the embryos of interest out of the agarose and put them in 1× Tricaine; keep mutants and phenotypically wild-type siblings separate.
2. Cull the embryos with a Tricaine overdose.

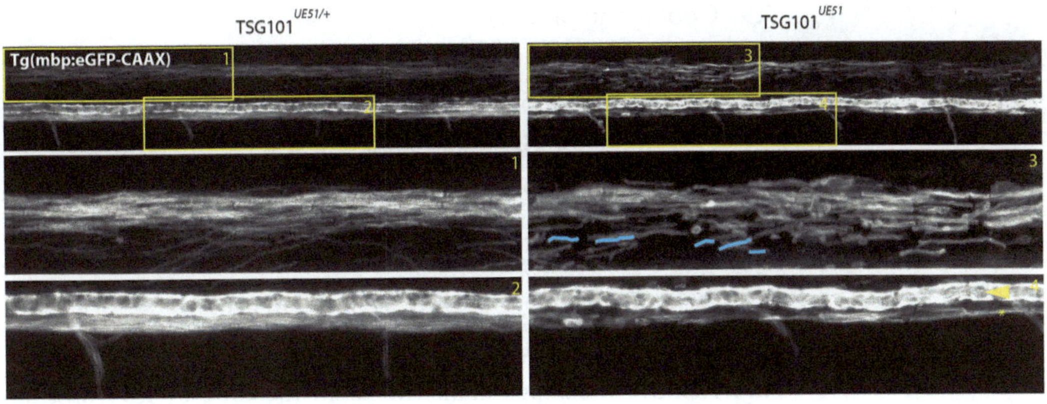

Fig. 5 Example of disrupted myelination in mutant tsg101^{UE51}, in which a cytosine to thymine transition in codon 14 of TSG101 changes an Arginine in a STOP. Tsg101^{UE51} mutants show mild hypomyelination in the dorsal spinal cord (box 3) compared to heterozygous siblings (box 1). Examples of shorter myelin sheaths are indicated in blue. The ventral spinal cord shows hypomyelination of the larger caliber axons of the ventral spinal cord (yellow star box 4), with the exception of the large Mauthner axon (yellow arrowhead box 4)

3. Collect the embryos in a 1.5 mL microcentrifuge tube; keep mutants and siblings separate.
4. Remove as much 1× Tricaine as possible.
5. Embryos can be liquid nitrogen frozen and stored at −80 °C until further use.
6. Genomic DNA is isolated from a pool of 100–150 embryos (*see* **Note 23**), using the Qiagen Puregene Core Kit A with adapted instructions as below (*see* **Notes 24** and **25**).
7. Thaw all tubes at room temperature.
8. For all tubes that contain mutant embryos divide 3 mL of kit Cell Lysis Solution over all tubes.
9. Pull the resulting content of all the tubes containing mutants in one 15 mL tube.
10. Repeat **steps 8–10** for the sibling embryos.
11. Add 15 μL kit Proteinase K (20 mg/mL) and invert 25×.
12. Incubate at 55 °C in the water bath overnight, invert periodically.
13. Let the lysate cool down at room temperature.
14. Add 15 μL kit RNase A Solution.
15. Mix by inverting 25×.
16. Incubate 2 h at 50 rpm in a 37 °C shaking incubator, invert every 10 min.
17. Cool the samples on ice for 3 min.
18. Add 1 mL Protein Precipitation Solution.

19. Cool 5 min on ice.
20. Vortex vigorously for 20 ss.
21. Cool on ice for 5 min.
22. Centrifuge for 10 min at 2000 × *g* at 4 °C (*see* **Note 26**).
23. Transfer the supernatant to a clean 15 mL tube containing 3 mL isopropanol, which is chilled on ice.
24. Mix by inverting 50×.
25. Incubate overnight at −20 °C in the freezer.
26. Centrifuge for 30 min at 2000 × *g* at 4 °C.
27. Remove the supernatant and drain the tube by inverting the tube on a clean piece of absorbent paper.
28. Add 3 mL of ice-cold 70% ethanol and invert 5× to wash DNA pellet.
29. Centrifuge for 3 min at 2000 × *g*.
30. Remove the supernatant.
31. Air-dry the pellet at room temperature.
32. Add 150 μL kit DNA Hydration Solution, pipette up and down, and transfer the total volume to a clean 1.5 mL tube.
33. Incubate at 65 °C for 1 h.
34. Incubate at room temperature overnight while shaking at 30 rpm.
35. Measure the DNA concentration and OD using a nanodrop (*see* **Note 27**).
36. Load 10 μL sample on a 1% agarose gel to check DNA quality and RNA contamination (*see* **Note 28**).

3.3.3 Whole Genome Sequencing

Whole genome sequencing (WGS) is a fast and cost-effective way to map the mutation causing the myelin phenotype of interest.

We will not discuss methods for WGS in this chapter. Various companies offer "next generation" sequencing on different plat forms. One can also elect to carry out whole exome sequencing at yet lower cost. This approach will also provide mapping and candidate mutation information.

3.3.4 Analysis of Whole Genome Sequencing Data

The method of assessing polymorphic loci at a genome-wide level simultaneously after WGS of recombinant animals is known as "mapping-by-sequencing." For an overview of the various methods we refer to [16]. In our forward genetic screen we have used variant discovery mapping (VDM), as the methods below describe.

There are a number of online data analysis platforms available for bioinformatics analysis after WGS. These platforms have incorporated workflows and produce mapping plots and annotated lists of variants. We made use of the galaxy interface (https://

usegalaxy.org), a free, web-based platform and used the CloudMap pipeline [19] https://usegalaxy.org/cloudmap

1. Sequence the nonmutant homozygous and heterozygous siblings of the mutants, and subtract the common variants from the variants identified in the pooled mutant sample. As a result, only the mutagen-induced SNPs will be mapped (*see* **Note 29**).
2. Advisable: Subtract SNPs from published wild-type genomes, making use of SNPFisher (http://snpfisher.nichd.nih.gov/snpfisher/snpfisher.html).
3. To narrow down the list of candidate mutations further, only select:
   (a) Homozygous variants within the mapping region.
   (b) The most likely type of mutations depending on the used mutagen.
   (c) Mutations that are predicted to actually have an effect on the gene product.
4. Advisable: Check the degree of conservation across species for that specific SNP using the UCSC genome browser (http://www.genome.ucsc.edu).

4 Notes

1. One can leave everything at the end of the procedure in the 1 M NaOH overnight, but it may be more practical to clean up after the 1 h of inactivation.
2. The ENU stock solution is very stable in the 10 mM acetic acid, but you should check the O.D. prior to use if it has been in storage for a significant period to test for possible degradation.
3. Although the ENU is very diluted when being measured, it should still be considered dangerous. Make sure therefore that you bring 1 M NaOH to the spectrophotometer, in the event of any spillage. It is better to take the diluted ENU to a nearby spectrophotometer than to bring the photometer to the site of more concentrated ENU, but make sure that the spectrophotometer is basically right beside the hood to avoid walking around with the cuvettes.
4. It is advisable to first select healthy male fish that breed well and fertilize good amounts of eggs prior to use. In order to do this wild-type male fish should be set up for breeding with the female fish that will be used for breeding to produce the F1 generation 2–3 weeks prior to ENU treatment.

5. For example, a final volume of 600 mL per 1 L breeding box with the breeding insert in can be used during the equilibration and ENU treatment phases, with five to six fish per tank. The breeding insert makes it easy to transfer fish from the equilibration/ENU tank to the recovery tanks with minimal disruption or transfer of liquid.
6. Equilibrate fish in these tanks for at least 1 h in the hood. Once the fish are in buffered fish water, keep noise to a minimum and avoid changing light levels. It is best to do this in the evening with little traffic through labs or in a very quiet corner of a lab.
7. The fish can be very sensitive during this time, so keep movement, light, noise, to an absolute minimum.
8. We have found leaving the fish in this lower dose of tricaine covered with the box overnight to be particularly effective at keeping the fish calm.
9. The number of mutations per male varies wildly, as might fecundity, etc. At this point it is fastest to first grow up a clutch of F1 fish from each male, by crossing to wild-type females of interest (Fig. 2). Once those animals are growing, start the test of mutagenesis efficiency.
10. Do not let too much time pass between inducing the mutations and outcrossing to generate F1 families; sperm stem cells might die or stop producing sperm cells due to the mutagenesis. As a result, only sperm stem cells that are not affected by the mutagen will produce sperm cells and the resulting families will not have any mutations.
11. From now on keep track of which F1 fish are derived from which P0 male. This is important when F3 embryos derived from different families have the same affected myelin phenotype and might have the same mutation. This is more likely when these families are derived from the same P0 male. It is advisable to start a database from this point to keep a good overview of the screen.
12. Before generating F2 families make sure the transgenic reporter line fish have good fluorescence. Since transgenes can be incorporated in the genome multiple times, the level of fluorescence can vary between fish.
13. It is advisable to separate the fluorescent positive and negative F2 fish for fluorescence; this will prevent that two fish that are both negative for the reporter are used to generate F3 embryos for screening. Keep both transgene positive and negative fish separate. Do not discard the negative fish since they might still carry a mutation of interest and can be crossed with fish that are

positive for the transgene, creating F3 offspring that will still have the transgene and can be screened.

14. When developmental stage should be determined very precisely use dividers and take those out the next morning. The fish will immediately start breeding and the fertilized eggs will start developing from that point.
15. For example, if there are six pairs created from F2 family #58, one could name them 58A-F.
16. Each tank should only contain one male and one female, as it is important to exactly determine which F2 fish are carriers of a mutation.
17. Keep the used F2 pairs separate and assign them a specific breeding number, for example, as in **Note 15**. This is crucial to know when F3 embryos derived from this clutch show a phenotype of interest and proves that these F2 fish are carriers of a mutation that affects myelination. These F2 fish will be used again in Subheading 3.3.1.
18. It is advisable to make a quick image of the myelin phenotype for administration purposes. This is especially convenient when a large-scale screen is carried out resulting in finding a great amount of phenotypes.
19. Discard families that give offspring with severe background mutations (e.g., heart edemas, malformations).
20. One of the parents must be negative for the transgene, to reduce the risk that the location of the transgene will be identified as the region in which the mutation resides.
21. For later subtraction (*see* Subheading 3.3.4) it is advisable to also isolate DNA from a pool of siblings that do not show the phenotype of interest.
22. Mutant embryos from multiple clutches can be combined to reduce the size of the candidate region where the mutation resides (Fig. 3b).
23. Including more mutants will reduce the candidate region in which the mutation of interest resides (Fig. 3) and result in the removal of more background mutations from the list of candidate mutations. Collect as many embryos as feasible, with a lower limit of 50 per pool.
24. Consult DNA preparation requirements with WGS company (Subheading 3.3.3).
25. Do not perform phenol extraction at any point because this interferes with the library preparation and cluster formation during whole genome sequencing, which will result in lower coverage.

26. The precipitated proteins should form a tight pellet. If this is not the case, put the sample on ice again for 5 min and repeat centrifugation.

27. In most cases sample requirements for OD are: OD260/280 = 1.8–2.2, OD260/230 > 2.0. However, one should consult the company where whole genome sequencing will be carried out for exact requirements.

28. A band at range > 10 kb should be visible. There should be no RNA which typically runs at 5 kb and 2 kb.

29. It is important that the subtracted variants are of high quality and sufficient sequencing depth in order to avoid falsely identified variants (e.g., sequencing errors).

References

1. Nusslein-Volhard C (2012) The zebrafish issue of development. Development 139:4099–4103. https://doi.org/10.1242/dev.085217
2. Monk KR, Naylor SG, Glenn TD, Mercurio S, Perlin JR, Dominguez C, Moens CB, Talbot WS (2009) A G protein-coupled receptor is essential for Schwann cells to initiate myelination. Science 325:1402–1405
3. Monk KR, Oshima K, Jörs S, Heller S, Talbot WS (2011) Gpr126 is essential for peripheral nerve development and myelination in mammals. Development 138:2673–2680. https://doi.org/10.1242/dev.062224
4. Lyons DA, Naylor SG, Scholze A, Talbot WS (2009) Kif1b is essential for mRNA localization in oligodendrocytes and development of myelinated axons. Nat Genet 41:854–858. https://doi.org/10.1038/ng.376
5. Cole KLH, Early JJ, Lyons DA (2017) Drug discovery for remyelination and treatment of MS. Glia 65(10):1565–1589. https://doi.org/10.1002/glia.23166
6. Lyons DA, Talbot WS (2015) Glial cell development and function in zebrafish. Cold Spring Harb Perspect Biol 7:a020586. https://doi.org/10.1101/cshperspect.a020586
7. Czopka T (2016) Insights into mechanisms of central nervous system myelination using zebrafish. Glia 64:333–349. https://doi.org/10.1002/glia.22897
8. de Bruijn E, Cuppen E, Feitsma H (2009) Highly efficient ENU mutagenesis in zebrafish. Methods Mol Biol 546:3–12. https://doi.org/10.1007/978-1-60327-977-2_1
9. Noveroske JK, Weber JS, Justice MJ (2000) The mutagenic action of *N*-ethyl-*N*-nitrosourea in the mouse. Mamm Genome 11:478–483
10. Lister JA, Robertson CP, Lepage T, Johnson SL, Raible DW (1999) Nacre encodes a zebrafish microphthalmia-related protein that regulates neural-crest-derived pigment cell fate. Development 126:3757–3767
11. Dutton KA, Pauliny A, Lopes SS, Elworthy S, Carney TJ, Rauch J, Geisler R, Haffter P, Kelsh RN (2001) Zebrafish colourless encodes sox10 and specifies non-ectomesenchymal neural crest fates. Development 128:4113–4125
12. Almeida RG, Czopka T, ffrench-Constant C, Lyons DA (2011) Individual axons regulate the myelinating potential of single oligodendrocytes in vivo. Development 138:4443–4450. https://doi.org/10.1242/dev.071001
13. Kelly PD, Chu F, Woods IG, Ngo-Hazelett P, Cardozo T, Huang H, Kimm F, Liao L, Yan YL, Zhou Y, Johnson SL, Abagyan R, Schier AF, Postlethwait JH, Talbot WS (2000) Genetic linkage mapping of zebrafish genes and ESTs. Genome Res 10:558–567
14. Postlethwait JH, Talbot WS (1997) Zebrafish genomics: from mutants to genes. Trends Genet 13:183–190
15. Schneeberger K (2014) Using next-generation sequencing to isolate mutant genes from forward genetic screens. Nat Rev Genet 15:662–676. https://doi.org/10.1038/nrg3745
16. Doitsidou M, Jarriault S, Poole RJ (2016) Next-generation sequencing-based approaches

for mutation mapping and identification in *Caenorhabditis elegans*. Genetics 204:451–474. https://doi.org/10.1534/genetics.115.186197

17. Eisen JS, Smith JC (2008) Controlling morpholino experiments: don't stop making antisense. Development 135:1735–1743
18. Hwang WY, Fu Y, Reyon D, Maeder ML, Tsai SQ, Sander JD, Peterson RT, Yeh J-RJ, Joung JK (2013) Efficient genome editing in zebrafish using a CRISPR-Cas system. Nat Biotechnol 31:227–229. https://doi.org/10.1038/nbt.2501
19. Minevich G, Park DS, Blankenberg D, Poole RJ, Hobert O (2012) CloudMap: a cloud-based pipeline for analysis of mutant genome sequences. Genetics 192:1249–1269. https://doi.org/10.1534/genetics.112.144204

Chapter 12

Manipulating Neuronal Activity in the Developing Zebrafish Spinal Cord to Investigate Adaptive Myelination

Jill M. Williamson, David A. Lyons, and Rafael G. Almeida

Abstract

In the central nervous system, oligodendrocyte-lineage cells and myelination can adapt to physiological brain activity. Since myelin can in turn regulate neuronal function, such "adaptive" myelination has been proposed as a form of nervous system plasticity, implicated in learning and cognition. The molecular and cellular mechanisms underlying adaptive myelination and its functional consequences remain to be fully defined, partly because it remains challenging to manipulate activity and monitor myelination over time in vivo at single-cell resolution, in a model that would also allow examination of the functional output of individual neurons and circuits. Here, we describe a workflow to manipulate neuronal activity and to assess oligodendrocyte-lineage cell dynamics and myelination in larval zebrafish, a vertebrate animal model that is ideal for live imaging and amenable to genetic discovery, and that has well-characterized neuronal circuits with myelinated axons.

Key words Oligodendrocyte, Myelin, Myelination, Neuronal activity, Neuronal circuits, Adaptive myelination, Activity-regulated myelination

1 Introduction

It is now abundantly clear that oligodendrocyte development and central nervous system (CNS) myelination are responsive to neuronal activity in vivo, not just in development but also throughout life. Many lines of evidence, including white matter imaging studies in humans (reviewed in [1]), sensory deprivation [2, 3] or stimulation [4, 5] paradigms and direct genetic [6], optogenetic [7] or chemogenetic [8] manipulation of neuronal activity in animal models have indicated that the proliferation, differentiation and survival of oligodendrocyte-lineage cells can be sensitive to brain activity. Furthermore, the formation and growth of myelin sheaths themselves are also sensitive to neuronal activity [7–11], in an axon or circuit-specific manner. The extent of myelination along an axon and its precise pattern are predicted to greatly regulate the neuron's conduction properties, such as conduction velocity and firing rate.

David A. Lyons and Linde Kegel (eds.), *Oligodendrocytes: Methods and Protocols*, Methods in Molecular Biology, vol. 1936, https://doi.org/10.1007/978-1-4939-9072-6_12, © Springer Science+Business Media, LLC, part of Springer Nature 2019

Thus, this "activity-regulated" or "adaptive" myelination is thought to represent a form of nervous system plasticity that enables vertebrates to adapt circuit output (and ultimately, behavior) to changes in neuronal activity driven by experiencing changes in the environment.

However, elucidation of the molecular mechanisms underlying axon-oligodendrocyte lineage cell communication is far from complete, and the exact effect of activity on oligodendrocyte-lineage cells seems to be greatly dependent on the particular stimulus and CNS region examined [12, 13]. Moreover, the relative contributions of oligodendrocyte or myelin adaptations and of axonal adaptations to neuronal activity are far from explained, and the functional consequences of physiological adaptations at the single-neuron or circuit level remain undefined [14]. Therefore, there is a need to simultaneously manipulate neuronal activity and assess oligodendrocyte and myelin adaptations in a living organism with well-characterized neuronal circuits and behavioral outputs to study the functional consequences of adaptive myelination.

Zebrafish are a well-established model for the study of developmental CNS myelination which offers not only great facility for live-imaging of oligodendrocytes and myelin at subcellular resolution (due to their rapid development, transparency, and existence of a repertoire of relevant fluorescent transgenic reporters) but also a genetically tractable platform for discovery and candidate-testing of molecular mechanisms of axon-glial communication [15, 16]. Furthermore, spinal cord myelination has been well characterized in developing zebrafish and proceeds in a stereotyped manner along well-defined axonal tracts that are individually identifiable between animals. For instance, the first spinal cord axons that become myelinated are first wrapped by oligodendrocytes around 55 h post-fertilization, and belong to a population of hindbrain neurons called reticulospinal neurons [17]. We recently discovered that the extent of reticulospinal axon myelination in the developing zebrafish is sensitive to synaptic vesicle release, since its abrogation with tetanus neurotoxin (TeNT) significantly reduced the number and length of their myelin sheaths [10]. Importantly, the connectivity [18] and functional outputs of circuits containing myelinated axons are well characterized in larval zebrafish: for instance, those of motor circuits that include reticulospinal axons and drive locomotive swimming behaviors such as escape responses or prey capture [19–22]. Indeed, pioneering functional imaging studies of neuronal activity in large brain volumes concomitant with behavioral outputs utilized the larval zebrafish model [23, 24]. Thus, larval zebrafish are a perfectly poised platform to address outstanding questions regarding adaptive myelination in a living vertebrate organism.

Here, we describe a workflow for delivering genetic, chemogenetic, and chemical regulators of neuronal activity, assessing their efficiency, and live-imaging oligodendrocyte development and myelination in the developing zebrafish spinal cord. In principle, this workflow can be adapted to other regulators of activity; refined to focus on particular aspects of oligodendrocyte biology; and extended to include subsequent functional and behavioral analysis.

2 Materials

2.1 Delivery of Neuronal Activity Regulators into Fertilized Zebrafish Eggs by Microinjection

1. P2-P10 pipettes and pipette tips.
2. mRNA or DNA plasmids (*see* Table 1).
3. Nuclease-free water.
4. Phenol red (0.5% solution).
5. Eppendorf tubes.
6. Ice.
7. Plastic petri dishes (50 mL).
8. Male and female adult zebrafish (*see* **Note 1**).
9. Embryo medium (10 mM HEPES-buffered E3 embryo medium; 5 mM NaCl, 0.17 mM KCl, 0.33 mM $CaCl_2$, 0.33 mM $MgSO_4$).
10. Plastic Pasteur pipettes.
11. Fine forceps.

Table 1
A selection of genetic tools to regulate neuronal activity

Regulator	Expression	Function
TeNT mRNA	Global	Inhibits all synaptic vesicle release by cleaving Vamp2
BoNT-B mRNA	Global	Inhibits all synaptic vesicle release by cleaving Vamp2
BoNT-C mRNA	Global	Inhibits all synaptic vesicle release by cleaving Syntaxin1 and Snap25
UAS:TeNT-FP plasmid	Mosaic	Inhibits synaptic vesicle release of individual neurons
UAS:BoNTB-FP plasmid	Mosaic	Inhibits synaptic vesicle release of individual neurons
UAS:K_{ir}2.1-FP plasmid	Mosaic	Decreases electrical excitability of individual neurons, by increasing potassium influx
UAS:TRPV1-FP plasmid	Mosaic	Increases electrical excitability of individual neurons *when the agonist capsaicin is administered*
UAS:TRPM8-FP plasmid	Mosaic	Increases electrical excitability *when the agonist menthol is administered*
UAS:TRPA1-FP plasmid	Mosaic	Increases electrical excitability *at 28.5 °C (but not 26.5 °C)*

12. Dissecting microscope.
13. Pneumatic Picopump.
14. Micromanipulator.
15. Micropipette Puller.
16. Thin-wall 1.0 mm glass capillaries.
17. Microloader pipette tips.
18. Stage micrometer slide.
19. Mineral oil.
20. Glass microscope slide.

2.2 Validation of Neuronal Activity Regulation

2.2.1 Ca^{2+} Imaging/SypHy Imaging

1. Mivacurium chloride 50× stock solution (25 mg/mL in water).
2. Embryo medium (*see* above).
3. Plastic petri dishes (50 mL).
4. Dissecting microscope.
5. Hot plate (set to 55 °C).
6. 1.2–2% low-melting-point agarose.
7. Glass pipette and pipette bulb.
8. Forceps.
9. Glass cover slips.
10. Glass microscope slides.
11. High-vacuum grease.

2.3 Bath Application Drug Treatments

1. Male and female adult zebrafish.
2. P2, P10, P20, and P1000 pipettes and pipette tips.
3. Embryo medium (*see* above).
4. Pronase 10× stock solution (5 mg/mL in water).
5. Plastic petri dishes (appropriate sizes) or multi-well plates (6–96 well plates).
6. Fine forceps.
7. Required drugs (*see* Table 2).

2.4 Mounting and Imaging Zebrafish Larvae

1. Embryo medium (*see* above).
2. Tricaine 25× stock solution (4 mg/mL in 20 mM Tris–HCl pH 7.5).
3. Plastic petri dishes (50 mL).
4. Dissecting microscope.
5. Hot plate (set to 55 °C).
6. 1.2–2% low-melting-point agarose.

Table 2
A selection of drug treatments to regulate neuronal activity

Treatment	Vehicle	Function
Pentylenetetrazole (PTZ)	n/a	GABA-A receptor antagonist
Tetrodotoxin (TTX) (*see* **Note 7**)	n/a	Voltage-gated Na+ channel blocker
Strychnine	n/a	Glycine receptor antagonist
CNQX	n/a	AMPA receptor antagonist
MK-801	n/a	NMDA receptor antagonist
MTEP	n/a	mGluR5 allosteric antagonist
Capsaicin	DMSO (up to 1%)	TRPV1-FP agonist—chemogenetic tool
Menthol	DMSO (up to 1%)	TRPM8-FP agonist—chemogenetic tool

7. Glass pipette and pipette bulb.
8. Fine forceps and microknives.
9. Glass cover slips.
10. Glass microscope slides.
11. High-vacuum grease.

3 Methods

3.1 Delivery of Neuronal Activity Regulators into Fertilized Zebrafish Eggs by Microinjection of mRNA/DNA

Genetically encoded neuronal activity regulators include suppressors of synaptic vesicle release such as the light-chain of tetanus [9, 10, 25–27] and botulinum [27] neurotoxins, suppressors of neuronal excitability such as the inward rectifying potassium channel Kir2.1 [26, 28, 29], and chemogenetic stimulators of activity based on transient receptor potential (TRP) channel regulation [30] (*see* Table 1). These regulators can be globally expressed (by injecting mRNA) or sparsely expressed (by injecting plasmid DNA, which yields mosaic expression in individual neurons) (Fig. 1). In this case, to drive neuronal expression, injected plasmids contain the regulator's coding sequence directly downstream of neuronal promoters, or of Upstream Activating Sequences (UAS), which require injection into transgenic eggs expressing the transcription factor Gal4 pan-neuronally or in specific neuronal subtypes. Gal4 specifically recognizes UAS and activates transcription of the downstream activity regulator. The coding sequence of these regulators can also be fused to that of a fluorescent protein (FP) for visual confirmation of the expressing cells. These manipulations can be performed in transgenic reporter lines (e.g., to visualize all OPCs, oligodendrocytes), or can be done concurrently with mosaic

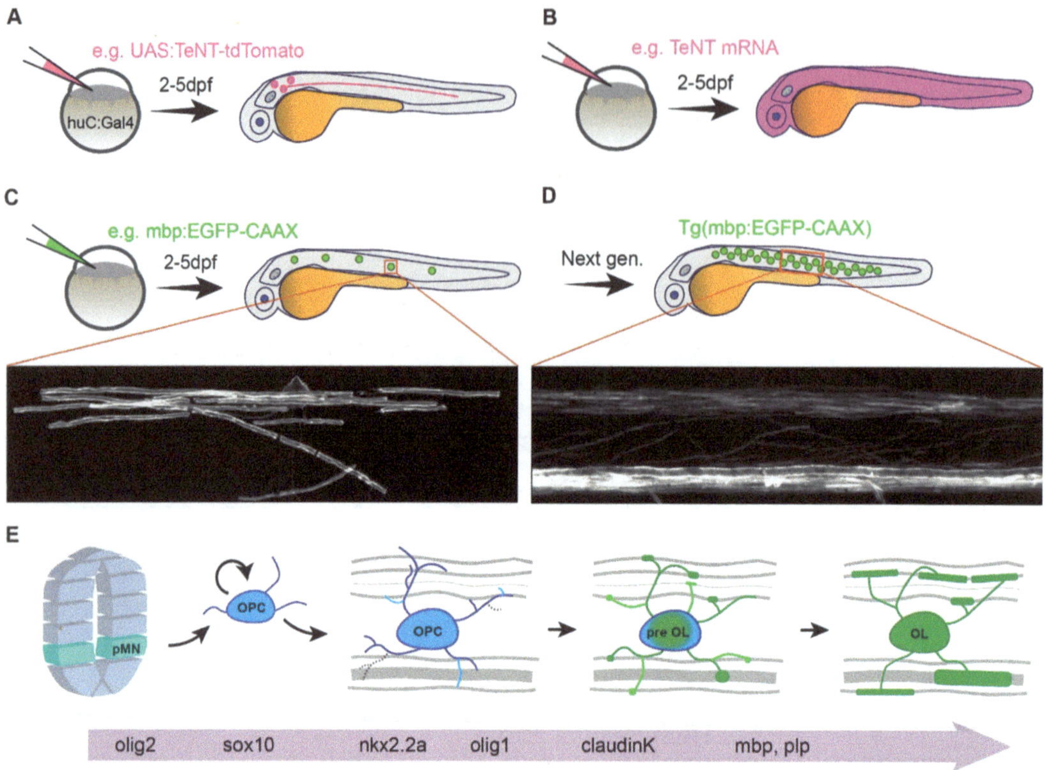

Fig. 1 Delivering neuronal activity regulators and assessing oligodendrocyte-lineage cell development in zebrafish larvae. Microinjection of DNA constructs (**a**) or mRNA (**b**) into 1-cell stage fertilized eggs enables transgenic regulators of activity to be expressed sparsely (**a**) or ubiquitously (**b**). Similarly, injection of oligodendrocyte or myelin reporter transgenes allows individual cell resolution of myelinating capacity in the injected larva (**c**), or global analyses of myelination (**d**) following raising injected larvae and identifying founders with stable germline transmission of the transgene. (**e**) Examples of promoters for reporting different stages of oligodendrocyte-lineage cell development

oligodendrocyte labeling to observe effects on individual oligodendrocytes.

1. Set up pair matings the day prior to performing injections.
2. Prepare injection solution on ice (*see* **Note 2**).
3. Obtain fertilized eggs from the pair matings. Injections must be performed within approximately 1 h of fertilization (1–4 cell stage) to maximize successful global mRNA expression/DNA integration.
4. Set up the injection station. We use the Pneumatic Picopump PV830 (World Precision Instruments Ltd.) fitted to a pressurized N_2 cylinder, and the Micromanipulator M3301R (World Precision Instruments Ltd.) with a standard dissecting microscope. As a guide, we use the following settings (which can be

adjusted as required): eject pressure = 20–30 psi, injection time = 10–50 ms.

5. Use a microloader pipette tip to fill a tapered glass needle (*see* **Note 3**) with 2–3 μL of injection solution and insert needle into injector. Use the micromanipulator to maneuver the needle into view under the microscope and orientate the needle to the desired angle (we typically use 45° angle relative to the microscope's stage).
6. Use a pair of fine forceps to break off the fine tip of the tapered needle—this opens the needle. Do not break off more than a third of the fine tapered tip to ensure the needle does not become too blunt.
7. Add a drop of mineral oil onto a stage micrometer slide, place this under the microscope and test inject once into the oil to confirm injection droplet size (*see* **Note 4**). The injection droplet size can be changed by either adjusting the injector settings (eject pressure, time) or increasing needle diameter by breaking off additional portions of the needle. Adjust these parameters until the desired injection droplet volume is achieved. We typically adjust the size of the droplet to 0.5 nL and inject up to 2 nL of solution per embryo.
8. Place a glass microscope slide into the lid of a plastic petri dish, and use a Pasteur pipette to gently line up fertilized eggs (1–4 cell stage) along the edge of the slide. Place this under the microscope.
9. Using the micromanipulator to move the needle to the first egg, insert the needle tip through the egg chorion and into the egg itself. Inject the desired volume of solution into the cell or yolk of each egg. To improve the efficiency of DNA integration and consistency between embryos, we always aim to inject into the one cell (not the yolk).
10. Store injected eggs in embryo medium in a petri dish in a 28.5 °C incubator. Embryos can be screened for expression of fluorescent reporters (if required) at 2–3 days post fertilization (dpf) under a fluorescence stereomicroscope, or by mounting embryos as per protocol 3.4 and screening under an epifluorescence microscope.

3.2 Validation of Neuronal Activity Regulators

3.2.1 Confirmation of Globally Reduced Neuronal Activity—Paralysis

We use embryo movement at 1 dpf to confirm global expression of inhibitory modulators (e.g., TeNT, BoNT). Unaffected embryos display a typical coiling movement in their chorions. Affected embryos should be completely paralyzed, i.e., display no coiling movement, which can be assessed in a quantitative manner (*see* example in ref. 31). Additionally, embryos >2 dpf will no longer exhibit escape response to touch. Since paralyzed embryos cannot properly hatch from their chorions, they must be dechorionated at 1–2 dpf.

3.2.2 Confirmation of Globally Reduced Neuronal Activity—Visual Background Adaptation

Global expression of activity blockers such as TeNT also suppresses visually evoked background adaptation in zebrafish larvae, whereby the bright light-evoked contraction of the black pigmented melanophores in the skin is blocked (*see* example in ref. 25). This can be easily assessed in a stereomicroscope.

3.2.3 Confirmation of Change in Neuronal Activity—Ca^{2+} Imaging

To perform Ca^{2+} imaging, neuronal activity regulators must be expressed in embryos with neuronally expressed genetically encoded Ca^{2+} indicator (GECI) such as GCaMP6fast.

1. Paralyze 3 dpf embryos in 0.5 mg/mL mivacurium chloride (a neuromuscular blocker) in embryo medium. This may take 5–10 min.
2. Embed embryos in drops of 1.3% low-melting-point agarose on a glass coverslip as per protocol 3.4, and use forceps to position embryo appropriately (*see* **Note 5**).
3. Draw a rectangular well using high-vacuum grease on a microscope slide, fill with 0.5 mg/mL mivacurium chloride in embryo medium, and place coverslip with agarose face-down into the well.
4. Identify neurons expressing the regulator of interest and record time-lapse movies of GECI fluorescence using an appropriate microscope. We have performed this experiment using an Olympus Revolution XDi spinning disk confocal microscope (for 20 min recordings) or a Zeiss Imager Z1 epifluorescence microscope (for 1 min recordings). For more on Ca^{2+} imaging in zebrafish, *see* refs. 32–34.
5. Analyze time-lapse movies using appropriate software, e.g., ImageJ.

3.2.4 Confirmation of Change in Synaptic Vesicle Release—SypHy Imaging

To confirm the effect of neuronal activity regulators, it is also possible to image synaptic vesicle release using the reporter SypHy, where pHluorin, a pH-sensitive GFP variant, is fused to synaptophysin [35]. In the acidic lumen of synaptic vesicles, SypHy's fluorescence is quenched, but upon vesicle release, pH increases and fluorescence is rapidly dequenched. We have previously expressed SypHy (under UAS regulatory control) sparsely in reticulospinal neurons and CoPA (Commisural Primary Ascending). A spinal interneurons, together with cytoplasmic fluorescent reporter tdTomato [10]. In immobilized larvae, we used the sparse tdTomato fluorescence to find these neurons' presynaptic terminals and define a region of interest to set up time-lapse imaging of the SypHy channel. We first photobleached background and residual SypHy membrane fluorescence in the region of interest and then imaged a small z-stack repeatedly in the SypHy channel, interspersing more rounds of bleaching if needed. Using a 20× objective and a Zeiss LSM880 with Airyscan equipped with a FAST module, we achieve imaging rates 0.02–2 Hz (1–2 Hz speeds achieved with FAST module). Follow protocol 3.2.3 or 3.4 to prepare samples for SypHy imaging (also *see* **Note 6**).

Plasmids containing UAS-controlled transgenes can be used to generate stable transgenic lines [36] and many cells will express the transgene when crossed with an appropriate neuronal:Gal4 transgenic line.

3.3 Delivery of Small-Molecule Neuronal Activity Regulators to Developing Zebrafish by Bath Application

These manipulations can be performed in transgenic reporter lines (e.g., to visualize all OPCs, oligodendrocytes), or can be done concurrently with mosaic labeling, e.g., to observe effects on individual oligodendrocytes. This protocol can also be used for chemogenetic experiments (e.g., delivery of the compounds that activate neurons expressing TRPV1/M8-FP).

1. Set up pair matings to obtain fertilized eggs the day before egg collection is required.
2. Collect fertilized eggs and store in embryo medium in a plastic petri dish at 28.5 °C.
3. Ensure all embryos have hatched from chorions by 2 dpf. Embryos can be manually dechorionated using fine forceps at 1–2 dpf, or chemically dechorionated using 0.5 mg/mL Pronase in embryo medium for 6 min followed by four gentle washes in fresh embryo medium.
4. Array embryos as needed into petri dishes or multi-well plates at 1–2 dpf. We use plastic 12-, 24-, or 96-well plates. In 96-well plates, we array up to three embryos per well in 150 μL embryo medium using wide-bore pipette tips (and later add 150 μL of the appropriate treatment to a final volume of 300 μL per well, **step 5**). Any inter-well spaces are also filled with embryo medium to reduce potential evaporation during treatments.
5. Prepare chemical treatment solution in embryo medium. Some drug treatments benefit from 1% DMSO (final concentration) to aid permeabilization into embryos; follow manufacturer's guidance for relevant vehicles and compound solubility.
6. Apply treatment to embryos when required and keep in a humidified chamber at 28.5 °C to prevent evaporation. The duration of the treatment will depend on the experiment. For example, a treatment from 2–4 dpf will affect neuronal activity during OPC specification, migration, differentiation, and myelination. The treatment may need to be refreshed daily.

To live-image the developing zebrafish spinal cord, we first immobilize and orient in agarose those larvae that we have selected for imaging. We find that mounting one larva per drop of agarose is

3.4 Mounting Zebrafish Larvae for Spinal Cord Live-Imaging

best for extended periods of imaging, but this can be increased to up to five larvae per agarose drop for screening purposes, followed by quick retrieval of selected larvae from the agarose.

1. At the appropriate stage of development, select morphologically normal larvae, place them in a petri dish containing 50 mL of embryo medium and anesthetize larvae by adding 2.1 mL of tricaine stock solution. For chemical treatments, the appropriate volume of tricaine stock solution can be added directly into the treatment wells. After 1–2 min, larvae should stop swimming or twitching. Check that heartbeat and blood flow are robust.
2. Microwave 1.2–1.5% low-melting-point agarose in a small glass vial for a few seconds until liquid. Higher concentrations can be used to ensure stability for extended periods of imaging, e.g., time-lapse experiments. Place on a hot plate set to 55 °C and let cool down until comfortable to hold.
3. Using a fire-polished glass pipette, carefully aspirate one larva and place it in the agarose vial, minimizing any excess medium going into the agarose. Swirl vial briefly to ensure larva is homogeneously covered by agarose.
4. Place a glass coverslip or glass-bottom petri dish on the stage of a dissection stereomicroscope. Use glass pipette to aspirate larva with some agarose, and pipette onto coverslip or dish in a single drop.
5. Before agarose starts to solidify (within 20 s), use fine forceps to adjust the larva to a lateral position, pressing gently with the forceps along the side of the larva to ensure it is as flat as possible against the coverslip or dish.
6. Let agarose solidify (around 5 min).
7. If using coverslip, create a high-vacuum grease square well on a glass slide with a plastic syringe. Well should be slightly smaller than the perimeter of the coverslip. Pipette a few drops of embryo medium with 1× tricaine to cover the well.
8. Carefully invert coverslip with solidified agarose drop and larvae over the well, ensuring larvae are in contact with the medium inside the well, and coverslip is sealed up against the grease wall.
9. If using glass dish, simply fill with embryo medium and 1× Tricaine.
10. The larva is now ready for imaging as per protocol 3.5.

We have been able to maintain larvae mounted in this manner alive and healthy for extended periods of imaging (~16 h). If needed, larvae can be retrieved from agarose using fine micro-knives, allowing repeated imaging over several days, or subsequent analysis or genotyping.

3.5 Imaging Oligodendrocyte-Lineage Cells and Myelination in the Zebrafish Spinal Cord

We typically obtain optical sections in z-stacks through the whole depth of the spinal cord or focused on individual cells using a point-scanning confocal microscope, which removes out of focus information and reveals cellular and subcellular structures in our region of interest. We use dry and water-immersion objectives between 20× and 63× with a high numerical aperture and sufficient working distance for imaging whole-mount zebrafish larvae. General steps include (1) finding region of interest using brightfield or fluorescence; (2) setting up appropriate laser lines and detection windows for excitation and emission of fluorophores; (3) rotating and centering region of interest using a fast scanning mode; (4) adjusting zoom and frame size to maximize efficiency of image collection (longer axis on horizontal plane, setting up vertical axis to smallest that will capture whole region of interest); (5) setting up the z-stack's first and last slices and step intervals; (6) adjusting scan speed and averaging to maximize signal-to-noise ratio; (7) ensuring use of full dynamic range of the image and preventing saturation of region of interest by regulating laser intensity and detector gain; (8) acquiring the z-stack once, or repeatedly for time-lapses. Given that individual researcher's imaging systems are variable, here we limit ourselves to general considerations related specifically to imaging oligodendrocyte-lineage cells and myelin in zebrafish larvae.

1. A comprehensive list of oligodendrocyte-lineage cell related transgenic zebrafish reporters is available in refs. 15, 37. The choice of reporter line and desired stage of oligodendrocyte development will determine the time of imaging: for instance, olig2:EGFP+; sox10:mRFP+ OPCs appear around 36–48 hpf, first in the anterior ventral spinal cord, and proceed toward the posterior spinal cord over the next hours. A subpopulation of OPCs also migrates dorsally at these stages. The first differentiated mbp:EGFP+ oligodendrocytes appear around 55–60 hpf, also in the anterior spinal cord, proceeding toward the posterior spinal cord.
2. For quantification of OPC number, we suggest transgenic reporter lines Tg(olig2:EGFP) and Tg(sox10:mRFP), the combination of which is only expressed by OPCs in the spinal cord [38]. Membrane-targeted mRFP in the Tg(sox10:mRFP) reporter line allows not only identification of *sox10*-expressing cells but also confirmation of the typical branched morphology of OPCs (other sox10+ expressing cells in the spinal cord include some neuron subtypes, with a morphology distinguishable from OPCs, and also lacking olig2:EGFP expression [38]). This combination will also allow unambiguous identification of OPC proliferation or oligodendrocyte apoptosis events in time-lapse movies without an additional proliferation marker [39]. From ~4.5 dpf, the high number of OPCs and

oligodendrocytes (which retain transgene expression) require higher magnification imaging (e.g., 63× objective) for more accurate counting through the z-stack.

3. For quantification of oligodendrocyte number, we suggest stable reporter line Tg(mbp:EGFP), and for global quantification of myelination, Tg(mbp:EGFP-CAAX), in which membrane-tethered EGFP labels all myelin [17]. We suggest imaging a length of spinal cord that includes a defined number of somites (e.g., 4–6 somites of the control spinal cord contain up to a hundred oligodendrocytes depending on stages analyzed) which are easy to locate consistently between individual larva in control and activity-regulated experimental conditions. We typically use a 20× objective for counting OPC or oligodendrocyte number, and a *z*-step of 1–2 μm ensure that each cell is sampled in principle at least twice. It is also possible to automate imaging and quantification of these reporters along the whole length of the spinal cord, which requires additional specialized equipment [40].

4. In addition to global analyses, zebrafish have the advantage of easily allowing mosaic, sparse expression of reporters in individual oligodendrocyte-lineage cells. This is useful for lineage-tracing or detailed morphological analysis of individual oligodendrocyte-lineage cells (e.g., quantification of the number and length of myelin sheaths that a single oligodendrocyte makes, i.e., its myelinating "capacity"). This can be achieved by microinjection of the plasmid containing the reporter transgene of interest, for instance mbp:EGFP-CAAX to label the membrane of individual oligodendrocytes, into fertilized eggs, as per protocol 3.1. It is also possible to achieve sparse labeling using the Gal4/UAS system, by combining a Gal4 driver line (e.g., sox10:KalTA4 for OPCs and mbp:Gal4VP16 for oligodendrocytes) with a UAS reporter line or injected plasmid (e.g., UAS:EGFP-CAAX) driving the reporter of interest. The repetitive UAS DNA is thought to be stochastically silenced epigenetically [41] (which can be overcome with other transgenic systems of transcriptional regulation [42]), and in our hands these larvae have a mosaic pattern of expression. Additionally, while less easy, it is also possible to achieve sparse labeling by transplanting individual cells at blastula stages from a donor transgenic embryo to a host embryo that does not contain the same reporter transgene [43]. For imaging of oligodendrocyte's myelinating capacity, we typically use either a 20× objective with a high zoom, or a higher-magnification objective (63×), and a 0.5–1 μm z-stack interval to ensure sufficient *z*-axis sampling of myelin sheaths. We suggest keeping track of the anterior-posterior and dorso-ventral location of each imaged cell between animals to ensure control and activity-regulated conditions are broadly comparable.

4 Notes

1. The transgenic fish crossed will depend on the experiment. For example, numerous transgenic lines expressing oligodendrocyte lineage fluorescent reporters are available [15], depending on which stage of oligodendrocyte development is to be analyzed. Additionally, UAS promoter-driven transgene plasmids must be injected into an appropriate Gal4-expressing transgenic background to facilitate expression of transgene in cells of interest. For example, to express UAS:TeNT-FP mosaically in random neurons one could inject UAS:TeNT-FP plasmid into Tg(HuC:Gal4) eggs (a transgenic line which expresses Gal4 protein pan-neuronally). Alternatively, one can inject neuronal-promoter-driven transgene plasmids (examples of pan-neuronal promoters include those of zebrafish gene *huC/elavl3* [44] and frog neural-specific beta tubulin *Xla. Tubb* [45]).
2. For mRNA injections, we typically inject 100–200 ng/μL in nuclease-free water. For DNA plasmid injections, we typically inject 5–15 ng/μL in nuclease-free water (concentration should be adjusted as required for each plasmid). We use the tol2kit to generate DNA plasmids, where the transgene of interest is flanked by tol2 transposon sites [36]; therefore we also include 25–50 ng/μL tol2 transposase mRNA to increase transgenesis efficiency. A higher [DNA] will be required if injecting non-tol2kit plasmids. Note that high [DNA] are toxic and will affect the viability of injected eggs. Phenol red can also be added to the injection solution (to 0.05% final concentration) to visualize injections more easily.
3. We use a P-97 Flaming/Brown Micropipette Puller (Sutter Instruments Co.) to pull tapered needles from thin-wall 1.0 mm glass capillaries. The following settings are used (but can be adjusted to alter the dimensions of the needle pulled): heat = 550, pull = 230, vel. = 50, time = 125.
4. We typically inject 1 nL of injection solution per egg, but around 0.5–4 nL can be injected. To calibrate needle for 1 nL injections, ensure one injection produces a droplet 0.1 mm in diameter (this generates a droplet of approximately 0.5 nL, using the equation: volume = $^4/_3\pi r^3$), then inject twice into each egg.
5. The orientation of the embryo will depend on which cells are being imaged. To image cells in the spinal cord embryos should be mounted laterally. To image cells in the brain embryos should be mounted dorsally (dorsal side facing coverslip).
6. For SypHy imaging, which we find to be especially sensitive to drift in the *z*-axis, we routinely use a glass-bottom dish and an

immersion objective, which we find minimizes drift compared to larvae mounted in an inverted coverslip.

7. Some small molecules benefit from being delivered by microinjection into the yolk of developing zebrafish. For instance, injection of 2 nL of 0.5 mM tetrodotoxin (TTX) directly into the yolk of 48 hpf larvae leads to their paralysis for 3 days [26].

References

1. Sampaio-Baptista C, Johansen-Berg H (2017) White matter plasticity in the adult brain. Neuron 96:1239–1251
2. Makinodan M, Rosen KM, Ito S et al (2012) A critical period for social experience-dependent oligodendrocyte maturation and myelination. Science 337:1357–1360
3. Hill RA, Patel KD, Goncalves CM et al (2014) Modulation of oligodendrocyte generation during a critical temporal window after NG2 cell division. Nat Neurosci 17:1518–1527
4. McKenzie IA, Ohayon D, Li H et al (2014) Motor skill learning requires active central myelination. Science 346:318–322
5. Hughes EG, Orthmann-Murphy JL, Langseth AJ et al (2018) Myelin remodeling through experience-dependent oligodendrogenesis in the adult somatosensory cortex. Nat Neurosci 21:696–706
6. Kougioumtzidou E, Shimizu T, Hamilton NB et al (2017) Signalling through AMPA receptors on oligodendrocyte precursors promotes myelination by enhancing oligodendrocyte survival. Elife 6:e28080
7. Gibson EM, Purger D, Mount CW et al (2014) Neuronal activity promotes oligodendrogenesis and adaptive myelination in the mammalian brain. Science 344:1252304
8. Mitew S, Gobius I, Fenlon LR et al (2018) Pharmacogenetic stimulation of neuronal activity increases myelination in an axon-specific manner. Nat Commun 9:306
9. Mensch S, Baraban M, Almeida R et al (2015) Synaptic vesicle release regulates myelin sheath number of individual oligodendrocytes in vivo. Nat Neurosci 18:628–630
10. Koudelka S, Voas MG, Almeida RG et al (2016) Individual neuronal subtypes exhibit diversity in CNS myelination mediated by synaptic vesicle release. Curr Biol 26:1447–1455
11. Krasnow AM, Ford MC, Valdivia LE et al (2018) Regulation of developing myelin sheath elongation by oligodendrocyte calcium transients in vivo. Nat Neurosci 21:24–28
12. Baraban M, Mensch S, Lyons DA (2016) Adaptive myelination from fish to man. Brain Res 1641:149–161
13. Nagy B, Hovhannisyan A, Barzan R et al (2017) Different patterns of neuronal activity trigger distinct responses of oligodendrocyte precursor cells in the corpus callosum. PLoS Biol 15:e2001993
14. Almeida RG, Lyons DA (2017) On myelinated axon plasticity and neuronal circuit formation and function. J Neurosci 37:10023–10034
15. Czopka T (2016) Insights into mechanisms of central nervous system myelination using zebrafish. Glia 64:333–349
16. Preston MA, Macklin WB (2015) Zebrafish as a model to investigate CNS myelination. Glia 63:177–193
17. Almeida RG, Czopka T, Ffrench-Constant C et al (2011) Individual axons regulate the myelinating potential of single oligodendrocytes in vivo. Development 138:4443–4450
18. Hildebrand DGC, Cicconet M, Torres RM et al (2017) Whole-brain serial-section electron microscopy in larval zebrafish. Nature 545:345–349
19. Lacoste AMB, Schoppik D, Robson DN et al (2015) A convergent and essential interneuron pathway for Mauthner-cell-mediated escapes. Curr Biol 25:1526–1534
20. Dunn TW, Gebhardt C, Naumann EA et al (2016) Neural circuits underlying visually evoked escapes in larval zebrafish. Neuron 89:613–628
21. Naumann EA, Fitzgerald JE, Dunn TW et al (2016) From whole-brain data to functional circuit models: the zebrafish optomotor response. Cell 167:947–960.e20
22. Severi KE, Portugues R, Marques JC et al (2014) Neural control and modulation of swimming speed in the larval zebrafish. Neuron 83:692–707
23. Ahrens MB, Li JM, Orger MB et al (2012) Brain-wide neuronal dynamics during motor adaptation in zebrafish. Nature 485:471–477

24. Lovett-Barron M, Andalman AS, Allen WE et al (2017) Ancestral circuits for the coordinated modulation of brain state. Cell 171:1411–1423.e17
25. Ben Fredj N, Hammond S, Otsuna H et al (2010) Synaptic activity and activity-dependent competition regulates axon arbor maturation, growth arrest, and territory in the retinotectal projection. J Neurosci 30:10939–10951
26. Hines JH, Ravanelli AM, Schwindt R et al (2015) Neuronal activity biases axon selection for myelination in vivo. Nat Neurosci 18:683–689
27. Sternberg JR, Severi KE, Fidelin K et al (2016) Optimization of a neurotoxin to investigate the contribution of excitatory interneurons to speed modulation in vivo. Curr Biol 26:2319–2328
28. Burrone J, O'Byrne M, Murthy VN (2002) Multiple forms of synaptic plasticity triggered by selective suppression of activity in individual neurons. Nature 420:414–418
29. Hua JY, Smear MC, Baier H et al (2005) Regulation of axon growth in vivo by activity-based competition. Nature 434:1022–1026
30. Chen S, Chiu CN, McArthur KL et al (2016) TRP channel mediated neuronal activation and ablation in freely behaving zebrafish. Nat Methods 13:147–150
31. Wei C, Thatcher EJ, Olena AF et al (2013) miR-153 regulates SNAP-25, synaptic transmission, and neuronal development. PLoS One 8:e57080
32. Leung LC, Wang GX, Mourrain P (2013) Imaging zebrafish neural circuitry from whole brain to synapse. Front Neural Circuits 7:76
33. Renninger SL, Orger MB (2013) Two-photon imaging of neural population activity in zebrafish. Methods 62:255–267
34. Muto A, Kawakami K (2016) Calcium imaging of neuronal activity in free-swimming larval zebrafish. Methods Mol Biol 1451:333–341
35. Granseth B, Odermatt B, Royle SJ et al (2006) Clathrin-mediated endocytosis is the dominant mechanism of vesicle retrieval at hippocampal synapses. Neuron 51:773–786
36. Kwan KM, Fujimoto E, Grabher C et al (2007) The Tol2kit: a multisite gateway-based construction kit for Tol2 transposon transgenesis constructs. Dev Dyn 236:3088–3099
37. D'Rozario M, Monk KR, Petersen SC (2017) Analysis of myelinated axon formation in zebrafish. Methods Cell Biol 138:383–414
38. Almeida RG, Lyons DA. Intersectional gene expression in zebrafish using the split KalTA4 system. Zebrafish 12:377–386
39. Almeida R, Lyons D (2016) Oligodendrocyte development in the absence of their target axons in vivo. PLoS One 11:e0164432
40. Early JJ, Cole KL, Williamson JM et al (2018) An automated high-resolution in vivo screen in zebrafish to identify chemical regulators of myelination. Elife 7:e35136
41. Akitake CM, Macurak M, Halpern ME et al (2011) Transgenerational analysis of transcriptional silencing in zebrafish. Dev Biol 352:191–201
42. Ghosh A, Halpern ME (2016) Transcriptional regulation using the Q system in transgenic zebrafish. Methods Cell Biol 135:205–218
43. Gansner JM, Dang M, Ammerman M et al (2017) Transplantation in zebrafish. Methods Cell Biol 138:629–647
44. Park HC, Kim CH, Bae YK et al (2000) Analysis of upstream elements in the HuC promoter leads to the establishment of transgenic zebrafish with fluorescent neurons. Dev Biol 227:279–293
45. Peri F, Nüsslein-Volhard C (2008) Live imaging of neuronal degradation by microglia reveals a role for v0-ATPase a1 in phagosomal fusion in vivo. Cell 133:916–927

Chapter 13

A Drug-Inducible Transgenic Zebrafish Model for Myelinating Glial Cell Ablation

Marja J. Karttunen and David A. Lyons

Abstract

To study cellular and molecular mechanisms of demyelination and remyelination in vivo, we developed a transgenic zebrafish line, Tg(mbp:mCherry-NTR), in which expression of the bacterial enzyme nitroreductase (NTR) is driven under the myelin basic protein promoter (mbp) and thus is expressed in myelinating glia. When NTR-expressing larvae are treated with the prodrug metronidazole, the reaction between NTR and Mtz results in a toxic metabolite which selectively kills NTR-expressing cells. Using the Tg(mbp: mCherry-NTR) line, we can ablate two-thirds of oligodendrocytes following a 2-day MTZ treatment. Demyelination is evident seven days later, and remyelination is observed 16 days after Mtz treatment. The Tg(mbp:mCherry-NTR) model can be used to image cell behavior during, and to test how genetic manipulations or chemical compounds regulate, demyelination and remyelination. In this chapter, we describe the methods we used to characterize the oligodendrocyte loss, demyelination and remyelination in the Tg(mbp:mCherry-NTR) model.

Key words Zebrafish, Demyelination, Transgenic line, Drug-inducible, Live imaging

1 Introduction

Myelin plays a number of essential roles in the vertebrate nervous system, from allowing axons to propagate electrical signals via saltatory conduction and protecting axons from damage, to providing them with metabolic support [1]. In the central nervous system (CNS) myelin is produced by specialized glial cells called oligodendrocytes. Given the numerous essential functions myelin plays, it is not surprising that impaired formation of myelin, damage to myelin, or loss of myelin is associated with a number of neurodevelopmental, neuropsychiatric, and neurodegenerative disorders, including leukodystrophies [2], autism [3], schizophrenia [4, 5], motor neuron disease [6], and multiple sclerosis [7].

A striking aspect of myelin is that, unlike other components of the nervous system, it can be regenerated following demyelinating insults. Myelin regeneration, known as remyelination, is carried out

David A. Lyons and Linde Kegel (eds.), *Oligodendrocytes: Methods and Protocols*, Methods in Molecular Biology, vol. 1936, https://doi.org/10.1007/978-1-4939-9072-6_13,

by newly generated oligodendrocytes, derived from resident oligodendrocyte precursor cells [8]. Importantly, remyelination is associated with reduced loss of axons [9] and an improved outcome for multiple sclerosis (MS) patients [10]. Animal models where demyelination is induced by administering toxins such as lysolecithin, ethidium bromide, or cuprizone (*see* Chapter 21) have further demonstrated that remyelination can restore conduction to axons and consequently motor function to animals [11, 12] and protect axons from degeneration [13, 14]. Thus, enhancing remyelination is a key therapeutic target in the effort to develop improved treatments for myelin disorders such as MS [8, 15].

Extensive research has used the aforementioned toxin-based models to study mechanisms of de- and remyelination and to identify factors that may regulate the success of remyelination. More recently, several mouse models have been published whereby targeted ablation of oligodendrocytes is achieved by conditional activation of diphtheria toxin [16–19] (*see* also Chapter 17). These too can be highly informative for studying aspects of demyelination, including the consequences of demyelination for axon health and the response of the immune system to demyelination.

However, it remains challenging to analyze the dynamic cellular interactions that occur following demyelination and during remyelination in mammalian models in vivo. For instance, very little is known about the effects of acute demyelination on the structure and composition of the underlying axon, or how various cells of the adaptive and innate immune system mediate and respond to demyelination, and influence remyelination. Such questions are of fundamental importance for designing effective interventions for demyelinating diseases. Thus, it would be of great benefit to have a system in which these questions could be addressed at the level of individual cells in vivo.

For this reason, we generated a system whereby we can specifically ablate mature oligodendrocytes in living zebrafish larvae, analogous to the Xenopus model described by Mannioui et al. (Chapter 14) (*see* also [20]). The larval zebrafish is an increasingly popular model to study developmental and regenerative events; in addition to the larvae being transparent and thus enabling live imaging, it is possible to obtain hundreds of embryos from single pair matings. These embryos develop rapidly such that by 5 days postfertilization (dpf) they have a functional and robustly myelinated nervous system. Furthermore, a growing battery of transgenic and genome-editing tools allows researchers to label cell types of interest, to turn on or off genes of interest, and to image the consequences in a living animal. Importantly, there is a well-defined population of large-caliber reticulospinal axons within the zebrafish spinal cord, the vast majority of which is known to be myelinated from approximately 4 dpf onward [21]. This is a pivotal consideration for studies of demyelination and remyelination, as it means that the majority large-caliber axons observed in the spinal cord following the demyelination method of this chapter can be regarded as truly demyelinated, rather than not-yet-myelinated.

A demyelinating fish model could make it possible to study demyelination and remyelination at a resolution that has thus far remained inaccessible in rodent models. For example, it could be possible to follow a single axon over time as it undergoes demyelination and remyelination, and relate its health status to its myelination status. Experimenters could also attempt interventions, both delaying and speeding up remyelination, and examine the effects of these on the axon's health, for example, in terms of organelle transport and mitochondrial activity. These are crucial issues to consider, as it is currently not known how soon following demyelination a single axon begins to show pathology, and how far the pathology can proceed before it becomes irreversible by remyelination.

Furthermore, larval zebrafish are increasingly used as tools for high-throughput chemical screening. Since embryos can be obtained in large numbers and develop outside the mother, it is relatively straightforward to deliver drug treatments to large cohorts of animals, and image the effects of these treatments on the phenotype of interest [22]. Our laboratory carries out large-scale chemical screens to identify modulators of developmental myelination [23]. However, since remyelination occurs in a distinctly different context from developmental myelination [8], a larval zebrafish in which demyelination can be induced may provide an opportunity to identify compounds that can regulate many biological events required for efficient remyelination in vivo, to complement the in vitro based assays that have identified compounds that promote oligodendrocyte differentiation [15].

Our zebrafish model of demyelination and remyelination makes use of the nitroreductase-metronidazole system [24, 25]. We generated a transgenic zebrafish line, Tg(mbp: mCherry-NTR) whereby the bacterial enzyme nitroreductase (NTR) is expressed under to control of the myelin basic protein (mbp) promoter [21], and thus expressed in mature myelinating oligodendrocytes (as well as Schwann cells, the myelinating cells of the peripheral nervous system). The NTR coding sequence is fused to that of the fluorescent protein mCherry, so that NTR-expressing cells can be easily identified within the animal. By itself, NTR expression does not affect the oligodendrocytes, but if the larvae are treated with the nontoxic prodrug metronidazole (Mtz), the NTR enzyme digests Mtz and converts it to a potent DNA interstrand cross-linking agent [25], causing cell death of the NTR-expressing cells.

In our system, we treat Tg(mbp:mCherry-NTR) larvae with Mtz from 5 to 7 dpf. This protocol consistently ablates two-thirds of oligodendrocytes while not harming the animals otherwise.

Myelin sheaths continue to deteriorate after the end of the treatment, such that 7 days later, demyelination of two-thirds of the reticulospinal axons in the spinal cord is observed by electron microscopy. This demyelination is still observed at 11 days post-treatment, but this is not accompanied by discernible loss of axons, or obvious axon pathology such as altered mitochondrial content. By 16 days after Mtz treatment, robust recovery has occurred, with myelin thickness restored to control levels. Oligodendrocyte numbers are not restored to control levels at this point. During the demyelinated phase, we observe a striking increase in microglia and macrophages in the spinal cord [26].

We propose that the Tg(mbp:mCherry-NTR) model of demyelination and remyelination will prove a valuable tool for studying demyelination and remyelination in vivo and for testing potential compounds for enhancing remyelination.

2 Materials

2.1 Induction of Demyelination in Larval Zebrafish

1. Female and male adult Tg(mbp:mCherry-NTR) zebrafish [26].
2. Embryo medium (10 mM HEPES-buffered E3 embryo medium; 5 mM NaCl, 0.17 mM KCl, 0.33 mM $CaCl_2$, 0.33 mM $MgSO_4$).
3. Disposable plastic petri dishes (50 mL).
4. Plastic Pasteur pipettes.
5. P20, P200, and P1000 pipettes.
6. Dimethyl sulfoxide (DMSO).
7. 5 mM solution of metronidazole in 1% DMSO and embryo medium
8. Tricaine (168 μg/mL 3-amino benzoic acid ethyl ester).
9. A fluorescent stereomicroscope (we use a Leica M165 FC).

2.2 Mounting Larvae and Live Imaging

1. 1.3% Low-melting-point agarose.
2. Glass coverslips.
3. Glass slides.
4. High-vacuum grease.
5. Confocal microscope (we use Zeiss LSM 710, 780, and 880 microscopes).

2.3 Preparation of Larvae for Transmission Electron Microscopy (TEM)

1. Tricaine (*see* above).
2. Primary fixative solution, consisting of:
   (a) 4% paraformaldehyde (PFA).
   (b) 2% glutaraldehyde.

   (c) 0.1 M sodium cacodylate, pH 7.4.
   (d) 18.2 mΩ deionized water.
3. 2 mL Eppendorf tubes or 200 μL PCR tubes.
4. Fine-pointed liquipettes.
5. Water bath at 15 ± 3 °C.
6. Microwave with Inverter technology.
7. Shaker.
8. Secondary fixative solution consisting of
   (a) 2% osmium tetroxide, aqueous solution.
   (b) 0.1 M imidazole.
   (c) 0.1 M sodium cacodylate, pH 7.4.
   (d) 18.2 mΩ deionized water.
9. Saturated aqueous solution of uranyl acetate.
10. 50%, 70%, 95% and 100% EM grade ethanol.
11. EM grade acetone.
12. EMBed 812 resin.
13. Silicone molds.
14. Oven that can be kept at 65 °C.
15. Razor blades.
16. Glass knives.
17. An ultramicrotome (we use a Reichert Jung Ultracut).
18. Diamond knife.
19. Chloroform.
20. A thin wooden stick with an eyelash glued to it.
21. Magic loop.
22. Toluidine blue O with 2% Borax.
23. Copper electron microscopy grids with hexagonal grid patterns (200 mesh grids).
24. Sato's lead stain, consisting of
   (a) 0.2 g lead citrate.
   (b) 0.15 g lead nitrate.
   (c) 0.15 g lead acetate.
   (d) 1 g sodium citrate.
   (e) 48.2 mL boiled and cooled 18.2 Ω deionized water.
   (f) 1.8 mL 5 M NaOH solution prepared using the boiled and cooled 18.2 Ω deionized water.
25. 1 mL syringes
26. Grid storage boxes.

2.4 Acquisition and Analysis of TEM Data

1. Electron microscope (we use a Jeol JEM-1400 Plus) with a digital camera.
2. ImageJ software.
3. Adobe Photoshop software.

2.5 Preparation, Staining, and Imaging of Cryosections

1. 4% paraformaldehyde (PFA).
2. Phosphate-buffered saline (PBS).
3. Eppendorf tubes or PCR strips.
4. Liquipettes.
5. 30% sucrose.
6. Optimal cutting temperature (OTC) medium.
7. 50 mL Falcon tubes.
8. 2-Methylbutane.
9. Dry ice.
10. Liquid nitrogen.
11. −80 °C freezer.
12. Cryostat (we use a Leica CM3050S).
13. Superfrost Plus slides.
14. Liquid Blocker super pap pen.
15. Hoechst dye, diluted 1:1000 in PBS.
16. Triton-X.
17. Fluoromount aqueous mounting medium.
18. Immersion oil.

3 Methods

All animal husbandry and experiments involving live animals must be performed according to institutional and national guidelines.

3.1 Ablation of Oligodendrocytes

Adult Tg(mbp:mCherry-NTR) animals may be incrossed, or outcrossed to other transgenic lines if labeling of other cell types is required by the experiment.

1. Set up pair matings to obtain fertilized eggs.
2. Collect eggs and keep in a 28.5 °C incubator in petri dishes containing embryo medium.
3. At 5 dpf, anesthetize larvae with tricaine diluted in embryo medium.
4. Identify those larvae that show strong expression of the mbp: mCherry-NTR transgene (as well as any other transgenes involved in the experiment) by manually sorting under a fluorescent stereomicroscope (*see* **Note 1**).

5. Prepare a 5 mM solution of metronidazole (Mtz) in 1% DMSO and embryo medium.
6. Place the selected larvae in 50 mL petri dishes containing either the metronidazole solution or only 1% DMSO (*see* **Note 2**).
7. Keep larvae in Mtz/DMSO in a 28.5 °C incubator for 48 h. After 24 h of treatment, replace the solutions with freshly prepared solutions.
8. At the end of the 48 h treatment, either proceed to experiment directly, or if a recovery period is required, transfer larvae into 1-L nursery tanks in a dedicated aquatic facility, and maintain them there for the requisite number of days.

3.2 Live Imaging

1. Anesthetize living zebrafish larvae in tricaine.
2. Embed larvae in a drop of 1.3% low-melting-point agarose and align for imaging (*see* **Note 3**).
3. As the agarose solidifies, create a well of high-vacuum grease on a glass slide by piping a rectangular shape on the slide. Fill this well with tricaine-containing embryo medium.
4. Press the coverslip with the embedded larvae onto the well, such that the larvae within the agarose are bathed in tricained embryo medium.
5. Image the embedded larvae on a confocal microscope (*see* **Note 4**).
6. When finished, detach the coverslip from the slide, add embryo medium onto the agarose drop, and use sharp blades to extract the larvae from the agarose. Either return larvae to petri dishes or tanks for subsequent experiments or, if the experiment is completed, cull by administering neat tricaine.
7. Carry out image analysis using ImageJ software (or similar), e.g., the counter plugin for counting cell bodies.

3.3 Preparation of Larvae for TEM

1. Place living larvae in either 2 mL Eppendorf tubes or PCR strips, depending on whether it is essential to identify individual larvae.
2. Terminally anesthetize larvae with tricaine.
3. Prepare primary fixative solution.
4. Remove tricaine and immediately replace it with primary fixative solution.
5. As soon as possible, perform microwave stimulation using a microwave with "Inverter" technology: 1 min at 100 W, 1 min out of the microwave, 1 min at 100 W, 1 min out. Then adjust the power setting to 450 W, and stimulate samples for 20 s, then take out for 20 s; repeat this five times. During all microwave stimulation, keep larvae in an ice bath of 15 °C ± 3 °C to prevent them from overheating (*see* **Note 5**).

6. Incubate samples in primary fix for 1–14 days at 4 °C.
7. When ready to continue processing, wash samples three times for 10 min in 0.1 M sodium cacodylate buffer on a shaker at gentle speed (~30 rpm).
8. Use sharp blades to dissect away the heads of the fixed larvae (*see* **Note 6**).
9. Prepare the secondary fixative by mixing equal volumes of 4% osmium tetroxide and a solution of 0.2 M imidazole and 0.2 M sodium cacodylate (*see* **Note 7**).
10. Immerse samples in secondary fixative and microwave-stimulate again: 1 min at 100 W, 1 min out, 1 min at 100 W, 1 min out. Change the power setting to 450 W. Stimulate samples for 20 s and take out for 20 s; repeat this five times.
11. Incubate the microwaved samples on a shaker at gentle speed at room temperature for at least 3 h.
12. Wash the samples in water five times for 10 min each.
13. Perform an en bloc stain by immersing samples in aqueous saturated uranyl acetate (8% w/v) and microwave-stimulating them (450 W for 1 min, out for 1 min, 450 W for 1 min).
14. Prepare an ethanol concentration gradient series (50%, 70%, 95%, and 100% EM grade ethanol) for the dehydration steps. Remove uranyl acetate and replace with 50% ethanol. Microwave-stimulate for 45 s at 250 W, and incubate on a shaker at room temperature for 10 min. Repeat for subsequent concentrations. At 100% ethanol, stimulate the samples for 1 min at 250 W, take out for 1 min and stimulate again at 250 W for 1 min. Then incubate for 10 min at room temperature on a shaker. Repeat three times.
15. Proceed to acetone dehydration: immerse samples in fresh EM grade acetone, stimulate at 250 W for 1 min, take out for 1 min, stimulate at 250 W for 1 min, and incubate on a shaker at room temperature for 10 min. Repeat three times.
16. During the acetone dehydration, prepare a pre-infiltration solution consisting of 50% acetone and 50% EMbed 812 resin. Following the last incubation period, remove the acetone and replace with the acetone-EMbed 812 solution. Leave overnight in room temperature on shaker.
17. The next day, replace the 50/50 solution with 100% EMbed 812 resin, and leave the samples to incubate in a fume hood overnight, with the lids of the tubes open to allow residual acetone to evaporate.
18. The next day, align samples in EMbed-filled silicone molds, ensuring that larvae lie parallel to the bottom of the mold (*see* **Note 8**).

19. Bake at 65 °C for at least 72 h to solidify the resin into blocks.
20. Trim the blocks using razor blades until the region of interest is reached. We trim starting at the tip of the tail, in an anterior direction, down to somite 16 (*see* **Note 9**).
21. From this region, use a diamond knife to cut 70 nm thick silver sections on an ultramicrotome. Flatten sections using a cotton bud dipped in chloroform, then lift them from the water using a magic loop and place on electron microscopy grids.
22. Stain sections on grids, first with a 1:1 solution of EM grade ethanol and saturated uranyl acetate. Stain for 5 min and wash with a running solution of 50% ethanol, 50% water for 30 s.
23. Stain sections with Sato's lead stain. Prepare this according to the recipe given above. Prior to combining the ingredients, heat the lead citrate on a foil boat placed on a hot plate in the fume hood, until it changes color from bright white to a dull gray. As soon as this color change occurs, remove lead citrate from the heat. Combine all ingredients in a beaker while stirring; solution should become clear. Aliquot the stain into 1 mL syringes, cap them with parafilm, and store protected from light at 4 °C.
24. Stain sections on grids with Sato's lead stain for 5 min, then wash under running water for 30 s and keep in storage boxes until imaging.
25. Acquire images using an electron microscope (*see* **Note 10**).
26. Quantify demyelination by counting the total numbers of axons, total numbers of myelinated axons, and the percentage of myelinated axons.

3.4 Preparation, Staining, and Imaging of Cryopreserved Tissue

1. Terminally anesthetize larvae in 4% PFA in PBS. Leave at 4 °C overnight.
2. Wash four times with PBS on a shaker for 5 min each.
3. Remove PBS and replace with 30% sucrose solution. Samples should float on the surface of the solution.
4. Leave at 4 °C overnight, or until samples have sunk to the bottom of the tubes.
5. Embed samples in silicone molds containing OTC medium.
6. During embedding, keep a Falcon tube containing 30 mL 2-methylbutane immersed in liquid nitrogen. Immediately after embedding, snap freeze each sample in this 2-methylbutane for 5 s.
7. Leave on dry ice until all samples have been snap frozen, and then detach block from silicone molds and transfer to labeled Eppendorf tubes.
8. Store blocks at −80 °C until cutting.

9. Cut 14 μm thick sections using a cryostat. We start collecting sections at somite 5 (*see* **Note 11**).
10. Collect sections onto Superfrost Plus slides.
11. In order to avoid sampling the same oligodendrocyte twice, be sure to leave at least 5 μm space between adjacent sections.
12. Store sections at −80 °C until staining and imaging.
13. Prior to imaging, perform a nuclear stain.
14. Prepare a 1:1000 dilution of Hoechst in PBS (*see* **Note 12**).
15. Draw a border around the sections on each slide using a Liquid Blocker pap pen.
16. Administer Hoechst solution to slides and leave, protected from light, for 5 min.
17. Wash 3× 5 min with 0.02% Triton-X in PBS, and 2× 5 min with PBS.
18. Mount slides using Fluoromount and a coverslip.
19. Image on a confocal microscope (*see* **Note 13**).

4 Notes

1. To align larvae for imaging, they should be laid on their sides on the coverslip and gently pressed against the coverslips along the length of their trunk. This ensures that they lay as close to the coverslip as possible.
2. Humanely cull the discarded larvae by placing them in neat tricaine.
3. Depending on the experiment, Mtz treatment can also be carried out in multiwell plates.
4. We typically image a four-somite stretch of the spinal cord to analyze the effects of metronidazole treatment on oligodendrocytes and myelin in Tg(mbp:mCherry-NTR) larvae.
5. The microwave stimulation is highly recommended, as a brief interval of microwaving has been found to markedly improve quality of zebrafish tissue fixation, although it is not known precisely what the underlying mechanism of this benefit is.
6. Cutting off the heads is considered to improve tissue permeation by the secondary fixative. You may also cut off the heads prior to primary fixation, but if you do so, be sure to proceed to fixation as soon as possible, before the tissue begins to degenerate.
7. Ensure that the pH of the latter solution is 7.4, as an altered pH will cause the osmium to precipitate out of solution and adversely affect fixation.

8. It is also recommended to include a printed label within the mold alongside the larva.
9. This is readily identifiable as location of the urogenital opening in the larva. This ensures that sections from different animals are always obtained from comparable regions.
10. We typically image an entire spinal cord, or a hemi-spinal cord, both at low magnification (1600×) and high magnification (4800–5700×). For high-magnification views of the entire spinal cord (or parts thereof), we align individual tiles using the Photomerge tool in Adobe Photoshop.
11. This region is readily identifiable using anatomical clues under a stereomicroscope, as it is the site immediately posterior to the swim bladder.
12. In order to visualize oligodendrocytes from cryosections, it is not necessary to use antibodies against oligodendrocyte antigens, as the endogenous mCherry signal from the mbp: mCherry-NTR transgene survives the fixation protocol reasonably well and can readily be used to label oligodendrocytes and myelin.
13. A 40× magnification allows a whole spinal cord to be captured in one image at good resolution. We image four sections per sample, count the number of oligodendrocytes from each, and take the mean of these to represent the number oligodendrocytes for that sample.

References

1. Nave K-A, Werner HB (2014) Myelination of the nervous system: mechanisms and functions. Annu Rev Cell Dev Biol 30:503–533. https://doi.org/10.1146/annurev-cellbio-100913-013101
2. Boespflug-Tanguy O, Labauge P, Fogli A, Vaurs-Barriere C (2008) Genes involved in leukodystrophies: a glance at glial functions. Curr Neurol Neurosci Rep 8:217–229
3. Pacey LKK, Xuan ICY, Guan S et al (2013) Delayed myelination in a mouse model of fragile X syndrome. Hum Mol Genet 22:3920–3930. https://doi.org/10.1093/hmg/ddt246
4. Hercher C, Chopra V, Beasley CL (2014) Evidence for morphological alterations in prefrontal white matter glia in schizophrenia and bipolar disorder. J Psychiatry Neurosci 39:376–385
5. Windrem MS, Osipovitch M, Liu Z et al (2017) Human iPSC glial mouse chimeras reveal glial contributions to schizophrenia. Cell Stem Cell 21:195–208.e6. https://doi.org/10.1016/j.stem.2017.06.012
6. Kang SH, Li Y, Fukaya M et al (2013) Degeneration and impaired regeneration of gray matter oligodendrocytes in amyotrophic lateral sclerosis. Nat Neurosci 16:571–579. https://doi.org/10.1038/nn.3357
7. Thompson AJ, Baranzini SE, Geurts J et al (2018) Multiple sclerosis. Lancet 391:1622–1636. https://doi.org/10.1016/S0140-6736(18)30481-1
8. Franklin RJM, Ffrench-Constant C (2017) Regenerating CNS myelin—from mechanisms to experimental medicines. Nat Rev Neurosci 18:753–769. https://doi.org/10.1038/nrn.2017.136
9. Kornek B, Storch MK, Weissert R et al (2010) Multiple sclerosis and chronic autoimmune encephalomyelitis. Am J Pathol 157:267–276. https://doi.org/10.1016/S0002-9440(10)64537-3
10. Patrikios P, Stadelmann C, Kutzelnigg A et al (2006) Remyelination is extensive in a subset of

multiple sclerosis patients. Brain 129:3165–3172. https://doi.org/10.1093/brain/awl217

11. Smith KJ, Blakemore WF, McDonald WI (1981) The restoration of conduction by central remyelination. Brain 104:383–404
12. Jeffery ND, Blakemore WF (1997) Locomotor deficits induced by experimental spinal cord demyelination are abolished by spontaneous remyelination. Brain 120:27–37. https://doi.org/10.1093/brain/120.1.27
13. Irvine KA, Blakemore WF (2008) Remyelination protects axons from demyelination-associated axon degeneration. Brain 131:1464–1477. https://doi.org/10.1093/brain/awn080
14. Mei F, Lehmann-Horn K, Shen Y-AA et al (2016) Accelerated remyelination during inflammatory demyelination prevents axonal loss and improves functional recovery. Elife 5: e18246. https://doi.org/10.7554/eLife.18246
15. Cole KLH, Early JJ, Lyons DA (2017) Drug discovery for remyelination and treatment of MS. Glia 65:1565–1589. https://doi.org/10.1002/glia.23166
16. Traka M, Arasi K, Avila RL et al (2010) A genetic mouse model of adult-onset, pervasive central nervous system demyelination with robust remyelination. Brain 133:3017–3029. https://doi.org/10.1093/brain/awq247
17. Traka M, Podojil JR, McCarthy DP et al (2015) Oligodendrocyte death results in immune-mediated CNS demyelination. Nat Neurosci 19:65–74. https://doi.org/10.1038/nn.4193
18. Pohl HBF, Porcheri C, Mueggler T et al (2011) Genetically induced adult oligodendrocyte cell death is associated with poor myelin clearance, reduced remyelination, and axonal damage. J Neurosci 31:1069–1080. https://doi.org/10.1523/JNEUROSCI.5035-10.2011
19. Locatelli G, Wörtge S, Buch T et al (2012) Primary oligodendrocyte death does not elicit anti-CNS immunity. Nat Neurosci 15:543–550. https://doi.org/10.1038/nn.3062
20. Kaya F, Mannioui A, Chesneau A et al (2012) Live imaging of targeted cell ablation in Xenopus: a new model to study demyelination and repair. J Neurosci 32:12885–12895. https://doi.org/10.1523/JNEUROSCI.2252-12.2012
21. Almeida RG, Czopka T, ffrench-Constant C, Lyons DA (2011) Individual axons regulate the myelinating potential of single oligodendrocytes in vivo. Development 138:4443–4450. https://doi.org/10.1242/dev.071001
22. MacRae CA, Peterson RT (2015) Zebrafish as tools for drug discovery. Nat Rev Drug Discov 14:721–731. https://doi.org/10.1038/nrd4627
23. Early JJ, Cole KL, Williamson JM et al (2018) An automated high-resolution in vivo screen in zebrafish to identify chemical regulators of myelination. Elife 7:4443. https://doi.org/10.7554/eLife.35136
24. Pisharath H, Parsons MJ (2009) Nitroreductase-mediated cell ablation in transgenic zebrafish embryos. Methods Mol Biol (Clifton, NJ) 546:133–143. https://doi.org/10.1007/978-1-60327-977-2_9
25. Curado S, Anderson RM, Jungblut B et al (2007) Conditional targeted cell ablation in zebrafish: a new tool for regeneration studies. Dev Dyn 236:1025–1035. https://doi.org/10.1002/dvdy.21100
26. Karttunen MJ, Czopka T, Goedhart M et al (2017) Regeneration of myelin sheaths of normal length and thickness in the zebrafish CNS correlates with growth of axons in caliber. PLoS One 12:e0178058. https://doi.org/10.1371/journal.pone.0178058

Chapter 14

Conditional Demyelination and Remyelination in a Transgenic *Xenopus laevis*

Abdelkrim Mannioui and Bernard Zalc

Abstract

Multiple sclerosis (MS) is the first cause of acquired disability progression in the young adult. Pathology of MS associates inflammation, demyelination, and neurodegeneration. The development of immunotherapies, by reducing the relapse rate, has profoundly impacted short-term prognosis and patients' quality of life. These anti-inflammatory medications, however, have not proven to be sufficient to prevent long-term disability progression, resulting from axonal transection and neuronal damage, consequences of prolonged demyelination. Promoting remyelination is therefore a key therapeutic strategy to limit handicap progression, and represent the major therapeutic challenge in MS. Here we present a simple, rapid, and cost-effective experimental model developed in *Xenopus laevis* to screen in vivo molecules promoting remyelination.

Key words Multiple sclerosis, Myelin, Tadpole, Transgenesis, Oligodendrocyte, MBP, GFP, Nitroreductase, Metronidazole

1 Introduction

During evolution, given the physical constrain imposed to the brain and spinal cord by the skull and vertebrae, acquisition of myelin has been the vertebrates solution to adapt to their dramatic increase in body length. By permitting a 50–100-fold acceleration of propagation of the nerve influx and providing the necessary metabolite support along very long axons, acquisition of myelin 425MY ago has indeed been an ideal adaptive solution to the large body size of vertebrates [1, 2]. In the central nervous system myelin sheath is synthesized by the oligodendrocyte. Loss of oligodendrocytes in diseases such as multiple sclerosis (MS) results in demyelination of axons, a situation which impaired propagation of nerve impulse (conduction block), causing an alteration of the normal function. MS is an autoimmune demyelinating disease affecting the young adult. In MS, it is generally accepted that demyelination process is the consequence of inflammatory attacks directed against this novel

David A. Lyons and Linde Kegel (eds.), *Oligodendrocytes: Methods and Protocols*, Methods in Molecular Biology, vol. 1936, https://doi.org/10.1007/978-1-4939-9072-6_14,

vertebrate evolutionary improvement. Furthermore, oligodendrocytes, and the myelin sheath itself, have been shown to provide metabolic support to the axons. Loss of the myelin sheath weakens the axon leading to axonal transection, the consequence of which is permanent handicap. For the past 15–20 years significant therapeutic progress has been made addressing the inflammatory component of MS. These treatments have proven efficient in diminishing the number of relapses, but, unfortunately, their outcome on the clinical evolution is less evident. This imposes the need for development of myelin repair strategies. One way of facilitating the repair of demyelinated lesions would be to promote endogenous oligodendrocyte development and their migration to the lesion sites and finally favor the successful re-ensheathing of myelin-striped axons by oligodendrocyte processes. The existing animal models of MS are not ideally suited to follow the process of demyelination and remyelination and there is a need to develop simple strategies adapted to larger scale screening of compounds favoring myelin repair [3]. Toxin-induced models of demyelination are not primarily intended as models that provide a disease facsimile of MS, but they allow the fundamental biology of remyelination to be explored without the confounding and complicating involvement of an autoimmune process [4]. Although rodents may be considered as the "standard" experimental model to investigate demyelination and remyelination, they are costly, time consuming, and not ideal for longitudinal investigation in the same animal.

Small organisms, such as amphibians, are increasingly being used in various stages of drug development processes and constitute a highly cost-effective alternative to mammals [5]. Zebrafish larvae like *Xenopus* tadpoles are transparent animals that develop outside the mother and are therefore particularly suited to investigate developmental processes. The use of transgenic larval zebrafish in a drug-reprofiling screen allows rapid in vivo screening and has been used successfully as an efficient way of identifying new indications for existing drugs [6–8]. To study myelination, demyelination, and remyelination we have produced a transgenic *Xenopus laevis* (*MBP-GFP-NTR*) designed to express in oligodendrocyte the fluorescent reporter GFP fused to *E. coli* nitroreductase (NTR), under the control of the 1.9 kb proximal portion of mouse *MBP* regulatory sequence. Cells expressing NTR are specifically ablated following treatment with metronidazole (a substrate of NTR) [9].

What is the advantage of the *Xenopus* model? In terms of myelin biology, the principle advantage of *Xenopus* resides in the fundamental difference in the molecular composition of myelin between fish and tetrapods. Fish, including zebrafish, use P0 as a major myelin protein both in the CNS and in the PNS. In contrast, starting with amphibians (including *Xenopus*), there has been a dramatic evolutionary change in myelin composition between

PNS and CNS. In tetrapods (including amphibians and mammals) the major myelin protein in the PNS remains P0, while in the CNS it is PLP and MBP. This event occurred in amphibians and has been maintained throughout evolution in mammals, including humans. This is well illustrated in our transgenic *Xenopus* since we are using an enhancer of mouse *MBP* to drive the transgene in the *Xenopus*, based on highly conserved sequence homology, from the *Xenopus* to human, of regulatory noncoding DNA portion upstream of the coding sequence. In this respect the large similarity of myelin proteins and sequences homologies among tetrapods justify the choice of *Xenopus* as an experimental animal model to investigate myelin biology.

We have previously shown that, as predicted, the GFP reporter is faithfully expressed in mature myelin-forming oligodendrocytes [9]. Furthermore, as illustrated here, treatment of *MBP-GFP-NTR* tadpoles with metronidazole induced a selective demyelination, reversible upon cessation of treatment. Thanks to GFP expression demyelination and remyelination could be monitored longitudinally by repeated examination of the same *Xenopus* embryo. This spontaneous remyelination is dramatically increased upon exposure of the *MBP-GFP-NTR Xenopus* tadpoles to candidate molecules to promote remyelination, a proof of concept that this conditional oligodendrocyte ablation model is ideally suited to screen for drugs favoring remyelination [10].

2 Materials

2.1 Xenopus laevis Holding Conditions

1. Water temperature range 20 ± 2 °C.
2. Room temperature 20 ± 2 °C.
3. pH: 7.5 (range 6.5–8).
4. Conductivity: 1600 μS (range 1200–1800).

2.2 Induction of Ovulation and Natural Mating

1. Female WT and transgenic male (*MBP-GFP-NTR*+/−).
2. Nets.
3. Human chorionic gonadotropin (HCG).
4. 1 mL syringes.
5. Aquarium.
6. Refrigerated Incubator or room temperature 20 °C.
7. Air pump with bubble stone.

2.3 Eggs Collection and De-Jellying

1. Plastic transfer pipettes.
2. Glass beaker 100 mL.
3. Petri dishes.

4. MMR 0.1×.
5. Cysteine 2% in MMR 0.1×.
6. Binocular microscope.
7. Timer.
8. Refrigerated Incubator 20 °C.
9. Gentamicin.

2.4 Rearing Xenopus Embryos

1. Aquarium.
2. Air pump with bubble stone.
3. Sera micron.
4. Purified water (after filtration and dechlorination of tap water).
5. Refrigerated Incubator 20 °C.

2.5 Staging and Sorting of GFP+ Embryos

1. Normal Table of *Xenopus laevis* by Nieuwkoop and Faber.
2. Macroscope Nikon AZ100.

2.6 Solutions and Preparation

1. Metronidazole solution: Metronidazole, Dimethyl sulfoxide (DMSO), Milli-Q water, in 1× MMR.
2. MMR (Marc's Modified Ringers) 10× stock solution (1 L): 58.44 g NaCl (1 M), 1.5 g KCl (20 mM), 1.2 g $MgSO_4$ (anhydrous) or 2.5 g $MgSO_4 \cdot 7H_2O$ (10 mM), 2.94 g $CaCl_2 \cdot 2H_2O$ (20 mM), 11.91 g Hepes (50 mM) pH 7.5–8, 2 mL 0.5 M EDTA (1 mM). Autoclave to sterilize.
3. Cysteine 2% (100 mL): 4 g L-cysteine powder, MMR 0.1×, pH 7.8–8.
4. MS-222 stock solution (5%): Tricaine methane sulfonate MS-222, Milli-Q water, sodium bicarbonate ($NaHCO_3$) to pH to 7.4.

2.7 Analyzing Experiments

1. Macroscope Nikon AZ100.
2. 6 well plate.
3. MS-222 0.1%.

3 Methods

3.1 Fertilization

In our colony, males and females *Xenopus* are maintained in separate tanks and kept heterozygous for the *MBP-GFP-NTR* transgene [9].

1. The evening before mating the females are primed by ip injection of 600 units of human Chorionic Gonadotropin and males received 10 units.

2. Set up single-pair mating in tanks containing frog system water maintained at 20 °C.
3. Allow frogs to mate for several hours.
4. After amplexus, when a sufficient number of eggs have been layed, remove the frogs to separate tanks and monitor fertilized embryos, i.e., eggs reaching stage 2 or more cells.
5. Fertilized eggs are de-jellied by adding cysteine (2%) for 10 min and rinsed three times in MMR 0.1×.
6. Living eggs are sorted out and maintained in groups of 20 in a Petri dish in MMR 0.1×.

3.2 Rearing and Staging of Embryo

Development of *Xenopus* embryo is temperature dependent. As mentioned above in our facility we keep the animals at 20 ± 2 °C on a 12-h light/12-h dark cycle (except during the demyelination treatment with metronidazole, which is conducted at 23 °C and in the dark).

1. *Xenopus* tadpoles were staged according to the normal table of *Xenopus laevis* (Daudin) of Nieuwkoop and Faber [11].

   (For digitized images and developmental data see <http://www.xenbase.org/anatomy/alldev.do>).
2. After reaching stage 48 embryos are fed twice a day with Sera® micron and water is changed every other day.

3.3 Transfer of Tadpoles

Transfer of tadpoles from one environment to another is by aspirating into a plastic Pasteur pipette, which tip is large enough to allow the tadpole to be aspirated before being transferred (Fig. 1).

3.4 Anesthesia

Tadpoles are transferred from their aquarium into either a 12 or 6 well culture dish (depending on the stage, i.e., the size of the animal) containing MS 222 (Tricaine 0.1%). Anesthesia is achieved when tadpoles stop swimming.

3.5 Evaluation of the Number of GFP+ Cells Per Optic Nerve

1. In each well, tadpoles are placed belly on the bottom. The only precaution is that the head should be as horizontal as possible so that optic nerve can easily be observed.
2. Put plate under the macroscope and identify the optic nerves. To identify the nerve, it may be easier to start from the eye, which is easily recognizable being the large dark structure of the lateral side of the head.
3. The optic nerve runs from the eye toward the brain (Fig. 2).
4. Next to the optic nerve muscles and vessels are located and under bright light you will see cells circulating through into the vessels.
5. As soon as you switch for blue light (fluorescence filters) green cells present in the optic nerve can be observed (Figs. 2 and 3).

Fig. 1 A stage 50 *Xenopus laevis* albinos tadpole in the tip of a plastic Pasteur pipette

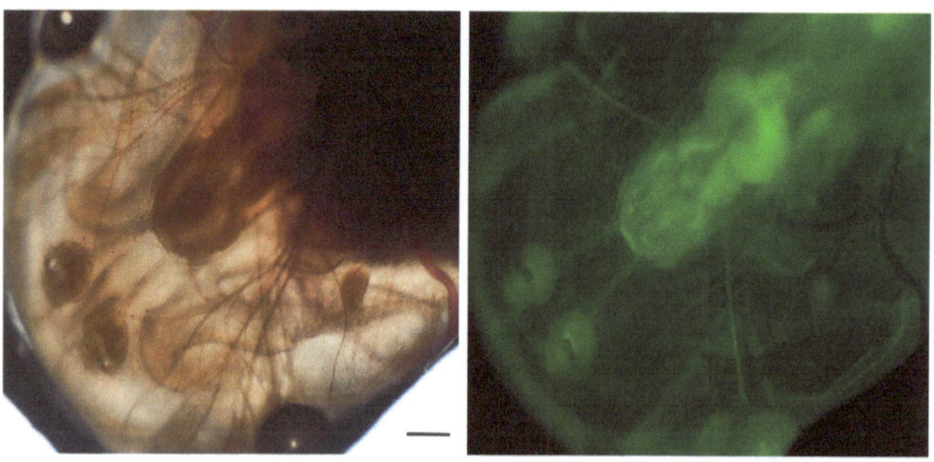

Fig. 2 Head of a stage 50 *MBP-GFP-NTR* tadpole viewed under a macroscope with bright light on the left and fluorescent light on the right. Even at this low magnification, myelinating oligodendrocytes are visible (bright green dots) in the optic nerves. In contrast, no fluorescent cells are detected in the olfactory nerves, which are never myelinated. Scale bar = 500 μm

6. The number of GFP expressing oligodendrocytes in each optic nerve is counted from the emergence of the nerve on the eye side toward the chiasm on the diencephalon side.
7. Since at these developmental stages the number of GFP+ cells per optic nerve varies between 25 ± 3 and 31 ± 2 (at stages

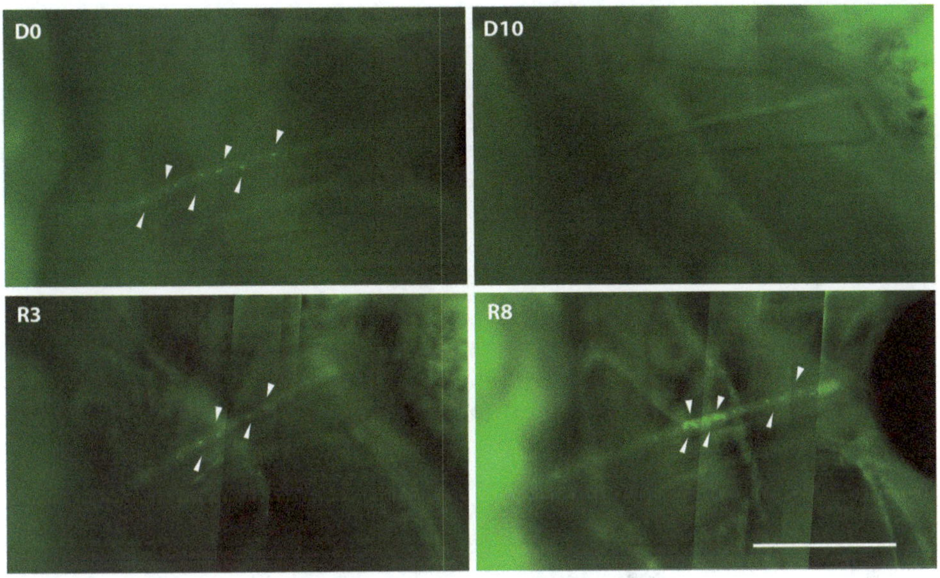

Fig. 3 Higher magnification of the optic nerve of stage 50 *MBP-GFP-NTR* tadpole viewed under a macroscope before demyelination (D0), after 10 days of treatment in metronidazole (10 mM) (D10) and at 3 (R3) and 8 (R8) days of recovery after cessation of the demyelinating treatment. White arrowheads point to some of the GFP+ oligodendrocytes. Note that at D10, oligodendrocytes are no longer detectable. Scale bar = 500 μm

52 and 55, respectively) the counting should be achieved in less than 5 min per tadpole, after which the animal is returned to normal water (*see* **Note 1**).

8. Each experiment requires a total of four episodes of counting. The number of GFP+ cells is counted before starting the demyelination (D0) at the end of demyelination period (D10), then after 3 and 8 days of recovery (R3 and R8, respectively) (Fig. 3).

3.6 Metronidazole Treatment

1. Metronidazole (10 mM, Fluka) is dissolved in filtered tap water containing 0.1% DMSO (Sigma Aldrich) and added in swimming water for 10 days.
2. Tadpoles are maintained in 600 mL of metronidazole solution (maximum 10 tadpoles/600 mL) at 23 °C in complete darkness (metronidazole is light-sensitive) and changed every 3 days throughout the duration of treatment, i.e., 10 days (*see* **Note 2**).
3. When tadpoles have reached stage 50–52, they are separated into three groups of six animals each (*see* **Note 3**).

   Group 1 are the controls: they are not submitted to metronidazole treatment, but the number of GFP+ cells is counted on D0, D10, R3, and R8. It is of note that during this period of 18 days the number of GFP+ cells is relatively stable and serves as reference for the two other groups (*see* **Note 1**).

Group 2 are the controls of spontaneous remyelination. Group 2 animals are demyelinated between D0 and D10 and at the end of metronidazole treatment they are returned in normal water for R3 and R8 counting episodes.

4. For regeneration experiments, metronidazole-treated animals are allowed to recover in either normal water or water containing compound(s) to be tested at different concentration in ambient laboratory lighting (12 h light 12 h dark).

3.7 Drug Treatment

1. Group 3 are the experimental tadpoles. Together with group 2 animals, group 3 embryos are exposed to metronidazole between D0 and D10 and then switched to normal water containing the molecule(s) to be tested or different concentrations of a given molecule. Each condition requires 6 tadpoles. The degree of remyelination of treated tadpoles, estimated by the number of GFP+ cells per optic nerve, is assayed after 3 (R3) and 8 (R8) days of treatment. In our model, drugs being tested are added directly in the aquarium water. In order to avoid using a too large quantity of the molecules to be tested, tadpoles are placed in a 50 mL beaker. Before running a complete dose response, in a preliminary experiment the concentration of the molecule to be tested is about ten times higher than the concentration used in vitro, for instance in an oligodendrocyte-neuron myelinating coculture system [12] (*see* **Note 4**).

3.8 Conclusion

Our *Xenopus* model should not be considered as a high-throughput screening tool, but rather as a medium-throughput screen. Indeed a skilled technician can easily screen ten different molecules at a time, within 13 days. High-throughput screens should be in vitro and candidate molecules identified in such in vitro tests need then be tested in vivo in our *Xenopus laevis* model. Our prediction is that starting from 10,000 molecules, chances are to identify between 100 and 500 hits using an in vitro high-throughput screen, which then could be tested in vivo in the transgenic *Xenopus* in approximately 3–8 months, respectively.

4 Notes

1. Under our experimental conditions it takes more than 2 weeks for tadpoles at stage 50 to reach stage 55. Therefore, the number of GFP+ cells per optic nerve is relatively stable during this period.
2. Treatment with metronidazole is dose and time dependent. When used at 15 mM a nearly complete demyelination is observed in 6 days, but this is at the expense of a larger number

of animals that will not survive the treatment. We prefer to use metronidazole at 10 mM for 10 days. In a recent publication nifurpurinol, another substrate for nitroreductase has been sucessfully used in the zebrafish at concentrations comprised between 1–4 μM [13]. We are in the process to test nifurpurinol in our Xenopus model.

3. When estimating the number of tadpoles necessary for an experiment, always add 10% more as a security since some animals may not survive during the whole course of the experiment, especially when comparing dose-response efficiency of drugs to be tested.
4. How to define the best concentration of a drug improving remyelination? After 10 days of demyelination, spontaneous remyelination is already observed at 3 days of recovery (R3) and is nearly complete after 8 days (R8). Therefore, in our hands the remyelination induced by different molecules tested (as assayed by the number of GFP+ cells per optic nerve) in comparison with controls was more evident at R3 than at R8 [10]. When, at a given concentration, remyelination induced by a molecule is statistically higher than the control at R3, it suggests that the effect is an acceleration of remyelination process. In contrast, an increase in remyelination at R8 would rather indicate the production of myelin on previously nonmyelinated axons, an effect which may not be desirable.

Acknowledgments

We are indebted to Drs. F. Kaya and A. Mazabraud for their contribution to the early steps of this work. This work was supported by Inserm, CNRS, UPMC, the program "Investissements d'Avenir" ANR-10-IAIHU-06, NeurATRIS, and ANR grant OLGA to B.Z.

References

1. Zalc B, Goujet D, Colman D (2008) The origin of the myelination program in vertebrates. Curr Biol 18:R511–R512
2. Salzer JL, Zalc B (2016) Myelination. Curr Biol 26:R971–R975. https://doi.org/10.1016/j.cub.2016.07.074
3. Miller RH, Fyffe-Maricich SL (2010) Restoring the balance between disease and repair in multiple sclerosis: insights from mouse models. Dis Model Mech 3:535–539. https://doi.org/10.1242/dmm.001958
4. Franklin RJM (2015) Regenerative medicines for remyelination: from aspiration to reality. Cell Stem Cell 16:576–577. https://doi.org/10.1016/j.stem.2015.05.010
5. De Smet F, Carmeliet P, Autiero M (2006) Fishing and frogging for anti-angiogenic drugs. Nat Chem Biol 2:228–229. https://doi.org/10.1038/nchembio0506-228
6. Buckley CE, Marguerie A, Roach AG et al (2010) Drug reprofiling using zebrafish identifies novel compounds with potential pro-myelination effects. Neuropharmacology 59:149–159
7. Fang Y, Lei X, Li X et al (2014) A novel model of demyelination and remyelination in a

GFP-transgenic zebrafish. Biol Open 4:62–68. https://doi.org/10.1242/bio.201410736

8. Kim S, Lee Y-I, Chang K-Y et al (2015) Promotion of remyelination by sulfasalazine in a transgenic zebrafish model of demyelination. Mol Cells 38:1013–1021. https://doi.org/10.14348/molcells.2015.0246
9. Kaya F, Mannioui A, Chesneau A et al (2012) Live imaging of targeted cell ablation in Xenopus: a new model to study demyelination and repair. J Neurosci 32:12885–12895. https://doi.org/10.1523/JNEUROSCI.2252-12.2012
10. Mannioui A, Vauzanges Q, Fini JB et al (2018) The Xenopus tadpole: an in vivo model to screen drugs favoring remyelination. Mult Scler 24:1421–1432. https://doi.org/10.1177/1352458517721355
11. Nieuwkoop PD (1994) Normal table of *Xenopus laevis* (Daudin). Garland Publishing Inc., New York. ISBN: 0-8153-1896-0
12. Lubetzki C, Demerens C, Anglade P et al (1993) Even in culture, oligodendrocytes myelinate solely axons. Proc Natl Acad Sci U S A 90:6820–6824
13. Bergemann D, Massoz L, Bourdouxhe J, Carril Pardo CA, Voz ML, Peers B (2018) Manfroid I. Nifurpirinol: A more potent and reliable substrate compared to metronidazole for nitroreductase-mediated cell ablations. Wound Repair Regen. Mar; 26(2):238–244. https://doi.org/10.1111/wrr.12633. Epub 2018 May 19

Chapter 15

Conditional Mutagenesis in Oligodendrocyte Lineage Cells

Sandra Goebbels and Klaus-Armin Nave

Abstract

Cell-type-specific gene targeting with the Cre/loxP system has become an indispensable technique in experimental neuroscience, particularly for the study of late-born glial cells that make myelin. A plethora of conditional mutants and Cre-expressing mouse lines is now available to the research community that allows laboratories to readily engage in in vivo analyses of oligodendrocytes and their precursor cells. This chapter summarizes concepts and strategies in targeting myelinating glial cells in mice for mutagenesis or imaging, and provides an overview of the most important Cre driver lines successfully used in this rapidly growing field.

Key words Transgenic mice, Cre, Cell-type-specific targeting, Oligodendrocyte, Myelin

1 Introduction

1.1 Genetics of Oligodendrocyte Lineage Cells

Mouse genetics has become a powerful technique to study oligodendrocyte function and myelination in vivo. Building on a rich body of work from the 1960s on naturally occurring mouse mutants with characteristic dysmyelinating phenotypes and illustrious names, such as *jimpy*, *rumpshaker*, *shiverer*, and *quaking* mice, oligodendrocyte and myelin research made early use of cDNA and genomic cloning techniques. This allowed researchers to work with genetically defined mouse mutants of cloned genes well before most other fields in the neuroscience. In the 1990s, the generation of targeted mutations in mice, by homologous genomic recombination and selection of correctly targeted embryonic stem (ES) cells [1] (schematically depicted in Fig. 1), became a revolutionary technique throughout biology, allowing the functional analysis of virtually any gene in vivo. Thus, a few years after its introduction as a technique [2], a plethora of newly cloned genes, predicted to be important for the development of oligodendrocyte lineage cells (OLC, in the following for oligodendrocyte precursor cells (OPC) and mature oligodendrocytes) and Schwann cells were successfully inactivated and their phenotypes dissected. Targeted genes

David A. Lyons and Linde Kegel (eds.), *Oligodendrocytes: Methods and Protocols*, Methods in Molecular Biology, vol. 1936, https://doi.org/10.1007/978-1-4939-9072-6_15,

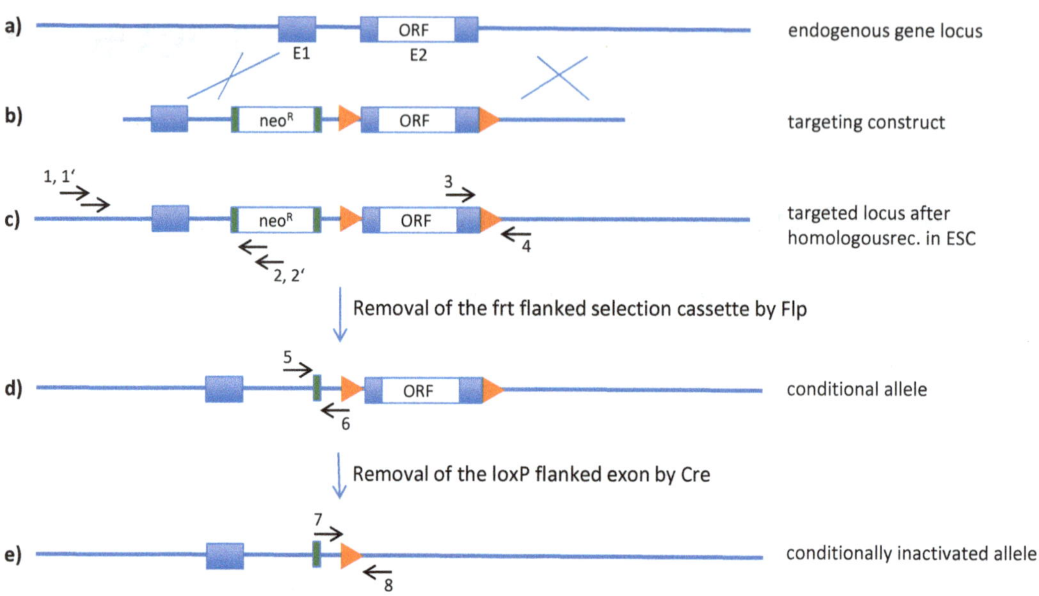

Fig. 1 Scheme for conditional targeting of a gene in mice. (**a**) Simplified genomic structure of the wild-type allele. The locus comprises two exons (blue boxes E1 and E2) with the entire coding region (ORF) located within E2. (**b**) The targeting construct harbors a frt-flanked neo^R selection cassette in intron 1 and the loxP-flanked exon 2. Frt sites are depicted as green rectangles, loxP sites as orange triangles. (**c**) Modified allele after homologous recombination in ES cells. Homologous recombinants are identified by nested PCR using primers #1 and #2, and #1′ and #2′, respectively. Integration of the 3′ loxP site is independently demonstrated by PCR with primers #3 and #4. (**d**) Allele after deletion of the neo^R selection cassette. Breeding of targeted mice with Flp-deleter mice can lead to mosaic offspring. Presence of the selection cassette may be addressed with primers #2 and #5, absence of the neo^R cassette with primers #5 and #6. (**e**) Cell-type-specific ablation of the modified allele in mice is demonstrated by PCR using primers #7 and #8

included those for myelin-specific structural proteins, such as MPZ [3], PLP1 [4], or CNP [5], and OLC-specific transcription factors, such as Olig 1 [6] or MYRF [7]. These added to the natural mouse mutants, often point mutations, for which gain- and loss-of-function effects were sometimes difficult to separate. Protocols for the positive/negative selection of mutant ES cells for germline transmission have been optimized [8] and the standard work flow in the authors' lab is detailed in [4, 9].

1.2 Principles of Conditional Mutagenesis in Mice

The ubiquitous inactivation of any widely expressed gene with a role in OLC can cause strong developmental defects and even early lethality in mice, which limits the value of "conventional" null mutants in the analysis of oligodendrocytes of the postnatal or adult nervous system. This is particularly true for genes whose knockouts die prematurely, such as the null mutants of the transcription factor (TF) genes *Olig2* [10], *Mrf* [7], or *Sox10* [11] and in the majority of cases, in which the targeted gene has vital roles outside the oligodendroglial cell lineage. This is also the case for TF genes, such as *Zfhx1b/Sip1* [12], *Yy1* [13], or *Hif1alpha* [14], for specific micro-RNAs and their processing enzymes, and for genes encoding growth factors, their receptors, and downstream effectors in signaling cascades.

To circumvent these problems, sequence-specific recombination systems, such as the "Flp/frt" system (derived from *Saccharomyces cerevisiae* [15]) or the more widely utilized "Cre/loxP" system (derived from the bacteriophage P1 [16]), have been established that enable a spatially and temporally controlled modification of target genes. Cre, a 38 kD *cy*clization *re*combinase, belongs to the integrase family of site-specific recombinases and catalyzes recombination at specific recognition sites, termed "loxP" sequences that comprise 2 palindromic 13 bp binding sites flanking a central asymmetric 8 bp "spacer region" (Fig. 2). The Cre/loxP technology can be used to carry out deletions, integrations, inversions, or translocations of double-stranded DNA fragments. However, the most widespread application of the Cre/loxP technology in mice is the cell-type-specific ("conditional") deletion of loxP-flanked ("floxed") target genes. For this purpose, two mouse lines are required that work in combination: (1) a Cre "driver" line that expresses Cre (or the CreER^{T2} variant, *see* below) under control of a cell-type-specific promoter and (2) a mouse line, in which the gene of interest has been "floxed," mostly by homologous recombination in ES cells. In the floxed line, an essential part of the targeted gene is flanked by two equally oriented, 34 bp long loxP sites (these sequences are absent from the normal murine genome). The position of both loxP sites is chosen such that their presence does not interfere with normal gene expression, typically in an intronic position. After crossbreeding the two mouse lines, all Cre-expressing cells in the double-transgenic offspring (as defined by the cell-type-specific promoter/enhancer driving the Cre transgene) will, in principle, recombine the floxed target gene, leaving a single loxP site behind. In mice homozygous for the floxed gene, this leads to the irreversible inactivation of gene function in a cell-type-specific manner.

Since the first demonstration of a cell-type-specific gene knockout in living mice, the ablation of a ubiquitously expressed DNA polymerase selectively from T-cells [17], a long list of conditional mouse mutants has demonstrated the usefulness of this approach, including for oligodendroglial and myelin-related research. For

13 bp 8 bp 13 bp

5′ ATAACTTCGTATA-ATGTATGC-TATACGAAGTTAT 3′

Fig. 2 Sequence of a wild-type loxP site. Please note that the bases in the 8 bp "spacer region" (except for the two in the middle) may differ in genetically engineered loxP variants

example, by inactivation of two ubiquitously expressed enzymes, PTEN and mTOR, selectively in cells of the oligodendroglial lineage, we and others have identified the PI3K/AKT/mTOR pathway as a physiological regulator of myelin membrane growth [18–20].

The Cre-loxP technology can also be used to turn on (or to turn off) regular transgenes generated by pronuclear injection. This approach has been helpful in mouse models of diseases that are caused by the overexpression of genes or transgenes with gain-of-function mutations. For example, a cell-type-specific "turn-off" strategy has been used to selectively remove an entire loxP-flanked transgene (encoding a mutant form of superoxide dismutase-1, SOD1), which is ubiquitously expressed, from OPC in the CNS. In this mouse model of amyotrophic lateral sclerosis (ALS), the loss of SOD1 toxicity from OPC was sufficient to delay disease onset [21], proving the role of OLC in a neurodegenerative disease. For "turning on" transgenes, most often a floxed "stop" cassette is placed upstream of a transgene that prevents functional translation of the RNA transcript. However, once the stop cassette is excised by Cre-mediated recombination, transgene expression, driven by a ubiquitously active promoter, is induced (*see* for example [22]). Importantly, upon Cre recombination the mouse genome is irreversibly changed. Experimental systems, in which transgene expression is switched "on" or "off" in a reversible manner, e.g., by the tetracycline regulated transactivator system [23, 24], have been discussed elsewhere [25].

1.3 Floxed Mice: Modifying Target Genes by Introducing loxP Sites

To generate mouse mutants with a Cre-modifiable target gene, essential regions of this gene need to be "floxed" in a way that the newly introduced loxP sequences do not interfere with gene regulation, mRNA splicing, or the protein coding region. Thus, loxP sites are typically placed into introns. However, also loxP insertions into 5′ or 3′ untranslated regions (UTRs) have been reported that did not interfere with wild-type-like expression of the target genes [26]. Since truncated proteins may still be functional or exert dominant-negative effects, floxing the first exon (s) of a gene with the ATG start codon is more likely to create a functional null allele. The generation of loxP-flanked alleles is commonly achieved by homologous recombination of a cloned genomic targeting construct that is transfected into embryonic stem cells (ESC) (Fig. 1). This "knock-in" strategy will be increasingly replaced by CRISPR/Cas9 nuclease guided gene editing of ESC or fertilized oocytes. Numerous protocols of Crispr/Cas9-based genome editing have been published recently, including the generation of conditional alleles [27–29].

Homologous recombination requires the cloning of a genomic targeting vector with a selection cassette, most commonly a neomycin resistance gene. In contrast to a conventional gene targeting construct, this neo^R cassette should be subsequently removed (also from intronic sites) because cryptic splice sites may affect gene expression and turn the floxed allele into a hypomorph or even into a null allele [26, 30]. Thus, strategies have been developed to remove such a cassette, either in vitro or in vivo, once the correctly targeted allele has been identified. To do so, the selection cassette is often flanked by "frt" sites (for 'flip recognition target'), the palindromic 34 bp sequence for the yeast Flp ("flippase") recombinase [15]. Subsequent breeding to "Flp-deleter" mice [15, 31], in which Flp is active in germ cells or at the zygote stage, results in the removal of a selection cassette in vivo and in the germline [30], as schematically shown in Fig. 1.

Alternatively, the selection cassette in a targeting construct can be made removable by flanking it with a third loxP sequence. Recombination would require crossbreeding to Cre-deleter mice, such as EIIa-Cre [32, 33], aiming for partial recombination of specifically those loxP sites that flank the selection cassette.

We note that these intercrosses produce often mosaic offspring. Frt sites will not be necessarily recombined in all cells of a mouse that has inherited both the frt-targeted allele and the Flp-deleter allele [30]. Dependent on the recombination status of the germline cells, its offspring will (or will not) transmit the desired loxP-flanked allele, i.e., the allele specifically lacking the frt-flanked selection cassette (*see* also Fig. 1). Likewise, crossing floxed mice with a mutant allele harboring 3 loxP sites to a Cre-deleter mouse, such as EIIa-Cre males, will produce mosaic offspring, in which the moderately efficient expression of Cre leads to different recombination events (i.e., recombination between loxP1/2, loxP2/3, loxP1/3, and no recombination at all) to a variable degree. Depending on the recombination status of the individual germ cells, further breeding of mosaic mice to wild type mice may yield all four types of targeted alleles in the next generation, and those mice that inherited the desired "selection cassette-removed-only allele" can be chosen for establishing the intended floxed mouse line. For this reason, mosaic offspring and all mice in the subsequent generation(s) have to be carefully checked for the segregation of the different loxP-flanked alleles and for the loss of the Cre or Flp genes. This requires a PCR analysis of genomic DNA using different sets of primers designed to detect all possible recombination events (*see* Fig. 1 for frt-flanked selection cassettes and [26, 34, 35] for the work flow of turning 3-loxP into 2-loxP mouse mutants).

The need for custom-made, self-designed floxed mutants has substantially decreased in recent years as a growing number of targeted ES cell clones and floxed mouse lines is made available to the scientific community by the International Knock-out Consortium (IKMC, www.knockoutmouse.org), which includes the European and the North American Conditional Mouse Mutagenesis Programs (EUCOMM, NorCOMM) [36]. In addition, more than 1270 floxed mouse strains are currently listed and made available by the Jackson Laboratory (https://www.jax.org/mouse-search). *See* **Notes 1** and **2** for additional transgenic mice of relevance to study oligodendrocyte biology.

1.4 Cre and Inducible CreER^T Mouse Lines for Oligodendrocyte Lineage Cells

Many genes have been identified that are predominantly expressed in oligodendrocytes and/or their precursor cells. Based on their spatiotemporal expression profile, the corresponding Cre (or CreER^T, *see* below) driver lines can be categorized by their expression in OPC, mature oligodendrocytes, or the entire oligodendroglial cell lineage. For example, OPC are most often targeted with Cre expressed under control of the Ng2 or Pdgfr alpha regulatory region. Oligodendrocytes can be targeted by Cnp, Plp1, or Tmem10 promoter-driven Cre-transgenes, and the entire lineage can be targeted with Sox10-Cre driver lines (*see* Table 1). Most mouse lines are also listed online by the Network Glia (https://www.networkglia.eu/tiermodelle) and by the Jackson Laboratory (https://www.jax.org/research-and-faculty/tools/cre-repository), including purchasing options.

Oligodendrocytes in the CNS are of different spatial origin, which suggests some heterogeneity also with respect to gene expression, development, and adult function [54, 66, 67]. Moreover, cortical and subcortical oligodendrocyte defects have been associated with diseases of the CNS, including schizophrenia, depression, autism, and Alzheimer's disease [68]. Thus, it may be desirable to target oligodendroglial genes in cortical and subcortical oligodendrocytes and spare oligodendrocytes in brain areas essential for basic motor functions (cerebellum, spinal cord). We thus added also forebrain-specific Cre driver lines into Table 1, which can be used to target relevant oligodendroglial subsets, for example, when the target gene itself is expressed only in oligodendrocyte lineage cells, or if the targeted transgene (which shall be turned on or off) is driven by an OLC-specific promoter.

A better temporal control over Cre-mediated genomic recombination has been achieved with an inducible system, in which Cre is expressed as a chimeric fusion protein harboring the ligand-binding domain of a modified estrogen receptor that is activatable by the administration of tamoxifen. Specifically, Cre (when fused to the G521R-mutated hormone-binding domain of the human estrogen receptor) is still responsive to the synthetic estrogen receptor ligand 4-hydroxytamoxifen (4-OHT) but not to endogenous 17β-estradiol (E2) [69, 70]. In the absence of tamoxifen, CreER^T is sequestered in the cytosol by binding to a heat-shock protein (Hsp90) and cannot enter the nucleus. However, upon binding of Tamoxifen (the prodrug) or 4-OHT (the active metabolite), CreER^T translocates into the nucleus and catalyzes recombination (Fig. 3). By this mechanism an extra level of temporal control is achieved in cell-type-specific gene targeting. The functionally improved version CreER^{T2} [71] is a triple mutant G400V/

Table 1
Selected genetically engineered mouse lines expressing Cre or CreER^T in the oligodendroglial lineage

Mouse line	Laboratory contact	Transgenic strategy	Side of recombination	References
OPC				
Pdgfra-CreER^T	W.D. Richardson w.richardson@ucl.ac.uk	PAC transgenic mice expressing a TM-inducible CreER^T under the control of the murine *Pdgfra* promoter/enhancer. The PAC comprises the entire *Pdgfra* gene (modified by fusion of the CreER^T coding sequence into the *Pdgfra* start codon), with 55 kb of 5′- and 74 kb of 3′ flanking sequences	OPC, few projection neurons in the forebrain (tested with TM application >P45)	[37]
Pdgfra-CreER^T	D. Bergles dbergles@jhmi.edu	BAC transgenic mice that express CreER^T under control of the murine *Pdgfra* promoter/enhancer. The BAC comprises the entire *Pdgfra* gene (with CreER^T being inserted behind the 5′UTR), with 71 kb of 5′- and 41 kb of 3′ flanking sequences	OPC, a subset of perivascular pericytes, ependymal cells in the choroid plexus (tested with TM application >P30)	[38]
Ng2-Cre	A. Nishiyama akiko.nishiyama@uconn.edu	BAC transgenic mice that express Cre under the control of the murine *Ng2* (*Cspg4*) promoter/enhancer. The BAC comprises the entire *Ng2* gene (with exon 1 being modified by the Cre coding sequence), with 60 kb of 5′- and 114 kb of 3′ flanking sequences	*Ng2*-expressing cells (incl. OPCs) in the CNS and other organs	[39]
Ng2-CreER^T BAC	A. Nishiyama akiko.nishiyama@uconn.edu	BAC transgenic mice that express a TM-inducible CreER^T under the control of the murine *Ng2* (*Cspg4*) promoter/enhancer. The BAC comprises the entire *Ng2* gene (with exon 1 being modified by the CreER^T coding sequence), with 60 kb of 5′- and 114 kb of 3′ flanking sequences	*Ng2*-expressing cells (incl. OPCs) in the CNS and other organs	[40]
Ng2-CreER^T	F. Kirchhoff frank.kirchhoff@uks.eu	Targeted insertion of CreER^T into the *Ng2* gene locus (by homologous recombination in ESC)	Ng2-expressing cells (incl. OPCs) in the CNS and other organs, few neurons (when tested with TM application >P30)	[41]
Ascl1-CreER^T	J.E. Johnson jane.johnson@utsouthwestern.edu	Random integration of a BAC (comprising the *Ascl1* gene and approx. 100 kb of 5′- and 200 kb of 3′-flanking sequences). The *Ascl1* coding sequence is replaced by CreER^T	Ventricular zone cells in certain domains of the developing nervous system (restricted to OL precursors when tested with TM application at E16)	[42]
OL				
Cnp-Cre[a]	K.-A. Nave nave@em.mpg.de	Targeted replacement of the *Cnp* coding region by Cre (by homologous recombination in ESC)		[5]

(continued)

Table 1
(continued)

Mouse line	Laboratory contact	Transgenic strategy	Side of recombination	References
			OL, SC, some OPC (also spinal MN and other CNS neurons when highly sensitive reporters are used)	
Mbp-Cre	M. Giovannini marco.giovannini@cephb.fr	Random integration of a transgenic construct harboring 1.3 kb of the 5′ flanking sequence of the murine *Mbp* gene directing Cre expression	Two lines (*Mbp*-Cre6; *Mbp*-Cre9) OL, testis, pituitary gland, heart and other organs. Recombination specificity in the brain not assessed at the cellular level	[43]
Mbp-Cre	M. Miura mmiura@nana.med.osaka-u.ac.jp	Random integration of a transgenic construct harboring 6.3 kb of the 5′ flanking sequence of the murine *Mbp* gene directing Cre expression	OL, mainly in white matter of the spinal cord	[44]
Mbp-CreER^{T}	A. Gow agow@med.wayne.edu	Random integration of a transgenic construct harboring 1.9 kb of the 5′ flanking sequence of the murine *Mbp* gene directing CreER^{T} expression	TM-induced recombination of floxed genes in OL	[45]
Plp-Cre	R. Kothary rkothary@ohri.ca	Random integration of a transgenic construct harboring 2.4 kb of 5′ flanking sequence, exon 1 (ATG mutated to CAG) and intron 1 of the murine *Plp1* gene to direct Cre expression	Widespread recombination in CNS, non-recombined cells are limited to vasculature	[46]
Plp-Cre	K.-A. Nave nave@em.mpg.de	Random integration of a transgenic construct harboring 3.7 kb of 5′ flanking sequence, exon 1 and intron 1 of the murine *Plp1* gene to direct Cre expression (Cre coding region (lacking the start ATG) was inserted in PLP exon 2; endogenous start ATG of PLP in exon 1 serves as an initiation codon)	Six lines with varying expression in OL, Purkinje cells and other neurons, and expression in other organs	[47, 48]
Plp-CreER^{T,b}	U. Suter usuter@cell.biol.ethz.ch	Random integration of a transgenic construct harboring 2.4 kb of 5′ flanking sequence, exon 1 (ATG mutated to CAG) and intron 1 of the murine *Plp1* gene to direct Cre expression	TM-induced recombination of floxed genes in OL, (OPCs), SC	[49]
Plp-CreER^{T,b}	B. Popko bpopko@neurology.bsd.uchicago.edu	Random integration of a transgenic construct harboring 2.4 kb of 5′ flanking sequence, exon 1 and intron 1 of the murine *Plp1* gene and the sequence of CreER^{T}	TM-induced recombination of floxed genes in OL and SC (virtually no recombination in OPCs) (proven for TM application >P30)	[50]
Mog-i-Cre	A. Waisman Waisman@uni-mainz.de	Targeted insertion of Cre into the first exon of the *Mog* gene (by homologous recombination in ESC)	TM-induced recombination of floxed genes in OL (and certain neuronal populations [51])	[52]

Tmem10/Opalin-Cre	B. Xiao bxiao365@yahoo.com	Targeted insertion of Cre into the ORF of the *Tmem10* gene (by homologous recombination in ESC)	OL; not in OPCs, astrocytes or microglia	[53]
OLC (OPC and OL)				
Olig2-Cre	R.J. Franklin rjf1000@cam.ac.uk W.D. Richardson w.richardson@ucl.ac.uk	Targeted introduction of Cre into the single exon of the *Olig2* gene (by homologous recombination in ESC)	OLC (close to 100% at OPC stage), motor neurons, subset of astrocytes	[54, 55]
Sox10-Cre[a]	W.D. Richardson w.richardson@ucl.ac.uk G. Koentges g.koentges@ucl.ac.uk	Random integration of a 170 kb PAC transgene	OLC and neural crest derived cell lineages	[56]
Sox10-iCreER^T	W.D. Richardson w.richardson@ucl.ac.uk	Random integration of a 127 kb PAC (harboring the entire *Sox10* gene) in which the *Sox10* ORF was replaced by an iCreER^T fusion gene	TM-induced recombination of floxed genes in OLC and neural crest derived cell lineages	[57]
Sox10-iCreER^T BAC	L. Dimou leda.dimou@uni-ulm.de	Random integration of a >210 kb BAC (harboring the entire *Sox10* gene) in which *Sox10* exon 3 was replaced by an iCreER^T fusion gene	TM-induced recombination of floxed genes in OLC and neural crest derived cell lineages	[58]
S4F:Cre[a]	A. McCallion andy@jhmi.edu	Random integration of a transgenic construct with the *Sox10* distal enhancer directing Cre expression	OLC and neural crest derived cell lineages	[59]
Olig1-Cre	Q.R. Lu Richard.Lu@cchmc.org C. Stiles stiles@dfci.harvard.edu D. Rowitch mailto:avid.Rowitch@ucsf.edu	Targeted replacement of the majority of the *Olig1* ORF by Cre (by homologous recombination in ESC). Two variants of the mouse line are available: frt-flanked neo cassette included [10] or removed (by breeding to a Flp-deleter mouse) [6]	Olig1-expressing cells of the CNS (incl. OL and >30% of GABAergic cortical interneurons)	[6, 10, 60]
Regional Cre drivers (for OLC-specific genes)				
FoxG1-IRES-Cre	D. Kawaguchi dkawaguchi@mol.f.u-tokyo.ac.jp	Targeted insertion of an IRES-Cre cassette into the 3′UTR of the *Foxg1* gene locus	Telencephalon	[61]
Emx-Cre[c]	K.R. Jones krjones@stripe.colorado.edu	Targeted insertion of an IRES-Cre cassette into the 3′UTR of the *Emx1* locus	Dorsal telencephalic pallium	[62]

(continued)

Table 1
(continued)

Mouse line	Laboratory contact	Transgenic strategy	Side of recombination	References
Emx-CreER^T	W.D. Richardson w.richardson@ucl.ac.uk	Random integration of a PAC transgene (approx. 180 kb in size), in which iCreER^T was fused into the translation initiation codon of the *Emx1* gene	Dorsal telencephalic pallium	[54]
Gsh2-iCre	W.D. Richardson w.richardson@ucl.ac.uk	Random integration of a PAC transgene (approx. 110 kb in size) in which iCre was fused into the translation initiation codon of the *Gsh2*-iCre gene	Lateral and/or caudal ganglionic eminences	[54]

Abbreviations used (in addition to gene names): *ATG* start codon, *BAC* bacterial artificial chromosome, *CreER*^{T} tamoxifen-inducible Cre recombinase, *ESC* embryonic stem cells, *iCre* optimized, codon improved version of Cre recombinase [63], *IRES* internal ribosomal entry site, *kb* kilobases DNA, *MN* motor neurons, *OL* oligodendrocytes, *OLC* oligodendrocyte lineage cells, *OPC* oligodendrocyte precursor cells, *ORF* open reading frame, *PAC* P1-derived artificial chromosome, *SC* Schwann cells, *TM* tamoxifen

[a]Cre expression in a variable subset of male germline cells reported: offspring of a Cre:floxed male may carry a ubiquitously recombined (i.e., deleted) floxed allele. It is suggested to pass the Cre-allele through the female germline

[b]Cre expression in a subset of male and (possibly) female germline cells reported: some offspring of a Cre::floxed male or female may carry the recombined floxed allele

[c]A modest level of TM-independent recombination of loxP-flanked genes has been reported [64, 65]: Vehicle-injected fl/fl::CreER /+ mice should be used as controls when utilizing this mouse line

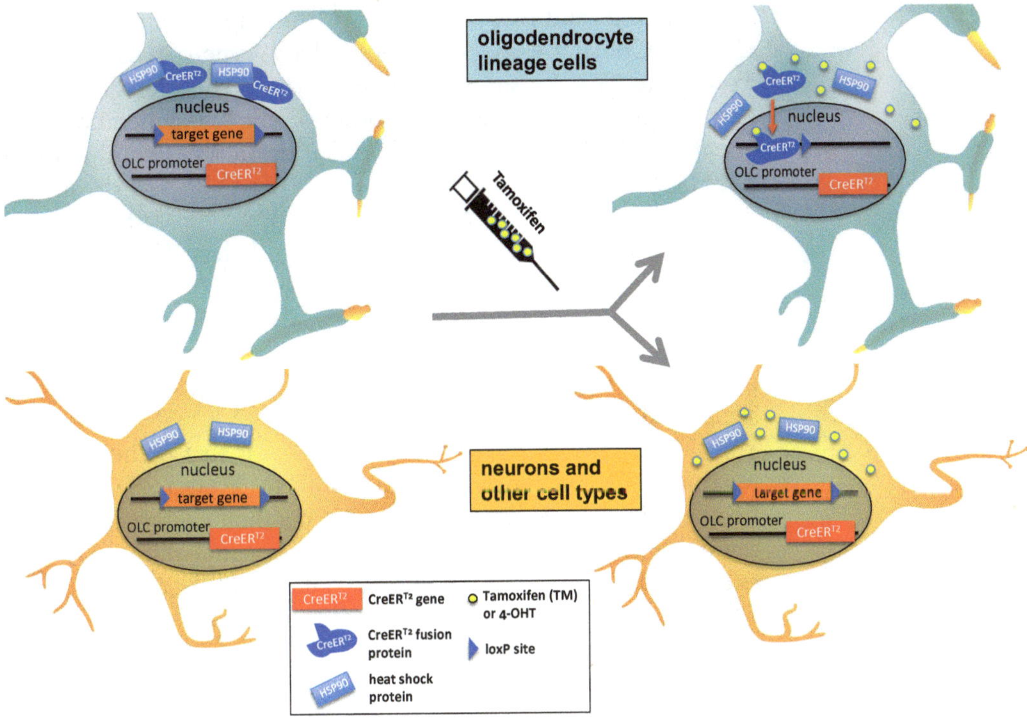

Fig. 3 Schematic depiction of the conditional inactivation of a loxP-flanked (floxed) target gene (orange) in oligodendrocyte lineage cells using Tamoxifen-dependent activation of CreER^{T2}. Note that the CreER^{T2} gene (red) is expressed only in oligodendrocytes (on top) but the encoded protein (blue) prevented from entering the cell nucleus by being bound to Hsp90, a heat-shock protein. Upon administration of tamoxifen (yellow), occupation of the hormone-binding site of CreER^{T2} causes a conformational change and the release of Hsp90, followed by nuclear entry of CreER^{T2} and genomic recombination

M543A/L544A of the ligand-binding domain that exhibits decreased recombinase activity in the nonactivated state and increased efficacy upon tamoxifen binding. Today, it is the predominantly used variant for the generation of tamoxifen-inducible Cre driver lines (Table 1).

1.5 Breeding Conditional Mouse Mutants

On average, and depending on strain, the life span of mice is about 2 years. Females become sexually mature at the age of 6–8 weeks and remain fertile for about 6 months. During that time, they can give rise to approx. 2–4 litters of 5–10 pups each. With 19 days of gestation and 3 weeks until weaning (and routine genotyping), the generation time of mice is about 3 months. Males should be used for breeding not before 8 weeks of age and remain fertile for approximately 12 months. Breedings are ideally set up with 1 male and 2 female mice that are only separated 1–2 days before delivery.

For an efficient and cost-saving breeding strategy, the first generation of mice, which are heterozygous for a floxed gene X (termed $X^{fl/+}$ in the following), should be intercrossed by brother/

sister matings and to a $Y^{Cre/+}$ (or $Y^{CreER/+}$) mouse that expresses Cre (or $CreER^{T2}$) under the desired cell-type-specific promoter from gene "Y." By simple Mendelian rules, the crossbreeding of heterozygous $X^{fl/+}$ mice to each other (in the "F0 generation") will generate 25% homozygous $X^{fl/fl}$ offspring in F1. Breeding of the same $X^{fl/+}$ heterozygotes to the $Y^{Cre/+}$ (or $Y^{CreER/+}$) driver mice will generate 25% $X^{fl/+}$::$Y^{Cre/+}$ (or $X^{fl/+}$::$Y^{CreER/+}$) offspring in F1 that are heterozygous for both mutant alleles. The subsequent mating of unaffected $X^{fl/+}$::$Y^{Cre/+}$ (or $X^{fl/+}$::$Y^{CreER/+}$) mice (F1) to the homozygous $X^{fl/fl}$ mice (F1, also unaffected) will already produce a F2 generation, in which 25% of the mice are "mutants" and possibly phenotypically affected, i.e., homozygous for two floxed alleles and heterozygous for the Cre driver ($X^{fl/fl}$::$Y^{Cre/+}$ or $X^{fl/fl}$::$Y^{CreER/+}$). Another 25% of mice are "controls" that are homozygous for the floxed gene but harbor no Cre allele ($X^{fl/fl}$::$Y^{+/+}$). However, by maintaining these F2 mice as future parental animals and mating them to each other, the generation of experimental mice and controls becomes more efficient: conditional mutants ($X^{fl/fl}$::$Y^{Cre/+}$ or $X^{fl/fl}$::$Y^{CreER/+}$) and controls ($X^{fl/fl}$::$Y^{+/+}$) will always be born in a 50/50 ratio. This breeding strategy can generally be applied if a $CreER^{T2}$ driver line is used, because the mutant animals ($X^{fl/fl}$::$Y^{CreER/+}$) are not burdened prior to Tamoxifen injection (Fig. 4). Obviously, for constitutive Cre driver lines this efficient breeding strategy is only suitable if the mutants ($X^{fl/fl}$::$Y^{Cre/+}$) are not burdened with a disabling phenotype or even infertile. In that case the parental $X^{fl/fl}$ mice should be intercrossed with unaffected double heterozygotes ($X^{fl/+}$::$Y^{Cre/+)}$, as suggested before for the generation of F2 mice.

The efficient breeding strategy will also not suffice if (1) Cre is a knock-in and haploinsufficiency of this locus matters, (2) when cytotoxic effects of Cre are suspected [72, 73] or (3) when the conditional deletion of only one of the floxed target alleles is already of phenotypic relevance. In these cases additional breedings have to be included in order to generate control genotypes, such as $X^{+/+}$::$Y^{Cre/+}$ (for toxicity and haploinsufficiency of the Cre-expressing locus) or $X^{fl/+}$::$Y^{Cre/+}$ ($X^{fl/+}$::$Y^{CreER/+}$) (for haploinsufficiency of the target locus).

In the unlikely event that conditional mutants ($X^{fl/fl}$::$Y^{Cre/+}$ or $X^{fl/fl}$::$Y^{CreER/+}$) are never born and embryonic lethality can be ruled out, one should consider that Cre and the floxed gene are located on the same chromosome. This can make it difficult, or virtually impossible, to obtain the required genotype. Here, using a different Cre driver line may be the simplest solution to the problem.

Floxed alleles can be easily converted to null alleles by crossbreeding to Cre-deleter strains, such as EIIa-Cre [32, 33] that recombine in the germline. Thus, two strategies for working with conditional mutants are followed by utilizing either homozygously floxed mice ($X^{fl/fl}$) or heterozygously floxed mice ($X^{fl/-}$) on the null mutant background. Both strategies have advantages and

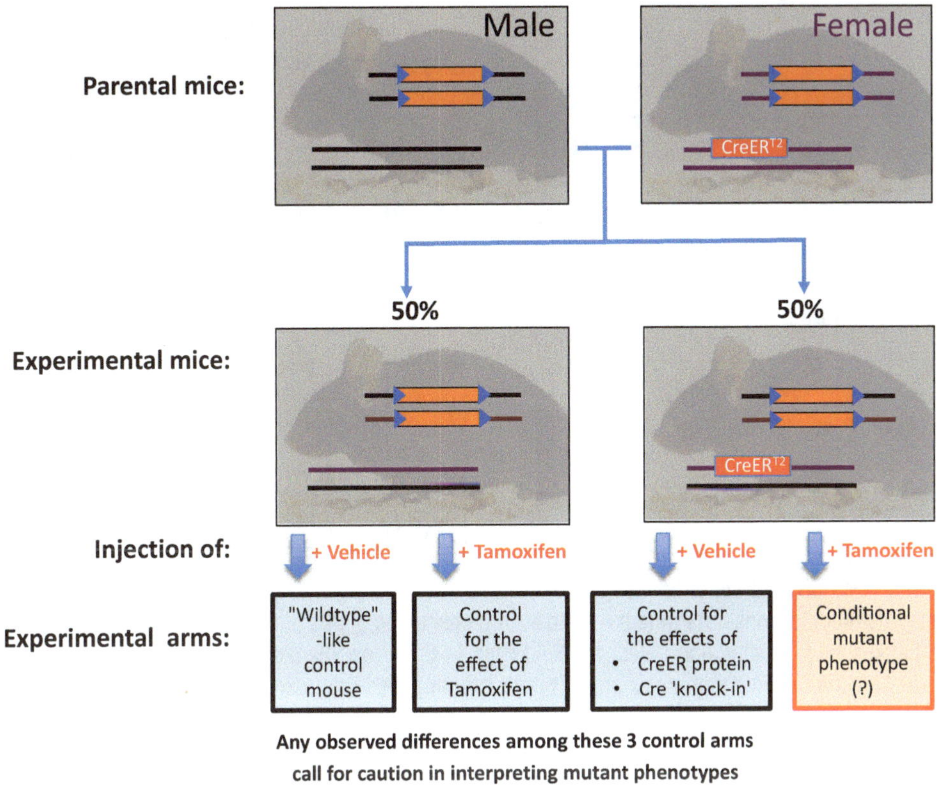

Fig. 4 Breeding scheme for the generation of conditional null mutants and controls. Having ensured that "floxed" mice are normally developed, the most convenient strategy is to generate and intercross parental animals of the genotypes fl/fl::+/+ and fl/fl::CreER^{T2}/+. This results in an equal number of offspring with the genotypes fl/fl::+/+ and fl/fl::CreER^{T2}/+. Injection of tamoxifen to the conditional mutants induces Cre-mediated recombination and thus cell-type-specific ablation of the gene of interest (orange frame). Experimental results from these animals should be carefully controlled by including three control conditions (black frames). First, vehicle treated fl/fl::CreER^{T2}/+ mice are used to control for effects of the Cre mutant allele, such as haploinsufficiency effects, Cre toxicity related issues or potential background recombination by non-induced Cre (in comparison to vehicle treated fl/fl::+/+ controls). Second, tamoxifen-treated fl/fl::+/+ animals without the Cre allele are used to determine the influence of Tamoxifen on the experimental paradigm

disadvantages. Since recombination of one allele is faster than of two, it may be more efficient to work on a heterozygous null mutant background, especially when Cre expression is low and/or transient. However, mice heterozygous for myelin genes can have relevant phenotypes [74, 75]. Thus, phenotypes caused by haploinsufficiency of the target gene may mask the conditional full mutant phenotype.

It is worth considering that Cre expression can sometimes be ectopically activated in the early embryo, which can give rise to fully recombined target genes in the germline. This has been reported for Sox10-Cre males [56] and is also sporadically evident in Cnp-Cre male mice [5] (Fig. 5, Table 1).

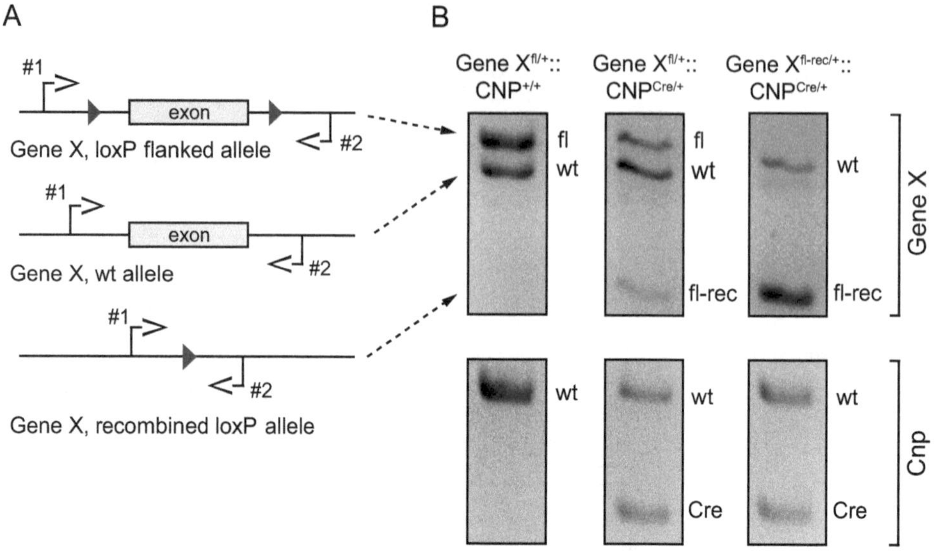

Fig. 5 Genomic PCR analysis of floxed mice by using a "two-Primer" strategy for the detection of short loxP-flanked gene modifications. (**a**) Primers #1 and #2 are positioned 5′ and 3′ from an essential exon of gene "X" that is flanked by loxP sites and detect the wild-type (+), the loxP-flanked (fl), and the recombined (fl-rec) allele. (**b**) Agarose gel electrophoresis on PCR products amplified from genomic tail tip DNA of mice with the genotypes *geneX*$^{fl/+}$::Cnp$^{+/+}$ (left panels); *geneX*$^{fl/+}$::Cnp$^{Cre/+}$ (middle panels); and *geneX*$^{fl\text{-}rec/+}$::Cnp$^{Cre/+}$ (right panels). The top row represents PCR products obtained from the targeted gene "X," the bottom row represents PCR products obtained from the Cnp allele. Please note that Cnp^{Cre} recombines in Schwann cells present in the tail tips. Thus, presence of the fl-rec allele is expected in all floxed mice that carry an additional Cnp^{Cre} allele (middle panels). In contrast, germline recombination, which fully converts the floxed allele into a ubiquitously present null allele, is indicated when (next to the wild-type allele) the floxed allele is not detectable, while the fl-rec product is present in high amounts (right panel). Please note that longer loxP-flanked sequences require a "three primer strategy" with, for example, an additional antisense primer being included that is located in the targeted exon. Here, this primer in combination with primer #1 would amplify a product from the wild-type and the floxed allele and primers #1 and #2 (more widely spaced from each other in this case) would only amplify a product after Cre-mediated recombination, since they would be spaced to much apart to amplify from the regular wild-type and the non-recombined floxed allele, respectively

1.6 Treatment Regimen for Tamoxifen-Induced CreER^{T2}-Mediated Recombination

Optimal protocols for tamoxifen-based CreER^{T2}-mediated recombination have to be empirically determined because the recombination efficiency depends on different variables, such as (1) the specifics of the "floxed" gene, including the methylation/acetylation state of the floxed gene (which defines the accessibility of loxP sites) and the length of the floxed DNA segment, (2) the type of applied drug, i.e., tamoxifen or 4-OHT, (3) the route of administration (i.p. injection, gavage) and the frequency and length of treatment, and (4) the variable of age, sex, and genetic background of the recipient mouse. We suggest the protocol in Subheading 1.8 as a starting point to assess the efficiency of tamoxifen-induced CreER^{T2}-mediated recombination in adult animals. An alternative is the application protocol proposed by the Jackson labs (https://www.jax.org/research-and-faculty/tools/cre-repository/tamoxifen).

1.7 Utilizing Cre-Indicator Mice

To assess cellular recombination specificity and efficiency of any Cre driver line, a plethora of "indicator" mice have been generated, in which (upon crossbreeding) Cre activates permanently the expression of reporter genes encoding fluorescent proteins or histochemically detectable enzymes. Note that the strength of the cellular signal does not reflect Cre expression, but rather the strength of the promoter that drives the reporter. Widely used are knock-ins into the constitutively active ROSA26 locus or transgenes expressed under control of ubiquitously active promoters, such as that of the chicken beta-actin gene. The Jackson laboratory alone lists more than 280 Cre "reporter" strains (https://www.jax.org/mouse-search). The general idea of this strategy is that upon Cre-mediated excision of a loxP-flanked "stop"-cassette, cells will express easily identifiable reporter genes encoding, for example, LacZ [76, 77], alkaline phosphatase [78], or fluorescent proteins such as EGFP [79, 80], EYFP [81, 82], ECFP [82], ZsGreen [81], an EGFP-tandem version of dimeric DsRed (=tdsRed [83]), or tdTomato [81]. Since Cre-mediated recombination is irreversible, reporter genes become permanently activated in Cre-expressing cells and all their descendants. This allows this strategy to be applied for cell lineage tracing and fate mapping analyses, which have been heavily used in the analysis of oligodendrocytes and the spatial and temporal analysis of their precursor cells [37, 84–88] (*see* **Note 3**).

In dual reporter lines, the excised "stop" sequence is by itself a reporter gene, encoding a different fluorophore. Upon Cre recombination, this leads to a "switch" of color, including a variable time in which both reporters may be detectable. Weakly expressed fluorescent reporters can sometimes be made visible by more sensitive immunostaining. An X-Gal staining protocol to analyze the expression of beta-galactosidase (lacZ) as a reporter is given in Subheadings 2 and 3.

For myelin research, membrane-tagged reporters are of special interest. The $ROSA^{mT/mG}$ reporter mouse line [89] expresses a ubiquitous cell membrane-localized tdTomato (mT) protein but, upon Cre-mediated recombination, a membrane-localized EGFP (mG). Similarly, $ROSA^{nT/nG}$ reporter mice switch the same reporter proteins, which are directed to the cell nucleus [90]. Other reporter mice express fluorescent proteins that are targeted to membranes by a GPI anchor (CAG::GFP-GPI) or myristoylation (CAG::myr-Venus) [91]. Note that in myelinating glial cells, fluorescent proteins are too bulky to be wrapped into a compacted myelin sheath [92], but rather demarcate cytosolic tubes, paranodes, and other non-compacted regions of the myelin sheath.

For the sparse-labeling of individual oligodendrocytes, any of the indicator mice discussed above can be used in combination with a CreER^{T2} driver line and a single low-dose injection (i.p.) of tamoxifen. Here, suggested doses range from 25 to 200 mg/kg [64]. Another application is the simultaneous activation of single reporter genes in multicolor fluorescent ("brain-bow") reporter mice that enable the single cell analysis of oligodendrocytes and their interactions in white matter tracts that are crowded with OLC [64, 93].

1.8 Technical Caveats

The Cre/loxP system is a powerful yet not unproblematic approach. One concern is that high-level expression of Cre itself may have off-target and thus unintended side effects on genomic integrity [94]. Thus, as a control group, the inclusion of mice with the Cre driver but without floxed alleles is highly recommended. Moreover, a randomly integrated Cre transgene may interfere with the expression of neighboring genes, a problem that can be reduced when only working with heterozygous Cre mice. While for targeted Cre "knock-in" mice the insertion site is well defined, haploinsufficiency may still occur and should be controlled for (by including non-Tamoxifen-induced CreER^{T2} mice or Cre mice without floxed alleles). We have, for example, replaced the coding region of Cnp by Cre, which has created a Cnp null allele, and heterozygous mice are fully normal and useful for the conditional deletion of floxed genes in oligodendrocytes. However, Cnp$^{Cre/+}$ mice develop a mild behavioral phenotype after 18 months of age with signs of neuroinflammation [74], which should be controlled for in any genetic experiment addressing brain aging. Expressing Cre downstream of an IRES (internal ribosomal entry site) may solve such problems [61].

Another problem, mostly affecting transgenic Cre lines, is the fact that promoters and regulatory elements are not always as cell-type-specific in development as anticipated from gene expression in the adult. For example, PLP1 is a highly cell-type-specifically expressed myelin protein, but 4 out of 6 Plp1-Cre driver lines generated in our lab showed a variable degree of recombination in neuronal subpopulations [47, 48]. This is most likely due to a low level of (transient) Cre expression already in neural stem cells. The same is true for Cnp-Cre mice, which are used to target oligodendrocytes, but also mark certain neuronal subtypes when used in combination with specific reporters [95, 96]. These off-target effects require a highly sensitive reporter line, in which the single target gene is very accessible and rapidly recombined. Thus, activation of a single reporter gene does not yet prove that any two floxed alleles of real target genes are likewise recombined.

However, off-target recombination illustrates a Cre-specific problem: the enzyme is effective in trace amounts and any recombination is irreversible. On the other hand, Cre-mediated recombination is often not fully penetrant, i.e., less than 100% of the envisioned target cell population can be recombined. In this case, recombination efficiency is not necessarily limited by the Cre driver, but by the floxed alleles, whose chromatin state and accessibility for Cre may prohibit recombination in some but not all cells. What defines the epigenetic state of these target cells is unknown.

As mentioned above, for some OLC-specific Cre driver lines (not only those generated by pronuclear injection) at least sporadic Cre expression has been detected in the male germline (less frequent in female germline cells), and is likely underdiagnosed (or not reported) in other lines. When unaware of this, the resulting transmission of unexpected null alleles can severely perturb the experiment. A suitable PCR strategy to identify (1) wild-type, (2) floxed, and (3) fully recombined (fl-rec) alleles is presented in Fig. 5. Note that if Cre is also expressed in Schwann cells, the fully recombined allele can always be detected in the genomic DNA obtained from tail biopsies (independent of germline recombination).

In experiments with CreER^{T2} driver lines one should keep in mind that a low level of recombination may occur in the absence of tamoxifen, because the fusion protein is not 100% excluded from the nucleus. This has been reported, e.g., for two widely used Plp1-CreER^{T2} lines [64, 65], and is possibly due to the strong activity of the Plp1 promoter in combination with an easily accessible target gene.

Improvements of the Cre/loxP technique hold the potential of enhancing the cellular specificity of Cre or speeding up conditional mutagenesis. One is based on a "Split-Cre" complementation system, in which two inactive Split-Cre fragments are expressed under different promoters. Here, Cre recombinase activity is only regained when both halfmers are expressed in the same cell [97, 98].

Alternatively, it may suffice and be much faster than crossbreeding to delete genes in mature oligodendrocytes of homozygously floxed mice by stereotactically injecting Cre-expressing viruses into the brain. For oligodendrocytes, adeno-associated viruses (AAV) of various serotypes that express Cre (or CreER^{T2}) under control of a 1.3 kb promoter fragment from the mouse Mbp gene are commercially available (Vector Biolabs).

2 Materials

2.1 X-Gal Staining

2.1.1 Avertin Solution

1. Dissolve 2 g 2,2,2-tribromoethanol in 2 mL *tert*-amyl alcohol and add 96 mL of warm ddH_2O (40 °C). Stir for 30 min on a magnetic stirrer.
2. Sterile filtrate through a 0.2 μm filter.
3. Store in the dark at 4 °C for up to 4 weeks.
4. Use 200 μL per 10 g of mouse weight.

2.1.2 4% Paraformaldehyde (PFA) in 0.1 M Phosphate Buffer Mix

1. 400 mL Na_2HPO_4 (0.2 M).
2. 100 mL NaH_2PO_4 (0.2 M).
3. 100 mL Formol (37%, filtered).
4. 400 mL ddH_2O.

2.1.3 X-Gal Solutions

1. X-Gal Solution A: 0.01% Sodium deoxycholate, 0.02% 4-Nonylphenyl-polyethylene glycol, 2 mM $MgCl_2$, 5 mM $K_3[Fe(CN)_6]$, 5 mM $K_4[Fe(CN)_6]$ mix and keep it the dark at room temperature.
2. X-Gal Solution B (X-Gal stock solution): dissolve 20 mg 5-bromo-4-chloro-3-indolyl-β-D-galactoside in 1 mL DMSO.

2.2 Extraction of Genomic DNA and Genotyping of *Cnp-Cre* Mutant Mice

1. 10× Modified Gitschier Buffer (MGB): 6.7 mL 1 M Tris–HCl pH 8.8 (f.c.: 670 mM), 1.66 mL 1 M $(NH_4)_2SO_4$ (f.c. 166 mM), 650 μL 1 M $MgCl_2$ (f.c. 65 mM), add ddH_20 to 10 mL.
2. Proteinase K stock solution: 10 mg/mL (in ddH_2O).
3. Triton X-100 diluted 1:10 in PCR grade ddH_2O.

2.3 Intraperitoneal Injection of Adult Mice with Tamoxifen

1. Tamoxifen.
2. Corn Oil.
3. Ethanol.
4. 1 mL syringe graded in 100 μL intervals.
5. 22–27 gauge needle.
6. Sharp container.
7. Chemical fume hood or biological safety cabinet.
8. Tissue lyser.

3 Methods

3.1 X-Gal Staining

1. Anesthetize mouse with Avertin-solution.
2. Perfuse mice through the left ventricle with 15 mL of Hank's balanced salt solution, followed by 50 mL of 4% paraformaldehyde in PBS.

3. Post-fix brain and/or spinal cord in 4% paraformaldehyde for 2 h.
4. Section in phosphate-buffered saline (PBS) on a vibratome at 20–200 μm.
5. Incubate sections in freshly prepared X-gal solution (600 μL of solution "B" added to 10 mL solution "A") in a humidified chamber (cell culture incubator) at 37 °C for up to 16 h in the dark. Check for staining under a microscope.
6. Post-fix in 4% paraformaldehyde for 30 min.
7. Wash sections two times in 1× PBS for 10 min each, and mount them in AquaPolymount or other suitable mounting media.

3.2 Extraction of Genomic DNA and Genotyping PCR for Cnp-Cre Mutant Mice

Extract genomic DNA from tail tips, mouths swaps, or ear punches by using a kit according to the manufacturer's protocols or by the following cheap "Quick DNA prep protocol."

All of the following steps are carried out in a 96 well plate using a shaking incubator and a water bath.

3.2.1 DNA Extraction

Quick DNA Prep Protocol

1. Prepare fresh 1× MGB (sufficient for 96 samples):
   (a) 2.2 mL 10× MGB.
   (b) 220 μL β-Mercaptoethanol (final conc. 1%).
   (c) 1.1 mL 10% Triton X-100 (final conc. 0.5%).
   (d) 18.48 mL PCR-grade ddH_2O.
2. Add 80 μL of 1× MGB to each tail/ear-punch/mouth swap.
3. Add 20 μL of proteinase K stock solution.
4. Digest tissues/cells for 1–2 h at 55°C while gently shaking.
5. Vortex samples briefly.
6. Quick spin down to collect all liquid.
7. Heat inactivate proteinase K by heating samples at 95°C for 5 min.

   (Important: traces of proteinase K suffice to inactivate Taq polymerase).
8. Spin down debris (5–10 min at 2000 × *g*).
9. Use 1–2 μL of supernatant for PCR.

3.2.2 Standard-PCR-Protocol for Genotyping Cnp-Cre Mice

1. Prepare PCR reaction mix:
   (a) 1 μL DNA (100–400 ng).
   (b) 0.5 μL Puro3 s primer (10 pmol/μL).

   (c) 0.5 μL Cnp-E3 s primer (10 pmol/μL).
   (d) 1.0 μL Cnp-E as (10 pmol/μL).
   (e) 2 μL dNTP mix (2 mM).
   (f) 2 μL 10× RedTaq buffer (Sigma).
   (g) 1 μL Red Taq polymerase (1 U/μL, Sigma).
   (h) Add ddH_2O.
   (i) 20 μL.
2. Use Thermocycling program:
   (a) 95 °C 3 min.
   (b) 50 °C 30 s.
   (c) 72 °C 60 s
   (d) 95 °C 30 s.
   (e) Back to 2 for 35 cycles.
   (f) 50 °C 60 s.
   (g) 72 °C 10 min.
   (h) Store at 4 °C.
3. Separate PCR products on agarose gels for visualization. A 700 bp product from the wild-type allele is amplified by Cnp-E3 s (5′-GCCTTCAAACTGTCCATCTC-3′) and Cnp-E as (5′-CCCAGCCCTTTTATTACCAC-3′). The targeted Cre-allele is positively identified with primers puro3 s (5′-CATAGCCTGAAGAACGAGA-3′) and Cnp-E as, yielding a 400 bp fragment.

3.3 Tamoxifen Induction Regimen

1. In a chemical fume hood: dissolve tamoxifen in corn oil at a concentration of 10 mg/mL by high speed shaking (e.g., in a tissue lyser) at 37 °C for 20 min. Tamoxifen is not soluble in water.
2. A Tamoxifen solution is sensitive to UV light and, similar to crystalline Tamoxifen powder, must be stored in the dark (foil-wrapped vial) at 4 °C.
3. Sanitize injection site with 70% ethanol.
4. Administer Tamoxifen on 2 × 5 consecutive days (with a pause of 2 days in between) with one intraperitoneal injection a day. A standard dose of 100 μL Tamoxifen solution contains 1 mg Tamoxifen per adult mouse and day, and is effective to induce Cre-mediated recombination (*see* **Notes 4–10**).

4 Notes

1. Several mouse lines have been generated that harbor floxed transgenes for experiments in optogenetics, i.e., the light-induced opening or closing of ion channels in Cre-expressing cells [99]. Cre-dependent optogenetic tool-mice will aid, e.g., in studies on activity-dependent myelination [100] when expressed in genetically defined neuronal subsets. This includes principal neurons and parvalbumin-positive interneurons, both of which can be myelinated within the cortex of mice [101].
2. Other floxed indicator strains expressing effective sensors or effectors that may be useful for research on OLC include:
   (a) Ai95D mice, expressing a Cre-dependent calcium indicator and fast variant of GCaMP6 [102].
   (b) R26:lacZbpA(flox)DTA and R26:eGFP(flox)DTA mice that can be used for Cre-mediated cell ablation [103, 104].
   (c) Floxed miRAP and floxed Ribo-Tag mice, which allow cell-type-specific profiling of miRNAs and translated mRNAs, respectively [105, 106].
   (d) PhAM^{floxed} (photo-activatable mitochondria) mice expressing the photo-convertible fluorescent protein Dendra2 for labeling and analysis of mitochondria, their fusion/fission and transport [107].
   (e) ROSA26-LSL-Cas9 mice for in vivo gene editing when combined with single guide RNAs and a source of Cre [108].
   (f) ATP-sensor ATeam1.03^{YEMK} mice to uncover energy deficits in axons with pathological myelination [114].
3. Recombination efficiencies of different reporter genes vary when tested in combination with the same Cre driver mouse (*see* for example [96]). Thus, reporter gene expression can be no proof that recombination of a floxed target gene has been efficient. Recombination must be additionally confirmed by other methods, such as PCR amplification of genomic DNA, Western blotting, in situ hybridization, or immunostaining.

Tamoxifen-induced recombination of target genes

4. Tamoxifen is a human carcinogen, teratogen, and mutagen. Review and follow the material safety data sheet and wear appropriate personal protective equipment. Pregnant women should not be exposed to or handle Tamoxifen in any form.
5. Tamoxifen is excreted by mice for up to 72 h after the last injection. Considered hazardous waste, bedding has to be

changed and adequately disposed 72 h after the last tamoxifen administration.

6. Tamoxifen-treated mice should not be housed with untreated animals, because unintended recombination can occur by the contamination with traces of tamoxifen during activities such as licking, grooming, or coprophagy.
7. Tamoxifen is transmitted transplacentally and via lactation, which can lead to CreER^{T2}-mediated recombination in embryos or early postnatal pups (*see* for example [49]).
8. Tamoxifen has considerable off-target effects that may confound the phenotypical analysis of floxed CreER^{T2} mutants [109]. It has, for example, been reported to impair consolidation and retrieval of memory in mice at doses relevant for inducing Cre-mediated recombination of floxed target genes [110]. Thus, inclusion of a tamoxifen-injected control group in the behavioral analysis of mice is highly recommended (Fig. 4).
9. Tamoxifen promotes the differentiation of OPC in vitro and in demyelinated lesions in vivo [111, 112] which again demonstrates the importance of including tamoxifen-injected experimental control groups (Fig. 4).
10. Ivermectin, a drug used to treat laboratory mice against ectoparasites, such as pinworms and fur mites, can cause unintended CreER^{T2}-mediated recombination of loxP sites [113]. Here, oral ivermectin treatment of parental mice expressing CreER^{T2} under the control of an ubiquitin promoter and YFP under Rosa26 regulatory elements induced CreER^{T2}-mediated recombination at loxP sites in T-cells of the offspring. Because of the blood brain barrier, the relevance of this finding to research on oligodendroglial cell lineages is yet unclear.

Acknowledgments

We thank Peter Brophy, Brian Popko, Dwight Bergles, Ueli Suter, Bill Richardson, Ori Peles, David Rowitch, and Richard Lu for personal communications on Cre driver lines, members of the Department of Neurogenetics for critical discussion, and Georg Wieser and Ulli Bode for help with the figures. Work in the authors' laboratories was supported by grants from the DFG (SPP 1757 to S.G. and K.A.N.) and by an European Research Council (ERC) advanced grant (to K.A.N.).

References

1. Capecchi MR (2005) Gene targeting in mice: functional analysis of the mammalian genome for the twenty-first century. Nat Rev Genet 6 (6):507–512
2. Thomas KR, Capecchi MR (1987) Site-directed mutagenesis by gene targeting in mouse embryo-derived stem cells. Cell 51 (3):503–512
3. Giese KP et al (1992) Mouse P0 gene disruption leads to hypomyelination, abnormal expression of recognition molecules, and degeneration of myelin and axons. Cell 71 (4):565–576
4. Klugmann M et al (1997) Assembly of CNS myelin in the absence of proteolipid protein. Neuron 18(1):59–70
5. Lappe-Siefke C et al (2003) Disruption of Cnp1 uncouples oligodendroglial functions in axonal support and myelination. Nat Genet 33(3):366–374
6. Xin M et al (2005) Myelinogenesis and axonal recognition by oligodendrocytes in brain are uncoupled in Olig1-null mice. J Neurosci 25 (6):1354–1365
7. Emery B et al (2009) Myelin gene regulatory factor is a critical transcriptional regulator required for CNS myelination. Cell 138 (1):172–185
8. Skarnes WC et al (2011) A conditional knockout resource for the genome-wide study of mouse gene function. Nature 474 (7351):337–342
9. Goebbels S et al (2006) Genetic targeting of principal neurons in neocortex and hippocampus of NEX-Cre mice. Genesis 44 (12):611–621
10. Lu QR et al (2002) Common developmental requirement for Olig function indicates a motor neuron/oligodendrocyte connection. Cell 109(1):75–86
11. Britsch S et al (2001) The transcription factor Sox10 is a key regulator of peripheral glial development. Genes Dev 15(1):66–78
12. Van de Putte T et al (2003) Mice lacking ZFHX1B, the gene that codes for Smad-interacting protein-1, reveal a role for multiple neural crest cell defects in the etiology of Hirschsprung disease-mental retardation syndrome. Am J Hum Genet 72(2):465–470
13. Donohoe ME et al (1999) Targeted disruption of mouse Yin Yang 1 transcription factor results in peri-implantation lethality. Mol Cell Biol 19(10):7237–7244
14. Kotch LE et al (1999) Defective vascularization of HIF-1alpha-null embryos is not associated with VEGF deficiency but with mesenchymal cell death. Dev Biol 209 (2):254–267
15. Dymecki SM (1996) Flp recombinase promotes site-specific DNA recombination in embryonic stem cells and transgenic mice. Proc Natl Acad Sci U S A 93(12):6191–6196
16. Sauer B, Henderson N (1988) Site-specific DNA recombination in mammalian cells by the Cre recombinase of bacteriophage P1. Proc Natl Acad Sci U S A 85(14):5166–5170
17. Gu H et al (1994) Deletion of a DNA polymerase beta gene segment in T cells using cell type-specific gene targeting. Science 265 (5168):103–106
18. Goebbels S et al (2010) Elevated phosphatidylinositol 3,4,5-trisphosphate in glia triggers cell-autonomous membrane wrapping and myelination. J Neurosci 30(26):8953–8964
19. Harrington EP et al (2010) Oligodendrocyte PTEN is required for myelin and axonal integrity, not remyelination. Ann Neurol 68 (5):703–716
20. Wahl SE et al (2014) Mammalian target of rapamycin promotes oligodendrocyte differentiation, initiation and extent of CNS myelination. J Neurosci 34(13):4453–4465
21. Kang SH et al (2013) Degeneration and impaired regeneration of gray matter oligodendrocytes in amyotrophic lateral sclerosis. Nat Neurosci 16(5):571–579
22. LoPresti P (2015) Inducible expression of a truncated form of tau in oligodendrocytes elicits gait abnormalities and a decrease in myelin: implications for selective CNS degenerative diseases. Neurochem Res 40 (11):2188–2199
23. Gossen M, Bujard H (1992) Tight control of gene expression in mammalian cells by tetracycline-responsive promoters. Proc Natl Acad Sci U S A 89(12):5547–5551
24. Gossen M et al (1995) Transcriptional activation by tetracyclines in mammalian cells. Science 268(5218):1766–1769
25. Schonig K, Freundlieb S, Gossen M (2013) Tet-transgenic rodents: a comprehensive, up-to date database. Transgenic Res 22 (2):251–254
26. Goebbels S et al (2005) Cre/loxP-mediated inactivation of the bHLH transcription factor gene NeuroD/BETA2. Genesis 42 (4):247–252
27. Nakagawa Y et al (2016) Ultra-superovulation for the CRISPR-Cas9-mediated production of gene-knockout, single-amino-acid-

substituted, and floxed mice. Biol Open 5 (8):1142–1148

28. Quadros RM et al (2017) Easi-CRISPR: a robust method for one-step generation of mice carrying conditional and insertion alleles using long ssDNA donors and CRISPR ribonucleoproteins. Genome Biol 18(1):92
29. Yang H, Wang H, Jaenisch R (2014) Generating genetically modified mice using CRISPR/Cas-mediated genome engineering. Nat Protoc 9(8):1956–1968
30. Meyers EN, Lewandoski M, Martin GR (1998) An Fgf8 mutant allelic series generated by Cre- and Flp-mediated recombination. Nat Genet 18(2):136–141
31. Farley FW et al (2000) Widespread recombinase expression using FLPeR (flipper) mice. Genesis 28(3–4):106–110
32. Holzenberger M et al (2000) Cre-mediated germline mosaicism: a method allowing rapid generation of several alleles of a target gene. Nucleic Acids Res 28(21):E92
33. Lakso M et al (1996) Efficient in vivo manipulation of mouse genomic sequences at the zygote stage. Proc Natl Acad Sci U S A 93 (12):5860–5865
34. Umans L et al (2003) Generation of a floxed allele of Smad5 for cre-mediated conditional knockout in the mouse. Genesis 37(1):5–11
35. Xu X et al (2001) Direct removal in the mouse of a floxed neo gene from a three-loxP conditional knockout allele by two novel approaches. Genesis 30(1):1–6
36. Ringwald M et al (2011) The IKMC web portal: a central point of entry to data and resources from the International Knockout Mouse Consortium. Nucleic Acids Res 39 (Database issue):D849–D855
37. Rivers LE et al (2008) PDGFRA/NG2 glia generate myelinating oligodendrocytes and piriform projection neurons in adult mice. Nat Neurosci 11(12):1392–1401
38. Kang SH et al (2010) NG2+ CNS glial progenitors remain committed to the oligodendrocyte lineage in postnatal life and following neurodegeneration. Neuron 68(4):668–681
39. Zhu X, Bergles DE, Nishiyama A (2008) NG2 cells generate both oligodendrocytes and gray matter astrocytes. Development 135 (1):145–157
40. Zhu X et al (2011) Age-dependent fate and lineage restriction of single NG2 cells. Development 138(4):745–753
41. Huang W et al (2014) Novel NG2-CreERT2 knock-in mice demonstrate heterogeneous differentiation potential of NG2 glia during development. Glia 62(6):896–913
42. Battiste J et al (2007) Ascl1 defines sequentially generated lineage-restricted neuronal and oligodendrocyte precursor cells in the spinal cord. Development 134(2):285–293
43. Niwa-Kawakita M et al (2000) Targeted expression of Cre recombinase to myelinating cells of the central nervous system in transgenic mice. Genesis 26(2):127–129
44. Hisahara S et al (2000) Targeted expression of baculovirus p35 caspase inhibitor in oligodendrocytes protects mice against autoimmune-mediated demyelination. EMBO J 19 (3):341–348
45. Gow A (2011) Using temporal genetic switches to synchronize the unfolded protein response in cell populations in vivo. Methods Enzymol 491:143–161
46. Michalski JP et al (2011) The proteolipid protein promoter drives expression outside of the oligodendrocyte lineage during embryonic and early postnatal development. PLoS One 6(5):e19772
47. Delaunay D et al (2008) Early neuronal and glial fate restriction of embryonic neural stem cells. J Neurosci 28(10):2551–2562
48. Delaunay D et al (2009) Genetic tracing of subpopulation neurons in the prethalamus of mice (Mus musculus). J Comp Neurol 512 (1):74–83
49. Leone DP et al (2003) Tamoxifen-inducible glia-specific Cre mice for somatic mutagenesis in oligodendrocytes and Schwann cells. Mol Cell Neurosci 22(4):430–440
50. Doerflinger NH, Macklin WB, Popko B (2003) Inducible site-specific recombination in myelinating cells. Genesis 35(1):63–72
51. Fruhbeis C et al (2013) Neurotransmitter-triggered transfer of exosomes mediates oligodendrocyte-neuron communication. PLoS Biol 11(7):e1001604
52. Hovelmeyer N et al (2005) Apoptosis of oligodendrocytes via Fas and TNF-R1 is a key event in the induction of experimental autoimmune encephalomyelitis. J Immunol 175 (9):5875–5884
53. Zou Y et al (2014) Oligodendrocyte precursor cell-intrinsic effect of Rheb1 controls differentiation and mediates mTORC1-dependent myelination in brain. J Neurosci 34(47):15764–15778
54. Kessaris N et al (2006) Competing waves of oligodendrocytes in the forebrain and postnatal elimination of an embryonic lineage. Nat Neurosci 9(2):173–179
55. Zawadzka M et al (2010) CNS-resident glial progenitor/stem cells produce Schwann cells as well as oligodendrocytes during repair of

CNS demyelination. Cell Stem Cell 6 (6):578–590
56. Matsuoka T et al (2005) Neural crest origins of the neck and shoulder. Nature 436 (7049):347–355
57. McKenzie IA et al (2014) Motor skill learning requires active central myelination. Science 346(6207):318–322
58. Simon C et al (2012) Sox10-iCreERT2: a mouse line to inducibly trace the neural crest and oligodendrocyte lineage. Genesis 50 (6):506–515
59. Stine ZE et al (2009) Oligodendroglial and pan-neural crest expression of Cre recombinase directed by Sox10 enhancer. Genesis 47 (11):765–770
60. Silbereis JC et al (2014) Olig1 function is required to repress dlx1/2 and interneuron production in Mammalian brain. Neuron 81 (3):574–587
61. Kawaguchi D et al (2016) Generation and analysis of an improved Foxg1-IRES-Cre driver mouse line. Dev Biol 412(1):139–147
62. Gorski JA et al (2002) Cortical excitatory neurons and glia, but not GABAergic neurons, are produced in the Emx1-expressing lineage. J Neurosci 22(15):6309–6314
63. Shimshek DR et al (2002) Codon-improved Cre recombinase (iCre) expression in the mouse. Genesis 32(1):19–26
64. Dumas L et al (2015) Multicolor analysis of oligodendrocyte morphology, interactions, and development with Brainbow. Glia 63 (4):699–717
65. Traka M et al (2016) Oligodendrocyte death results in immune-mediated CNS demyelination. Nat Neurosci 19(1):65–74
66. Crawford AH et al (2016) Developmental origin of oligodendrocyte lineage cells determines response to demyelination and susceptibility to age-associated functional decline. Cell Rep 15:761–773
67. Tripathi RB et al (2011) Dorsally and ventrally derived oligodendrocytes have similar electrical properties but myelinate preferred tracts. J Neurosci 31(18):6809–6819
68. Nave KA, Ehrenreich H (2014) Myelination and oligodendrocyte functions in psychiatric diseases. JAMA Psychiat 71(5):582–584
69. Brocard J et al (1997) Spatio-temporally controlled site-specific somatic mutagenesis in the mouse. Proc Natl Acad Sci U S A 94 (26):14559–14563
70. Feil R et al (1996) Ligand-activated site-specific recombination in mice. Proc Natl Acad Sci U S A 93(20):10887–10890
71. Feil R et al (1997) Regulation of Cre recombinase activity by mutated estrogen receptor ligand-binding domains. Biochem Biophys Res Commun 237(3):752–757
72. Forni PE et al (2006) High levels of Cre expression in neuronal progenitors cause defects in brain development leading to microencephaly and hydrocephaly. J Neurosci 26(37):9593–9602
73. Qiu L, Rivera-Perez JA, Xu Z (2011) A non-specific effect associated with conditional transgene expression based on Cre-loxP strategy in mice. PLoS One 6(5):e18778
74. Hagemeyer N et al (2012) A myelin gene causative of a catatonia-depression syndrome upon aging. EMBO Mol Med 4(6):528–539
75. Poggi G et al (2016) Cortical network dysfunction caused by a subtle defect of myelination. Glia 64(11):2025–2040
76. Akagi K et al (1997) Cre-mediated somatic site-specific recombination in mice. Nucleic Acids Res 25(9):1766–1773
77. Soriano P (1999) Generalized lacZ expression with the ROSA26 Cre reporter strain. Nat Genet 21(1):70–71
78. Lobe CG et al (1999) Z/AP, a double reporter for cre-mediated recombination. Dev Biol 208(2):281–292
79. De Gasperi R et al (2008) The IRG mouse: a two-color fluorescent reporter for assessing Cre-mediated recombination and imaging complex cellular relationships in situ. Genesis 46(6):308–317
80. Hartwich H, Satheesh SV, Nothwang HG (2012) A pink mouse reports the switch from red to green fluorescence upon Cre-mediated recombination. BMC Res Notes 5:296
81. Madisen L et al (2010) A robust and high-throughput Cre reporting and characterization system for the whole mouse brain. Nat Neurosci 13(1):133–140
82. Srinivas S et al (2001) Cre reporter strains produced by targeted insertion of EYFP and ECFP into the ROSA26 locus. BMC Dev Biol 1:4
83. Hasegawa Y et al (2013) Novel ROSA26 Cre-reporter knock-in C57BL/6N mice exhibiting green emission before and red emission after Cre-mediated recombination. Exp Anim 62(4):295–304
84. Clarke LE et al (2012) Properties and fate of oligodendrocyte progenitor cells in the corpus callosum, motor cortex, and piriform cortex of the mouse. J Neurosci 32 (24):8173–8185

85. Dimou L et al (2008) Progeny of Olig2-expressing progenitors in the gray and white matter of the adult mouse cerebral cortex. J Neurosci 28(41):10434–10442
86. Guo F et al (2010) Pyramidal neurons are generated from oligodendroglial progenitor cells in adult piriform cortex. J Neurosci 30 (36):12036–12049
87. Robins SC et al (2013) Evidence for NG2-glia derived, adult-born functional neurons in the hypothalamus. PLoS One 8(10):e78236
88. Tsoa RW et al (2014) Spatiotemporally different origins of NG2 progenitors produce cortical interneurons versus glia in the mammalian forebrain. Proc Natl Acad Sci U S A 111(20):7444–7449
89. Muzumdar MD et al (2007) A global double-fluorescent Cre reporter mouse. Genesis 45 (9):593–605
90. Prigge JR et al (2013) Nuclear double-fluorescent reporter for in vivo and ex vivo analyses of biological transitions in mouse nuclei. Mamm Genome 24:389–399
91. Rhee JM et al (2006) In vivo imaging and differential localization of lipid-modified GFP-variant fusions in embryonic stem cells and mice. Genesis 44(4):202–218
92. Aggarwal S et al (2011) A size barrier limits protein diffusion at the cell surface to generate lipid-rich myelin-membrane sheets. Dev Cell 21(3):445–456
93. Amitai-Lange A et al (2015) A method for lineage tracing of corneal cells using multi-color fluorescent reporter mice. J Vis Exp (106):e53370
94. Janbandhu VC, Moik D, Fassler R (2014) Cre recombinase induces DNA damage and tetraploidy in the absence of loxP sites. Cell Cycle 13(3):462–470
95. Genoud S et al (2002) Notch1 control of oligodendrocyte differentiation in the spinal cord. J Cell Biol 158(4):709–718
96. Tognatta R et al (2017) Transient Cnp expression by early progenitors causes Cre-Lox-based reporter lines to map profoundly different fates. Glia 65(2):342–359
97. Hirrlinger J et al (2009) Split-CreERT2: temporal control of DNA recombination mediated by split-Cre protein fragment complementation. PLoS One 4(12):e8354
98. Hirrlinger J et al (2009) Split-cre complementation indicates coincident activity of different genes in vivo. PLoS One 4(1):e4286
99. Madisen L et al (2012) A toolbox of Cre-dependent optogenetic transgenic mice for light-induced activation and silencing. Nat Neurosci 15(5):793–802
100. Gibson EM et al (2014) Neuronal activity promotes oligodendrogenesis and adaptive myelination in the mammalian brain. Science 344(6183):1252304
101. Micheva KD et al (2016) A large fraction of neocortical myelin ensheathes axons of local inhibitory neurons. Elife 5:e15784
102. Madisen L et al (2015) Transgenic mice for intersectional targeting of neural sensors and effectors with high specificity and performance. Neuron 85(5):942–958
103. Brockschnieder D et al (2006) An improved mouse line for Cre-induced cell ablation due to diphtheria toxin A, expressed from the Rosa26 locus. Genesis 44(7):322–327
104. Ivanova A et al (2005) In vivo genetic ablation by Cre-mediated expression of diphtheria toxin fragment A. Genesis 43(3):129–135
105. He M et al (2012) Cell-type-based analysis of microRNA profiles in the mouse brain. Neuron 73(1):35–48
106. Sanz E et al (2009) Cell-type-specific isolation of ribosome-associated mRNA from complex tissues. Proc Natl Acad Sci U S A 106(33):13939–13944
107. Pham AH, McCaffery JM, Chan DC (2012) Mouse lines with photo-activatable mitochondria to study mitochondrial dynamics. Genesis 50(11):833–843
108. Platt RJ et al (2014) CRISPR-Cas9 knockin mice for genome editing and cancer modeling. Cell 159(2):440–455
109. Jardi F et al (2017) A shortened tamoxifen induction scheme to induce CreER recombinase without side effects on the male mouse skeleton. Mol Cell Endocrinol 452:57–63
110. Chen D et al (2002) Tamoxifen and toremifene cause impairment of learning and memory function in mice. Pharmacol Biochem Behav 71(1–2):269–276
111. Barratt HE et al (2016) Tamoxifen promotes differentiation of oligodendrocyte progenitors in vitro. Neuroscience 319:146–154
112. Gonzalez GA et al (2016) Tamoxifen accelerates the repair of demyelinated lesions in the central nervous system. Sci Rep 6:31599
113. Corbo-Rodgers E et al (2012) Oral ivermectin as an unexpected initiator of CreT2-mediated deletion in T cells. Nat Immunol 13(3):197–198
114. Trevisol et al (2017) Monitoring ATP dynamics in electrically active white matter tracts. eLife 6. pii: e24241

Chapter 16

Recent Advances in Live Imaging of Cells of the Oligodendrocyte Lineage

Jaime Eugenin von Bernhardi and Leda Dimou

Abstract

Myelination is an important process that takes place also in the periphery during development and in the adulthood. Myelin serves as an electric isolator for axons, leading to a fast conduction of the action potential, and provides trophic support for the axon, both aspects highly important for the proper function of the nervous system. In the central nervous system, myelination starts shortly after birth and cells from the oligodendrocyte lineage tightly regulate this process during the whole life span. Initially, it was thought that under physiological conditions myelin generation only occurs in early postnatal development and that myelination stops at early adult ages. Historically, the process of myelination has mainly been studied in fixed tissue, and predominantly analyzed by electron microscopy, bringing valuable insights in the structure and distribution of myelin in the central nervous system. Nevertheless, the outdated notion of the static nature of myelin during adulthood was challenged in the past decades by the development of new techniques bringing in a new picture of a lively structure that is in constant remodeling under physiological and disease conditions. As fixed tissue can only provide information at a specific timepoint, the necessity of new techniques to study this process in vivo has become clear. In this chapter, we will review some of the latest techniques developed in order to study myelin and the oligodendrocyte lineage, as these cells are important for the formation and restructuration of the myelin. We will also introduce a protocol to prepare a cranial window to study NG2-glia (also known as oligodendrocyte progenitor cells) of the cerebral cortex in vivo, by 2-photon laser scanning microscopy. However, this technique can also be performed to study other cell populations or structures such as myelin, which will be discussed in this chapter as well. Despite being simple, this protocol has shown to be powerful to study the oligodendrocyte lineage and potentially is applicable to study myelin in vivo, which could turn into a key technique in the understanding of myelination and other functions that the oligodendrocyte lineage might have under physiological and disease conditions.

Key words Oligodendrocytes, NG2-glia, Progenitors, 2-Photon laser scanning microscopy, Imaging techniques, Mouse, Cranial window

1 Introduction

The term myelin was first coined in the mid-nineteenth century by the German pathologist Rudolf Virchow, who first observed through light microscopy that in some nerve fibers of the spinal cord there was a glistening, white, fatty substance inside them. The

David A. Lyons and Linde Kegel (eds.), *Oligodendrocytes: Methods and Protocols*, Methods in Molecular Biology, vol. 1936, https://doi.org/10.1007/978-1-4939-9072-6_16,

method used by Virchow revealed clear limitations in the study of myelin and, therefore, he derived inaccurate conclusions regarding its cellular localization. Later on, with the development of new imaging techniques, researchers discovered that myelin surrounded axons and was widely distributed in the peripheral and central nervous system (PNS and CNS, respectively). Nowadays, it is a general consensus that myelination of axons is a fundamental process for the proper function of the vertebrate nervous system, because it electrically isolates the axon, speeds up the action potential propagation, and provides trophic support to neurons (for review *see* [1]). Different imaging techniques had been essential in the initial study of myelin structure and organization in the brain, including methods such as immunofluorescence and electron microscopy.

Immunofluorescence studies have been useful in determining the general distribution of myelin in the immature and mature nervous system, mainly by using primary antibodies against major myelin proteins such as the proteolipid protein (PLP), the myelin associated-glycoprotein (MAG), and the myelin basic protein (MBP), providing an easy distinction between highly myelinated areas, known as white matter (WM), and less densely myelinated areas, for example, the gray matter (GM). However, only the great resolution provided by electron microscopy could reveal the fine structure of myelin sheaths, such as their multi-laminar organization around the axon [2]. It was only much more recently shown that myelin is a highly dynamic structure, both during development and in the adult brain, which can be regulated by experience and environmental interaction. This evidence raises a new view of the complex and dynamic functions of myelin in the nervous system, which techniques based on static analyses are insufficient to fully elucidate. The requirement of tissue fixation may introduce perturbation to the axon or myelin structure, which complicates the interpretation of data. Additionally, such techniques are restricted to the study of single timepoints and do not allow one to follow myelin dynamics during development, plasticity, or disease. Therefore, the development of new methods to study myelin and myelination in vivo has been critical in the last decades.

1.1 MRI and PET Techniques

In the field of clinical neuroimaging, Magnetic Resonance Imaging (MRI), in which signal intensities and image contrast are primarily determined by the spin density of water protons, has been widely used [3, 4]. Myelin, compared to other brain tissues, has a high lipid and low water content (~40% against ~80% of water content in other brain tissue [4]), whereby WM can be distinguished from GM areas of the brain by MRI. Thus, MRI became an important tool for studying WM/myelin in diseases of CNS, including multiple sclerosis (MS) [5]. Nevertheless, changes in water content can occur not only by changes in the myelin, but also by inflammation, gliosis, edema, and axonal loss [3], hence making it a nonspecific technique for myelin detection and therefore difficult to interpret changes of it in MRI scans.

Diffusion Tension Imaging (DTI), a form of MRI, has brought huge advances in the study of myelin in live healthy subjects, patients, and other nonhuman animal models. DTI takes advantage of the fact that water molecules diffuse differently in distinct brain tissues [6]. The diffusion of water in the WM is anisotropic (water diffuses more readily in one direction, along the axon, than in other directions), while in the GM and the cerebral spinal fluid it is isotropic (water diffuses equally in any direction). Most studies claim that the degree of anisotropy is directly associated to anatomical features of the WM, such as axon caliber, fiber density, and myelination. In other words, assuming that the axon and fiber characteristics remain constant, the degree of myelination will dictate the degree of anisotropy [6]. Interestingly, it has been shown that subjects training extensively for complex visual-motor tasks, such as juggling or piano playing, exhibit increases in the anisotropy of the corpus callosum [7, 8], which could be interpreted as an increase in myelin content in this area. When subjects stop their training for some weeks anisotropy decreased somewhat, pointing to changes of myelin toward basal conditions [8]. However, to date it has been difficult to use DTI to differentiate between changes to myelin and axons in health or disease.

An alternative method to study myelin in live subjects was developed more recently, employing Positron Emission Tomography (PET), which can provide a more quantifiable and myelin-specific approach. The physical principles of PET are based in the detection of radioactivity emitted after a radioactive tracer (in pM amounts) that is injected into a peripheral vein [9]. Because of radioactive decay, the injected substance starts emitting positrons that electrostatically interact with free electrons in the environment, leading to the formation of a pair of gamma ray photons that travel in opposite direction, a process known as annihilation [9, 10]. These photons can be detected by using pairs of collinearly aligned detectors in coincidence, which allow to spatially locate the signals in the body [9, 10]. The specificity of PET scanning resides in the usage of a radioactive tracer that can target a molecule of interest to a specific target tissue, and, in the case of the brain, this needs to be able to cross the blood–brain barrier (BBB). Notably, specific radiotracers for myelin have started to be developed within the last decade [11–14]. PET scanning has indeed been shown to be useful in observing disease progression, injury identification, and remyelination in MS. Nevertheless, it is important to consider that findings are tightly dependent on the WM injury size, due to the restriction in the spatial resolution of the scanner [13].

1.2 Label-Free In Vivo Microscopy Techniques

In recent years, development of nonlinear microscopy has enabled the label-free visualization of cells and tissues. Due to the structure, arrangement, and composition of myelin, it has been possible to create high-resolution in vivo images of myelinated axons within their tissue environment without an exogenous applied tracer or a reporter. These techniques could be potentially useful in the detection of myelin changes during pathology or experience-dependent plasticity over time.

Coherent anti-Stokes Raman scattering (CARS) microscopy, also known as CARS spectroscopy, is a variant of the Raman spectroscopy, which also uses the intrinsic molecular vibration to produce microscopic images [15, 16]. CARS can create an image from the CH_2 groups of lipid-rich samples. As myelin consists of oligodendrocytic membranes composed of ~70% lipids [16, 17], it contains a high density of CH_2 groups; whereby an image can be created from the large and directional resonant CARS signal from the CH_2 stretch (a change in the length of the molecular bond) vibration [17]. Notably, this technique also allows one to structurally characterize myelin, and detect nodes of Ranvier and the Schmidt-Lanterman incisures [17] as well as to measure myelin thickness [17] and composition.

Another label-free technique to image myelin is Third-Harmonic Generation (THG) microscopy. Higher harmonic generation is a nonlinear coherent scattering process, resulting from the phase matching and summation of light waves that are induced by inorganic and organic structures that possess specific physical properties, molecular arrangement, and order [18]. For myelin studies, THG signals have been of particular interest. These are generated when three photons at the fundamental frequency ω interact with a sample generating one emitted photon with a frequency 3ω and one-third the excitation photon wavelength, whereby the combined energy of these three photons is converted into one emitted photon with tripled energy [18–20]. This phenomenon happens at structural interfaces, such as at local transitions of the refractive index or the third-order nonlinear susceptibility χ^3 [19, 20], in which both correspond to the intrinsic physical property of materials. THG can happen at interfaces that are formed between aqueous interstitial fluids and lipid-rich structures or between water and large protein aggregates. Probably due to its multilayered nature, in which several interfaces between aqueous and lipid-rich membranes are present, myelin represents an extraordinary source of THG in the nervous system, which has been tested in both ex vivo and in vivo systems [21, 22]. It has been shown, e.g., in mouse cortical slices that high THG intensity signal corresponds to WM regions of the central nervous system, such as the mossy fibers in the cerebellum, the pons, the hindbrain, the corpus callosum in the cerebral cortex, and the spinal cord [21, 22].

More recently, Spectral Confocal Reflectance (ScoRe) microscopy has been developed in order to solely image myelinated axons in the mouse brain in vivo [23, 24]. ScoRe microscopy takes advantage of the high refractive index of the lipid-rich myelin, which reflection signals are obtained by the use of multiple confocal lasers. Through this process, different wavelengths show individual images of discontinuous myelin segments, but when merged, it is possible to recreate contiguous myelinated axons images. Despite the fact that the physical principals behind this phenomenon are not totally clear, multicolor reflection might be explained by thin-film interference. Similar to dichroic mirrors with alternating layers of optical coating with different refractive indices, myelin is composed of several layers of lipid-rich membranes, which possibly lead to constructive and destructive light interference, reinforcing certain reflected wavelengths and suppressing others. If this is the case, it is probable that factors, such as thickness and number of layers, may play a role in determining the wavelengths that are preferentially reflected, with further variability due to focal irregularities in myelin sheath thickness, lipid composition, and other local cellular variables.

Myelination is a process that occurs following oligodendrocyte lineage progression from oligodendrocyte progenitor cells (OPCs), also known as NG2-glia, into oligodendrocytes (OL), which produce the myelin that is wrapped around axons in the CNS. It appears clear, therefore, that the study of the NG2-glial and oligodendrocyte populations will be fundamental to fully understand the dynamics of myelination. Unfortunately, most MRI or PET machines have a restricted spatial resolution of ~200 μm and ~1 mm, respectively [21]; therefore, cellular and subcellular structures are not optically accessible, even though MRI has nowadays reached the capability of imaging in a resolution that can analyze brain cytoarchitecture [4, 25, 26]. Due to this limitation, we will not further discuss about these techniques in this chapter; however, we provide a table with a comparison between the different imaging techniques (Table 1). On the other hand, although label-free nonlinear microscopy techniques have an extraordinary resolution, to our knowledge, they have not yet been developed to specifically study either NG2-glia or OLs. Therefore, another approach is required for the study of the oligodendrocyte lineage on the cellular level.

1.3 Two-Photon Laser Scanning Microscopy

Two-photon laser scanning microscopy (2-PLSM) can be used for real-time imaging or time-lapse imaging of cells in the nervous system. This technique is based on the principles of the two-photon absorption and emission nonlinear process, theoretically developed by Maria Göppert-Mayer in her doctoral dissertation more than eight decades ago [27, 28]. In single photon absorption (like in confocal microscopy), a molecule absorbs an

Table 1
Comparison of different techniques for in vivo imaging of myelin and the oligodendrocyte lineage

Technique	Advantages	Disadvantages
MRI	• Complete brain image • Noninvasive	• Nonspecific • Low resolution relative to conventional microscopy
DTI	• Complete brain image • Functional imaging • Noninvasive	• Low resolution relative to conventional microscopy
PET scan	• Complete brain image • Functional imaging • Noninvasive	• Specific • Low resolution relative to conventional microscopy • Usage of radioactive isotopes
CARS microscopy	• High resolution (cellular level) • Invasive (requires window)	• Small imaging area
THG microscopy	• High resolution (cellular level) • Invasive (requires window)	• Small imaging area • No functional imaging possible
ScoRe microscopy	• High resolution (cellular level) • Requires a conventional confocal microscope with multi-lasers • Invasive (requires window)	• Small imaging area • No functional imaging possible
2-PLSM	• High resolution (cellular level) • Functional imaging • Invasive (requires window) • Low photobleaching • Low phototoxicity • Low thermal emission	• Small imaging area • Depth constrain (although it is improving)

excited photon and produces an emitted photon with a lower energy and longer wavelength. Interestingly, when two photons, each with half the energy necessary to promote the transition of the fluorophore from the ground state to its excited state, are simultaneously absorbed by the molecule (or in the case of the 2-PLSM within <1 fs delay), the emitted photon has higher energy and shorter wavelength than the one of the excited photon [27, 29]. This phenomenon is the key concept to understand the advantage of 2-PLSM as it allows one to perform in vivo imaging with a low energy and long wavelength light source, normally in the near-infrared spectrum (700–1400 nm). This principle allows to decrease the photo bleaching of the fluorophore, to reduce phototoxicity and to increase the tissue penetration (longer wavelengths are less scattered than shorter ones) leading to a superb resolution within the focal plane [30].

The use of 2-PLSM to study the oligodendrocyte lineage is well placed, due to the wide variety of available transgenic reporter mouse lines, which express fluorescent proteins like the green fluorescent protein (GFP) under the promoter of well-known molecular markers of the oligodendrocyte lineage. For example, transgenic reporter mice exist to label: (1) NG2-glia under the control of the proteoglycan NG2 [31, 32] or PDGFRα promoter [33], (2) mature oligodendrocytes under the PLP promoter [34], and (3) the complete oligodendrocyte lineage under the promoter of transcription factors like Sox10 [35] or Olig2 [36]. It has additionally been shown by in vivo brain imaging that oligodendrocytes can also be directly labeled by sulforhodamine 101 (SR101) [24], which previously had been reported to exclusively label astrocytes. Interestingly, more recent studies contradicted the idea that oligodendrocyte progenitor cells proliferate, migrate, and differentiate into oligodendrocytes only during development and could show that they also do so in the adult brain under physiological and pathological conditions [31, 34, 37–41]. Indeed, 2-PLSM has confirmed that NG2-glia constantly proliferate and differentiate into oligodendrocytes during the mice adulthood [31, 34]. Interestingly, proliferation and population size of NG2-glia are highly regulated by self-repulsion leading to a constant cell number under physiological conditions [31]. Notably, as soon as one NG2-glia differentiates or dies, a neighboring NG2-glia will proliferate and/or migrate into the NG2-glia-free area [31]. 2-PLSM in the field of the oligodendrocyte lineage is not uniquely centered in the mammalian brain, but also in other animal models, like the zebrafish larvae, which transparency and widely mutant lines make it an excellent model to study oligodendrocytes and their generation in the central nervous system [10–43].

In this chapter, we will describe a protocol to perform in vivo 2-PLSM or confocal imaging of the oligodendrocyte lineage in the mouse cerebral cortex through a cranial window. The methodology is mainly based on the protocol published in Bardehle et al. [44], although it has been modified here, and some extra considerations have been included in our methodology in order to study the oligodendrocyte lineage.

2 Materials

2.1 Reagents

1. Cortex buffer: NaCl (125 mM), KCl (5 mM), Glucose (10 mM), $CaCl_2$ (2 mM), $MgSO_4$ (2 mM) and HEPES (10 mM) dissolved in sterile ddH_2O and pH adjusted to 7.4.
2. Powder dental cement.
3. Povidone-iodine.
4. Paladur dissolvent.

5. Superglue.
6. Dextran-coupled Texas Red dye MW70000.

2.2 Equipment

1. Surgical tools.
2. 3–5 mm diameter glass coverslips.
3. Syringes.
4. Driller.
5. 2-Photon microscope with mouse adapter.

2.3 Drugs

2.3.1 Analgesia

Carprofen (Rimadyl) (4 mg/kg) diluted in sterile chloride sodium solution.

2.3.2 Anesthesia

Antagonizable Sleeping solution: midazolam (5 mg/kg of animal), medetomidine (0.5 mg/kg), and fentanyl (0.025 mg/kg) diluted in sterile sodium chloride solution. Other deep anesthesia can also be used.

2.3.3 Counter-Anesthesia

Awake solution: atipamezole (2.5 mg/kg of animal), flumazenil (0.5 mg/kg), and buprenorphine (0.1 mg/kg) diluted in sterile sodium chloride solution.

3 Methods

All experiments performed in animals, including the cranial window implantation and chronic imaging, must be conducted in accordance with institutional and governmental ethical guidelines and regulations in animal experimentation. Only after clarifying this point, you may proceed. A scheme of the operation procedure is shown in Fig. 1.

3.1 Procedure

1. Before anesthetizing the mouse, the optimal performance of the cranial window highly depends on a clean work space and sterile surgical tools; therefore, work surface and all surgical tools must be disinfected with 70–80% ethanol or equivalent disinfectant solutions beforehand.
2. Anesthetize the animal with an intraperitoneal (i.p.) injection of sleeping solution and carprofen (s.c., preferably 1 hour before anesthesia) as a long-term painkiller. After 5–10 min, confirm that the mouse is sleeping by checking its pinch-toe or pinch-tail reflex (*see* **Notes 1–3**).
3. Fix the mouse's head into a stereotaxic frame and make sure to have a rigid fixed animal. Once the animal head is fixed, the anesthetized mouse will keep its eyes open; apply ophthalmic ointment to the eyes to prevent corneal drying (*see* **Notes 4** and **5**).

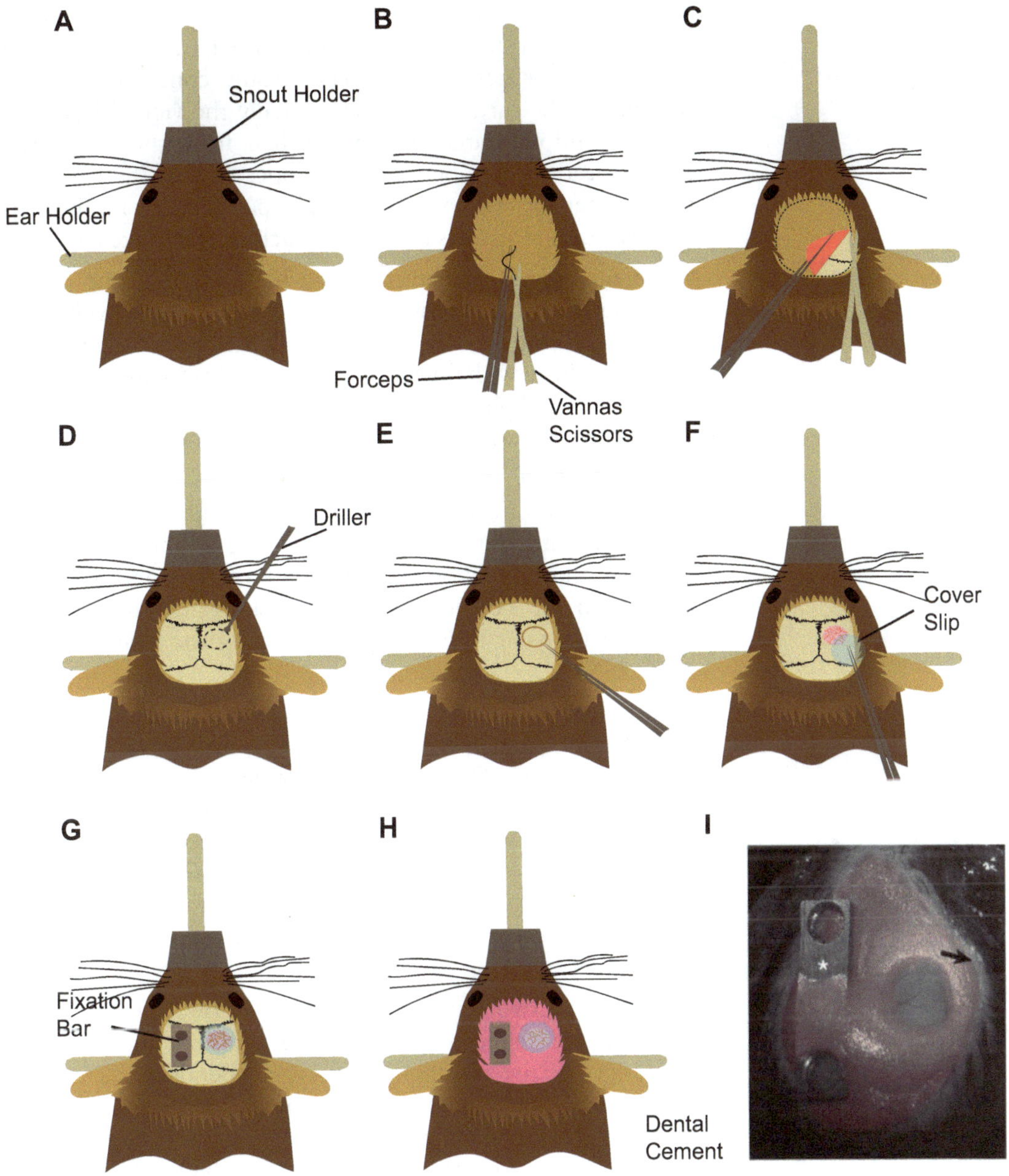

Fig. 1 Scheme representing a craniotomy. (**a**) Animal head is first placed and fixed in a stereotactic frame and (**b**) the head is shaved, exposing the head skin. An initial cut is done in the base of the head between the ears and afterward (**c**) the head skin is cut and removed exposing the skull. (**d**) A region of interest has to be selected and drilling has to be done in the skull in a circular fashion (**e**) until the bone is loosed and sequentially removed with the help of a sharp forceps. (**f**) A glass coverslip is placed on top of the craniotomy and (**g**) glued to the skull. Afterward, a fixation bar is placed in the contralateral hemisphere and finally (**h**) dental cement is applied over the skull and in the surrounding of the fixation bar and the coverslip. (**i**) Photograph of a cranial window with fixation bar and dental cement

4. Remove the hair from the area of the head where the craniotomy is going to be performed. Hair either can interfere with the surgery or get onto the craniotomy compromising the hygiene of the surgery or block partially the view field of the cranial window. Carefully trim the hair from the top of the mouse head located between the upper part of the neck and the area between the eyes. This can be performed with the help of a pair of fine scissors or a razor machine (*see* **Note 6**).
5. Once the animal is correctly fixed in the frame, one of the concerns must be to ensure the sterility of the area where the surgery will be performed. To disinfect the operated area, wipe the furless head with povidone-iodine, easily obtained in veterinary stores and pharmacies. Immediately after disinfecting the area, apply analgesic on the top of the mouse's head, to reduce the pain post-operation (*see* **Note 7**).
6. With the help of a scissor or a scalpel in your dominant hand and forceps in your non-dominant hand, lift the skin located between the more posterior segment of the ears. Make an incision in the bottom of the lifted skin, providing a small entrance for your scissors to keep cutting the head skin (*see* **Note 8**).
7. Remove the skin from the top of the mouse head and proceed to remove the skull periosteum with the help of an absorbant swab or a scalpel. The periosteum is a dense irregular connective tissue that surrounds most of the bones which is easily recognizable by its membrane-like appearance lying immediately over the skull. With the help of the scalpel, score the surface of the skull. This will create an irregular surface improving the adhesion of the glue and the dental cement applied later (*see* **Note 9**).
8. In literature, two types of imaging window can be found: (a) complete craniotomy plus cranial window and (b) skull thinning. The use and appropriate type depends on the experimental design and question (i.e., if you want to image a lesioned area or inject a viral vector, you need to open the skull in advance to perform the injury) (*see* **Notes 10** and **11**).
9. Drill in a ring shape of about 3–4 mm of inner diameter until the skull is thin in the edges of the circle. Carefully, with the help of forceps or the drill in off mode, gently press the upper part of the skull with drilled edges and check whether the circular piece of bone becomes loose (*see* **Note 12**).
10. Once the drilling is performed, add a drop of cortex buffer over the skull with drilled edges and with the help of thin-sharped tip forceps proceed to remove the skull (*see* **Note 13**).
11. In case of bleeding, flush with cortex buffer the blood until the bleeding stops completely (*see* **Note 14**).

12. Once the bleeding stops and after thorough washing with cortex buffer, place a 5 mm glass coverslip, which has to be previously cleaned with ethanol and washed out with cortex buffer, over the craniotomy (*see* **Note 15**).
13. With the help of a thin syringe needle push the glue around the edges of the coverslip carefully without glue getting inside the craniotomy or below the window (*see* **Notes 16** and **17**).
14. Add glue in the contralateral side of the craniotomy and place firmly a metal fixation bar, which will help to fix the head of the mouse during the imaging.
15. Mix the powder dental cement with the paladur dissolvent and carefully add the cement in the edges of the coverslip and the surrounding of the fixation bar. To avoid further infections of the exposed skull areas, add dental cement to cover them. Make a well-shaped structure with the dental cement around the coverslip, where you can fill up with immersion medium. Finally, wait until the cement dries and then you are ready to start the imaging.
16. Inject dextran-coupled Texas Red dye intravenously (i.v.), using the tail vein or retro-orbital vein [45], which will help to establish landmarks in order to find the area of imaging and follow a specific cell population during the whole experimental period.
17. During the imaging, the mouse must be fixed in an imaging frame and, again, a heating pad must be placed beneath the animal to keep the body temperature constant (*see* **Note 18**).
18. At the end of the experiment, administer i.p. "awake solution" and return the animal to its cage. Use a cage with a heating pad or place the complete cage on a heating pad to avoid hypothermia. Normally, a mouse can take several minutes to wake up and several hours to completely recover. Times may depend on, among other factors, the concentration of the drugs and the duration for which the animal was anesthetized.
19. Regular checking must be performed in between the time points of imaging. Check closely if there are any signs of infection, inflammation, or if the animal removes its cranial window (*see* **Notes 19–23**).

3.2 Final Remarks

The reasons to develop better techniques for in vivo imaging to study the oligodendrocyte lineage seem obvious. This glial population has shown to be highly dynamic under physiological conditions and to react to different stimuli, such as injury or neuronal activity, and during disease, such as multiple sclerosis, leaving behind the notion of static myelin in the CNS.

In this chapter, we have described a surgical procedure that can help to study the oligodendrocyte lineage in vivo with different types of microscopy that furthermore could be easily applied to several animal models (under physiological and pathological conditions). Nevertheless, we cannot totally ensure that the procedure itself does not lead to a reaction of this or another glial population. After experiments in our group and other research groups, it could however been shown that the craniotomy only leads to minimal or even no glial reaction [44]. In contrast, the literature has reported changes in the dendritic spine turnover depending on the type of the window (craniotomy vs. thinned skull), which has been attributed to the glial reaction to the surgery. As we cannot exclude any reaction, this reflects the importance of appropriate controls to study secondary effects of the cranial window in parallel to establishing the cranial window protocol.

The craniotomy is suited to 2-PLSM imaging, a powerful technique for in vivo imaging of cells in the CNS, because of its low phototoxicity and great power of penetration. Moreover, this method is convenient for studying the oligodendrocyte lineage due to the broad variety of well-established existing transgenic mouse lines, which allow the identification of different populations of the complete lineage. Nevertheless, it is also imperative to take in consideration that the number of reporter positive cells in constitutive and inducible reporter mouse lines which necessary depends on the recombination rate (reporter-positive cells overall the total population of cells of interest) of the used line. It must be kept in mind that in the case of a high proportion of reporter-positive cells, the discrimination of individual cells in a single focal plane could be difficult, while in the case of a low recombination rate, a selection of a specific subset of cells being labeled could lead to results not applicable to the complete cell population.

2-PLSM allows the performance of time-lapse imaging, following individual cells over long periods of time [31, 34], and of real-time imaging. However, as NG2-glia only move around 10 μm/week at a speed of 2 μm/day [31] resulting at a 60.9 nm/h (speed of 83.3 nm/h) a short-term real-time imaging would not be appropriate to observe the migration capability of NG2-glia. Even at a maximum speed of 250 nm/h (maximum speed reported in Hughes et al. [31]), no significant displacement of the cells could be observed in the real-time imaging. Nevertheless, it has been shown that NG2-glia and oligodendrocytes might have dynamic intracellular calcium levels, that are high enough to measure calcium spike signals in response to different factors and injury [46–48], making the study of the role of calcium signaling in the oligodendrocyte lineage in vivo by 2-PLSM possible. Several organic dyes, such as fura-2 or Oregon green, and fluorescent calcium indicator proteins, such as GCamp6 or RCamp2, could be used to study the oligodendrocyte calcium response to environmental changes.

As we previously described, the mammalian deep lying WM has a high population of oligodendrocytes in the CNS, but there is a lack of studies in a noninvasive in vivo model. Naturally, the main limitations of 2-PLSM and other imaging techniques is the scattering of photons in the heterogeneous brain tissue, which restricts the high-resolution imaging to superficial cortex layers or to cells located ~1 mm depth. To overcome this issue, several strategies have been implemented in other fields, such as microendoscopy, in which part of the cortex is removed and replaced with a metal cannula, which allows one to perform in vivo imaging in deeper structures [49, 50]. Nevertheless, we need to keep in mind that the surgical intervention in this methodology will naturally lead to inflammation and most probably would affect the behavior and dynamics of the oligodendrocyte lineage residing below. A more adequate strategy would probably be to apply further improvements of multi-photon microscopy, such as the three-photon microscopy. This increases the depth of optical images by raising the needed light source wavelength (~1700 nm) and has been shown to provide high-resolution, high-contrast images of the blood vessels up to ~1300 μm deep and neurons from hippocampal CA1 with good signal-to-noise ratio at ~1000 μm [45]. Therefore, this new development seems promising to overpass at least some of the limitations of the 2-PLSM.

It is important to highlight that any single microscopy modality, such as CARS, THG, ScoRe, confocal, and 2-PLSM, only delivers an incomplete picture of the tissue, restricted to either subcellular structures, molecular composition, or cellular organization. Interestingly, different multimodal methodologies have been developed, which combine confocal or 2-PLSM with other nonlinear microscopy techniques. It has been shown that different multiphoton microscopes can acquire images from THG and simultaneously from conventional fluorescent signals of the molecular reporter in the same sample [45]. Furthermore, it has been shown that in vitro calcium imaging by 2-PLSM can also detect a fluorescent marker and THG signals, providing both structural and functional information about the imaged cells [51].

Although CARS and THG have a great potential in myelin imaging, in general they require complex setups and laboratories are often not equipped with the required devices. On the other hand, although ScoRe microscopy does not use two-photon microscopy principles, it could be a better option, because it only requires a conventional confocal microscope using multiple lasers and in theory the depth penetration can be improved by using Ti: sapphire lasers, which produce light sources with longer wavelengths, or by using fiber-optic coupling [23]. Therefore, combining conventional confocal and ScoRe microscopy could bring better insights of how oligodendrocytes behave and myelinate over time. Even though ScoRe microscopy requires significantly low light levels than the ones used for conventional confocal fluorescence or 2-PLSM (on the range of 200–400 μW [23]), the potential phototoxicity and thermal injury generated by the technique must be studied in more depth, especially when combined with conventional confocal in vivo imaging.

4 Notes

1. It is important that the animal is unconscious during the surgery and the imaging (to avoid unnecessary pain and stress as well as artifacts due to breathing-associated movements).

   Subcutaneous injection of dexamethasone before the surgery can, in principle, be applied to reduce the inflammatory and swelling reaction of the brain. Dexamethasone is a synthetic corticosteroid, which has high affinity to the glucocorticoid receptor (GR). It has been shown that mice express GR in the oligodendrocyte lineage [52] and cortisol decreases the proliferation of NG2-glia in the adult rat [53]. Thus, we recommend not administering dexamethasone to mice for imaging of the oligodendrocyte lineage, although ultimately will depend in the experimental design and question.

   Isoflurane can also be used during the craniotomy and chronic imaging [54], having the benefit of a fast awakening of the animals after removal of the anesthetic. For induction 3% (wt/vol) isoflurane in 100% oxygen (0.5–1 L/min) and during surgery and imaging 1.5% (wt/vol) isoflurane in 100% oxygen (0.5–1 L/min) should be used.

   As isoflurane is a volatile drug, it must be used in well-ventilated areas.

2. If the animal wakes up during surgery, administer more sleeping solution by i.p. Anesthesia was probably not enough. Always administer anesthesia proportionally to the weight of the animal.

3. In case the mouse is anesthetized with sleeping solution longer than 3h, terminate the experiment and administer i.p. awake solution. Animal should be further imaged at another time. Prolonged exposure to the anesthetic can result in the death of the mouse. Other drugs, like isoflurane can give the researcher longer time periods to work.

4. If hair gets onto the surgery area, flush the hair away with saline or cortex buffer. If the hair sticks to the skull, try to remove it with the help of a scalpel or thin tip forceps. It is better to remove the hair at this time point, in order to avoid compromising the hygiene of the craniotomy.

5. If hairs get inside the craniotomy, flush and try to wash away the hairs with cortex buffer. If hair sticks to the cortex, try cautiously to remove it with a thin and small absorbant swab. The probability that your craniotomy hygiene is compromised increases the chances of having an infection and decreasing the view of field quality.
6. Anesthetized animals rapidly lose heat due to their large surface to body weight ratio. Therefore a pre-warmed heating pad (set at 35.5–37.5 °C considering that the mouse body temperature is around 36–39 °C), water blanket, or a thermal isolating material has to be located underneath the animal during the surgery and the imaging period. This will considerably improve the survival during surgery and the recovery of your mouse post-anesthesia.
7. Ethanol (70–80%) for disinfection can also be applied if the hair was not trimmed with an electric razor. Otherwise, it can cause irritation in the applied area and pain to the animal.
8. Take special care when you approach the skin close to the eyelids.
9. The improper removal of periosteum from the exposed skull or improper skull scoring can affect the stability of your cranial window, leading to the coverslip or the dental cement to become loose, thus making it easier for the mouse to remove it after the operation.
10. If a blood vessel located in the skull is damaged and starts bleeding during drilling, flush with buffer cortex until the bleeding stops. After stopping the bleeding, continue drilling. Next time, try to exert less pressure during the drilling or decrease the drilling speed.
11. If, during drilling, a fracture in the skull is generated and the driller penetrates through the skull, continue the procedure, remove the skull and evaluate the damage. If dura is compromised and there is an excess of bleeding, the mouse should not be used for imaging. Due to the fact that the oligodendrocyte lineage reacts to injury [31, 37, 55], results can be difficult to interpret. Next time, try to exert less pressure during the drilling or decrease the drilling speed.
12. The drilling is a critical step and it will determine the performance of the cranial window and the inflammatory response of the brain. Therefore, it is important not to exert excessive pressure over the skull during drilling. We recommend to gently lay the driller over the skull and keep the hand as steady as possible. Do not drill at very high speeds.

    The friction between the drill and the skull can potentially produce overheat on the drilled zone. Therefore, while drilling, constantly moisturize the skull with cortex buffer or saline solution and do not keep drilling in a single place for prolonged times (more than ~2 s).

We use a pressure hydraulic drill, but there are many other alternatives that can be used. Pneumatic dental drillers are commonly used in many laboratories due to the thin tip and their capacity of ejecting pressurized air, which helps to remove the skull spins released during the drilling. Biopsy punches are also used and are more comfortable in performing regular and size precise craniotomy, as well as reducing the risk of damaging the dura or the brain.

13. It is possible to thin a small circular-shaped area of the skull with a high-speed drill to a thickness of between ~15 and 50 μm [31, 56, 57]. This procedure is performed in a similar way to the method described above, but the difference is that the skull is not removed. The immersion medium is added on the thinned area of the skull and images can be acquired. Nevertheless, it has been shown that the remaining bone might affect the image quality of the deeper layers in the cortex (below ~100 μm) and reduce the depth capability of your preparation [57].

14. Cortex buffer must be at room temperature before applying over the brain in order to avoid cell cryodamage. It is very important to leave the brain surface completely untouched to avoid any reactive gliosis in the upper layers of the cortex. Many research laboratories use gelfoam in order to stop the bleeding. In our experience, gelfoam can generate reactive gliosis in the upper layers of the cortex due to the compression on the brain, making more difficult the interpretation of further results.

15. As alternatives for glass coverslip placement, you can also fit the glass coverslip within the craniotomy and subsequently exert a small pressure over the glass before gluing. This seems to increase the stability of the cranial window for chronic imaging. Nevertheless, this procedure might compromise the dura leading to a reactive gliosis and subsequent inflammation.

16. Exposure of the brain to glue can lead to a compromised dura and a subsequent inflammatory response. Furthermore, if the glue goes below the window it might affect the quality of the imaging.

17. If, after placing and gluing the coverslip, the brain bleeds, carefully remove the coverslip and rinse the craniotomy with cortex buffer. After the bleeding stops, place a new coverslip over the craniotomy. Blood will interfere with your imaging, blocking the light or will introduce a new element that will scatter the light.

18. Before starting the imaging, make sure that the animal is still anesthetized by checking pinch-toe or pinch-tail reflex.

Depending on the experiment, an imaging session can last from minutes to hours. If the animal wakes up, it might be distressed due to the fixation and the movement can add extra artifacts to the recording.

19. If the mouse removed the fixation bar, place a new bar in the space were the previous one was. In case it does not fit, with the help of a driller (dental cement gets extraordinary hard) trim the edges of the space where the previous bar was placed and then glue a new bar.
20. In case the mouse removed the cranial window, the solution depends on the institutional and governmental ethical guidelines and regulations in animal experimentation. Nevertheless, if the mouse just removed the dental cement and not the cranial window, prepare a new mix of dental cement and place it surrounding the cranial window. If the cranial window is also removed, inject i.p. sleep solution and place a new coverslip over the craniotomy. Check the craniotomy regularly to detect any sign of meningeal infection. Next time, try to have a better-fixed window by: (a) Properly removing the periosteum membrane. (b) Scoring the skull surface to have an irregular surface, which will improve the adherence of the dental cement. (c) Mix dental cement mixture with some drops of glue to improve the dental cement adherence.
21. In case the imaging gets worse in the subsequent days, clean the cranial window with 70–80% ethanol. Alternatively, carefully, replace the cranial window. The cranial window is exposed to the animals' environment; therefore, it can get dirty by the contact with other mice or elements in the cage.
22. In case white or white-yellow milky substance appears underneath the cranial window, this might be a sign of an infection and, therefore, an inflammatory reaction will take place. Discard the mouse. As we mentioned above, NG2-glia react to inflammation; therefore, the resulted images afterward might be complicated to interpret. Next time, make sure all surgical tools, coverslips, and workspace are cleaned with 70–80% ethanol and prepare a new sterile buffer cortex.
23. In case inflammation of the dura takes place and the images look blurry, wait 4–10 days until the inflammation finishes and start imaging again. If the inflammation does not disappear, discard the mouse.

References

1. Nave KA, Werner HB (2014) Myelination of the nervous system: mechanisms and functions. Annu Rev Cell Dev Biol 30:503–533. https://

doi.org/10.1146/annurev-cellbio-100913-013101

2. Sjostrand FS (1953) The lamellated structure of the nerve myelin sheath as revealed by high resolution electron microscopy. Experientia 9 (2):68–69
3. Alonso-Ortiz E, Levesque IR, Pike GB (2015) MRI-based myelin water imaging: a technical review. Magn Reson Med 73(1):70–81. https://doi.org/10.1002/mrm.25198
4. Watanabe T, Frahm J, Michaelis T (2016) In vivo brain MR imaging at subnanoliter resolution: contrast and histology. Magn Reson Med Sci 15(1):11–25. https://doi.org/10.2463/mrms.2015-0048
5. Paty DW, Oger JJ, Kastrukoff LF, Hashimoto SA, Hooge JP, Eisen AA, Eisen KA, Purves SJ, Low MD, Brandejs V et al (1988) MRI in the diagnosis of MS: a prospective study with comparison of clinical evaluation, evoked potentials, oligoclonal banding, and CT. Neurology 38(2):180–185
6. Alexander AL, Lee JE, Lazar M, Field AS (2007) Diffusion tensor imaging of the brain. Neurotherapeutics 4(3):316–329. https://doi.org/10.1016/j.nurt.2007.05.011
7. Bengtsson SL, Nagy Z, Skare S, Forsman L, Forssberg H, Ullen F (2005) Extensive piano practicing has regionally specific effects on white matter development. Nat Neurosci 8 (9):1148–1150. https://doi.org/10.1038/nn1516
8. Scholz J, Klein MC, Behrens TE, Johansen-Berg H (2009) Training induces changes in white-matter architecture. Nat Neurosci 12 (11):1370–1371. https://doi.org/10.1038/nn.2412
9. Bailey DL, Karp JS, Surti S (2005) Physics and Instrumentation in PET. In: Bailey DL, Townsend DW, Valk PE, Maisey MN (eds) Positron Emission Tomography. Springer-Verlag, London pp 13–39 https://doi.org/10.1007/1-84628-007-9_2
10. Ziegler S (2005) Positron emission tomography: principles, technology, and recent developments. Nucl Phys A 752:679–687. https://doi.org/10.1016/j.nuclphysa.2005.02.067
11. Wu C, Wang C, Popescu DC, Zhu W, Somoza EA, Zhu J, Condie AG, Flask CA, Miller RH, Macklin W, Wang Y (2010) A novel PET marker for in vivo quantification of myelination. Bioorg Med Chem 18(24):8592–8599. https://doi.org/10.1016/j.bmc.2010.10.018
12. Wang C, Wu C, Zhu J, Miller RH, Wang Y (2011) Design, synthesis, and evaluation of coumarin-based molecular probes for imaging of myelination. J Med Chem 54 (7):2331–2340. https://doi.org/10.1021/jm101489w
13. Stankoff B, Freeman L, Aigrot MS, Chardain A, Dolle F, Williams A, Galanaud D, Armand L, Lehericy S, Lubetzki C, Zalc B, Bottlaender M (2011) Imaging central nervous system myelin by positron emission tomography in multiple sclerosis using [methyl-(1)(1) C]-2-(4′-methylaminophenyl)-6-hydroxybenzothiazole. Ann Neurol 69(4):673–680. https://doi.org/10.1002/ana.22320
14. Stankoff B, Wang Y, Bottlaender M, Aigrot MS, Dolle F, Wu C, Feinstein D, Huang GF, Semah F, Mathis CA, Klunk W, Gould RM, Lubetzki C, Zalc B (2006) Imaging of CNS myelin by positron-emission tomography. Proc Natl Acad Sci U S A 103(24):9304–9309. https://doi.org/10.1073/pnas.0600769103
15. Cheng J, Xie X (2004) Coherent anti-Stokes Raman scattering microscopy: instrumentation, theory, and applications. J Phys Chem 108(3):827–840. https://doi.org/10.1021/jp035693v
16. Fu Y, Huff TB, Wang HW, Wang H, Cheng JX (2008) Ex vivo and in vivo imaging of myelin fibers in mouse brain by coherent anti-Stokes Raman scattering microscopy. Opt Express 16 (24):19396–19409
17. Wang H, Fu Y, Zickmund P, Shi R, Cheng JX (2005) Coherent anti-stokes Raman scattering imaging of axonal myelin in live spinal tissues. Biophys J 89(1):581–591. https://doi.org/10.1529/biophysj.105.061911
18. Weigelin B, Bakker GJ, Friedl P (2016) Third harmonic generation microscopy of cells and tissue organization. J Cell Sci 129 (2):245–255. https://doi.org/10.1242/jcs.152272
19. Barad Y, Eisenberg M, Horowitz M, Silberberg Y (1997) Nonlinear scanning laser microscopy by third harmonic generation. Appl Phys Lett 70(8):922–924. https://doi.org/10.1063/1.118442
20. Muller M, Squier J, Wilson KR, Brakenhoff GJ (1998) 3D microscopy of transparent objects using third-harmonic generation. J Microsc 191(3):266–274
21. Farrar MJ, Wise FW, Fetcho JR, Schaffer CB (2011) In vivo imaging of myelin in the vertebrate central nervous system using third harmonic generation microscopy. Biophys J 100 (5):1362–1371. https://doi.org/10.1016/j.bpj.2011.01.031
22. Witte S, Negrean A, Lodder J, Testa-Silva G, Mansvelder H, Louise Groot M (2011) Label-free live brain imaging and targeted patching with third-harmonic generation microscopy.

Proc Natl Acad Sci U S A 108(15):5970–5975. https://doi.org/10.1073/pnas.1018743108
23. Schain AJ, Hill RA, Grutzendler J (2014) Label-free in vivo imaging of myelinated axons in health and disease with spectral confocal reflectance microscopy. Nat Med 20 (4):443–449. https://doi.org/10.1038/nm.3495
24. Hill RA, Grutzendler J (2014) In vivo imaging of oligodendrocytes with sulforhodamine 101. Nat Methods 11(11):1081–1082. https://doi.org/10.1038/nmeth.3140
25. Duyn JH, van Gelderen P, Li TQ, de Zwart JA, Koretsky AP, Fukunaga M (2007) High-field MRI of brain cortical substructure based on signal phase. Proc Natl Acad Sci U S A 104 (28):11796–11801. https://doi.org/10.1073/pnas.0610821104
26. Chuang KH, Belluscio L, Koretsky AP (2010) In vivo detection of individual glomeruli in the rodent olfactory bulb using manganese enhanced MRI. NeuroImage 49 (2):1350–1356. https://doi.org/10.1016/j.neuroimage.2009.09.060
27. Göppert-Mayer M (1931) Über Elementarakte mit zwei Quantensprüngen. Ann Phys 9:273–294. https://doi.org/10.1002/andp.19314010303
28. Masters BR, So PT (2004) Antecedents of two-photon excitation laser scanning microscopy. Microsc Res Tech 63(1):3–11. https://doi.org/10.1002/jemt.10418
29. Svoboda K, Yasuda R (2006) Principles of two-photon excitation microscopy and its applications to neuroscience. Neuron 50 (6):823–839. https://doi.org/10.1016/j.neuron.2006.05.019
30. Denk W, Strickler JH, Webb WW (1990) Two-photon laser scanning fluorescence microscopy. Science 248(4951):73–76
31. Hughes EG, Kang SH, Fukaya M, Bergles DE (2013) Oligodendrocyte progenitors balance growth with self-repulsion to achieve homeostasis in the adult brain. Nat Neurosci 16 (6):668–676. https://doi.org/10.1038/nn.3390
32. Huang W, Zhao N, Bai X, Karram K, Trotter J, Goebbels S, Scheller A, Kirchhoff F (2014) Novel NG2-CreERT2 knock-in mice demonstrate heterogeneous differentiation potential of NG2 glia during development. Glia 62 (6):896–913. https://doi.org/10.1002/glia.22648
33. Kang SH, Fukaya M, Yang JK, Rothstein JD, Bergles DE (2010) NG2+ CNS glial progenitors remain committed to the oligodendrocyte lineage in postnatal life and following neurodegeneration. Neuron 68(4):668–681. https://doi.org/10.1016/j.neuron.2010.09.009
34. Hill RA, Patel KD, Goncalves CM, Grutzendler J, Nishiyama A (2014) Modulation of oligodendrocyte generation during a critical temporal window after NG2 cell division. Nat Neurosci 17(11):1518–1527. https://doi.org/10.1038/nn.3815
35. Simon C, Lickert H, Gotz M, Dimou L (2012) Sox10-iCreERT2: a mouse line to inducibly trace the neural crest and oligodendrocyte lineage. Genesis 50(6):506–515. https://doi.org/10.1002/dvg.22003
36. Dimou L, Simon C, Kirchhoff F, Takebayashi H, Gotz M (2008) Progeny of Olig2-expressing progenitors in the gray and white matter of the adult mouse cerebral cortex. J Neurosci 28(41):10434–10442. https://doi.org/10.1523/JNEUROSCI.2831-08.2008
37. Simon C, Gotz M, Dimou L (2011) Progenitors in the adult cerebral cortex: cell cycle properties and regulation by physiological stimuli and injury. Glia 59(6):869–881. https://doi.org/10.1002/glia.21156
38. Gibson EM, Purger D, Mount CW, Goldstein AK, Lin GL, Wood LS, Inema I, Miller SE, Bieri G, Zuchero JB, Barres BA, Woo PJ, Vogel H, Monje M (2014) Neuronal activity promotes oligodendrogenesis and adaptive myelination in the mammalian brain. Science 344(6183):1252304. https://doi.org/10.1126/science.1252304
39. Mangin JM, Li P, Scafidi J, Gallo V (2012) Experience-dependent regulation of NG2 progenitors in the developing barrel cortex. Nat Neurosci 15(9):1192–1194. https://doi.org/10.1038/nn.3190
40. Hines JH, Ravanelli AM, Schwindt R, Scott EK, Appel B (2015) Neuronal activity biases axon selection for myelination in vivo. Nat Neurosci 18(5):683–689. https://doi.org/10.1038/nn.3992
41. Mensch S, Baraban M, Almeida R, Czopka T, Ausborn J, El Manira A, Lyons DA (2015) Synaptic vesicle release regulates myelin sheath number of individual oligodendrocytes in vivo. Nat Neurosci 18(5):628–630. https://doi.org/10.1038/nn.3991
42. Czopka T (2016) Insights into mechanisms of central nervous system myelination using zebrafish. Glia 64(3):333–349. https://doi.org/10.1002/glia.22897
43. Czopka T, Ffrench-Constant C, Lyons DA (2013) Individual oligodendrocytes have only a few hours in which to generate new myelin

sheaths in vivo. Dev Cell 25(6):599–609. https://doi.org/10.1016/j.devcel.2013.05.013

44. Bardehle S, Kruger M, Buggenthin F, Schwausch J, Ninkovic J, Clevers H, Snippert HJ, Theis FJ, Meyer-Luehmann M, Bechmann I, Dimou L, Gotz M (2013) Live imaging of astrocyte responses to acute injury reveals selective juxtavascular proliferation. Nat Neurosci 16(5):580–586. https://doi.org/10.1038/nn.3371
45. Horton NG, Wang K, Kobat D, Clark CG, Wise FW, Schaffer CB, Xu C (2013) In vivo three-photon microscopy of subcortical structures within an intact mouse brain. Nat Photon 7(3):205–209. https://doi.org/10.1038/nphoton.2012.336
46. Butt AM (2006) Neurotransmitter-mediated calcium signalling in oligodendrocyte physiology and pathology. Glia 54(7):666–675. https://doi.org/10.1002/glia.20424
47. Hamilton N, Vayro S, Wigley R, Butt AM (2010) Axons and astrocytes release ATP and glutamate to evoke calcium signals in NG2-glia. Glia 58(1):66–79. https://doi.org/10.1002/glia.20902
48. Haberlandt C, Derouiche A, Wyczynski A, Haseleu J, Pohle J, Karram K, Trotter J, Seifert G, Frotscher M, Steinhauser C, Jabs R (2011) Gray matter NG2 cells display multiple Ca2+-signaling pathways and highly motile processes. PLoS One 6(3):e17575. https://doi.org/10.1371/journal.pone.0017575
49. Attardo A, Fitzgerald JE, Schnitzer MJ (2015) Impermanence of dendritic spines in live adult CA1 hippocampus. Nature 523 (7562):592–596. https://doi.org/10.1038/nature14467
50. Barretto RP, Ko TH, Jung JC, Wang TJ, Capps G, Waters AC, Ziv Y, Attardo A, Recht L, Schnitzer MJ (2011) Time-lapse imaging of disease progression in deep brain areas using fluorescence microendoscopy. Nat Med 17(2):223–228. https://doi.org/10.1038/nm.2292
51. Barille R, Canioni L, Rivet S, Sarger L, Vacher P, Ducret T (2001) Visualization of intracellular Ca2+ dynamics with simultaneous two-photon-excited fluorescence and third-harmonic generation microscopes. Appl Phys Lett 79(24):4045–4047. https://doi.org/10.1063/1.1425450
52. Matsusue Y, Horii-Hayashi N, Kirita T, Nishi M (2014) Distribution of corticosteroid receptors in mature oligodendrocytes and oligodendrocyte progenitors of the adult mouse brain. J Histochem Cytochem 62(3):211–226. https://doi.org/10.1369/0022155413517700
53. Alonso G (2000) Prolonged corticosterone treatment of adult rats inhibits the proliferation of oligodendrocyte progenitors present throughout white and gray matter regions of the brain. Glia 31(3):219–231
54. Holtmaat A, Bonhoeffer T, Chow DK, Chuckowree J, De Paola V, Hofer SB, Hubener M, Keck T, Knott G, Lee WC, Mostany R, Mrsic-Flogel TD, Nedivi E, Portera-Cailliau C, Svoboda K, Trachtenberg JT, Wilbrecht L (2009) Long-term, high-resolution imaging in the mouse neocortex through a chronic cranial window. Nat Protoc 4(8):1128–1144. https://doi.org/10.1038/nprot.2009.89
55. Dimou L, Gotz M (2014) Glial cells as progenitors and stem cells: new roles in the healthy and diseased brain. Physiol Rev 94 (3):709–737. https://doi.org/10.1152/physrev.00036.2013
56. Grutzendler J, Kasthuri N, Gan WB (2002) Long-term dendritic spine stability in the adult cortex. Nature 420(6917):812–816. https://doi.org/10.1038/nature01276
57. Isshiki M, Okabe S (2014) Evaluation of cranial window types for in vivo two-photon imaging of brain microstructures. Microscopy (Oxf) 63(1):53–63. https://doi.org/10.1093/jmicro/dft043

Chapter 17

The DTA Mouse Model for Oligodendrocyte Ablation and CNS Demyelination

Maria Traka

Abstract

Genetic mouse models facilitate investigation of mechanisms underpinning human diseases and aid the development of novel therapeutic treatments. To better understand the demyelination and remyelination processes in adult-onset demyelinating diseases like multiple sclerosis (MS), we have developed the DTA mouse model system that allows for the widespread ablation of the mature oligodendrocytes, resulting in demyelination throughout the central nervous system (CNS). Induction of oligodendrocyte death in young adult DTA mice causes extensive CNS demyelination that leads to a severe neurological disease, followed by a full recovery that is associated with extensive replenishment of oligodendrocytes and remyelination. Thus, the DTA mouse enables investigation of the mechanisms that promote remyelination in MS and other adult-onset demyelinating diseases. Approximately 30 weeks later, the recovered DTA mice develop a fatal secondary demyelinating disease that is mediated by autoimmune T cells. Therefore, the DTA mouse model is also ideal for elucidating the role of oligodendrocyte death in eliciting autoimmunity in MS. In this chapter we describe the methods we used to generate the DTA mouse model and to analyze both the primary and secondary demyelinating diseases in DTA mice.

Key words DTA, Oligodendrocyte death, Demyelination, Remyelination, Multiple sclerosis, Autoimmune disease

1 Introduction

Oligodendrocytes are essential for producing the myelin sheaths that facilitates the saltatory propagation of action potentials in the CNS. Furthermore, oligodendrocytes provide trophic support to axons through mechanisms that are not well understood [1, 2]. Multiple sclerosis (MS) is an autoimmune inflammatory demyelinating disease characterized by lesions formed in different CNS areas. In MS, autoimmune T cells and other immune cells infiltrate the CNS and attack the myelin and axons leading to neuronal dysfunction [3, 4]. The precise mechanisms that trigger MS, however, remain unknown [5]. It has been proposed that oligodendrocyte death, which occurs in new MS lesions in the absence of apparent signs of inflammation, could trigger

David A. Lyons and Linde Kegel (eds.), *Oligodendrocytes: Methods and Protocols*, Methods in Molecular Biology, vol. 1936, https://doi.org/10.1007/978-1-4939-9072-6_17,

autoreactivity against the myelin antigens leading to inflammation and demyelination in the CNS [6, 7].

To investigate the role of oligodendrocyte death in adult-onset demyelinating diseases like MS, we generated the DTA mouse model [8], in which we induce the death of mature oligodendrocytes in young adult *ROSA26-eGFP-DTA; PLP/CreER*^{T} (DTA) mice carrying the diphtheria toxin A-subunit gene (DT-A) [9] and expressing the tamoxifen inducible version of Cre recombinase (CreER^{T}) in oligodendrocytes [10]. DT-A induces cell death by catalyzing the inactivation of elongation factor 2 and inhibiting protein synthesis [11–13]. Tamoxifen injections in the DTA mice activate the Cre recombination at the loxP sites of the *ROSA26-eGFP-DTA* cassette and induce DT-A expression. This approach results in widespread death of oligodendrocytes starting at 7 days post-injection (dpi) throughout the CNS, with maximum cell loss occurring by 21 dpi. As a result, the DTA mice develop neurological symptoms such as ataxia and tremor that become severe by 35 dpi (peak of disease) correlating with demyelination and significantly impaired axonal conduction in the CNS [8]. The DTA mice recover significantly from their motor and axonal conduction defects by ~70 dpi, and they consistently showed replenishment of the missing oligodendrocytes and remyelination throughout the CNS [8].

Overall, the DTA mouse model shows that the CNS has a robust innate potential to remyelinate. Nevertheless, remyelination fails in MS leading to chronically demyelinated axons that eventually degenerate and contribute to disease progression. Therefore, enhancing the endogenous remyelination potential of the CNS [4] is a promising therapeutic approach to promoting functional recovery and axonal survival in multiple sclerosis.

The DTA mouse model is ideal for elucidating the mechanisms that could enhance the endogenous remyelination potential of the CNS and promote the functional recovery in MS as it specifically targets myelinating oligodendrocytes, without the confounding nonspecific effects present in the toxin, or in experimental autoimmune encephalomyelitis (EAE) and viral models [8].

The DTA mice that recover from oligodendrocyte loss, approximately 30 weeks later, develop a secondary fatal neurological disease characterized by seizures, ataxia, and weight loss, which gradually worsen over time [14]. To distinguish between the two diseases in the DTA mouse, we will refer to the primary disease as "acute" and the secondary disease as "late-onset." Time course analysis of the clinical symptoms in DTA mice during the acute up to the early stages of the late-onset disease is shown in Fig. 1 using the EAE-like clinical scoring system described in Table 1. The late-onset disease occurs without significant oligodendrocyte loss and it is characterized by widespread demyelination and axonal loss in the CNS, which is reminiscent of the progressive phase of MS

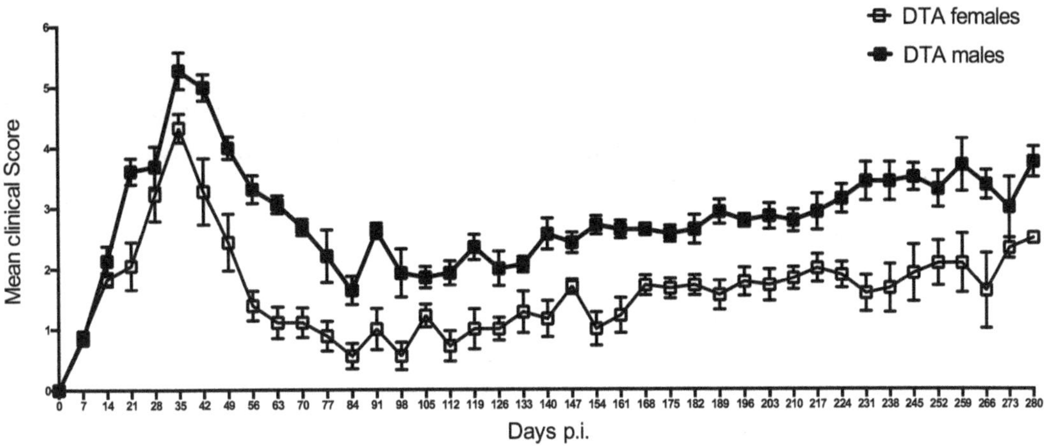

Fig. 1 Time course analysis of the disease symptoms observed in the male and female DTA mice during the acute disease and the early stages of the late-onset disease using the clinical scoring system outlined in Table 1. $N = 9$ mice in each group of males and females at 0 dpi. The number of DTA mice that survive during the late-onset disease is significantly reduced. Maria Traka and Brian Popko, unpublished data

Table 1
Clinical scoring of the tamoxifen-injected DTA mice during the acute disease and the early stages of the late-onset disease

Clinical score	Clinical symptoms
0	Mouse shows no symptoms of disease[a]
1	Mouse is slightly ataxic
2	Mouse shows moderate ataxia ± tremor + splayed gait
3	Mouse shows significant ataxia ± tremor + splayed gait
4	Mouse shows significant ataxia and stiff gait ± tremor and drags its hind limbs
5	Mouse shows significant ataxia, has a stiff gait and falls over on its side
6	Mouse is laterally recumbent and needs to be euthanized if it cannot move at all for humane reasons[b]

[a]Throughout the entire disease course, soft gel food is applied on the cage floor to help the DTA mice maintain a healthy weight

[b]At this stage, the mice are given subcutaneous (SQ) injections of 1 mL saline twice a day to prevent dehydration

[15]. CNS demyelination is also associated with local T cell infiltration and inflammation and with autoimmune MOG-specific T cell responses in the peripheral lymphoid organs, indicating that the late-onset disease is autoimmune T cell mediated. This hypothesis is supported by two main strands of evidence; first, adoptive transfer of the MOG-specific T cells into naïve *Rag1*-deficient mice results in mild neurological symptoms that associate with focal inflammatory lesions in the CNS matter; second, late-onset disease

symptoms were significantly attenuated in the DTA mice that were tolerized against MOG [14]. Furthermore, this proposed mechanism for the late-onset DTA disease supports the idea that oligodendrocyte death can induce late-onset CNS autoimmunity. Thus, the DTA mouse model offers the unique opportunity to investigate the processes that trigger autoimmunity in MS and screen for novel therapeutic approaches that could block this event and thereby prevent CNS damage. Here we include the protocols for inducing acute disease in DTA mice and the histological methods we used analyze the CNS defects in the acute and the late-onset disease.

2 Materials

2.1 Generation of the DTA Mice and Genotyping

1. Female and male *ROSA26-eGFP-DTA* and *PLP/CreER^T mice* (Jackson Laboratories).
2. Genotyping primers for *PLP/CreER^T* mice:

   Primer 1: 5′-GAT GTA GCC AGC AGC ATG TC-3′.
   Primer 2: 5′-ACT ATA TCC GTA ACC TGG AT-3′
3. Genotyping primers for *ROSA26-eGFP-DTA* mice:
   R26R1: 5′-AAA GTC GCT CTG AGT TGT TAT-3′.
   R26R2: 5′-GCG AAG AGT TTG TCC TCA ACC-3′.
   R26R3: 5′-GGA GCG GGA GAA ATG GAT ATG-3′.

2.2 Preparation of Tamoxifen and Injections of the DTA Mice

1. The 4-hydroxytamoxifen (≥70% Z isomer, *see* **Note 1**) is dissolved in a DMSO:ethanol:oil (4:6:90) mixture at a concentration of 8 mg/mL (*see* **Note 2**) as follows: Add DMSO (250 μL) and 100% EtOH (375 μL) to tamoxifen (50 mg) and mix well by vortexing until all the powder dissolves. Transfer the solution into a 15 mL tube and then add 5625 μL of sunflower seed oil and mix briefly by vortexing.
2. Sunflower seed oil.
3. 100% ethanol.
4. Dimethyl sulfoxide (DMSO).
5. 1 mL syringes and 25 G × 5/8 in. needles.
6. Ultrasonic bath.

2.3 Acute and Late-Onset Disease Monitoring

1. Rotating rod (Rotarod).
2. Gel food (clear H_20 76A diet gel).
3. Cage pad (ALPHA-Pad).
4. Saline solution (0.9% NaCl).

2.4 Histological Analysis of DTA Mice

2.4.1 Tissue Processing for Immunohistochemistry

1. ddH_2O.
2. 10 N NaOH.
3. 1 L 4% Paraformaldehyde (PFA) solution: Heat 500 mL of double distilled water (ddH_2O) in a glass beaker on a stir plate under a fume hood to 60 °C and add 40 g of PFA. Make sure that temperature does not exceed 60 °C. Carefully add a few drops of 10 N NaOH until the solution becomes clear and let it cool down. In a separate glass beaker dissolve 4.46 g NaOH and 19 g of sodium phosphate monobasic in 100 mL of ddH_2O on a stir plate and add to the PFA solution. Filter the solution and adjust the volume to 1 L by adding ddH_2O. Check the pH and adjust if necessary to pH 7.2–7.4 by adding a few drops of 5 N HCl. The solution can be stored at 4 °C for up to 1 week.
4. 10% sodium azide (NaN_3): Dissolve 10 g of sodium azide in 100 mL of ddH_2O in a glass beaker on a stir plate in a fume cupboard. Use protective gloves and lab coat since the sodium azide is highly toxic. Whenever possible, use a premixed stock solution. Store at 4 °C.
5. 30% sucrose solution (1 L): Dissolve 300 g of sucrose in 700 mL of ddH_2O in a glass beaker on a stir plate and add 100 mL of 10× phosphate buffered-saline (PBS), pH 7.4 and 10 mL of 10% sodium azide as preservative. Filter solution and adjust the volume to 1 L by adding ddH_2O. Store indefinitely at 4 °C.
6. Peristaltic pump.
7. Tissue-Tek Optimal Cutting Temperature compound (O.C.T., EMS).
8. Tissue-Tek cryomolds.
9. 18 G needles.

2.4.2 Immuno-histochemistry

1. Glass or plastic staining jars with cold (−20 °C) acetone.
2. Mouse monoclonal antibody CC-1 (Millipore Sigma, OP80).
3. Mouse monoclonal MBP antibody (SMI99, Biolegend).
4. Rabbit polyclonal MBP antibody (Abcam, cat. no. ab40390).
5. Rat monoclonal antibody CD11b (clone 5C6, Serotec).
6. Rat monoclonal CD3 antibody (17A2, Santa Cruz Biotechnology).
7. Anti-mouse Alexa Fluor 594 or 488 (Molecular Probes).
8. Anti-rabbit Alexa Fluor 594 or 488 (Molecular Probes).
9. Anti-rat Alexa Fluor 594 or 488 (Molecular Probes).
10. Blocking solution: 5% fetal calf serum (FCS) in 1× PBS with 0.1% Triton X-100.

11. Microcentrifuge.
12. Glycerol-based mounting media with DAPI.
13. Fluorescence microscope equipped with a digital camera.
14. Image J software (NIH).

2.4.3 Tissue Processing for Electron Microscopy

1. Fume hood.
2. Protective gloves.
3. Lab coat.
4. Safety goggles.
5. Face mask.
6. 2.5% Glutaraldehyde (GTA) and 4% PFA (EM fixative) solution (1 L): Heat 500 mL ddH_2O in a glass beaker on a stir plate under a fume hood to approximately ~60 °C and add 40 g of PFA. Make sure that temperature does not exceed 60 °C. Add carefully a few drops of 10 N NaOH until the solution becomes clear and let it cool down. In a separate glass beaker dissolve 4.46 g NaOH and 19 g of sodium phosphate monobasic in 100 mL of ddH_2O on a stir plate and add to the PFA solution. Filter the solution and add 32 of 10 mL-ampoules of 8% GTA. Adjust the volume to 1 L by adding ddH_2O. Check the pH and adjust if necessary to pH 7.2–7.4 by adding a few drops of 5 N HCl. The solution needs to be made fresh.
7. Small scissors.
8. Fine forceps.
9. Spatula.
10. 15 mL plastic tubes.
11. 0.2 M Millonig's phosphate buffer, pH 7.4 (premade, e.g., EMS).
12. Stereomicroscope.
13. Scalpel.
14. Ultramicrotome.
15. 2 mL clear glass vials.
16. 1 mL transfer pipettes.
17. Tube rotator.
18. 1% osmium tetroxide: mix equal volumes of 0.2 M Millonig's buffer and 2% osmium tetroxide solution. Work under a fume hood and handle with care since osmium is toxic.
19. 30%, 50%, 70%, 90%, 95%, and 100% ethanol.
20. Propylene oxide (PO).
21. Epon-812 mix as described in Table 2.
22. Disposable plastic tri-corn beaker.

Table 2
Reagents used for the EMBed-812 mix (100 mL)

Reagent	EMS cat. number	Weight (g)
EMBed-812	14,900	44.30
DDSA	13,710	33.20
NMA	19,000	20.10
DMP-30	13,600	1.67

23. Small vacuum chamber and 60 °C oven.
24. Sample labels.
25. Flat embedding molds.
26. Stereomicroscope.

2.4.4 Sectioning, Staining of Sections, and Imaging

1. Ultramicrotome.
2. Toluidine blue solution: 1 g of Toluidine Blue O, 2 g of sodium borate (Borax), 100 mL of ddH_2O. Filter with an 0.2 µm filter before use.
3. ddH_2O.
4. Xylene-based mounting medium.
5. Superfrost Plus Microscope Slides.
6. Prepare the Toluidine blue solution by dissolving 1 g of Toluidine Blue O and 2 g of sodium borate into 100 mL of ddH_2O.
7. Light microscope.
8. 200 mesh Cu grids.
9. 5% uranyl acetate solution: 2.5 g of uranyl acetate into 50 mL ddH_2O. Cover with foil and stir overnight. Add 10 drops of glacial acetic acid and store the solution for up to 6 months at 4 °C. Filter with an 0.2 µm filter before use.
10. Lead citrate solution: add sequentially the following chemicals in 30 mL of ddH_2O: 1.33 g of lead nitrate, 1.76 g of sodium citrate, dihydrate and 5 mL of 1 N NaOH. Stir the solution for 10 min to dissolve and adjust the volume to 50 mL with ddH_2O and store it for up to 6 months at 4 °C. Filter with an 0.2 µm filter before use.
11. Transmission Electron Microscope (TEM) equipped with a digital camera.

3 Methods

All experiments involving animals and animal tissues have to be performed according to the guidelines described in the animal use protocol that has been approved by the local institute.

3.1 Generation of the DTA Mice and Genotyping

Homozygous *ROSA26-eGFP-DTA* [9] females were generated by crossing the heterozygous *ROSA26-eGFP-DTA* mice. The resulting mice were genotyped by PCR using genomic DNA from mouse tail samples and the primers R26R1, R26R2, and R26R3. This breeding scheme produces approximately 50% *ROSA26-eGFP-DTA* heterozygous mice (an ~650 bp wild-type band and a 340 bp mutant band), 25% *ROSA26-eGFP-DTA* homozygous mice (a 340 bp band), and 25% wild-type mice (an ~650 bp band). The *PLP/CreER^T* mice were maintained by crossing *PLP/CreER^T* males with C57BL/6J females. The resulting mice were genotyped by PCR using genomic DNA from mouse tail samples and the primers Primer 1 and Primer 2. This breeding scheme produces approximately 50% *PLP/CreER^T* mice (a 532 bp band). The homozygous *ROSA26-eGFP-DTA* females (*see* **Note 3**) were crossed to *PLP/CreER^T* males to produce the double mutants *ROSA26-eGFP-DTA; PLP/CreER^T* (DTA mice).

3.2 Preparation of Tamoxifen and Injections of the DTA Mice

1. Sonicate the tamoxifen solution in water bath sonicator (without heat) until it is completely mixed (~30 min).
2. Protect the tamoxifen solution from light and keep it at 4 °C for 2 weeks maximum (*see* **Note 4**).
3. Mix briefly the tamoxifen solution and inject 5–7-week-old (*see* **Note 5**) male DTA mice intraperitoneally (IP) daily with a dose of 0.8 mg (100 μL)/day for 4 continuous days (*see* **Note 3**). Inject the female DTA mice IP daily with a dose of 0.8 mg (100 μL)/day for 3 continuous days (*see* **Note 3**).

3.3 Acute and Late-Onset Disease Monitoring

1. On the day of injection (0 dpi), weigh and score the DTA mice using the scoring system described in Table 1. Weigh and score the DTA mice at least once a week thereafter (*see* **Note 6**). Use *ROSA26-eGFP-DTA* littermate mice as control normal mice during the entire course of the acute and late-onset DTA disease.
2. At 0 dpi train both the DTA and the control mice to maintain themselves on the rotating rod (rotarod) at a constant speed (5 rpm) and monitor their motor performance once a week by recording the time (latency) that they spend on the rotarod in an accelerating speed mode (5–65 rpm) during 4 trial sessions, 5 min each. The motor performance of the DTA mice on the rotarod is always compared to control mice of the same gender.

During the peak of the disease (~35 dpi) and for the next couple weeks the DTA mice are significantly uncoordinated and they cannot stand on the rotarod. The DTA mice start recovering at ~49 dpi and their motor performance becomes nearly normal at ~77 dpi [8].

3. Start applying half a cup of gel food (clear H_20 76A diet gel) in the cage of the DTA mice at 0 dpi and replenish every few days to prevent significant weight loss. Once the DTA mice reach a clinical score of 3 or above then apply daily one cup of gel food.
4. Apply a cage pad (ALPHA-Pad) on the bottom of the cage on top of the bedding once the DTA mice reach a clinical score of 4 or above. This pad helps the mice that are at score 6 (on the side) to maneuver around by gripping the pad with their forefeet and drag themselves around better.
5. If the DTA mice become very sick and they cannot stand upright (score 6), they should receive subcutaneous (SQ) injections of 1 mL saline twice a day. Also, these mice are flipped from the side they are resting daily.
6. Upon recovering from oligodendrocyte loss, usually around 70 dpi (*see* **Note 7**), the DTA mice continue to be monitored at least once a week for late-onset disease symptoms such as seizures, weight loss, and motor defects using the scoring system shown in Table 1 (*see* **Note 8**) as well as rotarod analysis. Typically the late onset disease symptoms start around 280 dpi [14] (Fig. 1). Approximately 50% of the DTA mice are expected to die by 1 year post-activation [14] (*see* **Note 9**).

3.4 Histological Analysis of DTA Mice

3.4.1 Tissue Processing for Immunohistochemistry

Follow all appropriate precautions to prevent exposure to PFA, which is toxic, and its vapors: work under a fume hood and wear protective gloves, lab coat, safety goggles, and face mask during the PFA preparation, perfusion, and tissue processing. Discard any liquid and solid chemical waste according to local institute's regulations for waste disposal.

1. Mice are deeply anesthetized with approved anesthetic (e.g., Avertin) prior to the perfusion. Use pins to mount the animal's paws on a Styrofoam board that is placed within a receptive plastic container.
2. Adjust a needle in the plastic tube that is connected to a peristaltic pump operating at a velocity approximately 10 mL/min. Needle gauge depends on the mouse age (e.g., use 18 G for adults). Fill the tube with saline and turn pump off to avoid air bubbles.
3. Using small scissors and forceps open the mouse's chest cavity, expose the heart and insert the needle into the left ventricle,

start the pump and open the right atrium to flush out the blood with saline for a few minutes.

4. Stop the pump and transfer the plastic tube into the 4% PFA solution and perfuse the animal for 10 min. At the end the body should become stiff, which is an indication that the perfusion has been successful.
5. Using small scissors, forceps, and spatula, carefully dissect the brain and the spinal cord and the optic nerves from the perfused animal and place tissues into 4% PFA in a 15 mL plastic tube for 2 h at 4 °C for postfixation.
6. Wash the tissues three times for 5 min each with 1× PBS using 1 mL transfer pipettes.
7. Place tissues in cold (4 °C) 30% sucrose solution at 4 °C for 24–72 h or until the tissues sink in the bottom of the tube for cryoprotection.
8. On the day of the embedding, fill up a 250 mL glass beaker with cold (−20 °C) isopentane (or 2-methylbutane) and place it on a sheet of dry ice pellets inside a Styrofoam box.
9. Empty the tube with the sucrose solution and the tissues in a petri dish and carefully cut the brain into two mid-sagittal halves and the spinal cord horizontally into three smaller pieces of approximately 5 mm length at the cervical, lumbar, and thoracic level using a scalpel or a blade.
10. Wipe off the excess sucrose and embed each tissue carefully for cutting at the preferred orientation on a plastic cryomold filled with OCT media: usually sagittal or coronal for brain, horizontal for spinal cord segments, and longitudinal or horizontal for optic nerves.
11. Immerse the cryomold with the tissue in the isopentane bath kept cold in dry ice and let it freeze for a few minutes.
12. Store the tissue in −80 °C freezer until cutting.
13. Cut 10–12 μm thick cryosections using a cryostat at a temperature around −23 °C.
14. Store cryosections in −80 °C freezer until immunohistochemistry.

3.4.2 Immunohistochemistry

1. Thaw frozen sections at room temperature for approximately 10 min and place them in glass or plastic staining jars with cold (−20 °C) acetone, cover with lid and incubate slides for 10 min in −20 °C.
2. Wash slides two times with 1× PBS for 5 min each.
3. In the meantime prepare the blocking solution.

4. Apply the barrier pen around the tissue sections and place the slides in a humidified chamber. Add the blocking solution on the slides and incubate them for 1 h at room temperature (RT).
5. Dilute the antibodies to the appropriate concentration in the blocking buffer and spin down any debris using a microcentrifuge at ~12,000 × *g*.
6. Remove the blocking solution by tapping slides on paper tissue and place them back in the humidified chamber.
7. Cover the sections with the appropriate diluted primary antibody and incubate them overnight at 4 °C.
8. Wash slides three times with 1× PBS for 5 min each.
9. Prepare the secondary antibodies: Dilute the antibodies to the appropriate concentration in the blocking buffer and spin down any debris using a microcentrifuge at ~12,000 × *g*.
10. Place the slides in the humidified chamber, cover the sections with the appropriate diluted secondary antibody, and incubate them for 1–2 h at RT protected from light.
11. Wash slides three times with 1× PBS for 5 min each.
12. Remove the excess buffer from the slides by tapping them on paper tissue.
13. Apply a few drops of mounting media with DAPI and cover the sections with a glass coverslip.
14. Acquire images with a fluorescence microscope equipped with a digital camera, or similar.
15. Count the numbers of oligodendrocytes stained for CC-1 using the Image J Cell counter plugin, or similar.
16. Inspect myelin defects by MBP staining in different CNS areas such as cerebellar white matter, brain stem, and cervical spinal cord and analyze the degree of inflammation by looking for microglia/macrophages (CD11b-stained cells) and infiltrating T cells (CD3-stained cells) (*see* **Note 10**).

3.4.3 Tissue Processing for Electron Microscopy

Take all appropriate precautions to prevent exposure to toxic chemicals and their vapors used in this procedure: work under a fume hood and wear protective gloves, lab coat, safety goggles, and face mask during solution preparation, perfusion, and EM tissue processing. Discard any liquid and solid chemical waste according to local institute's regulations for waste disposal.

1. Prepare the 2.5% GTA and 4% PFA (EM fixative) solution (1 L).
2. Mice are deeply anesthetized with approved anesthetic (e.g., Avertin) prior to the perfusion, which is performed as described in Subheading 3.4.1 using the EM fixative solution.

3. Using the small scissors, forceps, and spatula, carefully dissect the brain, the optic nerves, and the spinal cord from the perfused animal and postfix them with the EM fixative in a 15 mL plastic tube for 2 h at 4 °C.
4. Transfer all the CNS tissues in 0.1 M Millonig's buffer (*see* **Note 11**) on a petri dish and dissect out the tissue samples under a stereomicroscope using fine forceps and a scalpel: a mid-sagittal slice of the cerebellum at the level of the vermis and cross-sectional slices from the corpus callosum at the level of the fornix, the brain stem at the level of the pons and the spinal cord segments at the cervical and lumbar enlargements. Trim all the tissue samples in ≤ 2 mm^3 blocks.
5. Transfer the tissue blocks in 2 mL clear glass vials and seal them with the closure during the EM processing.
6. Wash the samples three times for 30 min each with 0.1 M Millonig's buffer and use 1 mL transfer pipettes with thin ends for all subsequent washes. Rotate samples during all washes on a tube rotator.
7. Prepare 1% osmium tetroxide solution by mixing equal volumes of 0.2 M Millonig's buffer and 2% osmium tetroxide solution. Work under a fume hood and handle with care since osmium tetroxide is very toxic.
8. Remove the 0.1 M Millonig's buffer and add 1 mL of 1% osmium tetroxide solution in each vial. Make sure that the tissues are covered with liquid at all times and the vials are sealed. Let the samples incubate for 2 h at RT. The color of the tissue turns dark at this point.
9. Use wooden applicator sticks to transfer the tissue blocks to new vials filled with 0.1 M Millonig's buffer and wash the tissues three times for 10 min each under rotation.
10. Start the embedding procedure by dehydrating the samples using an ascending series of ethanol: 30%, 50%, 70%, 90%, 95%, and 100%. Incubate samples in each ethanol twice for 5 min each under rotation.
11. Incubate samples in pure 100% ethanol for 10 min under rotation. Open a new bottle of 100% ethanol for this step.
12. Replace the 100% ethanol with propylene oxide (PO) and seal the vials. Rotate the vials for 15 min and repeat this step once more.
13. In the meantime, prepare the Epon-812 mix as described in Table 2. Mix ingredients very well by using wooden applicator sticks in a disposable plastic tri-corn beaker and remove air bubbles using a small vacuum chamber or by placing the mix into a 60 °C oven for 10 min. Store the Epon-812 mix at 4 °C and use the same batch for the entire embedding procedure.

14. Prepare the PO:Epon-812 mix at 1:1 ratio in a plastic tube and mix well by vortexing briefly.
15. Prepare the PO:Epon-812 mix at 1:2 ratio in a plastic tube and mix well by vortexing briefly.
16. Replace the PO with the PO:Epon-812 mix (1:1) and incubate the samples for 2 h at RT under rotation.
17. Replace the PO:Epon-812 mix (1:1) with the PO:Epon-812 mix (1:2) and incubate the samples for 2 h at RT under rotation.
18. Replace the PO:Epon-812 mix (1:2) with the Epon-812 mix and incubate the samples overnight at RT under rotation.
19. The next day, bring the Epon-812 mix to RT and use wooden applicator sticks to transfer the samples into new vials filled with fresh Epon-812 mix. Incubate the specimens for 4 h at RT under rotation.
20. In the meantime, prepare the sample labels using a pencil or printer and place them in the right side (flat end) of each EM mold with the printed side facing up.
21. Add a thin layer of Epon-812 mix in the EM mold using a plastic 1 mL transfer pipette and orientate each tissue block carefully in the left side (pointed end) of each mold under a stereomicroscope. Orientate spinal cord, corpus callosum, optic nerve and brain stem for cross sectioning and cerebellum for sagittal sectioning. Fill up each mold with an adequate volume of Epon-812 mix.
22. Polymerize the molds with the samples on a flat surface at 60 °C for 48 h.

3.4.4 Sectioning, Staining of Sections, and Imaging

1. Cut 1 μm-thick (semithin) sections using an ultramicrotome, and transfer the sections using a steel loop into a drop of water on a slide and dry them on a slide warmer.
2. Stain the semithin sections in the toluidine blue solution for 1–2 min.
3. Rinse off gently the excess dye with ddH_2O.
4. Air-dry the slide and add a few drops of a xylene-based mounting medium and cover the semithin sections with a glass coverslip.
5. Observe the toluidine blue-stained semithin sections under a light microscope and use them as a guide for cutting the 70 nm-thick (ultrathin) sections.
6. Cut the ultrathin sections using an ultramicrotome and collect sections onto 200 mesh Cu grids.
7. Dry sections overnight before staining.

8. Stain grids with the 5% uranyl acetate solution for 15 min.
9. Wash the grids several times with ddH_2O.
10. Stain the grids with the lead citrate solution for 5 min.
11. Wash the grids several times with ddH_2O and let them dry.
12. Observe ultrathin sections stained with uranyl acetate and lead citrate under a transmission electron microscope and acquire five nonoverlapping images at 2900× magnification from the CNS white matter area of interest in both DTA mice and normal control mice (at least 3 mice/genotype) to be used for quantitative analysis of myelinated fibers.
13. Calculate the percentage of unmyelinated axons after counting the total number of axons with diameter larger than 0.32 μm in corpus callosum, 0.15 μm in optic nerve and usually 0.3 μm in other areas.
14. Calculate the *g* ratio (axonal diameter/fiber diameter) using ImageJ, or similar, to measure the axon diameter and the myelinated fiber diameter. *g*-Ratios are calculated for a minimum 40 myelinated axons/mouse as we previously described [14]. The presence of axons with thinner myelin (significantly increased *g* ratio than normal value) indicates remyelination.
15. Observe the toluidine blue-stained semithin sections under a light microscope and acquire nonoverlapping images at 40× (3 images) or 100× (4 images) from the CNS white matter area of interest. Use the images to count the mean axonal density in both DTA mice and normal control mice (at least 3 mice/genotype) by ImageJ or similar. The total number of axons is calculated by multiplying the mean axonal density by the total surface area of the corresponding toluidine blue-stained section of each CNS tissue measured by ImageJ [14].

4 Notes

1. It is important to use 4-hydroxytamoxifen (H-6278, ≥70% Z isomer) from Sigma. Using the ≥98% Z isomer (Sigma) will accelerate the disease course and increase the severity of the symptoms in the DTA mice leading them to early death.
2. This is a modified protocol for tamoxifen injections in DTA mice than originally reported [8]: The tamoxifen stock solution is prepared at the concentration of 8 mg/mL to inject the DTA mice at the lower dose of 0.8 mg/day and for a shorter period of time (4 consecutive days for the males and 3 consecutive days for the females). In the original protocol a 10 mg/mL stock solution of tamoxifen is used for injecting both male and female DTA mice at the dose of 1 mg/day for 5 consecutive

days. The modified protocol is recommended since it ameliorates the severity of the disease symptoms and reduces lethality during the acute disease in DTA mice.

3. The homozygous male *ROSA26-eGFP-DTA* are infertile and cannot be used as breeders.
4. Inspect the tamoxifen solution for precipitates and redissolve them with brief vortexing and sonication.
5. This tamoxifen protocol should not be used in young DTA mice (4 week-old or less) since they become very weak and die early due to a tamoxifen-associated toxicity.
6. If the mice lose more than 10% of their maximal weight, they should be euthanized (end point of disease).
7. When the DTA mice are injected at the age of 16 weeks or older, they show a slight delay of the disease onset (~1 week) and they never fully recover from the disease symptoms and usually die by 35 dpi (*Maria Traka and Brian Popko, unpublished data*).
8. The DTA mice display more severe tremors during the recovery stages of the acute disease than the early and peak stages of this disease.
9. Sudden noises and cage cleaning in the animal facility could trigger startle-evoked seizures and subsequent death in DTA mice during the late-onset disease.
10. Focal white matter lesions occur in DTA mice during the early stages (~280 dpi) of the late-onset disease, but they are rare. The focal lesions are negative for MBP staining and they frequently contain T cells stained by CD3 and a high density of microglia/macrophages stained by CD11b [14].
11. Sodium Cacodylate buffer 0.2 M, pH 7.4 can be used instead of 0.2 M Millonig's phosphate buffer, pH 7.4. Sodium Cacodylate buffer and its vapors are toxic and should be handled with care.

References

1. Nave KA, Trapp BD (2008) Axon-glial signaling and the glial support of axon function. Annu Rev Neurosci 31:535–561
2. Nave KA (2010) Myelination and the trophic support of long axons. Nat Rev Neurosci 11 (4):275–283
3. Trapp BD, Nave KA (2008) Multiple sclerosis: an immune or neurodegenerative disorder? Annu Rev Neurosci 31:247–269
4. Franklin RJ, Ffrench-Constant C (2008) Remyelination in the CNS: from biology to therapy. Nat Rev Neurosci 9(11):839–855
5. Nakahara J, Aiso S, Suzuki N (2010) Autoimmune versus oligodendrogliopathy: the pathogenesis of multiple sclerosis. Arch Immunol Ther Exp 58(5):325–333
6. Barnett MH, Prineas JW (2004) Relapsing and remitting multiple sclerosis: pathology of the newly forming lesion. Ann Neurol 55 (4):458–468
7. Stys PK, Zamponi GW, van Minnen J, Geurts JJ (2012) Will the real multiple sclerosis please stand up? Nat Rev Neurosci 13(7):507–514

8. Traka M, Arasi K, Avila RL, Podojil JR, Christakos A, Miller SD, Soliven B, Popko B (2010) A genetic mouse model of adult-onset, pervasive central nervous system demyelination with robust remyelination. Brain 133 (10):3017–3029
9. Ivanova A, Signore M, Caro N, Greene ND, Copp AJ, Martinez-Barbera JP (2005) In vivo genetic ablation by Cre-mediated expression of diphtheria toxin fragment A. Genesis 43 (3):129–135
10. Doerflinger NH, Macklin WB, Popko B (2003) Inducible site-specific recombination in myelinating cells. Genesis 35(1):63–72
11. Collier RJ (2001) Understanding the mode of action of diphtheria toxin: a perspective on progress during the 20th century. Toxicon 39 (11):1793–1803
12. Maxwell IH, Maxwell F, Glode LM (1986) Regulated expression of a diphtheria toxin A-chain gene transfected into human cells: possible strategy for inducing cancer cell suicide. Cancer Res 46(9):4660–4664
13. Palmiter RD, Behringer RR, Quaife CJ, Maxwell F, Maxwell IH, Brinster RL (1987) Cell lineage ablation in transgenic mice by cell-specific expression of a toxin gene. Cell 50 (3):435–443
14. Traka M, Podojil JR, McCarthy DP, Miller SD, Popko B (2016) Oligodendrocyte death results in immune-mediated CNS demyelination. Nat Neurosci 19(1):65–74
15. Bjartmar C, Wujek JR, Trapp BD (2003) Axonal loss in the pathology of MS: consequences for understanding the progressive phase of the disease. J Neurol Sci 206(2):165–171

Chapter 18

Human Glial Chimeric Mice to Define the Role of Glial Pathology in Human Disease

John N. Mariani, Lisa Zou, and Steven A. Goldman

Abstract

Human glial progenitor cells (hGPCs) can engraft, expand, and differentiate into functional oligodendrocytes and astrocytes when transplanted neonatally into murine hosts, in which they outcompete the host glial pool to ultimately colonize and dominate the recipient brains. When congenitally hypomyelinated mutants are used as hosts, the donor hGPCs generate myelinogenic oligodendrocytes as well as astrocytes, so that the recipient mice develop a largely humanized white matter, with entirely human-derived myelin. In addition, by neonatally engrafting hGPCs derived from patient- and disease-specific pluripotent stem cells, glial chimeric mice may be produced in which large proportions of all macroglial cells are not only human but also patient and disease specific. Human glial chimeric mice thus provide intriguing preparations by which to investigate the species-specific contributions of human glia to both cognition and human-selective neurodegenerative and neuropsychiatric diseases, as well as the potential for therapeutic glial cell replacement in these disorders. This review presents an overview of the uses, characteristics, and limitations of the human glial chimeric brain model, while providing a step-by-step protocol for the establishment of these mice.

Key words Glial progenitor cell, Oligodendrocyte progenitor cell, Stem cell, Cell transplantation, Mouse models

1 Introduction

Over the past two decades, we and others have explored the use of glial progenitor cells (GPCs) as engraftable cellular therapeutics for the treatment of glial disorders, in particular those characterized by oligodendrocyte loss and consequent demyelination (reviewed in [1, 2]). The intended therapeutic targets of this approach are broad, and include pediatric disorders of myelin formation and maintenance, such as the hereditary leukodystrophies and lysosomal storage disorders [3, 4], as well as the adult-onset disorders of myelin integrity, such as multiple sclerosis and the age-related white matter diseases of vascular insufficiency [5, 6]. With these therapeutic purposes in mind, a variety of increasingly refined methods have been established for isolating human glial progenitor

David A. Lyons and Linde Kegel (eds.), *Oligodendrocytes: Methods and Protocols*, Methods in Molecular Biology, vol. 1936, https://doi.org/10.1007/978-1-4939-9072-6_18,

cells, and for modeling their behavior and instructing their fates, both in vitro and in vivo, the latter via transplantation-based strategies.

Like their rodent counterparts [7], human GPCs may generate both astrocytes and oligodendrocytes [8], and in some contexts neurons as well [9, 10]; yet despite their multilineage competence, they have been most often referred to as oligodendrocyte progenitor cells (OPCs), and also as NG2 cells, the latter designation based on their expression of the NG2 chondroitin sulfate proteoglycan [11, 12]. While these terms are effectively synonymous, in this review these cells will be referred to as glial progenitor cells, GPCs. Human GPCs (hGPCs) can be reliably and efficiently isolated from human brain tissue by selecting for glial gangliosides recognized by the A2B5 antibody, with concurrent negative selection by depletion of polysialylated NCAM-defined neuroblasts [8, 10, 13]. They may also—and more specifically—be isolated on the basis of the CD140a-defined ectodomain of the PDGFα receptor, which includes all of the potentially oligoneogenic cells of the human brain [14]. Progressively more oligodendrocyte-biased phenotypes may be further selected, based upon the concurrent expression of the CD9 tetraspanin, or of the oligodendrocytic sulfatide recognized by the O4 antibody [15, 16]. Whether derived from adult or fetal human brain, each of these phenotypes has now proven efficient, in a variety of laboratories and model systems, in remyelinating structurally demyelinated or hypomyelinated tissue.

2 Experimental Background

2.1 Human GPCs May Be Transplanted to Restore Lost Myelin

Beginning with pioneering studies by Gumpel and Lachapelle, and later by Blakemore and colleagues [17–19], the ability of transplanted oligodendroglia and their progenitors to effect myelin repair has been studied in a number of model systems [20–23]; broadly reviewed in [24, 25]. Efforts focusing on the specific use of GPCs rather than their derived oligodendrocytes, and emphasizing the use of human GPCs in particular, are more recent [8, 26, 27]. Many of these studies have assessed the myelinogenic competence of GPCs in the shiverer mouse ($MBP^{shi/shi}$), a congenitally hypomyelinated mouse deficient in myelin basic protein [28, 29]. Using immunodeficient shiverer mice ($rag2^{-/-} \times MBP^{shi/shi}$) as hosts, we developed a multi-site delivery procedure by which donor GPCs could be introduced into the major presumptive white matter tracts of newborn mice [27]. This approach resulted in the widespread engraftment of donor hGPCs throughout the entire CNS, with infiltration of the forebrain, brainstem and cerebellum, and ultimately the spinal cord and roots (Fig. 1). The donor hGPCs then produced both oligodendrocytes and

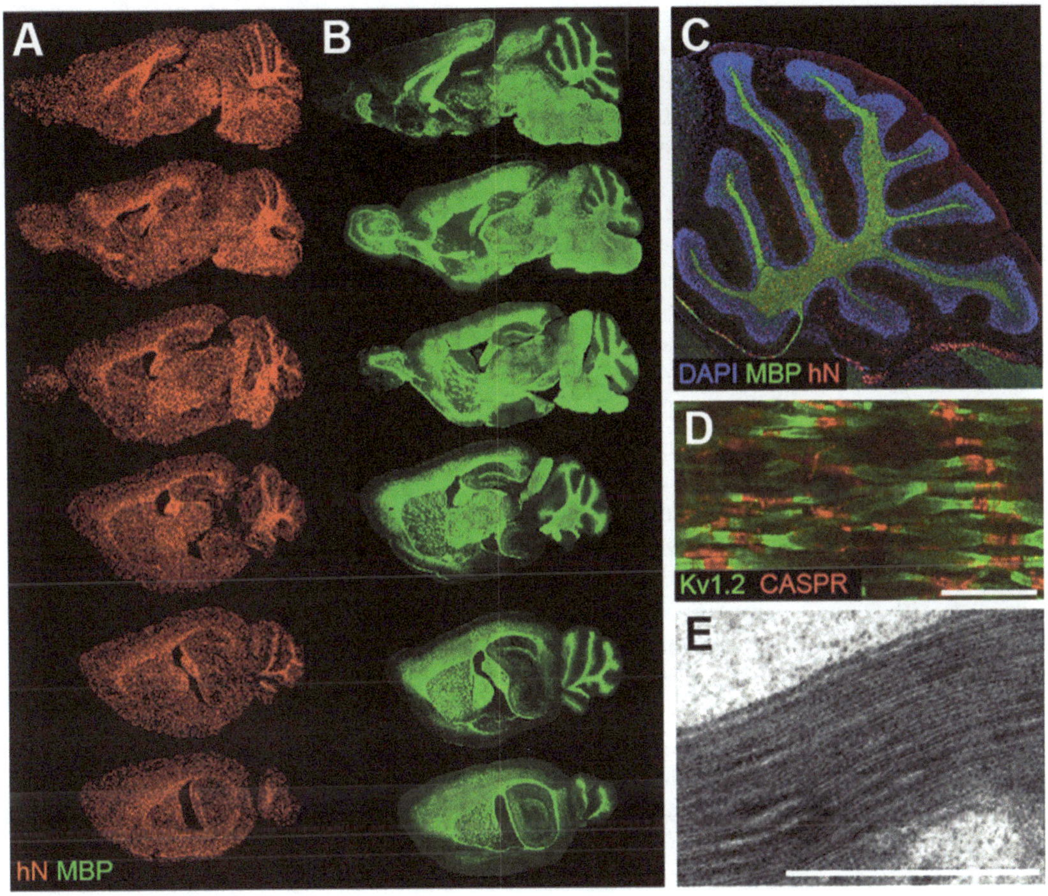

Fig. 1 Engrafted hGPCs chimerize and myelinate the congenitally unmyelinated shiverer brain. (**a–e**) Myelination of congenitally hypomyelinated shiverer mice by human fetal tissue-derived glial progenitor cells. (**a**, **b**) Representative sagittal images of an engrafted shi/shi × rag2$^{-/-}$ brain, sacrificed at 1 year of age. (**a**) Human donor cells identified by an anti-human nuclear antibody (hN; *red*). (**b**) Donor-derived myelin basic protein (MBP; *green*) in sections adjacent or nearly so to matched sections in **a**. All major white matter tracts heavily express MBP (which is all donor-derived in MBP-null shiverer mice). (**c**) Sagittal view through cerebellum of a year-old engrafted shi/shi × rag2$^{-/-}$ brain. All cells were stained with DAPI (*blue*); donor cells were identified by human nuclear antigen (hN, *red*), and donor-derived myelin by MBP (*green*). (**d**) Reconstituted nodes of Ranvier in the cervical spinal cord of a transplanted and rescued 1-year-old shi/shi × rag2$^{-/-}$ mouse, showing paranodal Caspr protein and juxtaparanodal potassium channel Kv1.2, symmetrically flanking each node. Untransplanted shiverer brains do not have organized nodes of Ranvier and, hence, cannot support saltatory conduction (Caspr, red; Kv1.2, green). (**e**) Electron micrograph of a 16-week-old shiverer mouse implanted perinatally with hGPCs shows a shiverer axon with a densely compacted myelin sheath. Scale: (**d**) 5 μm. (**e**) 1 μm. Figure adapted from ref. 61

astrocytes, resulting in the dense and efficient central myelination of these otherwise unmyelinated mice. This in turn was associated with their substantially prolonged survival, with frank rescue and phenotypic recovery of a large minority [27]. Indeed, whereas untreated shiverers invariably die by 20–21 weeks of age, a fraction of neonatally engrafted mice achieved normal life spans of over 2 years, suggesting the potential power of this approach toward therapeutic remyelination.

In the shiverer model, donor hGPCs mature in a highly context-dependent fashion, such that those donor cells that engraft presumptive white matter develop as myelinogenic oligodendrocytes and fibrous astrocytes. In contrast, those cells that invade the cortical and subcortical gray matter and differentiate do so as astrocytes [8]. Moreover, the differentiation of hGPCs differs depending upon whether the cells have been transplanted into normally myelinated or congenitally hypomyelinated recipients: in normally myelinated hosts, transplanted hGPCs colonize both the gray and white matter, ultimately differentiating as astrocytes or remaining as glial progenitors [30]. In contrast, hGPCs transplanted into shiverer mice first preferentially expand in the callosal and capsular white matter, giving rise therein to new oligodendrocytes as well as astrocytes and new hGPCs (Figs. 1 and 2).

Yet even more stunning than their broad colonization, the implanted hGPCs not only integrate into the recipient brains, but over time predominate; they first displace and then ultimately *replace* the resident glial populations of their hosts. In both shiverers and myelin wild-type recipients, colonization of the cortical and subcortical gray by migrating hGPCs is followed first by their selective expansion, and then by the inexorable displacement and in situ death of resident murine glial progenitors [30]. As a result of this apparent competition between resident mouse GPCs and the invading human GPCs, by a year of age the donor hGPCs are typically distributed in a relatively uniform manner throughout the forebrain white and gray matter, while mouse GPCs are scarce, and often absent from the now human glial-dominant forebrains. Importantly, this competitive advantage of human over murine glial progenitors is evident in wild-type as well as in shiverer mice; in both recipient environments, the human glial progenitors typically outcompete their murine counterparts.

2.2 Human Donor Glial Progenitors Dominate the Competition

The selective expansion of the human glial population in the mouse appears at least in part to be a product of the more sustained proliferation of the transplanted human GPCs, which appear to retain cell-autonomous regulation of expansion, and expand in numbers by over 40-fold in the corpus callosum alone over the first year of life [27]. Yet the advantage of human GPCs in the mouse environment is clearly based on more than preferential expansion, since the murine GPCs are completely replaced by their human counterparts over time. Rather, the process of hGPC colonization seems overtly competitive. Human GPCs typically expand outward from their periventricular and callosal points of introduction, in advancing waves that seem to both repulse and overwhelm resident murine progenitors, which proceed to die,

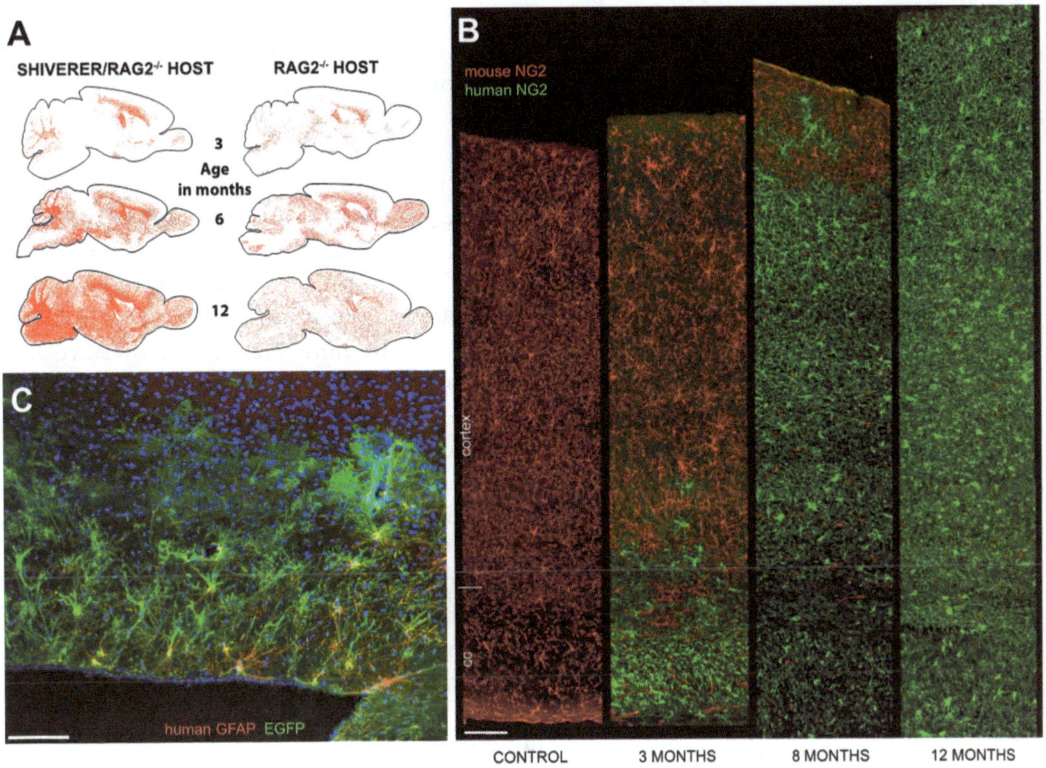

Fig. 2 Human GPCs outcompete and ultimately replace resident mouse GPCs. (**a**) hGPCs neonatally transplanted into either congenitally hypomyelinated shiverer × rag2$^{-/-}$ (*left* column), or into normally myelinated rag2$^{-/-}$ (*right*) mouse brain, disperse and expand broadly throughout the brain as a function of age. hGPCs reach higher density in white matter than gray matter of the hypomyelinated shiverer (*left*), in contrast to their relatively uniform distribution in normally myelinated brain (*right*). Red dots indicate individual human donor GPCs, as labeled by human nuclear antigen (*left* and *right* columns). (**b**) Progressive replacement of host mouse GPCs, by donor hGPCs, as identified by species-specific antibodies against NG2, compared to an unengrafted control mouse (*left-most* column). By 1 year, hGPCs have replaced mouse GPCs throughout the entire depth of the cortex. (**c**) EGFP-tagged human astrocytes in the corpus callosum, cortex, and fimbria in a 2-year-old rag2$^{-/-}$ mouse. The hGPCs were labeled with an EGFP-expressing lentivirus (*green*) in vitro, prior to transplantation. Dorsal is toward the top and rostral to the left in this sagittal section. (EGFP, *green*; human-specific GFAP, *red*). Scale: 100 μm. (**a**, **b**) Adapted from ref. 30

largely in situ (Fig. 2). The human GPCs ultimately attain a relatively uniform distribution, achieving apparent contact inhibition in a manner analogous to that reported developmentally by Bergles and colleagues [31]. While the molecular basis for the competitive dominance of the human GPCs is unknown, several recent studies have identified differential expression of both MYC and hedgehog-dependent pathways as contributing to clonal dominance during early ontogeny [32–34]. Further assessment of differential gene expression by mouse and human GPCs in vivo may permit the discovery of similar regulators of competition that favor the relative dominance of human over mouse GPCs when the two are in direct competition, a decidedly unnatural situation that might nonetheless provide us important insights into the differential expansion of favored glial cell populations in the developing human brain, and the signals that determine which among competing populations is ultimately dominant.

Regardless of its molecular basis, the domination of the host brains by human GPCs leads to the slow but inexorable glial humanization of these brains, as mature astrocytes undergo presumably normal turnover in adulthood, with astrocytic replacement from now-humanized resident progenitor pools. This process results in the substantial astrocytic humanization of these rodent brains, first by fibrous astrocytes of the white matter, and then by protoplasmic astroglia of both cortical and subcortical gray matter [30, 35]. In both shiverer and myelin wild-type hosts, the proportion of human astrocytes thus increases monotonically as a function of time, and is matched by a corresponding decrease in the proportion of mouse astrocytes. Over time, these brains thus become chimeric for human astrocytes, as well as for human glial progenitors, and in shiverer mice, this process includes replacement of host oligodendrocytes by their human counterparts as well. As a result, in 1-year-old shiverer mice neonatally engrafted with human GPCs, essentially all resident glial progenitor cells, all oligodendrocytes, and large proportions of all astrocytes are of human origin (Figs. 1 and 2).

2.3 Human Glia Maintain Uniquely Hominid Features in Chimeras

Human astrocytes are larger and more complex than those of rodents; human cortical astrocytes can exhibit over triple the diameter, and over tenfold the number of terminal processes, as rodent astrocytes [36, 37]. Human astrocytes have both structural features and functional competencies unique to hominids, and exhibit a range of astrocytic pleomorphism without precedent in infraprimate mammals. On that basis, we have postulated that the functional roles of glia have expanded during evolution, and especially so with the appearance of hominids [36, 37]. These evolutionary changes are of particular interest because astrocytes have been shown to play vital roles in information processing within the CNS [38, 39]. Astrocytes are required for synaptogenesis and maintenance of synaptic density [40], and a number of specific astrocytic modulators of synaptic plasticity have been identified, including the glypicans [41], TNFα [42], and the neuroligins [43], among others. Importantly, these ligands may be differentially expressed by human astroglia, potentially offering advantages to human astroglia in the regulation of synaptic plasticity, relative to infraprimate glia [37].

On the basis of these observations, we asked if the greater structural complexity of human astrocytes relative to those of rodents might be accompanied by functional differences. In particular, we considered whether human glial chimeric mouse brains, with their substantial colonization by human astroglia and their progenitors, might manifest functional distinctions from wild-type mice, and if so whether these functional differences might reflect aspects of human cognitive evolution [44]. This possibility was anticipated by the observation that human astrocytes propagate Ca^{2+} waves significantly more rapidly than rodents [37]. To define the human-selective contributions of astrocytic complexity to network function, we therefore assessed the behavior of human glial chimeras to both matched unengrafted and allografted mice. To that end, hGPCs, pre-biased in vitro to astrocytic phenotype, were transplanted into neonatal immunodeficient mice, thereby establishing human glial chimeras with especially large complements of human astrocytes as well as GPCs. By 7–10 months of age, the *majority* of all forebrain OPCs and astrocytes in these mice were typically of human origin [44]. The engrafted human glia appeared to mature in a cell-autonomous fashion, in that the diameter, domain size, and morphology of human astrocytes in the chimeric mouse brain each approximated that of astrocytes in the normal adult human brain.

2.4 Human Glial Chimeras for Assessing the Glial Contribution to Network and Cognitive Function

Since human glia appeared to maintain the cell-autonomous size, complexity, and domain architecture of human astrocytes in the mouse environment, together with Maiken Nedergaard's group we asked if human glial chimeric brains exhibited increased synaptic plasticity. To that end, we compared the threshold for inducing hippocampal long-term potentiation (LTP) in chimeric mice with that of littermate controls. We found that high-frequency stimulation significantly potentiated the field EPSP slopes of neurons in the chimeric hippocampi, and indeed did so to a larger degree and for a substantially longer period, than in unengrafted littermate controls. Thus, human glial chimeras manifested substantially facilitated LTP [44]. Importantly, such stable, long-lasting changes in synaptic function are thought to be involved in learning and memory, of which LTP is typically considered an in vitro surrogate. Since hippocampal LTP was enhanced in chimeric mice, we next asked whether the human glial chimeras might have cognitive advantages over their unengrafted or mouse GPC-allografted counterparts. We found that the human glial chimeras indeed performed better than control mice across a variety of learning tasks, which included auditory fear conditioning, novel object and place recognition, and Barnes maze navigation. In all of these tests—but not in any test of social interactivity or primary perception—the human glial chimeras performed better and acquired new causal associations more quickly than did murine-allografted or untransplanted controls [44]. In short, they learned more rapidly, and at least along the axes of the tests performed, could be defined—if colloquially so—as smarter. As such, these glial chimeras may provide us a viable—if provocative—model by which to evaluate the species-specific contributions of human glia to human cognition.

2.5 hiPSC-Derived Glial Chimeras as Models of Human Genetic and Inflammatory CNS Disease

Most of the studies described thus far used hGPCs derived from brain tissue, of both adult and fetal origin. Yet GPCs and their derived astrocytes and oligodendrocytes may also be derived from pluripotential stem cells (PSCs) [45–47]. Bipotential oligodendrocyte-astrocyte progenitors may now be produced in quantity from both embryonic stem cells (hESCs) and induced pluripotent cells (iPSCs), typically in 3–4 months, a time course similar to that of their appearance from neural stem cells in early development [15, 16, 48]. The cells produced by these protocols are as efficient at myelinogenesis in vivo as their tissue-derived counterparts, myelinating most axons within the hypomyelinated shiverer forebrain, and rescuing a substantial fraction of neonatally engrafted shiverer mice from early death [47] (Fig. 3).

Importantly, while these protocols for producing hGPCs from pluripotent stem cells were developed to permit the production of hGPCs in clinically useful quantities, they also enabled the establishment of chimeric mice using glia derived from patient- and disease-specific PSCs. These in turn may be used to assess not only the in vivo pathology of disease-derived glia but also the extent to which that glial pathology may contribute to disease pathogenesis. By way of early example, Fossatti and colleagues produced hGPCs from iPSCs derived from patients with primary progressive multiple sclerosis (PPMS), and then transplanted the immature oligodendroglia derived from these cells into shiverer mice, in which the cells differentiated as myelinogenic cells that ensheathed resident axons [15, 49]. While the mice did not manifest any stigmata of PPMS, the authors were nonetheless able to establish mice stably chimerized with PPMS-derived oligodendroglia, thereby providing a potential in vivo model for future studies of genetic vulnerability to PPMS. More broadly, this study established the potential value of mice chimerized with patient-specific hGPCs for identifying susceptibility factors for autoimmune inflammatory diseases of the CNS.

2.6 Human Glial Chimeras as Hosts for Infections Unique to the Human Brain

Human glial chimeras may have applications beyond that of modeling the species-specific role of human glia in neurological function and dysfunction. In particular, these mice may permit us to better assess the biology of human-selective infectious and inflammatory diseases of the CNS. For instance, a number of pathogenic viruses of the human brain are gliotropic, and some are specifically human in their species selectivity. Viruses such as the JCV polyomavirus, the cause of progressive multifocal leukoencephalopathy (PML), and human herpesvirus-6, another prominent viral encephalitis of immunocompromised individuals, have never been amenable to experimental study in vivo, since these are human glial-specific pathogens [50]. On that basis, we asked whether human glial chimerization might permit these animals to be infected with human gliotropic viruses in vivo, and if so whether these infected

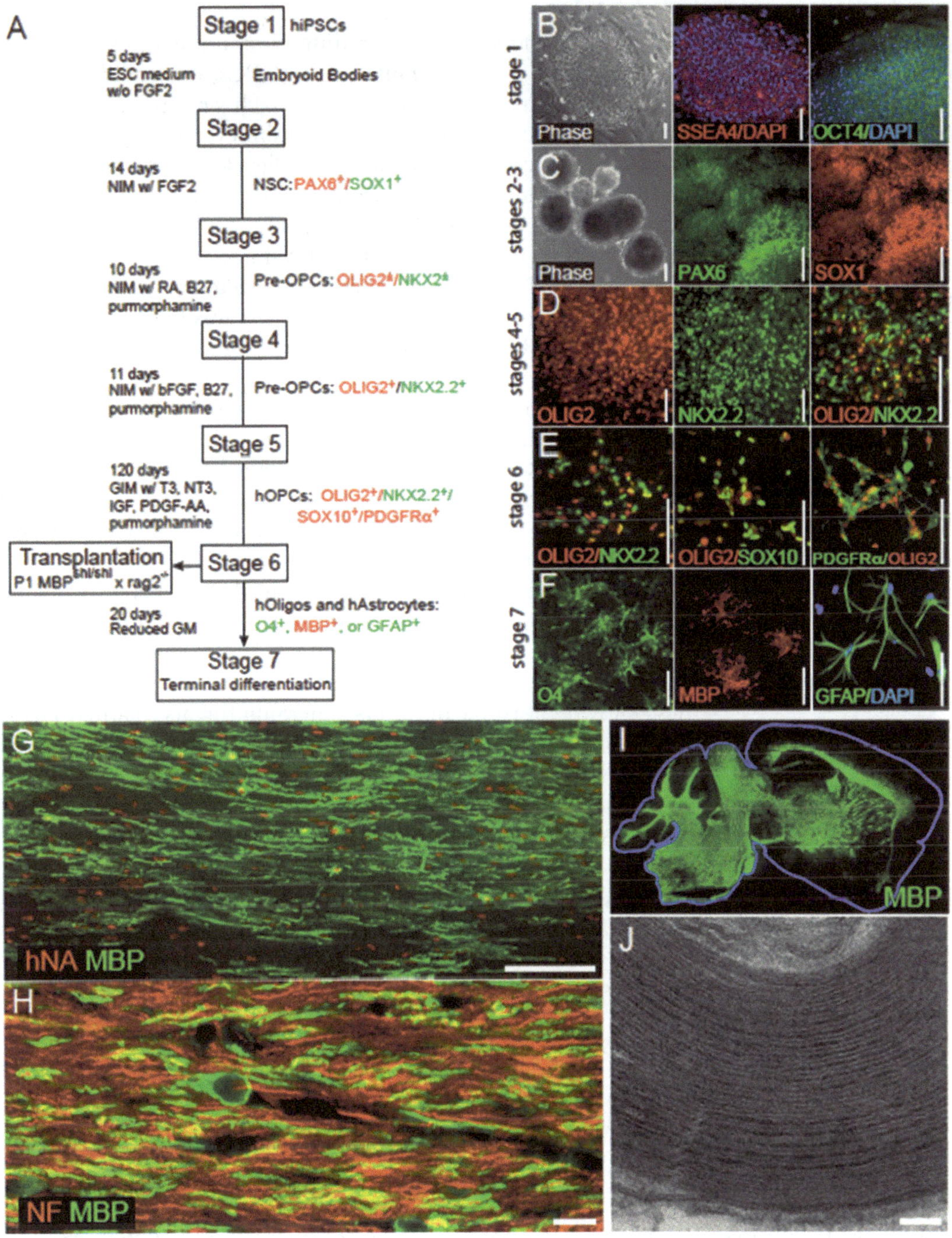

Fig. 3 Human pluripotent stem cell-derived GPCs can remyelinate dysmyelinated hosts. (**a**) This schematic outlines the multi-stage protocol by which GPCs, oligodendrocytes and astrocytes may be generated from human pluripotent stem cells, whether embryonic stem cells (hESCs) or induced pluripotent stem cells (iPSCs). (**b–f**) Show representative images taken at serial stages of glial differentiation, with the serial expression of selected marker proteins noted at each stage. (**g**) 3 months after neonatal transplant into hypomyelinated shiverer mice, human induced pluripotent cell (hiPSC)-derived GPCs have matured as myelinating, myelin

human glial chimeras might develop pathology replicating that of humans. By neonatally engrafting immunodeficient myelin-deficient rag2$^{-/-}$ × *shi/shi* mice with hGPCs, we established mice with a fully humanized white matter, which were then infected by intracerebral injection of JCV [51]. Within several weeks thereafter, the JCV-infected mice manifested oligodendrocytic loss with demyelination and astroglial proliferation, replicating thereby the cardinal features of human PML (Fig. 4).

We further noted that in the JCV-infected human glial chimeras, that hGPCs and astrocytes were infected more rapidly than were oligodendrocytes, and viral replication was noted primarily in astrocytes and hGPCs rather than in oligodendrocytes, which instead exhibited viral T antigen-associated apoptotic death. By establishing human glial chimeras in wild-type rather than hypomyelinated hosts—which produced mice colonized with human astrocytes and GPCs but not oligodendrocytes—we then established that human astroglia were *sufficient* for JCV spread, and that astrocytes, rather than oligodendrocytes, were the principal targets for, and reservoir of, infection in vivo [51].

This study established the utility of glial chimeras for studying the biology of human gliotropic viruses in vivo, and promises to be the first of many using human glial-transplanted animals to investigate human-specific brain infections in vivo. In particular, beyond these known gliotropic disorders, emerging threats that may target neural stem and progenitor cell populations as well as their derived glia, such as the Zika virus, Chikungunya virus, CMV, and HIV, among others, might also be especially amenable to in vivo investigation using humanized glial chimeras. Indeed, our JCV-infected humanized mice revealed the limits of our knowledge of these human gliotropic encephalitides, and hence the promise afforded by human glial chimeras for fundamental mechanistic discovery as well as therapeutic modeling of these enigmatic and often dire disorders.

Fig. 3 (continued) basic protein (MBP)-expressing oligodendrocytes (MBP, *green*; human nuclear antigen, *red*). (**h**) The hiPSC-derived oligodendrocytes ensheath mouse axons (neurofilament, *red*; MBP, *green*). (**i**) Human iPSC-derived oligodendrocytes can myelinate the entire brain of shiverer mice, which do not otherwise express MBP (*green*). (**j**) The myelin generated by hiPSC oligodendrocytes is ultrastructurally normal, exhibiting major dense lines and thick myelin sheaths. The use of such serial and distinct stages of growth factor exposure, paired with more extended periods of differentiation, have led to the production of highly enriched populations of human GPCs, that are highly efficient at myelinogenesis in vivo while manifesting no evident tumorigenesis. Scale: (**b–e**) 100 μm; (**f**) 25 μm; (**g**) 100 μm; (**h**) 10 μm; (**j**) 100 nm. Figure adapted from refs. 47 and 61

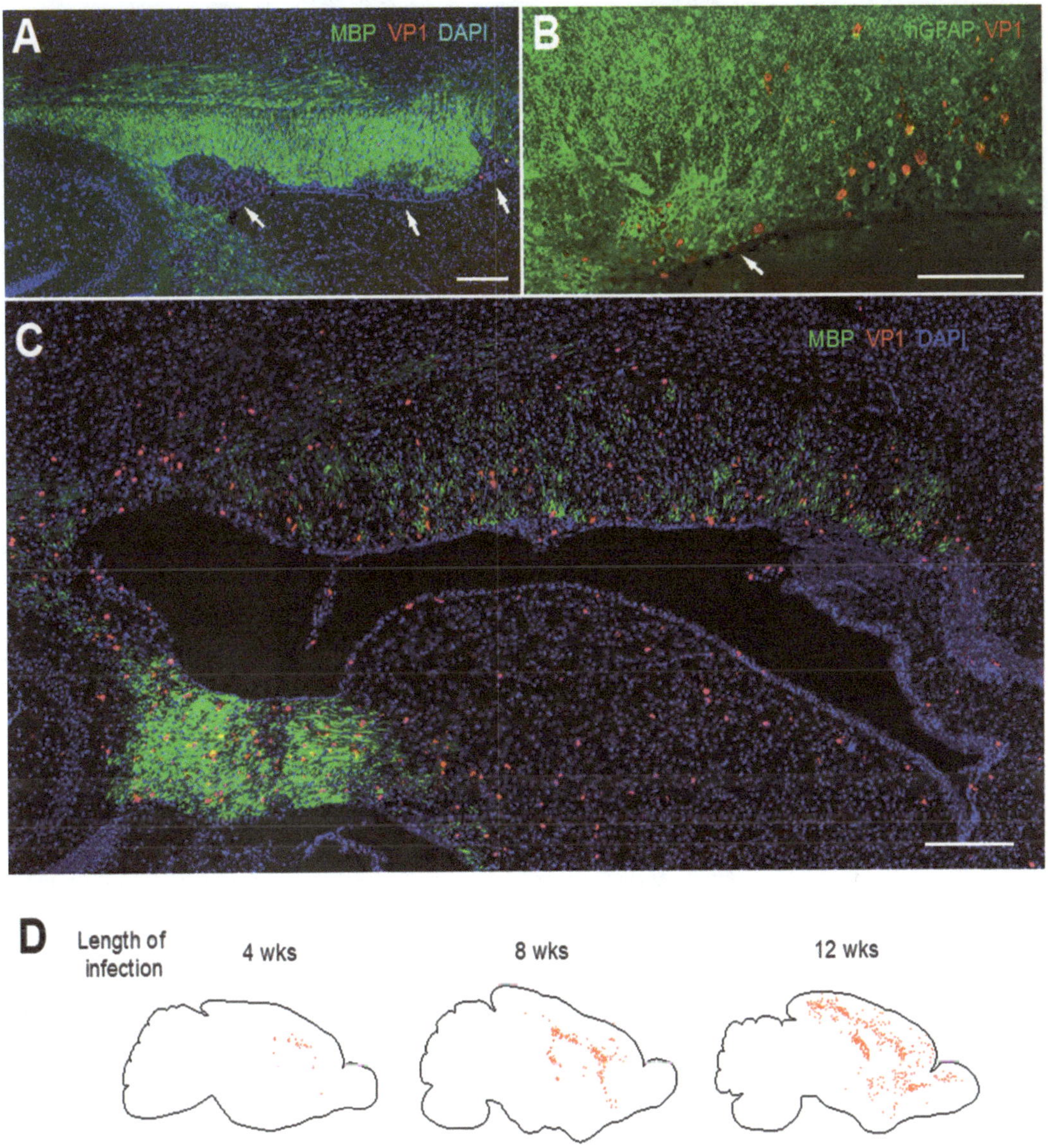

Fig. 4 Human glial chimeric mice can develop a prototypic human demyelinating viral encephalitis. Infection by the human-specific gliotropic JC polyomavirus (JCV) was observed and followed in human glial chimeras, after intracerebral viral injection at a single locus. Viral spread in vivo was tracked by immunostaining for the late JC viral VP1 antigen, as a function of time after infection. In these chimeric shiverers, essentially all myelinating oligodendrocytes, as well as most GPCs and large numbers of astrocytes, are human, and only human cells are infected by this virus. (**a**, **b**) By 4 weeks after infection, focal regions of demyelination (**a**, *arrows*) and infection-associated astrogliosis (**b**, *arrow*) are noted in the forebrain white matter of human glial chimeras, here abutting the callosal wall of the lateral ventricle. (**c**) By 11 weeks after infection, diffuse hypomyelination of the corpus callosum and capsules was noted. (**d**) Sagittal sections of an infected chimera at each of three timepoints; individual JCV-infected $VP1^+$ cells are dot-mapped onto the schematic. Infected human cells became progressively more widespread with time, to include much of the forebrain white matter by 12 weeks post-infection, with marked cortical spread by that point as well. Scale: 200 μm (**a**, **c**); 100 μm (**b**). Figure adapted from ref. 51

2.7 Humanized Glial Chimeras as Models of Neurodegenerative and Neuropsychiatric Disorders

Going forward, human glial chimeras may find great value in defining the relative contributions of glial pathology to diseases of the nervous system for which the relative roles of neurons and glia remain unclear or explored. In particular, the neurodegenerative and neuropsychiatric disorders involve both neuronal and glial pathology, and yet have traditionally been considered primary neuronal disorders; evident glial pathology such as hypomyelination has been assumed, often with little evidence, to have been secondary to neuronal loss. A host of neurodegenerative disorders, including ALS, Huntington disease, Alzheimer's disease, and others, have been reported to have significant glial involvement, and yet the specific contributions of glial dysfunction to disease pathogenesis have been difficult to establish. In vivo chimeras may allow causal glial pathology to be better defined in these disorders, and ultimately to be more specifically targeted. For instance, Benraiss and colleagues used mice rendered chimeric for mutant huntingtin-expressing GPCs to identify a critical role for glial pathology in the progression of Huntington disease [52]. Similarly, Windrem and colleagues used glial chimeras produced from iPSC GPCs derived from patients with childhood-onset schizophrenia, to define a causal role for glial dysfunction in schizophrenia [53]. Remarkably, the disease-derived glial chimeric mice manifested not only the hypomyelination typically associated with childhood-onset schizophrenia, but also aspects of its associated behavioral pathology, attesting to the fidelity with which the humanized chimeras can recapitulate aspects of disease phenotype in vivo. More broadly, these studies demonstrated that—at least in neonates—normal human GPCs could replace disease-derived GPCs in vivo; by so doing, these findings served to expand the potential scope of cell replacement therapy beyond isolated dysmyelination, to include treatment of the neurodegenerative and neuropsychiatric disorders—conditions that just a few years ago would not even remotely have been considered potential targets of cell therapy [1].

2.8 The Future

Human glial chimeras are already proving of great utility in defining the glial contributions to neurological disease, and the specific role of human glia in both cognition and cognitive pathology. Yet they may also prove of great value in screening agents, both small molecules and biologics, whose effects on glial cells require assessment, or that might be specifically directed at glial molecular targets. Indeed, as chimerization of experimental animals with multiple phenotypes becomes possible—with pluripotent stem cell-derived human microglia [54–56] and vascular cells, human peripheral immune cells [57], and even with human neurons [58–60]—and as our ability to produce these other phenotypes on a disease-specific basis becomes more mature—one can readily envision the real-time assessment of the interactions of multiple neural phenotypes in vivo, in a largely humanized system. This capability should allow the deployment and benefit of far more precise and informative animal models of human neurological disease than any yet available.

To capitalize upon these opportunities, we offer the following protocols, and welcome any questions from those engaged in the deployment and use of this technology.

The Method: How to Make a Chimera

3 Materials

3.1 Preparation of Cells

1. Cell Scraper.
2. 2× scalpels—size 11.
3. HBSS, no calcium, no magnesium (HBSS−/−) at 25 °C—At least 12 mL per 6 well plate of adherent cells.
4. 100 μm cell strainer.
5. Dissection microscope.
6. 10 cm cell tissue culture treated plate.
7. 2× 15 mL tubes.
8. 50 mL tube.
9. Hemocytometer.
10. 15 mL tube per 6-well plate.
11. Centrifuge.
12. 0.2 mL PCR tube.
13. HBSS, no calcium, no magnesium (HBSS−/−) at 4 °C—15 mL.
14. 1.5 mL tube with screw top.
15. Ice Bucket.
16. 20/200 μL pipette closely calibrated to the Hamilton syringe to be used for transplantation.

3.2 Transplantation of hGPCs into Neonatal Mice

1. HBSS−/− at 4 °C—15 mL.
2. 70% ethanol.
3. Stereotaxic frame with digital micrometer (or other apparatus to secure neonatal mouse head).
4. Heating Block.
5. Dissection microscope.
6. Fiber optic light.
7. 10 μL Hamilton Gastight syringe with attached micropipette.
8. 3× 10 mL syringe.

9. 2× 0.22 μm syringe filter.
10. 3× 30 gauge needle.
11. 3× 26 gauge needle.
12. Sterile cotton swab.
13. Sterile sheet.
14. Sterile gauze.
15. Alcohol prep pads.
16. Povidone-iodine prep pad.
17. Paper towel.
18. P1000 pipet tip box.

4 Methods

4.1 Preparation of Cells

Adherent hESC or hIPSC hGPCs must first be detached and large colonies dissected.

4.1.1 Detachment and Dissection of hESC or hIPSC-Derived hGPCs

1. Remove media from wells to be used.
2. Add 1 mL of HBSS−/− to each well.
3. Using cell scraper, gently detach cells into HBSS−/−.
4. Tilt plate and ensure cells are suspended in HBSS−/− by scraping into pool of liquid.
5. Pipette out cells into 15 mL tube.
6. Repeat **steps 2–5** to ensure maximal recovery of cells.
7. Allow cells to settle in tube.
8. Remove supernatant leaving ~200 μL of media on top.
9. Pipette up and down with a p1000 pipette ~5 times to break up clusters.
10. Transfer cells to the middle of a 10 cm tissue culture treated plate (*see* **Note 1**).
11. Examine cells under a dissecting microscope in a sterile hood.
12. Dissect larger cell aggregates until they are <100 μm in size with two size 11 scalpels.
13. Collect cells in a 15 mL tube.
14. Wash with HBSS−/− as needed to collect all cells from the plate. Ideally until total collected volume = 12 mL total.
15. Pre-wet 100-μm cell strainer with 1 mL of HBSS−/− on 50 mL tube.
16. Transfer 12 mL of cells through 100-μm cell strainer.
17. Add 1 mL of HBSS−/− to tube cells were in to wash and then pass them through the cell strainer.

4.1.2 Washing and Resuspension of Cells

If using hESC or hIPSC-derived hGPCs, begin at **step 2**. Sorted human fetal cells can either be used directly after sorting (FACS or MACS), or can be left on ultra-low attachment plates overnight in glial media.

1. For sorted fetal cells, transfer cells in media to a 15 mL tube and add HBSS−/− to 14 mL.
2. Remove 10 μL and count cells on hemocytometer.
3. Centrifuge cells for 10 min at 200 × *g*.
4. Carefully remove supernatant.
5. Resuspend in 100 μL HBSS−/−.
6. Transfer cell suspension to 0.2 mL pcr tube.
7. Wash 15 mL tube with 100 μL of HBSS−/−.
8. Transfer wash to 0.2 mL pcr tube.
9. Centrifuge for 10 min at 200 × *g*.
10. Completely remove supernatant.
11. Resuspend cells in HBSS−/− to achieve desired concentration for injections (e.g., 100,000 cells/μL) (*see* **Note 2**).
12. Place pcr tube in 1.5 mL tube and keep on ice until ready to use.

4.2 Transplantation of Cells

Transplantation of cells into neonatal mice is best accomplished at P1.

4.2.1 Equipment Preparation

1. Place sterile sheet where stereotaxic apparatus will be set up.
2. Set up stereotaxic apparatus under dissection microscope.
3. Set up fiber optic lighting to illuminate injection area.
4. Spray down working area with 70% ethanol and wipe down to dry.
5. Fill one 10 mL syringe with 70% ethanol.
6. Fill one 10 mL syringe with HBSS−/−.
7. Attach a 0.22 μm syringe filter to each syringe.
8. Attach 26G needle to each syringe filter.
9. Set heating block to 45 °C.
10. Sterilize P1000 pipet tip box with 70% ethanol and place on top of heating block.
11. Moisten sterile gauze with sterile HBSS−/− and place into box.
12. Fasten Hamilton syringe on stereotactic apparatus.

13. Flush Hamilton syringe micropipette with 1 mL of 70% ethanol by inserting filled ethanol syringe into the top of the Hamilton syringe.
14. Flush Hamilton syringe with 1 mL of HBSS−/− as you did in **step 12**.
    (a) If air bubbles persist in syringe, flush again with HBSS−/−.
15. Place plunger into Hamilton syringe taking care that air bubbles do not form.
16. Depress plunger to 4 μL mark.
17. Using a cotton swab, remove excess HBSS−/− from the tip of the Hamilton syringe needle. Use a downward motion as to not damage the needle.
18. Draw 1 μL of air into the syringe to create a separation air bubble.

4.2.2 Animal Preparation

Carry out these procedures within a biosafety hood; practice proper sterile techniques.

1. Remove top of cage.
2. Place a couple of food pellets at the opposite end of the cage from the nest as to lure the mother away.
3. When the mother leaves the nest, remove half of the pups placing them in the humidified warming chamber.
   (a) Never remove more than half of the pups at a time as to not stress the mother.
4. Remove one pup from humidified warming chamber.
5. Wipe pup down with povidone-iodine.
6. Place pup in a glove-barriered hole in the ice within ice bucket.
7. Cover the pup with ice and wait 2–4 min for the pup to be properly anesthetized.
8. Remove the pup from the ice and wipe the pup down with an ethanol wipe.
9. Place the pup into the stereotactic device securing their head.

4.2.3 Loading of Cells

1. Triturate the cell suspension gently using the P20 pipette to homogenize; avoid creating air bubbles.
2. Withdraw an appropriate volume for your injection type into the Hamilton syringe:
   (a) 2 μL for forebrain injection.
   (b) 3 μL for forebrain + hindbrain injection.
3. Return tube of cells to ice while injecting each mouse.

4.2.4 Injection of Cells in the Forebrain

1. Remove the 30 g needle from its sheath to bend the needle tip to a 90° angle.
2. Using the bent needle, create holes in the skull at these four sites (*see* **Note 3**):
   (a) Front: 1.0 mm posterior from bregma, ±1.0 mm from the midline.
   (b) Back: 2.4 mm posterior from bregma, ±1.0 mm from the midline.
3. Remove bent needle and insert injection needle into the punctured hole stopping at the level of the skull (*see* **Note 4**).
4. Lower the needle to the following depths:
   (a) Anterior: 1.0 mm.
   (b) Posterior: 0.9 mm.
5. Inject 0.5 μL in each site. Each site should be injected in 1 s.
6. Keep needle in the brain for 5 s following the injection and then retract it (*see* **Note 5**).

4.2.5 Injection of Cells in the Hindbrain

If significant engraftment of the hindbrain is desirable (e.g., for enhanced survival of the shiverer mouse), adding a fifth injection site into the hindbrain is advised.

1. Reposition pup in the stereotactic apparatus so that its head is down at a 90° angle to its body and secure its head.
2. Locate the cisterna magna, a soft spot below the base of the skull resembling a shallow circle.
3. Create a hole at the midline using the same technique as before with the bent needle.
4. Remove the bent needle and lower the injection needle into the punctured hole stopping at the level of the skull.
5. Lower the needle to a depth of 1.1 mm.
6. Inject 1 μL over the course of 2 s.
7. Allow the needle to remain in the brain for 5 s following the injection.
8. Retract the injection needle (*see* **Note 5**).

4.2.6 Recovery of Pups and Clean Up

1. Remove pup from the injection holder.
2. Wipe pup down with alcohol prep pad.
3. Place the pup in the humidified recovery chamber.
4. When all pups are awake within the humidified chamber, return them to their nest while the mother is away.
5. Remove Hamilton syringe plunger.
6. Flush the Hamilton syringe with 1 mL HBSS−/−.

7. Flush the Hamilton syringe with 1 mL of 70% ethanol.
8. Attach 26G needle to empty 10 mL syringe and fill with air.
9. Flush out the remaining ethanol.
10. Use the cotton swab to remove any additional fluid from the tip of the pipette.
11. Unmount the Hamilton syringe.

Utilizing this protocol will maximize the migration, survival, and engraftment of transplanted hGPCs. They will remain as hGPCs or differentiate into astrocytes or myelinating oligodendrocytes in a context-dependent manner. Time following injection before significant engraftment is reached will vary depending on mouse model and population of cells injected.

5 Notes

1. Dissection of clusters will be more easily achieved if the volume of liquid transferred to the plate is low enough to maintain surface tension. Avoid transferring bubbles, as they will interfere with visualization.
2. Depending upon the composition of cellular populations, the mouse model, and the speed of delivery, the number of cells injected will vary. Ideally, a volume of more than 2 μL into the neonatal forebrain should not be exceeded, as this may lead to irreparable tissue damage and disruption of the host tissue architecture.
3. While pressing down on the skull with the bent syringe, the tissue will depress. The skull has been punctured once it pops back up. Do not go deeper than this, as that will risk tissue damage.
4. Verification that the needle is penetrating the brain can be obtained by confirming that the skull does not depress as the needle is lowered. If it does snap back, then the skull puncture may not be sufficiently large. Step 2.4.2 may then need to be repeated, or adjustment of the syringe site may be necessary to assure proper injection of cells directly into the brain at the proper depth.
5. In rare cases, backflow may be observed. Nonetheless, in these instances, the vacuum created by removing the needle will often pull the cells back into the intended location.

References

1. Goldman SA (2016) Stem and progenitor cell-based therapy of the central nervous system: hopes, hype, and wishful thinking. Cell Stem Cell 18:174–188. https://doi.org/10.1016/j.stem.2016.01.012
2. Goldman SA (2017) Progenitor cell-based treatment of glial disease. Prog Brain Res 231:165–189. https://doi.org/10.1016/bs.pbr.2017.02.010
3. Goldman SA (2011) Progenitor cell-based treatment of the pediatric myelin disorders. Arch Neurol 68:848–856. https://doi.org/10.1001/archneurol.2011.46
4. Osorio MJ, Goldman SA (2016) Glial progenitor cell-based treatment of the childhood leukodystrophies. Exp Neurol 283:476–488. https://doi.org/10.1016/j.expneurol.2016.05.010
5. Franklin RJM, Goldman SA (2015) Glia disease and repair-remyelination. Cold Spring Harb Perspect Biol 7:a020594. https://doi.org/10.1101/cshperspect.a020594
6. Goldman SA, Nedergaard M, Windrem MS (2012) Glial progenitor cell-based treatment and modeling of neurological disease. Science 338:491–495. https://doi.org/10.1126/science.1218071
7. Raff MC, Miller RH, Noble M (1983) A glial progenitor cell that develops in vitro into an astrocyte or an oligodendrocyte depending on culture medium. Nature 303:390–396
8. Windrem MS, Nunes MC, Rashbaum WK et al (2004) Fetal and adult human oligodendrocyte progenitor cell isolates myelinate the congenitally dysmyelinated brain. Nat Med 10:93–97. https://doi.org/10.1038/nm974
9. Belachew S, Chittajallu R, Aguirre AA et al (2003) Postnatal NG2 proteoglycan-expressing progenitor cells are intrinsically multipotent and generate functional neurons. J Cell Biol 161:169–186. https://doi.org/10.1083/jcb.200210110
10. Nunes MC, Roy NS, Keyoung HM et al (2003) Identification and isolation of multipotential neural progenitor cells from the subcortical white matter of the adult human brain. Nat Med 9:439–447. https://doi.org/10.1038/nm837
11. Nishiyama A, Lin XH, Giese N et al (1996) Co-localization of NG2 proteoglycan and PDGF alpha-receptor on O2A progenitor cells in the developing rat brain. J Neurosci Res 43:299–314. https://doi.org/10.1002/(SICI)1097-4547(19960201)43:3<299::AID-JNR5>3.0.CO;2-E
12. Nishiyama A, Komitova M, Suzuki R, Zhu X (2009) Polydendrocytes (NG2 cells): multifunctional cells with lineage plasticity. Nat Rev Neurosci 10:9–22. https://doi.org/10.1038/nrn2495
13. Roy NS, Wang S, Harrison-Restelli C et al (1999) Identification, isolation, and promoter-defined separation of mitotic oligodendrocyte progenitor cells from the adult human subcortical white matter. J Neurosci 19:9986–9995
14. Sim FJ, McClain CR, Schanz SJ et al (2011) CD140a identifies a population of highly myelinogenic, migration-competent and efficiently engrafting human oligodendrocyte progenitor cells. Nat Biotechnol 29:934–941. https://doi.org/10.1038/nbt.1972
15. Douvaras P, Wang J, Zimmer M et al (2014) Efficient generation of myelinating oligodendrocytes from primary progressive multiple sclerosis patients by induced pluripotent stem cells. Stem Cell Reports 3:250–259. https://doi.org/10.1016/j.stemcr.2014.06.012
16. Piao J, Major T, Auyeung G et al (2015) Human embryonic stem cell-derived oligodendrocyte progenitors remyelinate the brain and rescue behavioral deficits following radiation. Cell Stem Cell 16:198–210. https://doi.org/10.1016/j.stem.2015.01.004
17. Blakemore WF, Crang AJ, Franklin RJ (1990) Transplantation of glial cell cultures into areas of demyelination in the adult CNS. Prog Brain Res 82:225–232
18. Gumpel M, Lachapelle F, Gansmuller A et al (1987) Transplantation of human embryonic oligodendrocytes into shiverer brain. Ann N Y Acad Sci 495:71–85
19. Lachapelle F, Gumpel M, Baulac M et al (1983) Transplantation of CNS fragments into the brain of shiverer mutant mice: extensive myelination by implanted oligodendrocytes. I. Immunohistochemical studies. Dev Neurosci 6:325–334. https://doi.org/10.1159/000112359
20. Archer DR, Levén S, Duncan ID (1994) Myelination by cryopreserved xenografts and allografts in the myelin-deficient rat. Exp Neurol 125:268–277. https://doi.org/10.1006/exnr.1994.1029
21. Archer DR, Cuddon PA, Lipsitz D, Duncan ID (1997) Myelination of the canine central nervous system by glial cell transplantation: a model for repair of human myelin disease. Nat Med 3:54–59. https://doi.org/10.1038/nm0197-54

22. Duncan ID (2005) Oligodendrocytes and stem cell transplantation: their potential in the treatment of leukoencephalopathies. J Inherit Metab Dis 28:357–368. https://doi.org/10.1007/s10545-005-7058-z
23. Yandava BD, Billinghurst LL, Snyder EY (1999) "Global" cell replacement is feasible via neural stem cell transplantation: evidence from the dysmyelinated shiverer mouse brain. Proc Natl Acad Sci U S A 96:7029–7034
24. Ben-Hur T, Goldman SA (2008) Prospects of cell therapy for disorders of myelin. Ann N Y Acad Sci 1142:218–249. https://doi.org/10.1196/annals.1444.014
25. Franklin RJ, Ffrench-Constant C (2008) Remyelination in the CNS: from biology to therapy. Nat Rev Neurosci 9:839–855
26. Windrem MS, Roy NS, Wang J et al (2002) Progenitor cells derived from the adult human subcortical white matter disperse and differentiate as oligodendrocytes within demyelinated lesions of the rat brain. J Neurosci Res 69:966–975. https://doi.org/10.1002/jnr.10397
27. Windrem MS, Schanz SJ, Guo M et al (2008) Neonatal chimerization with human glial progenitor cells can both remyelinate and rescue the otherwise lethally hypomyelinated shiverer mouse. Cell Stem Cell 2:553–565. https://doi.org/10.1016/j.stem.2008.03.020
28. Popko B, Puckett C, Lai E et al (1987) Myelin deficient mice: expression of myelin basic protein and generation of mice with varying levels of myelin. Cell 48:713–721
29. Readhead C, Popko B, Takahashi N et al (1987) Expression of a myelin basic protein gene in transgenic shiverer mice: correction of the dysmyelinating phenotype. Cell 48:703–712
30. Windrem MS, Schanz SJ, Morrow C et al (2014) A competitive advantage by neonatally engrafted human glial progenitors yields mice whose brains are chimeric for human glia. J Neurosci 34:16153–16161. https://doi.org/10.1523/JNEUROSCI.1510-14.2014
31. Hughes EG, Kang SH, Fukaya M, Bergles DE (2013) Oligodendrocyte progenitors balance growth with self-repulsion to achieve homeostasis in the adult brain. Nat Neurosci 16:668–676. https://doi.org/10.1038/nn.3390
32. Amoyel M, Bach EA (2014) Cell competition: how to eliminate your neighbours. Development 141:988–1000. https://doi.org/10.1242/dev.079129
33. Amoyel M, Simons BD, Bach EA (2014) Neutral competition of stem cells is skewed by proliferative changes downstream of Hh and Hpo. EMBO J 33:2295–2313. https://doi.org/10.15252/embj.201387500
34. Clavería C, Giovinazzo G, Sierra R, Torres M (2013) Myc-driven endogenous cell competition in the early mammalian embryo. Nature 500:39–44. https://doi.org/10.1038/nature12389
35. Goldman SA, Nedergaard M, Windrem MS (2015) Modeling cognition and disease using human glial chimeric mice. Glia 63:1483–1493. https://doi.org/10.1002/glia.22862
36. Oberheim NA, Wang X, Goldman S, Nedergaard M (2006) Astrocytic complexity distinguishes the human brain. Trends Neurosci 29:547–553. https://doi.org/10.1016/j.tins.2006.08.004
37. Oberheim NA, Takano T, Han X et al (2009) Uniquely hominid features of adult human astrocytes. J Neurosci 29:3276–3287. https://doi.org/10.1523/JNEUROSCI.4707-08.2009
38. Araque A, Parpura V, Sanzgiri RP, Haydon PG (1999) Tripartite synapses: glia, the unacknowledged partner. Trends Neurosci 22:208–215
39. Kang J, Jiang L, Goldman SA, Nedergaard M (1998) Astrocyte-mediated potentiation of inhibitory synaptic transmission. Nat Neurosci 1:683–692. https://doi.org/10.1038/3684
40. Ullian EM, Sapperstein SK, Christopherson KS, Barres BA (2001) Control of synapse number by glia. Science 291:657–661. https://doi.org/10.1126/science.291.5504.657
41. Allen NJ, Bennett ML, Foo LC et al (2012) Astrocyte glypicans 4 and 6 promote formation of excitatory synapses via GluA1 AMPA receptors. Nature 486:410–414. https://doi.org/10.1038/nature11059
42. Stellwagen D, Malenka RC (2006) Synaptic scaling mediated by glial TNF-alpha. Nature 440:1054–1059. https://doi.org/10.1038/nature04671
43. Stogsdill JA, Ramirez J, Liu D et al (2017) Astrocytic neuroligins control astrocyte morphogenesis and synaptogenesis. Nature 551:192–197. https://doi.org/10.1038/nature24638
44. Han X, Chen M, Wang F et al (2013) Forebrain engraftment by human glial progenitor cells enhances synaptic plasticity and learning in adult mice. Cell Stem Cell 12:342–353. https://doi.org/10.1016/j.stem.2012.12.015
45. Hu B-Y, Du Z-W, Zhang S-C (2009) Differentiation of human oligodendrocytes from

pluripotent stem cells. Nat Protoc 4:1614–1622. https://doi.org/10.1038/nprot.2009.186
46. Izrael M, Zhang P, Kaufman R et al (2007) Human oligodendrocytes derived from embryonic stem cells: Effect of noggin on phenotypic differentiation in vitro and on myelination in vivo. Mol Cell Neurosci 34:310–323. https://doi.org/10.1016/j.mcn.2006.11.008
47. Wang S, Bates J, Li X et al (2013) Human iPSC-derived oligodendrocyte progenitor cells can myelinate and rescue a mouse model of congenital hypomyelination. Cell Stem Cell 12:252–264. https://doi.org/10.1016/j.stem.2012.12.002
48. Stacpoole SRL, Spitzer S, Bilican B et al (2013) High yields of oligodendrocyte lineage cells from human embryonic stem cells at physiological oxygen tensions for evaluation of translational biology. Stem Cell Reports 1:437–450. https://doi.org/10.1016/j.stemcr.2013.09.006
49. Fossati V, Douvaras P (2014) Generating induced pluripotent stem cells for multiple sclerosis therapy. Regen Med 9:709–711. https://doi.org/10.2217/rme.14.63
50. Haley SA, Atwood WJ (2014) An animal model for progressive multifocal leukoencephalopathy. J Clin Invest 124:5103–5106. https://doi.org/10.1172/JCI79186
51. Kondo Y, Windrem MS, Zou L et al (2014) Human glial chimeric mice reveal astrocytic dependence of JC virus infection. J Clin Invest 124:5323–5336. https://doi.org/10.1172/JCI76629
52. Benraiss A, Wang S, Herrlinger S et al (2016) Human glia can both induce and rescue aspects of disease phenotype in Huntington disease. Nat Commun 7:11758. https://doi.org/10.1038/ncomms11758
53. Windrem MS, Osipovitch M, Liu Z et al (2017) Human iPSC glial mouse chimeras reveal glial contributions to schizophrenia. Cell Stem Cell 21:195–208.e6. https://doi.org/10.1016/j.stem.2017.06.012
54. Abud EM, Ramirez RN, Martinez ES et al (2017) iPSC-derived human microglia-like cells to study neurological diseases. Neuron 94:278–293.e9. https://doi.org/10.1016/j.neuron.2017.03.042
55. Douvaras P, Sun B, Wang M et al (2017) Directed differentiation of human pluripotent stem cells to microglia. Stem Cell Reports 8:1516–1524. https://doi.org/10.1016/j.stemcr.2017.04.023
56. Pandya H, Shen MJ, Ichikawa DM et al (2017) Differentiation of human and murine induced pluripotent stem cells to microglia-like cells. Nat Neurosci 20:753–759. https://doi.org/10.1038/nn.4534
57. Araínga M, Su H, Poluektova LY et al (2016) HIV-1 cellular and tissue replication patterns in infected humanized mice. Sci Rep 6:23513. https://doi.org/10.1038/srep23513
58. Brüstle O, Choudhary K, Karram K et al (1998) Chimeric brains generated by intraventricular transplantation of fetal human brain cells into embryonic rats. Nat Biotechnol 16:1040–1044. https://doi.org/10.1038/3481
59. Chen C, Kim W-Y, Jiang P (2016) Humanized neuronal chimeric mouse brain generated by neonatally engrafted human iPSC derived primitive neural progenitor cells. JCI Insight 1:e88632. https://doi.org/10.1172/jci.insight.88632
60. Keyoung HM, Roy NS, Benraiss A et al (2001) High-yield selection and extraction of two promoter-defined phenotypes of neural stem cells from the fetal human brain. Nat Biotechnol 19:843–850. https://doi.org/10.1038/nbt0901-843
61. Osorio MJ, Goldman SA (2016) Cell therapy for pediatric disorders of glia. In: Tuszynski MH (ed) Translational neuroscience. Springer, Boston, MA, pp 275–296

Chapter 19

A Method to Detect Cytochrome *c* Oxidase Activity and Mitochondrial Proteins in Oligodendrocytes

Graham R. Campbell and Don J. Mahad

Abstract

Cytochrome *c* oxidase or mitochondrial respiratory chain complex IV is where over 90% of oxygen is consumed. The relationship between complex IV activity and mitochondrial proteins, which provides a guide to understanding the mechanisms in primary mitochondrial disorders, has been determined by histochemistry (complex IV activity) and immunohistochemistry in serial sections. In the central nervous system (CNS), mitochondrial activity and immunoreactivity have been determined in populations of cells in serial sections as capturing cells in more than one section is difficult. In this report, we describe a method to determine complex IV activity in relation to mitochondrial proteins at a single-cell level in the CNS. We performed complex IV histochemistry and immunohistochemistry consecutively in snap-frozen sections. Although the product of complex IV histochemistry reduces the sensitivity of standard immunohistochemistry (secondary antibody and ABC method), the biotin-free Menapath polymer detection system enables mitochondrial proteins to be detected following complex IV histochemistry. The co-occurring chromogens may then be separately visualized and analyzed using multispectral imaging. Our technique is applicable for exploring mitochondrial defects within single cells, including oligodendrocytes, in a variety of CNS disorders and animal models of those diseases.

Key words Mitochondria, Cytochrome *c* oxidase, Neurodegeneration, Immunohistochemistry and histochemistry

1 Introduction

1.1 Mitochondrial Respiratory Chain

Oligodendrocytes lacking cytochrome *c* oxidase (COX), or mitochondrial respiratory chain complex IV, are recognized as glycolytic cells that appear to metabolically support myelinated axons, in vivo [1]. Complex IV is the terminal complex of the electron transport chain and the site where over 90% of inhaled oxygen is consumed [2]. Complex IV activity may be determined and densitometrically analyzed at a single-cell level in snap-frozen tissue sections using complex IV histochemistry [3–5], which correlates well with bio-

David A. Lyons and Linde Kegel (eds.), *Oligodendrocytes: Methods and Protocols*, Methods in Molecular Biology, vol. 1936, https://doi.org/10.1007/978-1-4939-9072-6_19, © Springer Science+Business Media, LLC, part of Springer Nature 2019

chemically analyzed activity in tissue homogenates [6]. Complex IV deficiency is a well-recognized feature in patients with primary mitochondrial dysfunction due to defects in mitochondrial DNA (mtDNA) and/or nuclear DNA [2, 7–9].

The mitochondrial respiratory chain complexes are made up of multiple subunits, which may be immunohistochemically detected in snap-frozen as well as fixed tissue sections [10–12]. The pattern of mitochondrial respiratory chain complex subunit immunoreactivity in relation to mitochondrial activity provides a guide to the possible underlying mechanism of mitochondrial dysfunction [11]. The subunit I of complex IV (COX-I) is a major component of the catalytic core as it contains three of the four oxygen binding sites (heme *a* and Cu_B-heme a_3 binuclear centers) [13, 14]. Although the catalytic subunits of complex IV may be absent in some cases when the biochemical defect is genetically determined, in other cases complex IV deficiency occurs despite immunohistochemically detectable COX-I [11]. For example, point mutations of mtDNA may impair complex IV activity without substantially reducing the steady state levels of COX-I [15]. Furthermore, complex IV deficiency due to modifications, rather than loss, of COX-I has been reported in animal models and isolated mitochondria [16, 17].

To date, the relationship between complex IV activity and mitochondrial respiratory chain complex subunits has been histologically determined in serial sections [4, 11, 12, 18]. Single muscle fibers and individual colon crypts are particularly suitable for studying mitochondrial activity in relation to genetic defects, mRNA expression, and mitochondrial protein immunoreactivity because of their uniform composition in serial tissue sections [11, 12, 18]. In the central nervous system (CNS), the relationship between mitochondrial activity and protein immunoreactivity has been explored using populations of cells in serial sections rather than within single cells due to the difficulty in capturing cells in serial sections [12, 19]. The importance of determining mitochondrial activity and protein immunoreactivity at a single-cell level in CNS tissue sections is highlighted by the fact that the level (heteroplasmy) of mtDNA defects, and the resulting protein and activity, may vary between cells [20, 21]. By combining complex IV histochemistry (activity) with immunohistochemical detection of COX-I and porin (a voltage gated anion channel present in all mitochondria [22]) we describe a technique to analyze complex IV activity relative to mitochondrial proteins within single cells in snap-frozen tissue sections.

1.2 Detection of Mitochondrial Activity and Mitochondrial Protein Immunoreactivity in Serial Sections from Cases with Mitochondrial DNA Defects

We used serial sections of colon crypts from an aging control, where mtDNA mutations have been reported [18, 23], and muscle from a case with a single deletion of mtDNA to detect complex IV and complex II activity by histochemistry as well as COX-I (a catalytic subunit of complex IV) and porin (a marker of mitochondria) by immunohistochemistry. The lack of complex IV activity with intact activity of complex II, which is entirely encoded by nuclear DNA, is a well-recognized histopathological feature of primary mtDNA disorders [12].

The colon crypts deficient in complex IV activity lack COX-I immunoreactivity in serial sections, indicating the loss of activity following the loss of catalytic subunit in cases with biochemically significant mtDNA defects [12]. However, the complex IV-deficient structures contain mitochondria, as judged by complex II activity and porin immunoreactivity in serial sections [24]. When complex IV and complex II histochemistry assays are performed consecutively (complex IV/complex II histochemistry), complex IV-active and complex IV-deficient structures are apparent in brown and blue, respectively [24]. The unmixed multispectral images of complex IV/complex II histochemistry and pseudo-colored immunofluorescence-like image show an apparent loss of staining for complex II activity where complex IV activity is detected, which is due to the inability to deposit reduced NBT crystals and/or reduce NBT by complex II via the intermediate electron donor, phenazine methosulfate. Complex IV/complex II histochemistry on a muscle section shows fibers with and without complex IV activity, which is genetically determined [24]. The serial sections show a striking decrease in COX-I subunit in those fibers lacking complex IV activity [24] despite the presence of mitochondria as judged by porin immunoreactivity [24]. As in colon crypts, the product of complex IV histochemistry interferes with the detection of complex II in muscle fiber that contain complex IV activity [24].

1.3 The Product of Complex IV Histochemistry Impairs the Detection of Mitochondrial Proteins by Standard Immuno-histochemistry and Immunofluorescent Histochemistry

We used hippocampal sections from a case with multiple deletions of mtDNA [25] to explore the relation between complex IV activity and mitochondrial protein immunoreactivity in the CNS. There are complex IV-active and -deficient [24] cells, morphologically consistent with neurons and containing NeuN labeling nuclei (not shown), in the snap-frozen tissue sections. In a serial section fixed in PFA, COX-I immunoreactivity is present within a proportion of cells [24], which is consistent with the positive correlation between complex IV activity and COX-I reported in primary mitochondrial disorders [12]. When a serial section is subjected to complex IV/complex II histochemistry there are cells lacking complex IV activity but with complex II activity [24].

To detect mitochondrial proteins in relation to complex IV activity at a single-cell level we performed complex IV histochemistry and immunofluorescent histochemistry consecutively [24]. The mitochondria in cells lacking complex IV activity are evident by the immunofluorescent labeling of porin subsequent to complex IV histochemistry. Intriguingly, neither porin nor COX-I is detectable in the population of cells with complex IV activity, which is not due to the lack of mitochondria or the catalytic subunits of complex IV.

To explore the relationship between the formation of complex IV histochemical reaction product and mitochondrial protein immunoreactivity, we used serial sections of snap-frozen rat spinal cord and performed time course experiments with complex IV histochemical assay [4, 10] prior to subjecting the same sections to standard immunohistochemistry using secondary antibody and ABC method with Vector® SG (gray) as the chromogen [24]. The formation of complex IV histochemical reaction product beyond the 20 min time point prevents the immunohistochemical detection of COX-I [24] and porin (data not shown). The unmixing of the multispectral images and quantitation of immunoreactivity using Nuance system identifies a striking loss of COX-I immunoreactivity at and beyond the 20 min time point [24].

1.4 Immunohistochemical Detection of Mitochondrial Proteins Subsequent to Complex IV Histochemistry

In order to explore the apparent loss of mitochondrial immunoreactivity following complex IV histochemistry we used the biotin-free Menapath X-Cell plus polymer detection system, instead of the secondary antibody and ABC method, to detect the primary mouse monoclonal antibodies. The biotin-free Menapath polymer detection system has a greater sensitivity compared with the avidin biotin amplification method as indicated by the optimum primary antibody dilution of 1:3200 for porin instead of 1:400 necessary for the secondary antibody and ABC method. Porin is detectable in rat spinal cord ventral horn cells subsequent to complex IV histochemistry with the use of Menapath polymer detection system [24]. The quantitation of immunoreactivity (intensity and area above a fixed threshold) using Nuance system shows that porin immunoreactivity in the entire ventral horns detected using the biotin-free Menapath polymer detection system is not affected by the product of complex IV histochemistry [24].

In comparison to porin, COX-I immunoreactivity following complex IV histochemistry is partial and not apparent in a number of ventral horn cells with abundant complex IV activity despite the increased sensitivity of the biotin-free method. The impact of complex IV histochemistry on subsequent detection of porin and COX-I using the biotin-free method is similar in non-CNS tissue such as muscle and colon crypts to rat spinal cord ventral horns.

1.5 Analysis of Complex IV Activity in Relation to Mitochondrial Protein Immunoreactivity at a Single-Cell Level

We validate the combined histochemical and immunohistochemical method to detect complex IV activity and mitochondrial protein immunoreactivity at a single-cell level using a specific inhibitor of complex IV, sodium azide [26], and different reaction times of complex IV histochemistry [4].

As expected, the incubation of spinal cord sections with sodium azide leads to a global inhibition of complex IV [24]. The reduction in complex IV activity by sodium azide is apparent within ventral horn cells abundant in mitochondria as illustrated by the composite of pseudo-colored unmixed multispectral images of complex IV histochemistry (Fig. 6 in [24]) and porin immunohistochemistry. When ventral horn cells abundant in mitochondria were outlined by a fixed threshold of porin immunoreactivity, there is a significant ($p < 0.001$) reduction in densitometrically determined complex IV activity as well as the ratio between complex IV activity and porin immunoreactivity following exposure to sodium azide [24]. The sodium azide-mediated reduction in complex IV activity is also apparent in the regions of cells that bear COX-I immunoreactivity using the biotin-free method. Interestingly, the inhibition of complex IV by sodium azide restores the ability to immunohistochemically detect COX-I following complex IV histochemistry in ventral horns cells from naïve rat spinal cords, using the secondary antibody and ABC as well as biotin-free methods.

In the time course experiments the ratio between complex IV activity and porin immunoreactivity at a single-cell level increases, exponentially, when ventral horn cells were identified by the abundance of mitochondria [24]. The increase in mitochondrial-activity-to-protein ratio is also apparent in the regions of cells that bear COX-I immunoreactivity using the biotin-free method.

In summary, we describe a method to determine mitochondrial activity in relation to mitochondrial protein immunoreactivity at a single-cell level in the CNS. The use of Menapath detection system overcomes the difficulties in detecting mitochondrial proteins following histochemistry and, together with multispectral imaging capabilities of Nuance system, offers a solution to the difficulties in capturing single CNS cells in serial sections. Our method is applicable to investigate oligodendrocytes in a variety of CNS disorders as well as animal models.

2 Materials

2.1 Tissue Acquisition

1. Isopentane.
2. Human colon, muscle, and brain.
3. Cryostat.
4. Rats.

5. Isoflurane.
6. Rinse buffer: 2000 U/L heparin, 0.025% lidocaine, 0.002% $NaNO_2$, and 0.02 M *N*-2-hydroxyethylpiperazine-NO-2-ethanesulfonic acid (HEPES; pH 7.4).
7. Dissection tools.

2.2 Histochemistry

1. Complex IV medium: 100 μM cytochrome *c*, 4 mM diaminobenzidine tetrahydrochloride (DAB·4HCl), 20 μg/mL catalase in 0.1 M phosphate buffer (pH 7.0).
2. Complex II/Succinate dehydrogenase (SDH) medium: 130 mM sodium succinate, 200 mM phenazine methosulfate, 1 mM sodium azide, 1.5 mM nitroblue tetrazolium (NBT) in 0.1 M phosphate buffer, pH 7.0.
3. PBS.
4. Ethanol 70%, 95%, 100%.
5. Histoclear.
6. DPX®.

2.3 Immunohistochemistry

1. 4% paraformaldehyde (PFA).
2. Tris-buffered saline Tween-20 (TBST).
3. 70% methanol, 95% methanol and 100% methanol.
4. 0.3% H_2O_2.
5. Mouse monoclonal antibodies raised against porin (IgG_{2b}) and COX-I (IgG_{2a}).
6. 1% normal goat serum.
7. Avidin-biotin-horseradish peroxidase complex (ABC).
8. Antibodies against porin and COX-I.
9. ABC solution for 30 min.
10. Alexia fluor 568.
11. Streptavidin-Alex fluor 488.
12. Menapath X-Cell Plus HRP Polymer detection system.
13. Vector® SG.
14. Histoclear.
15. DPX®.

3 Methods

3.1 Human Tissue

1. Snap-freeze samples of human colon, muscle, and brain in isopentane at −160 °C.
2. Obtain the colon sample from colonic mucosa at least 12 cm from the edge of the lesion following resection of a colon tumor.

3. The snap-frozen blocks are used to obtain cryostat sections (15 μm), which are air-dried for 60 min at room temperature prior to performing histochemistry and/or immunohistochemistry.

3.2 Rat Spinal Cord

1. Anesthetize naïve Sprague Dawley rats with isoflurane and perfuse with rinse buffer, according to standard procedures.
2. Dissect the spinal cord and snap-freeze in isopentane for cryostat sectioning.

3.3 Complex IV and/or Complex II Histochemistry

In serial sections, complex IV and complex II histochemistry was performed separately to detect cytochrome *c* oxidase and succinate dehydrogenase activity, respectively. Complex IV activity was detected using a well-established histochemical technique [10].

1. Apply complex IV medium (*see* Subheading 2.2) to sections for 50 min at 37 °C.
2. In a serial section, detect complex II activity using succinate dehydrogenase (SDH) medium. Incubate sections for 45 min at 37 °C.
3. Wash two times in PBS.
4. Dehydrate in grades of ethanol 70%, 95%, and 100% (5 min).
5. Clear in Histoclear (5 min).
6. Mount in DPX®.

3.4 Sequential Complex IV/Complex II Histochemistry

Complex IV histochemistry and complex II histochemistry was applied sequentially to the same section.

1. Subject a snap-frozen cryostat section to complex IV medium.
2. Wash two times in PBS.
3. Expose the same tissue section to complex II medium (*see* **Note 1**).
4. Incubate snap-frozen air-dried tissue sections for 30 min in 1 mM sodium azide. To inhibit complex IV (*see* **Note 2**).
5. Add complex IV medium.
6. Incubate a control section in PBS for 30 min prior to applying the complex IV medium.

3.5 Immunohistochemistry

1. Allow sections to dry from −80 °C storage.
2. Fix in 4% paraformaldehyde (PFA) for 10 min at 4 °C.
3. Permeabilize in solutions of Tris-buffered saline Tween-20 (TBST), then 70% methanol, and then 95% methanol with 0.3% H_2O_2 for 10 min each and 100% methanol for 20 min followed by rehydration.
4. Two methods of primary antibody detection were used (*see* **Note 3**).

3.5.1 Method 1

1. Apply blocking serum of 1% normal goat serum for 30 min for those sections to be visualized using the avidin-biotin-horseradish peroxidase complex (ABC) method. Incubate the primary antibodies at concentrations of 1:400 for porin and 1:100 for COX-I for 90 min.
2. Apply a secondary anti-mouse antibody (Vector Laboratories) 1:200 for 30 min.
3. Wash with TBST.
4. Add the ABC solution for 30 min (*see* **Note 4**).

3.5.2 Method 2

1. Use the Menapath X-Cell Plus HRP Polymer detection system (A. Menarini Diagnostics, Wokingham, UK) without a blocking step and an optimum concentration of 1:3200 for both porin and COX-I primary antibodies.
2. Detect staining using the polymer kit as per manufacturing guidelines (A. Menarini Diagnostics, Wokingham, UK).
3. Use the Vector® SG as the HRP substrate.
4. Dehydrate the sections in grades of ethanol 70%, 95%, and 100% twice.
5. Apply Histoclear.
6. Mount in DPX®.

3.6 Combining Complex IV Histochemistry with Immunohistochemistry

1. After the complex IV histochemistry (Subheading 3.3).
2. Wash off Excess medium with PBS.
3. Carry out immunohistochemistry with chromogens or fluorochromes (*see* Subheading 3.5) beginning with fixation in 4% PFA solution.
4. Mount the sections subjected to both complex IV histochemistry and immunohistochemistry in aqueous medium.

3.7 Multispectral Imaging and Densitometric Analysis of Multiple Chromogens

1. Obtain Bright field images of the two chromogens (diaminobenzidine tetrahydrochloride and Vector® SG) using the Nuance imaging system (CRi, Woburn, MA) (*see* **Note 5**).
2. Determine global densitometric values of complex IV activity and immunoreactivity in the entire ventral horns by manually outlining the ventral horns in unmixed images. Outline the single cells with abundant mitochondria, which are mostly neurons based on morphology and NeuN labeling (not shown) in ventral horns of rat spinal cords by setting a fixed threshold on porin or COX-I immunoreactivity in the sections subjected to immunohistochemistry only.
3. Keep the threshold of porin or COX-I immunoreactivity constant.

4. Use threshold to outline the cells abundant in mitochondria in the unmixed images of sections subjected to both complex IV histochemistry and immunohistochemistry. Record the mean signal and the area of immunoreactivity within each cell.
5. Copy the outline of cells, based on the immunoreactivity above the threshold, on to the unmixed images of complex IV activity.
6. Record the mean signal intensity of complex IV histochemistry (activity) within each corresponding cell.

4 Notes

1. Following the complex IV/complex II histochemical reaction any cell lacking complex IV activity will be blue, formed by the presence of NBT crystals and the lack of DAB·4HCl. Brown staining indicates complex IV activity.
2. Sodium azide is a specific inhibitor of cytochrome *c* oxidase [26].
3. The mouse monoclonal antibodies raised against porin (IgG_{2b}) and COX-I (IgG_{2a}) were purchased from Molecular Probes® (Invitrogen Ltd., Paisley, UK).
4. For double immunofluorescent labeling subtype specific secondary antibodies from Molecular Probes® (Invitrogen Ltd., Paisley, UK) were used, one biotinylated and the other conjugated to Alexia fluor 568, followed by streptavidin-Alex fluor 488 for 30 min.
5. The Nuance imaging system (CRi, Woburn, MA), which uses liquid crystal tunable filters to acquire multispectral images and deconvolution methods to separate or "unmix" the original double-labeled image to reveal unmixed images (in gray scale) of each chromogen [27]. The unmixed images may then be used to quantitate each chromogen, apply pseudo-color and combined into a composite that appears like an immunofluorescently labeled image [27].

References

1. Fünfschilling U, Supplie LM, Mahad D et al (2012) Glycolytic oligodendrocytes maintain myelin and long-term axonal integrity. Nature 485:517–521. https://doi.org/10.1038/nature11007
2. DiMauro S, Schon EA (2003) Mitochondrial respiratory-chain diseases. N Engl J Med 348:2656–2668. https://doi.org/10.1056/NEJMra022567
3. Dimlich RV, Showers MJ, Shipley MT (1990) Densitometric analysis of cytochrome oxidase in ischemic rat brain. Brain Res 516:181–191
4. Johnson MA, Bindoff LA, Turnbull DM (1993) Cytochrome c oxidase activity in single muscle fibers: assay techniques and diagnostic applications. Ann Neurol 33:28–35. https://doi.org/10.1002/ana.410330106
5. Seligman AM, Karnovsky MJ, Wasserkrug HL, Hanker JS (1968) Nondroplet ultrastructural demonstration of cytochrome oxidase activity

with a polymerizing osmiophilic reagent, diaminobenzidine (DAB). J Cell Biol 38:1–14

6. Hevner RF, Liu S, Wong-Riley MT (1995) A metabolic map of cytochrome oxidase in the rat brain: histochemical, densitometric and biochemical studies. Neuroscience 65:313–342
7. Betts J, Lightowlers RN, Turnbull DM (2004) Neuropathological aspects of mitochondrial DNA disease. Neurochem Res 29:505–511
8. Shoubridge EA (2001) Nuclear genetic defects of oxidative phosphorylation. Hum Mol Genet 10:2277–2284
9. Tanji K, Bonilla E (2000) Neuropathologic aspects of cytochrome C oxidase deficiency. Brain Pathol 10:422–430
10. Johnson MA, Kadenbach B, Droste M et al (1988) Immunocytochemical studies of cytochrome oxidase subunits in skeletal muscle of patients with partial cytochrome oxidase deficiencies. J Neurol Sci 87:75–90
11. Rahman S, Lake BD, Taanman JW et al (2000) Cytochrome oxidase immunohistochemistry: clues for genetic mechanisms. Brain 123 (Pt 3):591–600
12. Tanji K, Bonilla E (2008) Light microscopic methods to visualize mitochondria on tissue sections. Methods 46:274–280. https://doi.org/10.1016/j.ymeth.2008.09.027
13. Fontanesi F, Soto IC, Barrientos A (2008) Cytochrome c oxidase biogenesis: new levels of regulation. IUBMB Life 60:557–568. https://doi.org/10.1002/iub.86
14. Taanman JW, Williams SL (2001) Assembly of cytochrome c oxidase: what can we learn from patients with cytochrome c oxidase deficiency? Biochem Soc Trans 29:446–451
15. Pye D, Kyriakouli DS, Taylor GA et al (2006) Production of transmitochondrial cybrids containing naturally occurring pathogenic mtDNA variants. Nucleic Acids Res 34:e95–e95. https://doi.org/10.1093/nar/gkl516
16. Elfering SL, Haynes VL, Traaseth NJ et al (2004) Aspects, mechanism, and biological relevance of mitochondrial protein nitration sustained by mitochondrial nitric oxide synthase. Am J Physiol Heart Circ Physiol 286: H22–H29. https://doi.org/10.1152/ajpheart.00766.2003
17. Martin LJ, Liu Z, Chen K et al (2007) Motor neuron degeneration in amyotrophic lateral sclerosis mutant superoxide dismutase-1 transgenic mice: mechanisms of mitochondriopathy and cell death. J Comp Neurol 500:20–46. https://doi.org/10.1002/cne.21160
18. Taylor RW, Barron MJ, Borthwick GM et al (2003) Mitochondrial DNA mutations in human colonic crypt stem cells. J Clin Invest 112:1351–1360. https://doi.org/10.1172/JCI19435
19. Tanji K, Kunimatsu T, Vu TH, Bonilla E (2001) Neuropathological features of mitochondrial disorders. Semin Cell Dev Biol 12:429–439. https://doi.org/10.1006/scdb.2001.0280
20. Sciacco M, Bonilla E, Schon EA et al (1994) Distribution of wild-type and common deletion forms of mtDNA in normal and respiration-deficient muscle fibers from patients with mitochondrial myopathy. Hum Mol Genet 3:13–19
21. Taylor RW, Turnbull DM (2005) Mitochondrial DNA mutations in human disease. Nat Rev Genet 6:389–402. https://doi.org/10.1038/nrg1606
22. Mannella CA (1998) Conformational changes in the mitochondrial channel protein, VDAC, and their functional implications. J Struct Biol 121:207–218. https://doi.org/10.1006/jsbi.1997.3954
23. Greaves LC, Preston SL, Tadrous PJ et al (2006) Mitochondrial DNA mutations are established in human colonic stem cells, and mutated clones expand by crypt fission. Proc Natl Acad Sci U S A 103:714–719. https://doi.org/10.1073/pnas.0505903103
24. Mahad DJ, Ziabreva I, Campbell G et al (2009) Detection of cytochrome c oxidase activity and mitochondrial proteins in single cells. J Neurosci Methods 184:310–319. https://doi.org/10.1016/j.jneumeth.2009.08.020
25. Reeve AK, Krishnan KJ, Elson JL et al (2008) Nature of mitochondrial DNA deletions in substantia nigra neurons. Am J Hum Genet 82:228–235. https://doi.org/10.1016/j.ajhg.2007.09.018
26. Keilin D, Hartree EF (1945) Purification and properties of cytochrome c. Biochem J 39:289–292
27. Mansfield JR, Hoyt C, Levenson RM (2008) Visualization of microscopy-based spectral imaging data from multi-label tissue sections. Curr Protoc Mol Biol Chapter 14:Unit 14.19–14.19.15. https://doi.org/10.1002/0471142727.mb1419s84

Chapter 20

Transmission Electron Microscopy of Oligodendrocytes and Myelin

Marie-Theres Weil, Torben Ruhwedel, Martin Meschkat, Boguslawa Sadowski, and Wiebke Möbius

Abstract

In this chapter, we describe protocols to study different aspects of oligodendrocytes and myelin using electron microscopy. First, we describe in detail how to prepare central nervous system tissue routinely by perfusion fixation of the animal and conventional embedding in Epon resin. Then, we explain how, with some modifications, chemically fixed tissue can be used for immunoelectron microscopy on cryosections. Chemical fixation and Epon embedding can also be applied to purified myelin to assess the quality of the preparation. Furthermore, we describe how cryopreparation by high-pressure freezing can be used to study the fine structure of myelin in nerve, brain, and spinal cord tissue. The differences in the structural appearance of oligodendrocytes and myelin between cryopreserved and conventionally processed samples are compared using representative images. Since primary cultured oligodendrocytes are used to study structure and function in vitro, we provide protocols for chemical fixation and Epon embedding of these cultures. Finally, we explain how the cytoskeleton of cultured oligodendrocytes can be visualized by using transmission electron microscopy on platinum-carbon replicas. In this chapter, we provide a wide range of protocols that can be applied to shed light on the different biological aspects of myelin and oligodendrocytes.

Key words Electron microscopy, Immunolabeling, Oligodendrocytes, Cytoskeleton, Cryopreparation, Central nervous system, Myelin, Chemical fixation

1 Introduction

With the pace of instrumental development, imaging techniques are increasingly gaining importance, especially for the investigation of central nervous system biology and function. In combination with genetic tools and in vivo imaging techniques, electron microscopy adds valuable ultrastructural information to the related functional readouts. For the acquisition of ultrastructural data, several preparation strategies can be considered depending on the biological question. To assess general changes in cellular structures associated with the phenotype in a given animal model, chemical

David A. Lyons and Linde Kegel (eds.), *Oligodendrocytes: Methods and Protocols*, Methods in Molecular Biology, vol. 1936, https://doi.org/10.1007/978-1-4939-9072-6_20,

fixation by perfusion and conventional Epon embedding are recommended [1]. This standard technique provides results regarding, e.g., quantification of myelin thickness and serves as basis for a more refined working hypothesis.

In the course of a study, more questions will arise that require assessment of the localization of a protein within a cell or tissue. Apart from the fact that one can learn a lot about the function of a protein by knowing its localization, the protein of interest might also be changed in abundance or localization in the mutant model. Immunoelectron microscopy is a suitable tool to study such changes [2]. Here, the antigen is visualized by gold-conjugated antibodies or protein A in the context of the subcellular environment. This so-called reference space adds valuable information, since changes occurring in mutant animals often affect not only the localization of the protein of interest but also the size and shape of organelles. Immunoelectron microscopy has always been a popular and powerful tool in cell biology, but relies on the quality of the available antibodies [3, 4]. These are crucial for the success of this technique, but often antibodies that work reliably with other techniques fail to work for immune-electron microscopy. Therefore, it is one of the most important tasks today to develop better and more reliable probes.

Apart from these in situ visualization strategies, electron microscopy can also complement biochemical analyses. To study the composition of myelin in transgenic mutants or different model organisms, this lipid-rich structure can be isolated biochemically due to its low density using density gradient centrifugation. To prove the purity of the myelin preparation, the content can be analyzed by electron microscopy. After chemical fixation and embedding, contaminating components of a myelin pellet, such as attached axonal membranes and mitochondria, can be identified. This helps in interpreting data, e.g., from proteome analysis.

Conventional sample preparation relies on chemical fixation. The disadvantage of this method is that the buffer conditions, such as pH value and osmotic strength, can never comply with the requirements of all components in a tissue or even in a cell. Therefore, fixation artifacts are almost inevitable and especially obvious in myelin of chemically fixed rodent tissue as lipid-rich structures are difficult to fix. To be able to discriminate effects on the myelin structure caused by the mutation in the animal model from fixation artifacts or subtle changes in the myelin architecture, a different approach in the sample preparation is required. By the application of high-pressure freezing (HPF), living samples are physically immobilized in milliseconds by freezing at approx. 2000 bar, and then later subjected to dehydration and chemical fixation during substitution of the ice with an organic solvent at low temperature [5–9]. By the application of 2000 bar during freezing, ice crystal formation is suppressed resulting in the formation of so-called amorphous ice. In such a vitrified sample, the ultrastructure is preserved in a close-to-native state for an optimal sample preparation. Since the invention of this technique almost 50 years ago, we gained a much more detailed and complete view on the structure of living matter, especially for myelin of mice [10]. For the freeze substitution procedure, a plethora of published protocols exists. The protocol described here worked very well for mouse and rat myelin. For other tissues, other protocols might be more suitable and require testing.

Hypotheses generated from in vivo studies often require in vitro validation. For this purpose, primary oligodendrocyte cultures are useful tools. Here, we describe how such cultures can be embedded for electron microscopic inspection. In thin section electron microscopy, the architecture of the cytoskeleton is difficult to study since fibers can hardly be traced over longer distances. To overcome this disadvantage, a whole mount technique using platinum-carbon replicas [11] is described at the end of this chapter, which is also suitable to study the cytoskeleton in primary oligodendrocyte cultures.

2 Materials

2.1 Chemical Fixation and EPON Embedding

1. Mice or rats to be processed for electron microscopy.
2. Anesthesia for mice; Avertin: 2 g 2,2,2,-Tribromoethanol in 2 mL tertiary amyl-alcohol, add 96.5 mL H_2O pre-warmed to 37 °C, stir for 30 min, filter with 0.2 μm filter, store 5 mL aliquots at −20 °C in the dark (light sensitive!).
3. Anesthesia for rats; 100 mg/mL ketamine (10 mL stock solution), stored at 4 °C, 20 mg/mL xylazine (50 mL stock solution), stored at 4 °C.
4. Fixative in phosphate buffer according to Karlsson-Schultz [12]: For 200 mL working solution: take 0.36 g $NaH_2PO_4 \cdot H_2O$, 3.1 g $Na_2HPO_4 \cdot 2H_2O$, 1 g NaCl, add 130 mL distilled water, 20 mL 25% glutaraldehyde and 50 mL 16% formaldehyde (final concentration: 109.5 mM $NaH_2PO_4 \cdot H_2O$, 93.75 mM $Na_2HPO_4 \cdot 2H_2O$, 86.2 mM NaCl, 2.5% glutaraldehyde, 4% paraformaldehyde). Filter with 0.2 μm filter to remove any particles. The salt solution can also be prepared as 2× stock in advance and stored at room temperature (RT). Fixatives are always added fresh prior to use.
5. 25% glutaraldehyde.
6. 16% paraformaldehyde (PFA) stock solution: Dissolve 16 g PFA in 90 mL double-distilled water upon heating and stirring at 65 °C. Afterward, add droplets of 5 N NaOH until solution becomes clear. Adjust volume to 100 mL. Filter cold solution

with 0.2 μm filter. Aliquot and store PFA at −20 °C. Thaw aliquots in a water bath at 60 °C for ~15 min until the solution becomes clear.

7. HBSS.
8. Styrofoam board.
9. Butterfly needle.
10. Perfusion peristaltic pump with tubing and adaptor.
11. Schott bottles or falcons.
12. Dissection tools (forceps, scissors, clamps, needles).
13. Glass vials or falcons to store the tissue.
14. Brain slicer.
15. Platinum double-edged razor blades.
16. Leica Vibratome VT1200S.
17. Platinum-coated double-edged razor blades.
18. Plastic Pasteur pipettes.
19. Super glue.
20. Hollow biopsy punch of different diameters depending on the size of the structure of interest.
21. Cutting mat.
22. Fine brush.
23. Forceps.
24. Tea filter bags.
25. Small segment or 8-segment baskets.
26. Basket lids.
27. Lynx Tissue Processor EM sample holder.
28. Lynx Tissue Processor.
29. 0.2 M phosphate buffer (PB): Prepare solution A containing 27.6 g/L $NaH_2PO_4 \cdot H_2O$ and solution B containing 35.7 g/L $Na_2HPO_4 \cdot 2H_2O$. Mixtures of these two solutions will determine the pH: 20 mL of solution A mixed with 80 mL of solution B will result in 0.2 M phosphate buffer with pH 7.3.
30. 2% osmium tetroxide (OsO_4) in 0.1 M PB: Dissolve 2 g OsO_4 in 50 mL distilled water and dilute 1:1 with 0.2 M PB.
31. Acetone
32. Epon resin: 21.4 g glycid ether 100, 14.4 g 2-dodecenylsuccinic acid anhydride (DDSA), 11.3 g methylnadic anhydride (MNA). Stir the solution for 10 min and then add 0.84 mL 2,4,6-Tris(dimethyl-amino-methyl-)-phenol (DMP-30) stirred for additional 20 min before use.

33. Embedding molds or gelatine capsules and Parafilm™-covered glass slides.

Caution: Osmium tetroxide, uranyl acetate, and some components of the Epon resin are extremely toxic or carcinogens and should be handled only under the fume hood with great care and appropriate personal protection (nitrile gloves, safety goggles, and lab coat).

2.2 Chemical Fixation and Embedding for Immunoelectron Microscopy

1. Animals to be processed for electron microscopy.
2. Karlsson-Schultz phosphate buffer for immunoelectron microscopy: Final concentration: 109.5 mM $NaH_2PO_4{\cdot}H_2O$, 93.75 mM $Na_2HPO_4{\cdot}2H_2O$, 86.2 mM NaCl, 0.25% glutaraldehyde, 4% paraformaldehyde. Filter with 0.2 μm filter to remove any particles. The salt solution can also be prepared as 2× stock in advance and stored at RT. Fixatives are always added fresh prior to use.
3. 25% glutaraldehyde
4. 16% PFA stock solution.
5. HBSS.
6. Butterfly needle.
7. Perfusion pump with tubing and adaptors.
8. Schott bottles or falcons.
9. Glass vials.
10. Cutting mat.
11. Fine brush.
12. Forceps.
13. Toluidine blue.
14. 10% gelatine (for bakery).
15. Ice bucket.
16. 2.3 M sucrose in PB.
17. Whatman™ filter paper.
18. Aluminum pins.
19. Cryovials.
20. Rubber-coated forceps.
21. Liquid nitrogen.
22. Leica UC6 cryo-ultramicrotome.
23. Cryoimmuno diamond knife 35°.
24. Pick up solution: 1% methylcellulose, 1.15 M sucrose.
25. Perfect loop™.
26. Carbon and Formvar-coated grids.

27. Parafilm™.
28. Plastic Pasteur pipettes.
29. PBS.
30. Glycine.
31. Bovine serum albumin.
32. Primary antibody (if possible raised in rabbit).
33. Rabbit secondary antibody, if the primary was not raised in rabbit.
34. Protein A gold, conjugated with 10 nm or 15 nm gold particles.
35. Neutral uranyl acetate: 2% uranyl acetate, 150 mM oxalic acid. Adjust the pH to 7 with 25% ammonium hydroxide.
36. Methylcellulose-uranyl acetate mix: dilute 4% uranyl acetate 1:10 in 2% methylcellulose.
37. Ice box.
38. Metal plate.
39. Filter paper.
40. Loops.
41. Rack to dry the loops.
42. Gridbox.

2.3 Chemical Fixation and Embedding of Myelin Preparations for Conventional Electron Microscopy

1. Myelin preparations.
2. Karlsson-Schultz phosphate buffer: 109.5 mM $NaH_2PO_4 \cdot H_2O$, 93.75 mM $Na_2HPO_4 \cdot 2H_2O$, 86.2 mM NaCl, 2.5% glutaraldehyde, 4% paraformaldehyde. Adjust pH to 7.4. Filtered with 0.2 μm filter. Salt solution can be prepared as 2× stock in advance and stored at RT. Fixatives are added fresh prior to use.
3. Tabletop centrifuge.
4. 1.5 mL Eppendorf tubes.
5. 2.5% low-melting agarose.
6. Razor blade or scalpel.
7. Needle.
8. Cutting mat.
9. Forceps.
10. Segment baskets.
11. Basket lids.

1. Ice box with ice-cold PBS.
2. Scissors for cutting skin and bones.

2.4 Cryopreparation by High-Pressure Freezing

3. Fine spring scissors.
4. Scalpels.
5. Brain slicer.
6. Razor blades.
7. Gelatine (for bakery).
8. Superglue.
9. Leica vibratome VT1200S.
10. 20% poly-vinyl-pyrrolidinone (MW = 10,000; PVP; in PBS)
11. Hollow biopsy punch of 2 mm diameter.
12. Specimen carrier, Type A and B.
13. Specimen holder cylinder half and middle plate.
14. Filter paper.
15. 1-Hexadecene.
16. Leica HPM100.
17. Rubber-coated forceps.
18. Perforated cryovials.
19. Styropor box filled with liquid nitrogen.
20. Tannic acid.
21. Glass-distilled acetone.
22. Rubber-coated forceps.
23. Plastic Pasteur pipettes.
24. Leica aluminum tins as containers equipped with disposable insets (flow-through rings).
25. Conical closures (used as disposable lids).
26. Leica AFS 2.
27. Osmium tetroxide.
28. 5% uranyl acetate in methanol.
29. Ice bucket.
30. Epoxy resin (Epon) (*see* recipe).

2.5 Monolayer Embedding of Oligodendrocytes

1. Cultured primary cells in small dishes (optimal: 3 cm dishes, no well plates).
2. 0.1 M Phosphate buffer (PB, *see* recipe above).
3. 1% OsO_4 (in PB).
4. Uranyl acetate.
5. Tungstophosphoric acid hydrate.
6. Epon resin.
7. Gelatine capsules.

8. Formvar-coated 100 mesh hexagonal carbon grids (*see* [13, 14] for preparation).
9. Gridbox.

2.6 Cytoskeleton Visualization of Oligodendrocytes

1. Cells cultured on glass coverslips.
2. Serum-free DMEM.
3. PEM buffer: 100 mM PIPES, 1 mM EGTA, 1 mM $MgCl_2$, pH 6.9.
4. Triton X-100.
5. Polyethylene glycol (35 kDa).
6. Glutaraldehyde.
7. Tannic acid.
8. Uranyl acetate.
9. Critical point dryer, e.g., Leica EM CPD 300.
10. Ethanol.
11. Sputter coater, e.g., Leica EM ACE600.
12. Needle.
13. 12 well cell culture plate.
14. Hydrofluoric acid.
15. Household detergent.
16. Perfect loops™.
17. Formvar-coated 100 mesh hexagonal carbon grids (*see* [13, 14] for preparation of formvar films).
18. Gridbox.

3 Methods

3.1 Chemical Fixation and EPON Embedding

Electron microscopy by chemical fixation (Fig. 1) has proven to be a useful tool to study the ultrastructure of oligodendrocytes (Fig. 2) and myelin in the central nervous system. With the help of conventional electron microscopy, alteration in the myelin ultrastructure of mutant mice (Fig. 3) could be revealed, such as changes in myelin thickness, altered paranodes or adaxonal space.

For this purpose, it is necessary to perfuse the animal with aldehyde-based fixatives. After postfixation in aldehyde-containing fixative, the dissected optic nerve or a segment of the spinal cord can be embedded directly. In contrast to that, when brain vibratome sections need to be cut, a region of interest is selected, excised, and then embedded. Sometimes, spinal cord samples also require vibratome sectioning. If an embedding machine is used, depending on the size of the sample, they are processed in small segment baskets of suitable size. To avoid loss of samples during

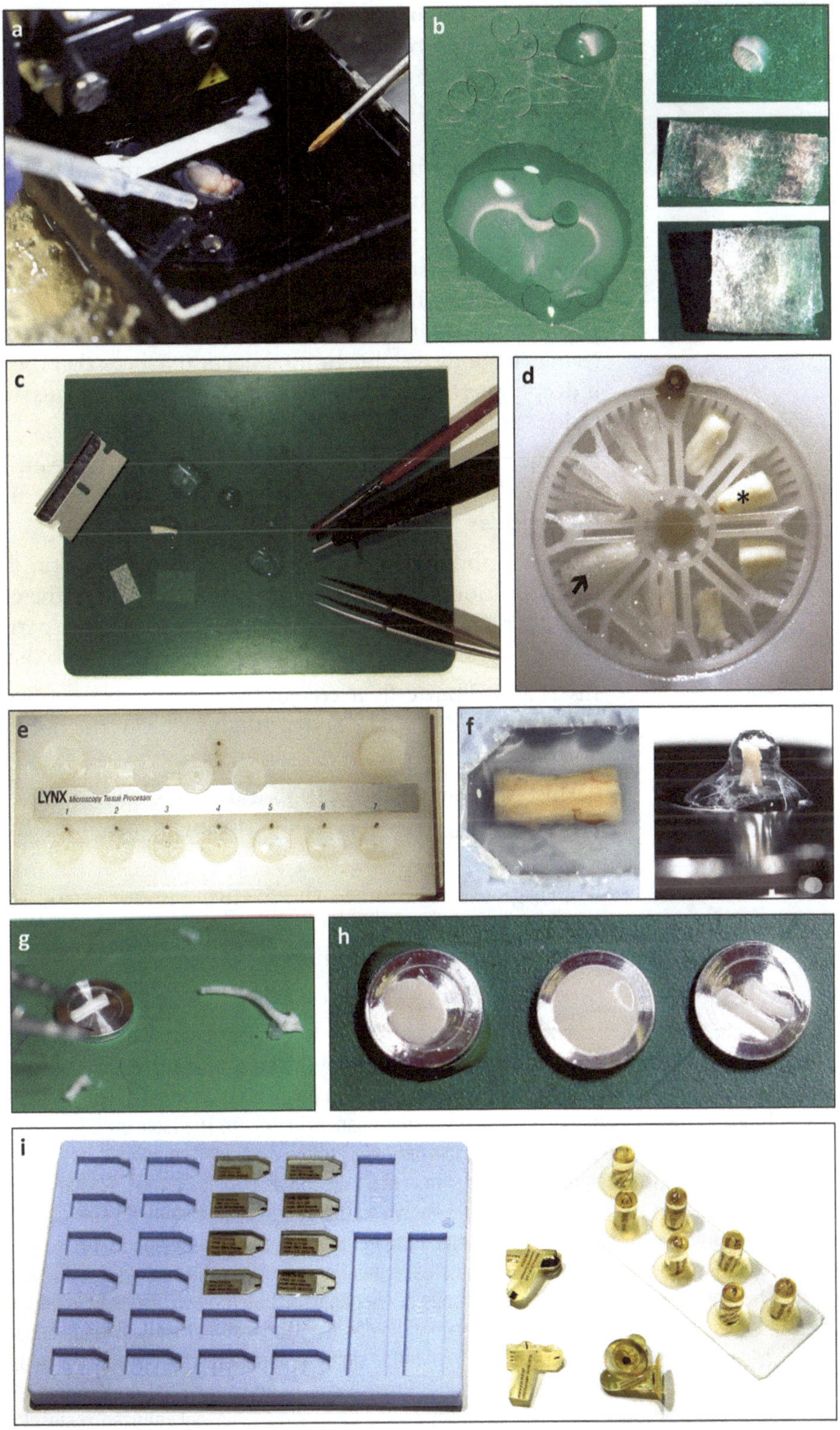

Fig. 1 Preparations strategies for transmission electron microscopy. (**a**) For conventional fixation of brain samples, the animal is perfused and after postfixation, vibratome sections are cut. A fine brush and a Pasteur pipette are used to handle the vibratome sections. A similar procedure is used to prepare fresh brain and

processing, small samples, such as optic nerves or punched tissue pieces, have to be wrapped in small pieces of tea filter paper. The embedding, including the contrasting and infiltration with the Epon resin, can be done manually or automatically.

3.1.1 Perfusion of the Animal for Conventional Fixation of Brain and Spinal Cord Samples

1. Prepare the perfusion pump by attaching the tubing and the adaptor. Flush the system with HBSS. Avoid introducing air bubbles into the system.
2. Anesthetize the animals. For mice, use 0.2 mL avertin per 10 g body weight by intraperitoneal (i.p.) injection. For rats, anesthetize with ketamine (60 mg/kg body weight) and xylazine (8 mg/kg body weight) by i.p. injection. Check that the animal is deeply anesthetized by pinching its paw and making sure that it does not twitch. The anesthesia should last 1–2 h.
3. Put the animal with the back down on a Styrofoam board. Stretch the limbs away from the body and pin them down with needles. Wet the fur with 70% ethanol.
4. Open the abdominal cavity with a transversal incision. Make a lateral incision in the diaphragm from one edge of the chest to the other. Cut laterally along the chest and flip that part of the rib cage up. Fix it with needles to gain good access to the heart.
5. Insert a 27G butterfly needle in the left ventricle and fix it with clamps or needles to the Styrofoam board. Cut the right atrium immediately.
6. Perfuse the animal with HBSS until the liver becomes yellow (10 mL) at very low speed of perfusion (1 mL/min, *see* **Note 1**).
7. Exchange the solution to the fixative without introducing air bubbles and perfuse for 10 min with fixative. The animal will twitch and then become stiff.

Fig. 1 (continued) spinal cord slices for high-pressure freezing. (**b**) The area of interest is punched out using a hollow biopsy punch (2 mm diameter). To avoid sample loss upon processing, the tissue sample is packed in tea filter paper by folding over twice from each side (*see* right panels in (**b**)). (**c**) The tools used for punching the area of interest in brain vibratome sections are displayed on a cutting mat. (**d**) The sample has to be placed with the flat side on one of the 8-segment baskets shown in on the left side (black arrow). Tissue pieces, such as spinal cord, can be placed directly into the basket shown on the right side (asterisk). (**e**) The loading station for preparing the baskets used for processing of the samples in an embedding machine, such as the Lynx Tissue Processor. (**f**) Embedding of spinal cord samples in gelatine for vibratome sectioning and subsequent cryopreparation by high-pressure freezing. (**g**) The optic nerve has to be cut into two pieces to fit into a carrier. The nerve piece is placed with fine forceps into specimen holder. (**h**) Specimen carrier with loaded spinal cord, rostral corpus callosum, and optic nerve samples (from left to right). (**i**) Embedding of optic nerve samples and spinal cord pieces in the flat, blue embedding molds. Brain tissue is embedded flat on Parafilm™-coated glass slides with Epon-filled gelatine capsules (*see*n on the right side). The polymerized resin blocks are shown in the middle (spinal cord on the top, optic nerve on the left and brain tissue on the right side)

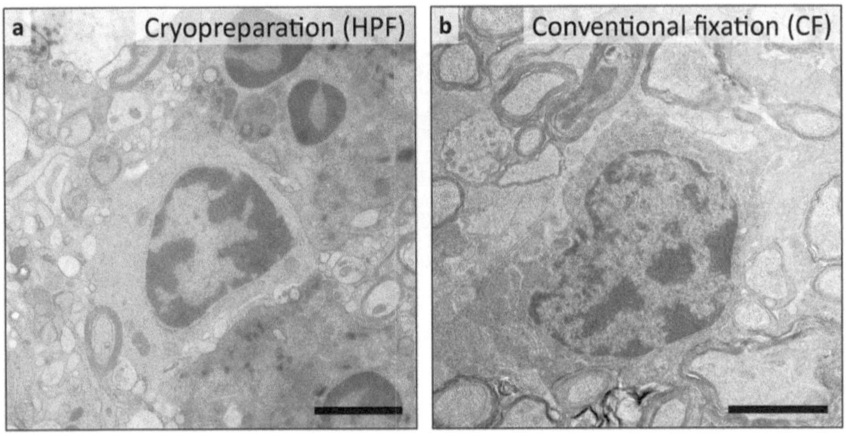

Fig. 2 Oligodendrocytes in the mouse corpus callosum. (**a**) and (**b**) Oligodendrocytes are shown in cryopreserved and conventionally fixed samples. (**a**) In the cyroprepared sample, the nucleus contains a lot of heterochromatin and is surrounded by a narrow rim of pale cytoplasm. The free ribosomes and microtubules are abundant in the cytoplasm, but no glycogen or glial fibrillary acidic protein bundles can be detected. (**b**) In conventional prepared samples, the cytoplasm of oligodendrocytes appears darker than the cytoplasm of other cell types. The oligodendrocyte nucleus shows a patchy pattern of hetero- and euchromatin with a large proportion of condensed chromatin. Sometimes, the fine processes can be found in attachment to myelin sheaths. Scale bar 2 μm

8. Dissect the brain immediately as described below and immerse the tissue in the same fixative overnight. The spine is roughly dissected as described in **step 3** and immersed in fixative before dissecting the spinal cord.

3.1.2 Preparation of Brain Samples After Conventional Fixation

1. Wet the skin of the head with 70% ethanol. Cut the skin with a scalpel from the nose to the neck and flip it to either side.
2. In case the optic nerves should be prepared as well, insert small scissors carefully into the eye socket behind the eye ball and cut the optic nerve. Avoid stretching the nerves by pulling the eyes.
3. Cut the skull with scissors from the brainstem to the coronal suture.
4. Perform a lateral cut to either side and remove that part of the skull.
5. Cut the skull from the coronal suture to the ears along the brain.
6. Cut the bone in between the eyes and remove the skull by flipping it to either side.
7. Remove the brain from the skull by inserting angled forceps beneath the brain and gently pulling it out. The optic nerves should remain attached to the brain.

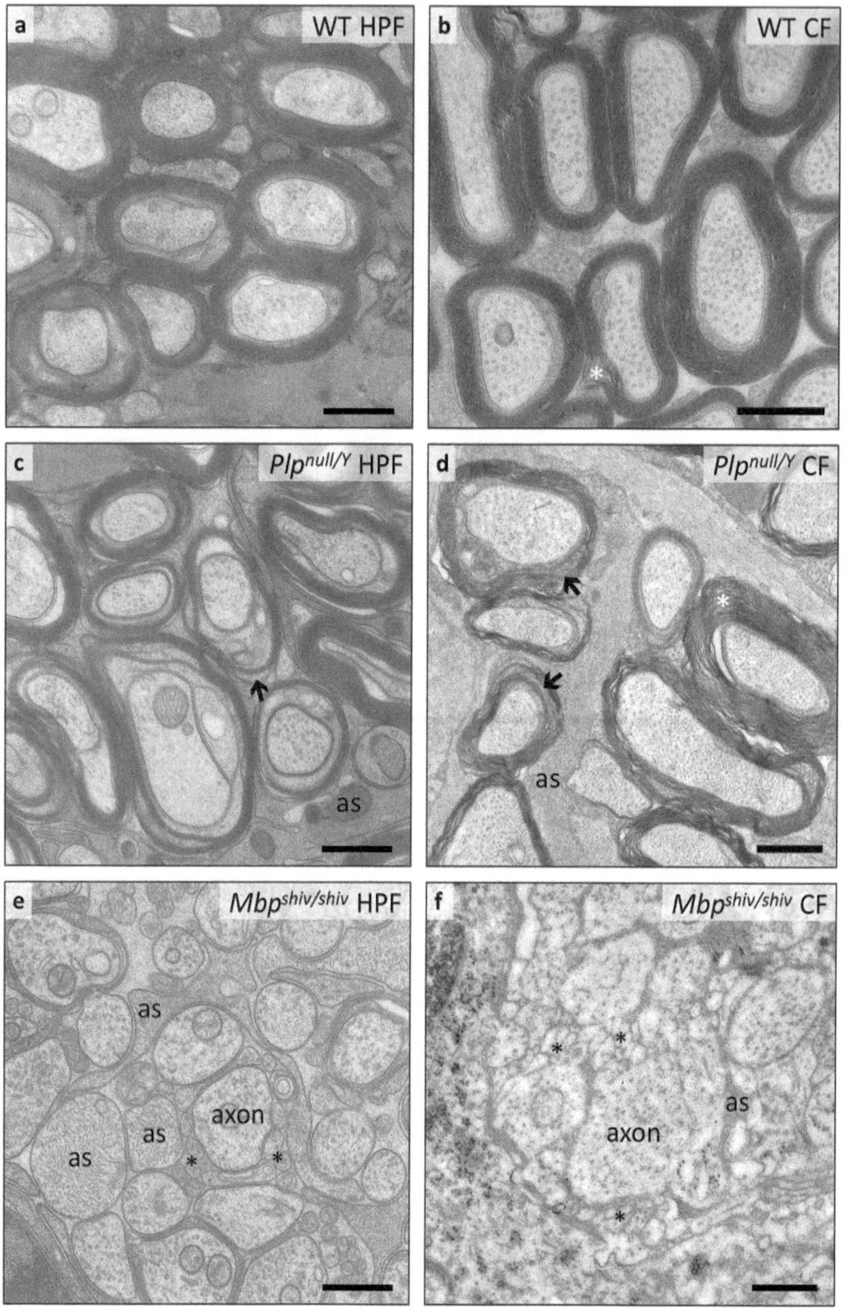

Fig. 3 Mouse optic nerve from different myelin mutants in transmission electron microscopy using cryopreparation and conventional fixation. (**a**) and (**b**) Representative electron micrographs of the optic nerve from wild type (WT), (**c**) and (**d**): proteolipid protein (PLP)-deficient ($Plp^{null/y}$) and (**e**) and (**f**): myelin basic protein (MBP)-deficient mice (*shiverer*, $Mbp^{shiv/shiv}$) prepared by cryopreparation (HPF, left panel) and conventional fixation (CF, right panel). Cryopreparation results in well-preserved myelin with smooth and tightly attached myelin lamellae. In conventional fixation, a typical artifact is the loosening of the myelin membranes in focal areas (*see* white asterisk in (**b**) and (**d**)). In $Plp^{null/y}$ mice, large cytosolic channels within single myelin membranes appear frequently (black arrow). Due to the cryopreparation that preserved the sample in a close-

8. Continue fixation of the brain by immersion in Karlsson-Schultz fixative overnight at 4°C. For prolonged storage, replace the fixative by 1% PFA in PB.
9. In order to embed the region of interest, cut 1 mm slices with a brain slicer matrix and razor blades or prepare vibratome sections of the fixed brain as described in the next steps.
10. For vibratome slices, cut the brain with a razor blade sagittal along the sagittal suture or coronal depending on region of interest. To speed up processing time, the tissue can be trimmed in such a way that the part of the tissue that is not of interest is removed beforehand.
11. Use superglue to mount the flat side of the tissue block on the vibratome plate. Wait 30 s for the glue to harden. Attach the plate to bottom of the cutting chamber filled with ice-cold PBS. Attach a platinum double-edged razor blade to the vibratome head and set the cutting window in such a way that it contains the whole tissue.
12. Cut the brain in 200 μm thick slices using an amplitude of 1 mm and speed of 1 mm/s with the cortex facing the razor blade first. Gently hold the top of the tissue with a fine brush and push a stream of buffer with a plastic Pasteur pipette onto the razor blade (Fig. 1a).
13. Select two sections containing the region of interest and transfer them to a cutting mat with a fine brush. Punch out a piece of tissue containing the area of interest using a hollow biopsy punch (Fig. 1b and c).
14. If an embedding machine, such as the Lynx II, is used, the punch of brain tissue is packed in a small piece of tea filter paper like a small packet so that the tissue is contained within the folded filter paper (Fig. 1b). Put the flat side of the tea bag into the 8-segment basket filled with 0.1 M phosphate buffer (PB) and cover it with the basket lid after filling up all segments (Fig. 1d and e). The filter paper packet will prevent the loss of small samples, such as optic nerves and tissue pieces by floating out of the basket segments (*see* **Note 2**).
15. The tissues can then either be processed automatically with an embedding machine or manually by incubating the samples with the different solutions. The protocols used for either automated or manual embedding are shown in Tables 1a and 1b. The steps are the same, but the incubation times vary.

Fig. 3 (continued) to-native state, the cytosolic channels are more apparent than in conventional fixed samples. Astrocytic processes (as) are also visible. In *shiverer* mice ($Mbp^{shiv/shiv}$), compaction of myelin is impaired due to the lack of MBP. Thus, instead of myelin lamellae microprocesses are formed (black asterisk). These microprocesses are only clearly visible in cryofixed samples. as, astrocyte, scale bar: 500 nm

Table 1a
Automated embedding protocol for conventional electron microscopy

Solution	Duration	Temperature
PB buffer	10 min	4 °C
PB buffer	10 min	4 °C
PB buffer	10 min	4 °C
2% OsO_4	4 h	4 °C
PB buffer	10 min	RT
PB buffer	10 min	RT
PB buffer	10 min	RT
30% acetone	20 min	RT
50% acetone	20 min	RT
70% acetone	20 min	RT
90% acetone	20 min	RT
100% acetone	10 min	RT
100% acetone	10 min	RT
100% acetone	10 min	RT
100% acetone	10 min	RT
acetone: Epon mix (2:1)	2 h	RT
acetone: Epon mix (1:1)	2 h	RT
acetone: Epon mix (1:2)	4 h	RT
Fresh pure Epon	4 h	RT

16. In brief, for automated embedding, wash the samples with 0.1 M PB three times for 10 min at 4°C. After postfixation and staining in 2% OsO_4 in 0.1 M PB for 4 h at 4 °C, wash the tissue again with 0.1 M PB for 10 min at RT. Dehydrate the samples with increasing concentrations of acetone in water (30, 50, 70 and 90%) for 20 min each. Incubate the samples in 100% acetone for 10 min four times. Infiltrate the tissue with increasing concentrations of Epon resin mixed with acetone (2:1, 1:1 for 2 h each and 1:2 for 4 h) prior to incubation with pure Epon resin for 4 h.
17. After infiltration with Epon resin, remove the samples from the baskets and the tea filter packets. Place the samples in

Table 1b
Protocol for manual embedding for conventional electron microscopy

Solution	Duration	Temperature
PB buffer	15 min	RT
PB buffer	15 min	RT
PB buffer	15 min	RT
2% OsO_4	4 h	4 °C
PB buffer	15 min	RT
PB buffer	15 min	RT
PB buffer	15 min	RT
50% acetone	20 min	RT
70% acetone	20 min	RT
90% acetone	20 min	RT
100% acetone	20 min	RT
100% acetone	20 min	RT
100% acetone	20 min	RT
100% acetone	20 min	RT
acetone: Epon mix (2:1)	2 h	RT
acetone: Epon mix (1:1)	2 h	RT
acetone: Epon mix (1:3)	Overnight	RT
Fresh pure Epon	2 h	RT

embedding molds (for optic nerves and spinal cord) or flat on Parafilm™-covered glass slides with Epon-filled gelatine capsules upside down on the samples (Fig. 1i and *see* **Note 3**).

18. Polymerize the resin for 24 h at 60 °C.
19. Boil resin blocks with gelatine capsules in water to remove the gelatine.

3.1.3 Preparation of the Spinal Cord After Conventional Fixation

1. For the dissection of the spinal cord after the perfusion of the animal, lay down the animal with the back facing up on a Styrofoam plate. Wet the skin with 70% ethanol. Cut the skin along the spinal cord and flip to either side.

2. Cut away all the muscle tissue and the organs attached to the spine. Leave some parts of the rib cage attached to the spine to facilitate the dissection of the spinal cord.
3. After prolonged fixation in Karlsson-Schulz fixative, pin the extracted spine to the Styrofoam plate using needles. Gently hold the upper side of the spine and insert strong, fine spring scissors on one side of the spinal cord. Cut a piece of the bone and then insert the scissor on the other side of the spinal cord and continue cutting the spine. Proceed to expose the spinal cord by alternately lifting one part of the spine and cutting the bones around the exposed spinal cord.
4. Use forceps to lift the cervical part of the spinal cord and remove it from the underlying spine. Gently pull the spinal cord up and cut the attached dorsal root ganglia.
5. Put the extracted spinal cord on a petri dish and cut a piece of tissue containing the area of interest. Small pieces of the spinal cord, as well as vibratome sections of the spinal cord, can be processed for conventional electron microscopy. In general, tiny samples are better infiltrated by chemicals and resins than large tissue blocks. Therefore, thin vibratome sections are recommended. In some cases, however, a larger piece of spinal cord is required for analysis. Vibratome sections (*see* protocol above subheading 3.1.2, **step 14** onward) are packed into tea filter packets and placed into 8-segment baskets for embedding (as shown for brain samples in Fig. 1b, *see* **Note 2**). Wrapping into tea filter packets prevents bending or loss of the small tissue slices during the embedding process. Larger pieces of spinal cord can directly be transferred into a small segment basket (Fig. 1c and d) and be processed for embedding as described above.
6. After dehydration and resin infiltration, vibratome sections are placed flat on Parafilm™-covered microscope glass slides (Fig. 1i and *see* **Note 3**). Then, Epon-filled gelatine capsules are placed upside down on top of the samples. Spinal cord pieces are placed into embedding molds and covered with Epon resin (Fig. 1i). Include small paper labels for identification of the samples into the resin. The blocks are polymerized at 60 °C for 24 h.

3.1.4 Preparation of the Optic Nerve After Conventional Fixation

1. Kill the animal (mouse or rat) by cervical dislocation. In case of rats, or according to local animal handling regulations, the animal needs to be anesthetized before.
2. Cut off the head of the animal.
3. Make an incision with a small, sharp scissor behind the eyes. The eye ball should loosen from the eyecup and the end of the optic nerve attached to the eye ball should be cut.

4. Wet the skin with 70% ethanol. Cut the skin with a scalpel from the nose to the neck and flip it to either side.
5. Cut the skull with strong scissors from the brainstem to the coronal suture.
6. Cut a lateral cut to either side and remove the skull.
7. Cut the skull from the coronal suture to the ears along the brain.
8. Cut the bone in between the eyes and remove the skull by flipping to either side.
9. Remove the brain from the skull by inserting angled forceps beneath the brain and gently pulling the brain out.
10. Flip the brain over without stretching the optic nerve (*see* **Note 4**).
11. Cut the optic nerve at the chiasm where it is attached to the brain.
12. Pick up the optic nerves at the chiasm with angled forceps and pull them up gently.
13. Place the optic nerve into Karlsson-Schulz fixative overnight (or at least for 2 h).
14. Pack the optic nerve in tea filter packets and place in 8-segment baskets (as shown for brain sample in Fig. 1b and d). Close with lid and process for embedding as indicated in Table 1a or 1b.
15. Embed the samples in embedding molds (Fig. 1i), include a label indicating the experimental conditions and polymerize the blocks for 24 h at 60 °C.

3.2 Chemical Fixation and Embedding for Immunoelectron Microscopy

A powerful tool to study the subcellular localization of proteins is immunoelectron microscopy. Here, antibodies directed against the protein of interest are employed to identify its localization in relation to the ultrastructure of the tissue. Since the antigenicity of a protein is negatively influenced by the strength and duration of the chemical fixation, care has to be taken regarding the concentration of the fixative and the incubation time. Therefore, the animals have to be perfused with a fixative containing 4% PFA and 0.25% glutaraldehyde (or less). Dissected tissue should be exposed to fixatives only for a limited time of 2–4 h or maximal overnight (depending on the sample size), and then transferred into 1% PFA that is suitable for prolonged storage at 4 °C. Small tissue pieces (such as optic nerve or vibratome sections of the spinal cord or brain) are then infiltrated with 2.3 M sucrose as a cryoprotectant for 20–22 h. The samples are then mounted on aluminum pins and frozen in liquid nitrogen. After cutting cryosections according to Tokuyasu, the sections are first blocked and then stained with primary antibodies to detect the protein of interest. The sections are then incubated with either gold-conjugated Protein A that binds to rabbit antibodies or other secondary gold-conjugated antibodies. The antibody-gold complexes are then fixed and the samples are contrasted in a two-step protocol with neutral uranyl acetate first and secondly with uranyl acetate and methylcellulose (for further reading, *see* [13–15]). At the electron microscope, the subcellular localization of the protein is apparent as black dots on the background of the general ultrastructure of the tissue(Fig. 4).

1. Perfuse the animal as described in Protocol 1 with a Karlsson-Schultz phosphate buffer for immunoelectron microscopy containing 0.25% glutaraldehyde or less (*see* **Note 1**).
2. After removal of the tissue, store the samples in 1% PFA at 4 °C until further processing.
3. Cut vibratome sections of the spinal cord and brain at 200 μm thickness (*see* protocol 1 for details).
4. For brain and spinal cord sections, infiltrate the vibratome sections with 2.3 M sucrose for 20–22 h at 4 °C on a rotating wheel. For optic nerve, stain the nerve briefly in 1% Toluidine blue and embed it in small blocks of 10% gelatine prior to infiltration with sucrose.
5. Place a small petri dish on a big petri dish filled with ice. Add a drop of 2.3 M sucrose on the dish and transfer the section or the optic nerve-gelatine blocks into it.
6. Use a scalpel to cut a small piece containing the area of interest from the vibratome sections (size around: 1 mm × 1 mm).
7. Mount the tiny tissue piece or optic nerve-gelatine block on an aluminum pin with a droplet of sucrose. Remove carefully all the excessive sucrose with Whatman™ filter paper.
8. Freeze the pin with the attached sample in liquid nitrogen using rubber-coated forceps. Samples can be stored in liquid nitrogen or processed for cutting immediately.
9. Trim the block and cut ultrathin cryosections with a Leica UC6 cryo-ultramicrotome equipped with a cryoimmuno diamond knife and a trimming diamond knife (Diatome, Biel, Switzerland). Use 1.15 M sucrose and 1% methylcellulose to pick-up and mount the sections on formvar- and carbon-coated copper grids using a Perfect loop™. Store samples on a paper-covered glass slides in a petri dish at 4 °C.
10. To perform immunolabeling of the sections with gold-conjugated Protein A, cover the bench top with Parafilm™ and place droplets of the incubation solutions with a plastic Pasteur pipette directly on the Parafilm™. Always float one grid upside down on each droplet. Move the grids from droplet to droplet with forceps by touching the grid only at the rim.

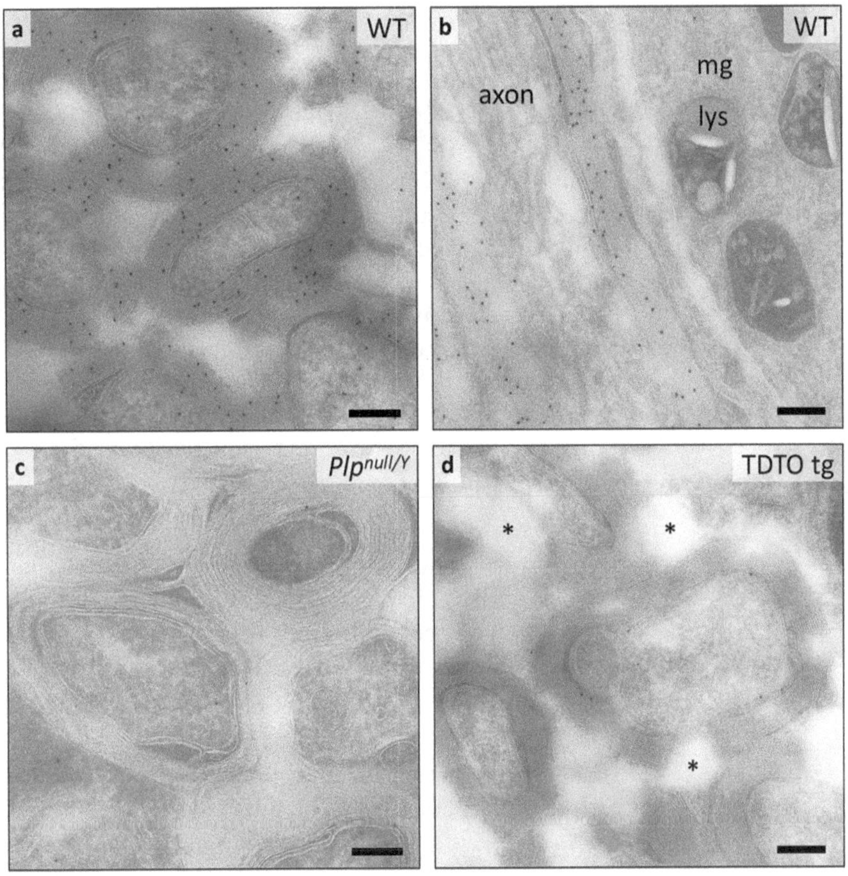

Fig. 4 Cryo-Immunoelectron microscopy of mouse optic nerve. (**a**) and (**b**) Immunogold labeling of proteolipid protein (PLP) in optic nerve cryosections of adult WT mice using a polyclonal rabbit antibody directed against PLP and protein A-gold. In panel (**b**), the nerve is sectioned longitudinally. Here, an adjacent microglia cell (mg) containing numerous lysosomes (lys) can serve as an internal negative control for the labeling. (**c**) Immunogold labeling of PLP on optic nerve sections of a PLP-deficient mouse (Plp$^{null/Y}$) that serves as a negative control. A few gold particles represent background. (**d**) shows labeling of td Tomato in a transgenic reporter mouse (TDTO tg) expressing this reporter protein only in oligodendrocytes. This protein is expressed in the cytosol and appears in the outer and inner tongue of the myelin sheath. Here we used a polyclonal antibody directed against RFP (Rockland, US, # 600–401-379) and protein A-gold. The sample was fixed with only 4% formaldehyde. This is insufficient to stabilize the lipid-rich myelin structure resulting in loss of material from the section (indicated by black asterisks). lys, lysosomes; mg, microglia cell, scale bar: 200 nm

Make sure to clean the forceps with water and tissue paper in between different samples and steps (*see* **Note 5**).

11. Wash off the pick-up solution from the grids by incubation in droplets of PBS for 2 min. Repeat this step twice.
12. Incubate with PBS containing 0.1% glycine to block free aldehyde groups from fixative for 2 min. Repeat step twice.
13. Block the samples using 1% BSA in PBS for 3 min.

14. Incubate the sections with the primary antibody of choice (in appropriate dilution) in 1% BSA in PBS for approximately 1 h. The optimal dilution and incubation time has to be determined beforehand. Usually, the concentration of antibody solution has to be higher than for immunohistochemistry. Antibody concentrations between 0.2 and 2 μg/mL should be tested. If the antibody concentration is unknown, a dilution series has to be performed. For the incubation with the primary antibody, droplets of 5–7 μL of antibody solution are used per grid. As a negative control, the primary antibody is omitted, and the grid is floated only on blocking solution. Cover the droplets during the antibody-incubation with a flat box or petri dish to avoid evaporation of the antibody solution and deposition of dirt particles on the samples (*see* **Note 6**).
15. Wash the grids in 0.1% BSA in PBS for 2 min. Repeat this step four times.
16. If the primary antibody is not derived from rabbit, incubate the samples with a suitable bridging antibody. This rabbit antibody has to bind antibodies derived from the species of the primary antibody. Float the grids on the secondary antibody for 30 min. Then, wash the grids in 0.1% BSA in PBS for 2 min. Repeat this step four times.
17. Incubate the samples with gold-conjugated Protein A (CMC, Utrecht, Netherlands) at the indicated dilution for 20 min.
18. Wash the grids in PBS for 2 min. Repeat this step four times.
19. Fix the antibody-gold-complexes and denature the Fc-domain of the antibody using 1% glutaraldehyde in PBS for 5 min.
20. Wash the sections with distilled water for 1 min. Repeat this step nine times.
21. Stain the samples with neutral uranyl acetate for 5 min at RT.
22. Continue staining and protecting the samples with a layer of methylcellulose by incubation with 0.4% uranyl acetate in 2% methylcellulose for 5 min on ice (*see* **Note** 7). Use a metal plate covered with Parafilm™ for a flat surface.
23. After the incubation of the samples with methylcellulose-uranyl acetate mix, take a stainless steel wire loop of 3 mm diameter and insert it into the droplet underneath the grid in such a way that the grid resides in the middle of the loop and gently lift the loop upward. Dry the grid by striking the loop over Whatman™ filter paper until no more liquid is released.
24. Place the loops upright into a rack and let the grids dry for ~20 min. Place a box above the loops to avoid contamination with particles in the air.

25. To remove the grids from the loops, place the loop with the grid upside down on a filter paper. Insert forceps in between the grid and the edge of the loop and keep repeating this step in a rotating motion around the grid. This way the grid can be released from the loop without breaking the methylcellulose layer.
26. Store the grid in a grid box until imaging.

3.3 Chemical Fixation and Embedding of Myelin Preparations for Conventional Electron Microscopy

Due to the high lipid content of myelin, it can be purified biochemically via density gradient centrifugation [16]. Often not only Western Blotting, but also electron microscopy is performed to assess the purity of the preparation. The following protocol allows an easy and rapid quality control of sample preparation and gives first insights into the ultrastructure of myelin preparations in mutant animals as well (Fig. 5).

1. Centrifuge myelin preparations in a 1.5 mL Eppendorf tube at maximum speed in an Eppendorf centrifuge for 10 min.
2. Fix the pellet with Karlsson-Schultz fixative for 1 h at RT.
3. Cover the pellet with 2.5% low-melting agarose (pre-warmed to 60 °C) and place on ice for several minutes to allow the agarose to solidify.
4. Cut the bottom of the Eppendorf tube with a scalpel or a razor blade.
5. Remove the pellet from the remaining tube with a needle.
6. Cut the pellet in half on a cutting mat and place in a small segment basket. Cover with basket lid.
7. Perform the embedding protocol as described in protocol 1.
8. Embed the samples by placing the flat side of the pellet on a Parafilm™-coated glass slide and cover it with an Epon-filled gelatine capsule.
9. After polymerization for 24 h at 60 °C, remove the gelatine by boiling the samples in water for 15 min.

3.4 Cryopreparation by High-Pressure Freezing

The state-of-the-art preservation of the ultrastructure of nervous tissue employs high-pressure freezing of the tissue sample. This method of cryopreservation quickly and physically immobilizes the sample in a close-to-native state. The tissue is fully hydrated as the water within the sample is rapidly transformed into amorphous ice. In contrast to chemical fixation, the samples are not fixed with aldehydes, but subjected to freezing in live state. During freeze substitution, the ice is replaced by organic solvents, while fixing and contrasting the samples at low temperatures.

Therefore, fresh tissue needs to be extracted and either a piece of tissue (e.g., optic nerve) or a piece of a vibratome slice of the brain or spinal cord is cryopreserved (Fig. 1 f–h). For that, the

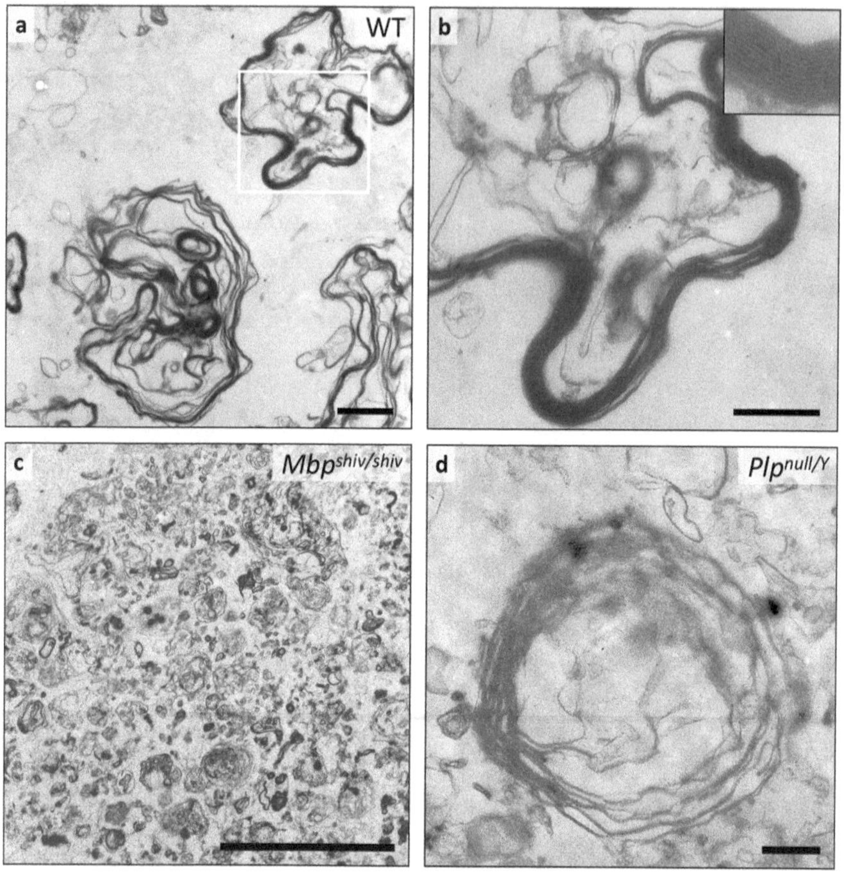

Fig. 5 Myelin preparations of different myelin mouse mutants. (**a**) Myelin preparation from WT mice. (**b**) Area indicated by a white box in (**a**) at higher magnification. Myelin lamellae are clearly visible in the preparation (*see* also inset in (**b**)). (**c**) The overview image of the myelin preparation from *shiverer* mice ($Mbp^{shiv/shiv}$) shows that compact myelin cannot be detected. (**d**) In contrast to this, the myelin from PLP-deficient mice ($Plp^{null/Y}$) is still compacted. Scale bars: 1 μm in (**a**), 500 nm in (**b**) and (**d**), 2 μm in (**c**)

sample needs to be loaded in the specimen carrier prior to freezing with liquid nitrogen at 2000 bar in a Leica HPM100 (Fig. 1g and h) (for further reading, *see* [17]). The subsequent freeze substitution is carried out in a semiautomated fashion in a Leica AFS 2. The dehydration and fixation is performed at −90 °C with different cocktails containing tannic acid followed by osmium tetroxide and uranyl acetate to stabilize and contrast the specimen. Thereafter, the samples are embedded in resin after infiltration with increasing amounts of Epon (in acetone, modified from [18], Fig. 1i).

3.4.1 High-Pressure Freezing of Fresh Dissected Tissue

1. Cool down the high-pressure freezer (e.g., Leica HPM100) according to the manufacturer's instructions with liquid nitrogen until the internal storage dewar is full. Refill the machine after 20 min again and allow the machine to cool down for another 30 min before some test freezing cycles with empty

assemblies of middle plate and cylinder halves (without carrier). Check that all freezing parameters are in the normal range such as: rise time: 4.1 ms, shift p/T: 1.3 ms, hp dwell: 376.5 ms, and cooling rate: 21505 °C/s.

2. Kill the animal (mouse, rat or other) by cervical dislocation (in case of rats, anesthetize the animal before).
3. Cut off the head of the animal.

3.4.2 Cryopreservation of the Optic Nerve

1. Dissect the optic nerve fresh as described in protocol 1 (*see* **Note 4**).
2. Place the optic nerve in ice-cold PBS. Transfer the nerve into drop of 20% PVP (in PBS) on the cutting mat with a fine brush or angled forceps.
3. Cut the nerve with a scalpel in ~2 mm long pieces.
4. Submerge the specimen carrier in 20% PVP and transfer the tissue into the carrier by placing the specimen carrier underneath the tissue and moving it up. Make sure to avoid touching the sample. If necessary only touch the rim of the tissue. For the optic nerve, two pieces can be processed in one specimen carrier (Fig. 1g and h). For tiny nerves of young animals, the 0.1 mm carrier (Type A) can be used, for all other samples the 0.2 mm carrier (Type A, other side) is preferred. Always use the smallest carrier possible, since smaller samples with less space freeze better than bulky samples.
5. Remove putative air bubbles, blot the bottom side of the carrier containing the sample on filter paper and transfer it into the middle plate on top of the lower cylinder half with the sample facing upward. Wet the second carrier (0.3 mm, Type B, flat side facing down onto the sample) with hexadecene and close the specimen carrier.
6. Remove excessive liquid with filter paper.
7. Close the specimen sandwich with the upper cylinder half.
8. Freeze the sample.
9. Remove the samples from the sample dewar and transfer the carriers into a Styrofoam box with liquid nitrogen (*see* **Note 8**).
10. Remove the carrier from the middle plate. Remove the lid of the carrier.
11. Transfer the carrier with the frozen tissue with rubber-coated forceps into perforated cryovials under liquid nitrogen atmosphere. The cryovials should be perforated in the upper half of the vials to prevent floating in liquid nitrogen. Make sure to label the tubes with the sample details.
12. Store the samples in liquid nitrogen until freeze substitution.

3.4.3 Cryopreservation of the Brain

1. Dissect the brain fresh according to the protocol for fixed tissue as described in protocol 1.
2. Wash the brain in ice-cold PBS and put it on a petri dish.
3. Cut the brain with a razor blade sagittal along the sagittal suture or coronal depending on region of interest. To speed up processing time, the tissue can be trimmed in such a way that only the relevant brain area remains.
4. Mount the flat side of the brain on the vibratome plate using superglue. Wait for the glue to harden and then attach the plate to bottom of the cutting chamber filled with ice-cold PBS. Cut the brain in 200 μm thick slices using an amplitude of 1 mm and speed of 0.6–0.7 mm/s with the cortex facing the razor blade first (*see* Fig. 1a).
5. Select two sections containing the region of interest that are of good quality (equal thickness, no cuts or damages in the area of interest). Transfer them to a petri dish and then to the cutting mat using a fine brush.
6. Punch out a piece of tissue containing the area of interest using a 2 mm hollow biopsy punch.
7. Load the tissue in the specimen carrier as described in Subheading 3.4.2, **steps 4–12** for the preparation of the optic nerve (Fig. 1h) and proceed as described in the following steps of the protocol.

3.4.4 Cryopreservation of the Spinal Cord

1. Dissect the spinal cord freshly as described above for fixed tissue.
2. Put the extracted spinal cord on a petri dish and cut a piece of tissue of 5 mm length containing the area of interest.
3. Immerse the spinal cord in Karlsson-Schultz fixative for 1–5 min to fix the rim of the tissue.
4. Put a drop of 10% gelatine (pre-warmed to 45 °C) in an embedding mold in an ice bucket. Gently blot the spinal cord on tissue paper to remove excess liquid and transfer it into the gelatine. Adjust the orientation according to the purpose, e.g., for longitudinal sections lay the spinal cord down flat longitudinally, for cross-sections of the spinal cord put the sample upright. Allow the gelatine to harden for 30 s to 1 min (Fig. 1f).
5. Cover the tissue with gelatine solution and wait for the gelatine to solidify.
6. Remove the block of gelatine.
7. Use superglue to attach the block of gelatine containing the tissue on the vibratome plate (Fig. 1f).
8. Continue to cut and freeze the tissue as described above.

3.4.5 Freeze Substitution of the High-Pressure Frozen Samples

1. Fill the Leica AFS 2 with liquid nitrogen until the internal dewar is full. Precool the chamber to −90 °C.
2. Prepare 0.1% tannic acid fresh in glass-distilled acetone.
3. Equip the Leica-provided aluminum tins with disposable insets (called flow-through rings) for ten specimens and label the individual flow-through rings with notches made using a razor blade. Pipette the tannic acid solution into the tins until the flow-through rings are half-full. Freeze the tannic acid / acetone mix in liquid nitrogen. Avoid spilling of liquid nitrogen into the tins.
4. Remove the sample from the liquid nitrogen container and put them into a Styrofoam box filled with liquid nitrogen.
5. Take out the specimen carrier from the cryovials and transfer them upside down with rubber-coated forceps into the insets under liquid nitrogen atmosphere (*see* **Note 8**).
6. After all samples are loaded into the inserts, transfer the tins into the Leica AFS 2. Cover the tins then with disposable plastics lids (conical closures).
7. Start the freeze substitution according to the protocol shown in Table 2 in a semiautomated fashion. The optic nerves and the spinal cord samples are incubated with the tannic acid for 100 h, whereas brain tissue is only incubated with tanning agent for 24 h (*see* **Note 9**).
8. After the incubation with the tannic acid solution, the samples are washed with acetone. Aspirate the solution with a plastic Pasteur pipette and quickly replace it with acetone precooled to −90 °C. Perform four washes of 30 min each and make sure that the samples are always covered with liquid.
9. During the fourth wash, prepare the contrasting cocktail containing 2% OsO_4 and 0.1% uranyl acetate (from a 5% stock solution in methanol) in 25 mL acetone. Handle the chemicals under the fume hood with complete personal protection (lab coat, safety goggles, mask). Dissolve 0.5 g OsO_4 crystals in 25 mL acetone in a 50 mL Falcon tube with plastic Pasteur pipettes and then add 500 μL 5% uranyl acetate. Transfer the cocktail into a closed glass vial and then into the AFS to allow the contrasting solution to cool down. When the fourth wash is completed, incubate the samples in the OsO_4 and uranyl acetate mix according to the protocol shown in Table 2 (7 h at −90 °C, 14 h temperature raise to −20 °C with an increment of 5 °C/h, 16 h at −20 °C followed by a 2.4 h temperature raise of 10 °C/h to 4 °C). The waste (including the washing steps) should be handled carefully and disposed of according to the appropriate regulations.

Table 2
Automated freeze substitution protocol (modified from [16])

Solution	Duration	Temperature
0.1% tannic acid in acetone	24 h (for brain samples) 100 h (for optic nerves and spinal cord samples)	−90 °C
Ice-cold acetone	4 × 30 min	−90 °C
2% osmium tetroxide, 0.1% uranyl acetate in acetone	7 h 14 h 16 h 2.4 h 1 h	−90 °C Warm up to −20 °C, in 5 °C/h −20 °C Warm up to 4 °C, in 10 °C/h 4C
Cold acetone	4 × 30 min	4C
Cold acetone	1 h	RT
Acetone: Epon mix (2:1)	2 h	RT
Acetone: Epon mix (1:1)	2 h	RT
Acetone: Epon mix (1:2)	2 h	RT
90% Epon in acetone	Over night	RT
Fresh pure Epon	Over night	RT

10. When the freeze substitution protocol reached 4 °C, remove the tins from the AFS and place on ice under the fume hood. Wash out the contrasting cocktail four times with acetone. Then, allow the samples to warm up to RT.
11. Infiltrate the tissue with increasing concentrations of Epon in acetone (2:1, 1:1, 1:2 acetone/Epon) by incubation for 2–3 h for each step.
12. Incubate the samples with 90% Epon in acetone at RT overnight.
13. Infiltrate the samples with fresh Epon resin for 4 h.
14. Remove the flow-through rings from the tins and the samples from the rings using a needle and forceps (*see* **Note 3**).
15. To embed the optic nerve, place the tissue in flat embedding molds. For cross-sections of the spinal cord and the brain, place the samples on Parafilm™-covered glass slides with Epon-filled gelatine capsules turned upside down on the sample. Include labels in the resin to identify the samples (Fig. 1i).
16. Polymerize the Epon resin at 60 °C for 24 h.
17. Remove the gelatin capsule by boiling the resin blocks in water for 15 min.

3.5 Monolayer Embedding of Oligodendrocytes

With the help of this method, cultured cells can be fixed and embedded in Epon to assess their morphology on the ultrastructural level. After cutting the Epon blocks either parallel to the culture substrate or in transverse orientation, the samples can be imaged at the transmission electron microscope. With this method, the morphology of subcellular structures, such as organelles, can be examined from "top-view" (Fig. 6) or "side view." Hence, the morphology of the whole cells cannot be assessed with this method.

1. Grow primary cells in small cell culture dishes (optimal: 3 cm dishes).
   (a) **Note**: Do not use multi-well plates as they would need to be broken apart during the sample preparation.
2. Fix the cells with Karlsson-Schultz fixative for 15 min at RT.
3. Wash the cultured cells thrice with 0.1 M phosphate buffer (PB). All steps are carried out in the cell culture dish directly (*see* **Note 10**).
4. Fix the cells with 1% OsO_4 in 0.1 M PB at 4 °C for 1 h.
5. Wash the samples three times with double-distilled water.
6. Dehydrate with 50% ethanol for 10 min, repeat this step.
7. Incubate samples with 70% ethanol for 10 min, repeat this step.
8. Stain the cultured cells with a freshly prepared mix of 1.5% uranyl acetate and 1.5% tungstophosphoric acid hydrate in 70% ethanol for 90 min at RT.
9. Dehydrate with 90% ethanol for 10 min, repeat this step.
10. Incubate samples with 99.9% ethanol for 5 min, repeat this step five times.

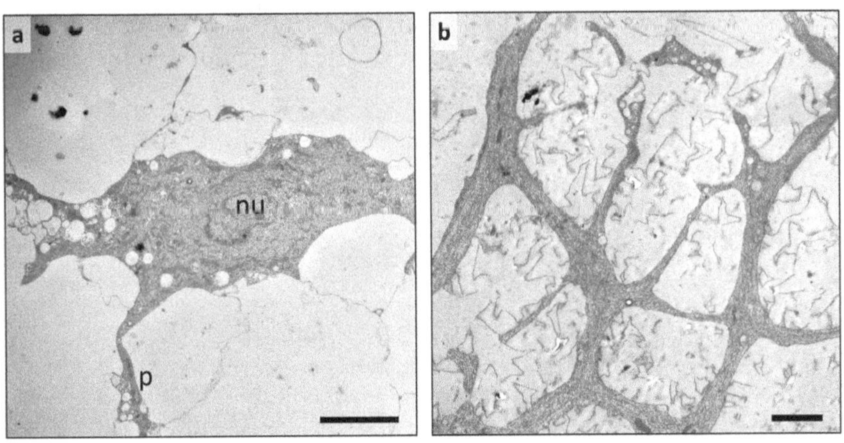

Fig. 6 Monolayer embedding of primary mouse oligodendrocytes in culture. (**a**) Representative image of a mature primary mouse oligodendrocyte (5 days in vitro) with the nucleus (nu) and primary processes (p). (**b**) Branching processes in the periphery of the cell. Scale bars: 5 μm in (**a**) and 2 μm in (**b**)

11. Infiltrate the cells with increasing concentrations of Epon in ethanol (1:2, 1:1, 2:1 Epon/ethanol) by incubation for 30 min for each step at RT.
12. Incubate the samples in pure Epon resin for 1 h at RT.
13. Remove as much Epon resin from the cell culture dish as possible.
14. Fill gelatine capsules with Epon resin and include a label indicating all relevant information for the experiment. Put the capsule upside down on the monolayer.
15. Polymerize the resin at 60 °C for 24 h.
16. The capsules need to be removed by applying a bit of force. The monolayer covers the smooth surface of the block.

3.6 Cytoskeleton Visualization of Oligodendrocytes

The cytoskeleton of the oligodendrocytes defines its shape, endocytosis and intracellular transport. The distribution of actin and microtubule bundles can easily be visualized in fluorescence microscopy, but to examine the ultrastructural architecture of the cytoskeleton the resolution of an electron microscope is needed (Fig. 7). This brings about the problem that cytoskeletal elements are hardly contained over longer distances within the volume of an ultrathin section and cannot be imaged in total.

Here, we present a "whole mount technique," where the cultured cells are examined as a whole [11, 19]. Therefore, the cells are fixed, and all lipids and cytosolic proteins are extracted with detergents. The remaining cytoskeleton is then fixed and contrasted during dehydration. After critical point drying, the cells are then sputter coated. The resulting replicas are subsequently detached from the coverslips and mounted on grids. Often, the remaining cell material and cytoskeleton around the nucleus is too dense to be examined by electron microscopy, whereas the peripheral areas of the cells can be imagined at high resolution to obtain insight into the higher-ordered structures of the cytoskeletal filaments [11, 19] (Fig. 7).

1. Grow cells in small cell culture dishes (optimal: 3 cm dishes).
2. Wash the cultured cells three times with pre-warmed 0.1 M PB. All steps are carried out in the cell culture dish (*see* **Note 10**).
3. Extract lipids and proteins using one of the following protocols for 5–7 min at RT.
   - *General cytoskeletal visualization:* 1% Triton X-100 and 2% PEG in PEM buffer.
   - *Visualization of labile, easily extractable components:* 0.5% Triton X-100 and 0.25% glutaraldehyde in PEM buffer.
4. Wash twice for 1 min in PEM buffer.

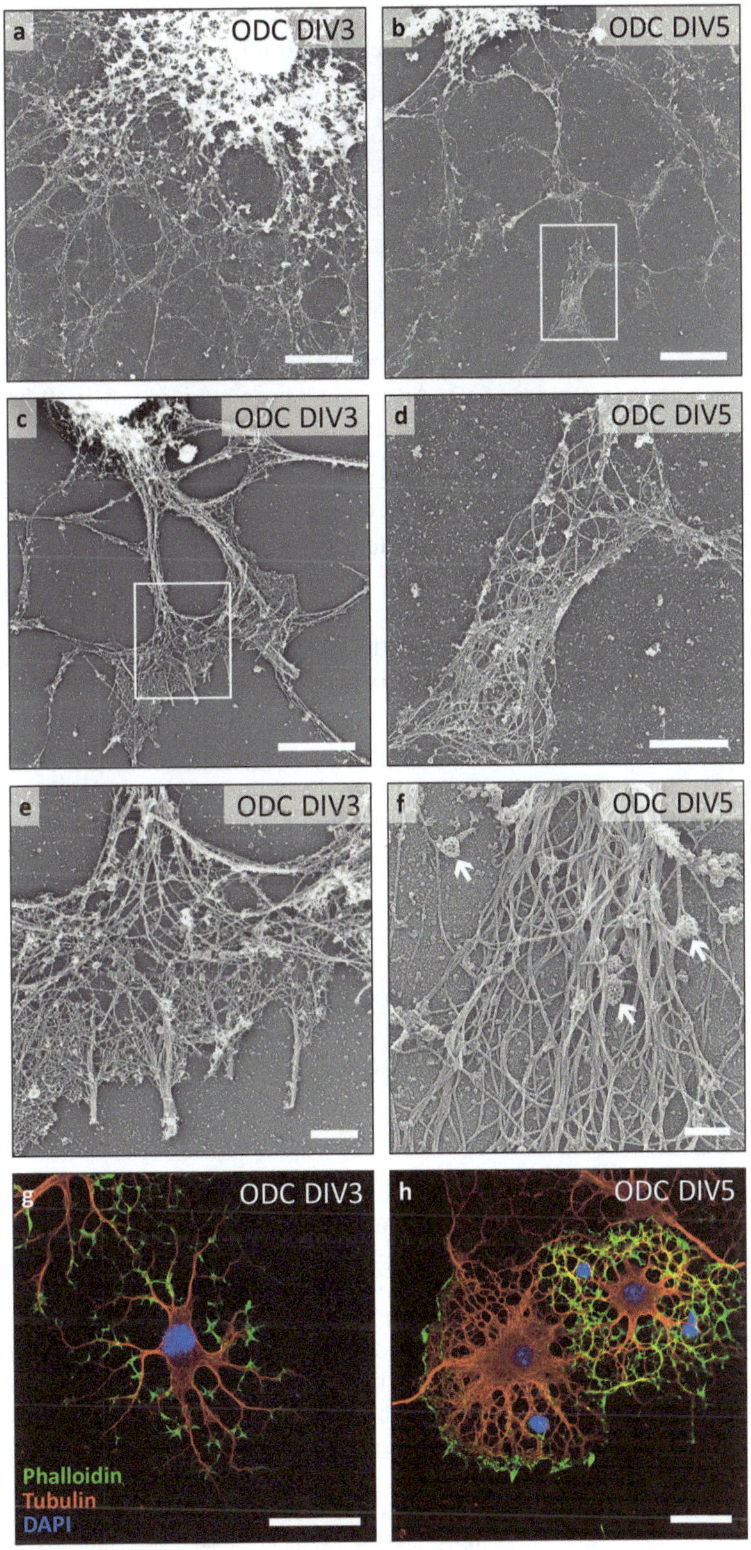

Fig. 7 Cytoskeletal visualization of primary mouse oligodendrocytes in culture. (**a**), (**c**) and (**e**): Representative images of replicas generated from primary mouse

5. Fix the cytoskeletal proteins using 2% glutaraldehyde in PEM buffer for 20 min at RT.
6. Incubate with freshly prepared 0.1% tannic acid in PEM buffer.
7. Wash three times with double-distilled water. Then, incubate with water for 5 min at RT to remove the tannic acid completely before contrasting.
8. Contrast the samples with 0.1% uranyl acetate in water for 20 min at RT.
9. Incubate with increasing concentrations of ethanol for 10 min at each step: 30, 50, 70, 90% and anhydrous ethanol.
10. Add contrast by staining with 0.2% uranyl acetate in 100% ethanol for 20 min at RT.
11. Wash three times with 100% ethanol.
12. The samples need to be transferred into the coverslip-sample holder of the automated critical point dryer. Keep track of the sample position in the holder (Fig. 8a). Then, transfer the samples to the critical point dryer (e.g., Leica EM CPD 300). Set the device according to the manufacturer's instructions to perform 12 exchanges with liquid CO_2 at 5–10 °C prior to critical point drying (slow heating, slow speed during gas out).
13. Stick the dried glass coverslips on sticky notes (label them with the sample name) and adhere the sticky notes with double-sided tape onto the sample table in the sputter coater, such as Leica EM ACE600 (Fig. 8b and *see* **Note 11**).
14. Deposit 2.7 nm platinum at 1 mPa vacuum during 45° angle rotation (*see* manufacturer's protocol for setting up the protocol for sputter coating). Monitor the thickness of the coating with a quartz. Set the working distance to 45 mm and the sample height to 2 mm.
15. Deposit 5 nm carbon layer at 90°.

Fig. 7 (continued) oligodendrocytes at day 3 and (**b**), (**d**) and (**f**): 5 in vitro (DIV3 and DIV5 respectively) at different magnifications. The area in the white box is magnified in the panel below. The white arrows mark clathrin-coated pits and allow quantification of endocytosis. (**g**) Fluorescent confocal images of younger (**h**) and more mature primary mouse oligodendrocytes stained for actin with phalloidin in green and for tubulin in red, whereas the nuclei of the cells are shown in blue by a DAPI stain. The actin filaments are generally found at the leading edge of the cells and the processes, whereas tubulin is only found in the processes. The underlying architecture of the cytoskeleton cannot be determined with these confocal images and hence, replicas can reveal the ultrastructure of the cytoskeleton. Scale bars: 5 μm in (**a**) and (**b**), 1 μm in (**c**) and (**d**) and 250 nm in (**e**) and (**f**), 25 μm in (**g**) and (**h**)

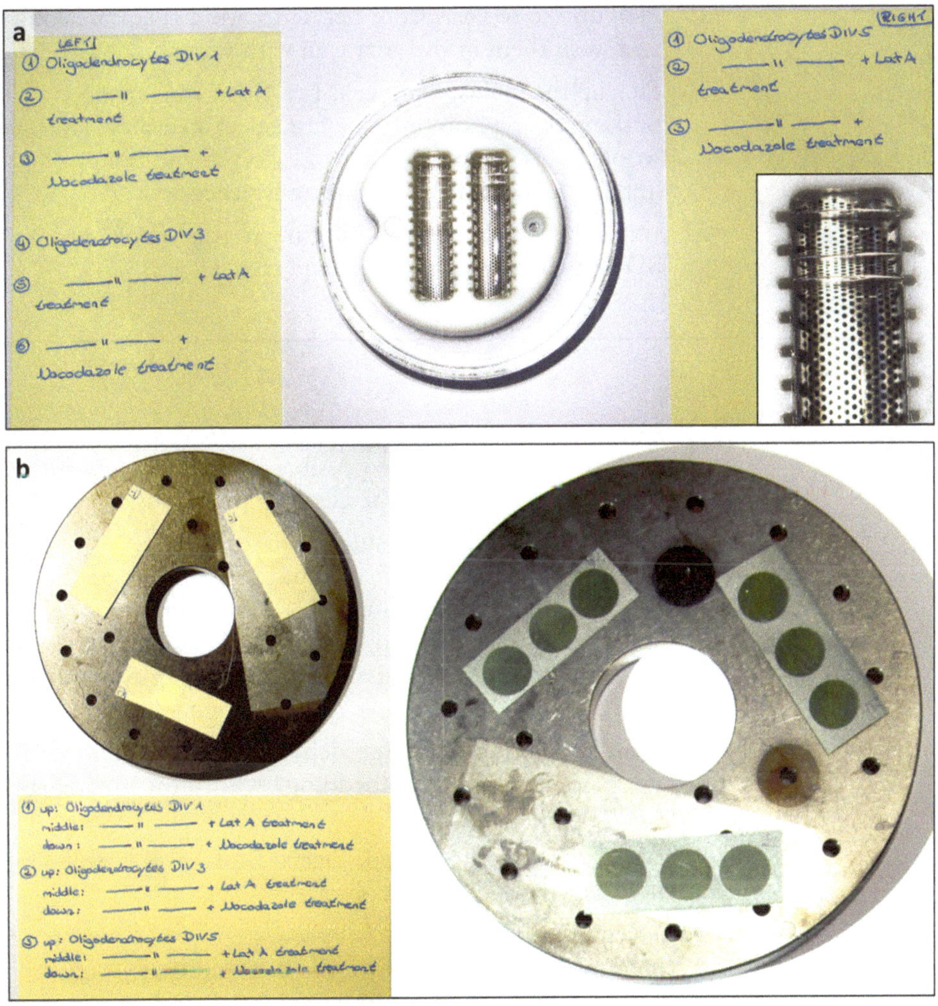

Fig. 8 Critical point drying and platinum-carbon coating for generation of replicas for cytoskeletal imaging in transmission electron microscopy. (**a**) Image of the coverslip-sample holder of the critical point dryer immersed into anhydrous ethanol. The 12 mm coverslips are placed in the holder and the sample position is noted down. The inset on the right side shows in detail how the coverslips are positioned in the holder. (**b**) The dried glass coverslips are placed with double-sided tape onto the sample table of the sputter coater. The image shows the coverslips before (left) and after sputter coating (right) with platinum-carbon

16. Scratch horizontal and vertical lines into the coating of the coverslips in such a way that the resulting pieces of replicas fit on a 3 mm copper grid (*see* **Note 12**).
17. Float the coverslips on 10% hydrofluoric acid in a 12 well cell culture plate. The coverslips will sink down to the bottom of the plate and the replicas will float.
18. Use a Perfect loop™ to transfer the replicas to the next well filled with water and a droplet of highly diluted detergent.

19. Pick up the replicas from below to avoid them breaking apart and wash them in the next well with water (*see* **Note 13**).
20. Pick up the replicas with cells (*see* **Note 14**) on copper grids by pushing the replicas onto the grids with an eyelash attached to a wooden stick. Dry the grid and store them in a gridbox until imaging. No further contrasting is necessary.
21. Image by TEM. The resulting micrographs will appear with a 3D aesthetics after inverting the contrast.

4 Notes

1. Make sure to perfuse the animals at low speed to keep the ultrastructure as intact as possible. A good perfusion is critical for a good outcome of the electron microscopy.
2. Make sure to pack the tissue well in tea bag filter paper in such a way that the tissue cannot slip out of the baskets during transfer between different solutions (Fig. 1b). Furthermore, the flat end of the packet should be facing the wall of the 8-segment basket to generate a flat tissue for embedding as this avoids loss or crumbling of the sample (Fig. 1d).
3. Handle the samples after dehydration and infiltration with Epon resin (prior to embedding) with great care as they can easily break.
4. When dissecting the optic nerve fresh, avoid stretching the nerve as it induces artifacts.
5. Always clean forceps in between steps when performing an immunolabeling to avoid cross-contamination between samples and dirt particles on the sections.
6. Include a no-primary antibody control to detect nonspecific binding of the bridging antibody and Protein A.
7. Do not process more than 6–8 samples at a time as otherwise the methylcellulose layer becomes too thick, and hence makes it more difficult to image since the electron beam has to pass through.
8. Make sure to only touch the samples with forceps that are precooled to liquid nitrogen temperature as this avoids thawing, and hence freeze damage of the samples.
9. Do not let the samples fall dry during the freeze substitution.
10. Do not let the cells dry out.
11. Do not affix the coverslips too strongly on the sample table for the sputter coater as they will break upon detachment.

12. When edging the replicas, the squares in between the lines should not become too small as the replicas will break easily, when transferring them between the wells.
13. Try to immerse and remove the replicas with the loop on an angle as they will break into small pieces otherwise.
14. Replicas that contain cells can be identified due to the reflection of the cells in a binocular microscope.

References

1. Möbius W, Cooper B, Kaufmann WA et al (2010) Electron microscopy of the mouse central nervous system. Methods Cell Biol 96:475–512
2. Möbius W, Posthuma, G (2018) Sugar and ice: Immunoelectron microscopy using cryosections according to the Tokuyasu method. Tissue Cell (in press) https://doi.org/10.1016/j.tice.2018.08.010
3. Griffiths G, Lucocq JM (2014) Antibodies for immunolabeling by light and electron microscopy: not for the faint hearted. Histochem Cell Biol 142:347–360
4. Hacker C, Lucocq JM (2014) Analysis of specificity in immunoelectron microscopy. Methods Mol Biol 1117:315–323
5. Dahl R, Staehelin LA (1989) High-pressure freezing for the preservation of biological structure: theory and practice. J Electron Microsc Tech 13:165–174
6. Vanhecke D, Graber W, Studer D (2008) Close-to-native ultrastructural preservation by high pressure freezing. Methods Cell Biol 88:151–164
7. Dubochet J (2007) The physics of rapid cooling and its implications for cryoimmobilization of cells. Methods Cell Biol 79:7–21
8. Dubochet J, Zuber B, Eltsov M et al (2007) How to "read" a vitreous section. Methods Cell Biol 79:385–406
9. Korogod N, Petersen CC, Knott GW (2015) Ultrastructural analysis of adult mouse neocortex comparing aldehyde perfusion with cryo fixation. elife 4:e05793
10. Möbius W, Nave K-A, Werner H (2016) Electron microscopy of myelin: structure preservation by high-pressure freezing. Brain Res 164:92–100
11. Svitkina T (2007) Electron microscopic analysis of the leading edge in migrating cells. Methods Cell Biol 79:295–319
12. Karlsson U, Schultz RL (1965) Fixation of the central nervous system for electron microscopy by aldehyde perfusion. J Ultrastruct Res 12:160–186
13. Peters PJ, Pierson J (2008) Immunogold labeling of thawed cryosections. Methods Cell Biol 88:131–149
14. Slot JW, Geuze HJ (2007) Cryosectioning and immunolabeling. Nat Protoc 2:2480–2491
15. Webster P, Webster A (2014) Cryosectioning fixed and cryoprotected biological material for immunocytochemistry. Methods Mol Biol 1117:273–313
16. Norton WT, Poduslo SE (1973) Myelination in rat brain: method of myelin isolation. J Neurochem 21:749–757
17. Kaech A, Ziegler U (2014) High-pressure freezing: current state and future prospects, Methods Mol Biol 1117:151–171
18. Siksou L, Rostaing P, Lechaire J-P et al (2007) Three-dimensional architecture of presynaptic terminal cytomatrix. J Neurosci 27:6868–6877
19. Resch GP, Urban E, Jacob S (2010) The actin cytoskeleton in whole mount preparations and sections. Methods Cell Biol 96:529–564

Chapter 21

Toxin-Based Models to Investigate Demyelination and Remyelination

Christopher E. McMurran, Chao Zhao, and Robin J. M. Franklin

Abstract

Clinical myelin diseases, and our best experimental approximations, are complex entities in which demyelination and remyelination proceed unpredictably and concurrently. These features can make it difficult to identify mechanistic details. Toxin-based models offer lesions with predictable spatiotemporal patterns and relatively discrete phases of damage and repair: a simpler system to study the relevant biology and how this can be manipulated. Here, we discuss the most widely used toxin-based models, with a focus on lysolecithin, ethidium bromide, and cuprizone. This includes an overview of their respective mechanisms, strengths, and limitations and step-by-step protocols for their use.

Key words Demyelination, Remyelination, Animal models, Toxin, Lysolecithin, Lysophosphatidylcholine, Ethidium bromide, Cuprizone

1 Introduction

The causes of demyelinating disease are many, and the consequences can be devastating for a patient. Several toxins have been reported to cause clinical demyelination in humans, including sodium cyanate and the antiseptic hexachlorophene [1]. However, these cases are rare when compared to the autoimmune diseases multiple sclerosis and neuromyelitis optica, or even genetic mutations as occur in Pelizaeus-Merzbacher disease or adrenoleukodystrophy. Toxin-based models do not attempt to replicate the complex pathogenesis of these diseases, but rather provide a reductionist setting to study certain aspects, where kinetics are predictable and confounding variables minimized. This approach has yielded substantial insight into the biology of demyelination and remyelination, with much relevance to human disease.

1.1 Types of Toxin-Based Model

Due to the diversity in the range of toxins used to cause demyelination and their modes of delivery, this chapter will focus on three of the most common models. Focal injection of two substances

David A. Lyons and Linde Kegel (eds.), *Oligodendrocytes: Methods and Protocols*, Methods in Molecular Biology, vol. 1936, https://doi.org/10.1007/978-1-4939-9072-6_21,

(lysolecithin and ethidium bromide) will be discussed, as will administration of cuprizone in the diet. Other toxins reported to cause primary demyelination in animal models include 6-aminonicotinamide [2], antibodies to myelin proteins (e.g., [3].), and bacterial endotoxin [4], and these are briefly summarized in Table 1.

1.1.1 Lysolecithin

Lysolecithins (lysophosphatidylcholines, LPCs) are hydrolysis products of phospholipids, which are found naturally in cell membranes and have physiological roles in mediating the phagocytosis of apoptotic cells [9]. The demyelination literature typically refers to the singular *lysolecithin* (lysophosphatidylcholine), which is in fact a mixture of lysolecithins with different fatty acid tails, prepared by the enzymatic action of phospholipase A_2 on a natural substrate rich in phospholipids such as egg yolk (for example: Sigma L4129). Focal injection of lysolecithin was first shown to cause CNS demyelination in the mouse spinal cord in the 1970s [10] and subsequent studies have shown similar lesions in other species including rat [11] and rabbit [12], and other white matter tracts such as the corpus callosum [13] and caudal cerebellar peduncle (CCP) [14].

Lysolecithin selectively destroys myelin, and kills the majority of oligodendrocytes [15] in an ellipsoid region spanning several millimeters. Following injection, there is marked inflammation in the core of the lesion with astrogliosis at the border [16]. Axonal injury can be more pronounced than in other toxin-based models. Remyelination occurs extensively and rapidly in young animals (within 1 month in the young adult mouse [16]) and is mainly carried out by new oligodendrocytes, generated from oligodendrocyte progenitor cells (OPCs) [17]. Remyelination by Schwann cells can also occur in the center of a lesion, particularly if it is large or there is excessive astrocyte loss. In older animals the regeneration of myelin sheaths occurs more slowly [18].

The demyelinating action of lysolecithin is likely a combination of two effects (Fig. 1a). Firstly, being made up of amphipathic molecules, concentrated lysolecithin can act as a detergent, compromising the integrity of cell membranes [19]. Oligodendrocytes are particularly vulnerable to this effect, which may be in part because other cell types are able to metabolize lysolecithin, limiting its concentration in their membranes [20]. Secondly, lysolecithin can signal directly to immune cells, being a chemoattractant signal for monocytes and lymphocytes [21, 22] and direct activator of monocytes, macrophages, microglia, and lymphocytes [23–25]. Thus, the toxin may directly activate immune cells to strip axons of their myelin, which is already compromised by the detergent action. Evidence for an immune-mediated component is that lysolecithin-induced demyelination is reduced when immune signaling molecules are neutralized [26], or in mice devoid of T

Table 1
Overview of toxin-based models for demyelination/remyelination

Toxin	Focal/Systemic	Mechanism	Principal uses
Lysolecithin	Focal	– Detergent effect on oligodendrocytes – Activates immune cells	– Most commonly used focal model, with very well-defined kinetics – Electrophysiological recordings in a specific location
Ethidium Bromide	Focal	– Intercalates in DNA – Affects all nuclei in lesion, sparing axons	– Remyelination in the relative absence of astrocytes – Lesions in CCP can be combined with an in-dwelling catheter for infusion
Cuprizone	Systemic	– Disturbs copper homeostasis, causing metabolic stress – Oligodendrocytes are especially vulnerable and undergo apoptosis	– No surgical procedures required – Remyelination with intact blood-brain barrier [5] – Similarities with pattern III lesions in multiple sclerosis [6]
6-aminonicotinamide	Focal, with systemic effects on gray matter	– Likely to involve vitamin B_3 (nicotinic acid) antagonism causing gliotoxicity [7]	– Remyelination in the relative absence of astrocytes *(Largely superseded by other models)*
Anti-galactocerebroside + complement	Focal	– Antibody against a major constituent of myelin causes selective fixation of complement	– Highly axon-sparing lesion
Bacterial endotoxin (lipopolysaccharide)	Focal	– Direct Toll-Like Receptor signaling to oligodendrocytes and astrocytes [8] – Activates immune cells	– Focal model of immune-mediated demyelination – Similarities with pattern III lesions in multiple sclerosis [4]

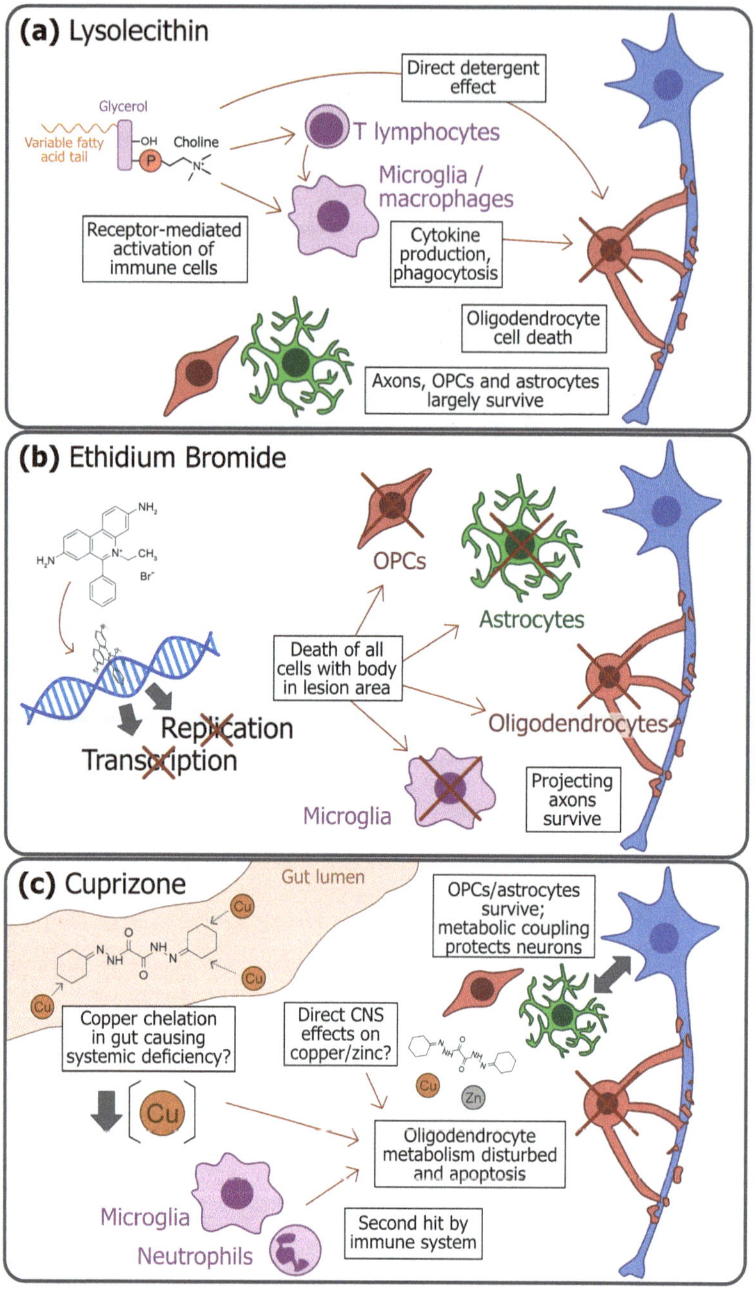

Fig. 1 Schematic overview of the mechanisms by which lysolecithin (**a**), ethidium bromide (**b**), and cuprizone (**c**) cause demyelination

cells [27]. Despite these roles in the pathology, macrophages, microglia, and T cells are also necessary for efficient remyelination [28–30]. Demyelination does not depend on $CXCR2^+$ neutrophils, which are essential in the cuprizone model [31].

By virtue of its relatively high selectivity to oligodendrocytes, lysolecithin can also be used to study demyelination and remyelination in organotypic slice culture. Slices of cerebellum several hundred micrometers thick are most commonly used, as their largely self-contained neural networks allows good survival of myelinated Purkinje cells ex vivo [32]. This model has been successful in both rat [32] and mouse tissue [33] and has proved useful for screening compounds suspected to influence remyelination in a shorter timescale than for live animals [33–35].

1.1.2 Ethidium Bromide

Ethidium bromide (EB) is a planar molecule that can intercalate between base pairs in double-stranded DNA [36] to inhibit transcription and replication. This property has made it useful as a veterinary cytotoxic drug to treat trypanosomiasis [37], as well as for visualization of nucleic acids for research purposes. Focal demyelination by EB injection into a white matter tract was first demonstrated in the spinal cords of cats in the 1980s [38] and has since become well established in mice and rats, with other locations including the CCP white matter tract [14] and the gray matter of the hippocampus [39].

Injected EB is cytotoxic to any cell with a cell body in the lesion area (Fig. 1b). This will include oligodendrocytes and astrocytes [38] as well as OPCs [40], but axons projecting from distant nuclei are spared, providing especially high doses are not used. This means that EB and lysolecithin lesions differ in some crucial respects. EB lesions tend to be larger and remyelination occurs more slowly, likely due to the death of progenitor cells and astrocytes/microglia, which must be recruited from surrounding intact tissue [14]. The death of astrocytes also results in an increased contribution of Schwann cells to remyelination, as astrocytes inhibit OPCs from adopting a Schwann cell fate [41, 42]. Clearance of myelin debris is frequently delayed in EB lesions, contributing to a slower rate of remyelination [43, 44], a probable consequence of microglial death and an impaired inflammatory response. Thus, comparisons between EB and lysolecithin lesions have helped to disentangle some of the important roles that astrocytes and microglia play in remyelination.

1.1.3 Cuprizone

In contrast to the focal injection of lysolecithin or ethidium bromide, the copper chelator cuprizone (biscyclohexanone oxalyldihyrazone) is administered to animals in their diet. This abrogates the need for surgical equipment and expertise, making it the toxin-based model of choice for many groups studying remyelination. Since the 1960s, cuprizone has been known to cause demyelination in mice [45] though subsequent studies using a range of species, strains, ages, and doses showed much variability in the degree of demyelination. To achieve consistency and reproducibility, most

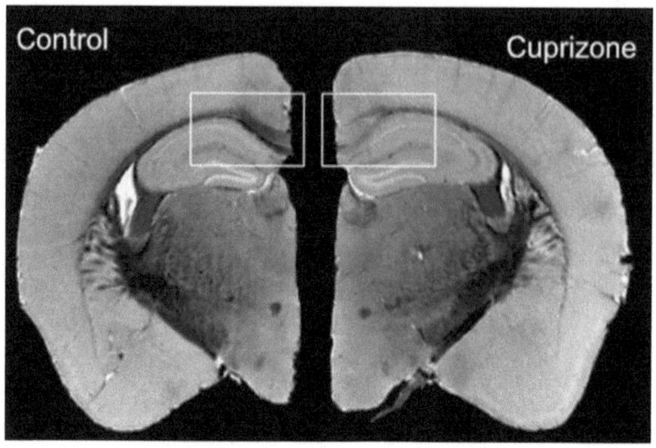

Fig. 2 T2*-weighted gadolinium-enhanced Magnetic Resonance Image (MRI) of brain hemispheres from a mouse fed a 0.2% cuprizone diet for 5 weeks (right) and a control animal (left). Myelin appears dark. The cuprizone-treated mouse shows marked demyelination in the corpus callosum (white boxes), seen here at its caudal end

experimenters feed a 0.2% cuprizone diet to young adult male C57BL/6 mice for a period of 4–6 weeks, which has been shown to give a predictable pattern of demyelination with the scope for transgenic studies [46].

Demyelination occurs most notably in the corpus callosum, but also at other sites including the hippocampus, external capsule, cerebellar peduncles, and cerebral cortex [47]. Callosal demyelination becomes gradually apparent from 3 weeks of treatment, and is accompanied by extensive inflammation and astrogliosis [46]. The corpus callosum is most affected toward its caudal end (Fig. 2), thus it is important that equivalent locations are selected for meaningful comparison [47]. Remyelination overlaps with continued demyelination: at 3 weeks OPCs begin to accumulate [48] and mRNA for myelin proteins is detected from week 5 [49]. Often the mice are returned to their regular diet for 1–2 weeks to allow a period of remyelination in the absence of active demyelination. Remyelination is substantial after a 3 weeks recovery, with restoration of the perinodal architecture [50], although this may not entail complete functional regeneration as axonal degeneration and locomotor impairment can reappear several months later [51].

Cuprizone has a high affinity for copper, an important cofactor for metabolic reactions including oxidative phosphorylation in mitochondria. However, the exact mechanism by which it causes demyelination remains a matter of controversy [52]. One hypothesis suggests that the toxin itself remains in the gut, where it chelates copper and brings about a systemic deficiency and failure of copper-dependent processes [53, 54]. An alternative theory proposes that cuprizone can distribute into the plasma and CNS, where it locally disrupts the homeostasis of copper and other metals [55, 56]. In any case, dietary cuprizone treatment leads to disturbances in cellular metabolism and selective apoptosis of oligodendrocytes, which are particularly vulnerable due to their high metabolic rate and low levels of antioxidants (Fig. 1c) [56, 57]. OPCs and astrocytes are largely spared, the latter of which are also thought to protect neurons through metabolic coupling [52]. However, it is important to note that cuprizone is a systemic toxin and will likely have off-target effects in the CNS as well as other tissues—notably the liver [58].

An initial phase of early oligodendrocyte death [59] is exacerbated by a second hit from the immune system. Full demyelination depends on the actions of neutrophils [31] and microglia [60] but not lymphocytes [61]. However, as with lysolecithin and EB lesions, the innate immune system also plays a vital role in remyelination, by clearing debris and secreting growth factors [30, 62, 63].

Aspects of the cuprizone paradigm can be modified to suit the specific needs of an experiment. A common version is the "chronic" cuprizone model, whereby mice receive a 12-week course of cuprizone and show impaired remyelination with more axonal damage [64, 65]. In another variation, mice receive the mTOR inhibitor rapamycin, which blocks OPC differentiation and delays remyelination until the end of cuprizone treatment [66]. This allows separation of the demyelination and remyelination phases, as seen in lysolecithin or EB lesions, though the effect of rapamycin on the microglial and astrocyte responses remains to be fully characterized.

1.2 Why Use Toxin-Based Models to Study Demyelination and Remyelination?

Toxin-based models are just one strategy to model myelin diseases in vivo, and should therefore be considered alongside other approaches. Alternatives include autoimmune models such as Experimental Autoimmune Encephalomyelitis (EAE) [67, 68], and infectious models such as Theiler's Murine Encephalomyelitis (TMEV) [69, 70], which are generally better representations of the effector aspects of multiple sclerosis. Genetic models have also furthered our understanding, for example, the effects of mutations that cause myelin disease can be modeled in transgenic mice [71].

A key strength of toxin-based models is their predictability, both in their temporal and spatial patterns. Lysolecithin and EB lesions demonstrate a consistent sequence of events with daily resolution, giving a clear background on which to study factors that may influence the rate of demyelination or remyelination. Cuprizone also has a predictable course, albeit over a longer timescale. This is in contrast to EAE, where the sites of demyelination will vary between animals, and lesions of different ages will occur in the same animal, complicating interpretation. Focal lesions of specific white matter tracts can also be favorable for animal welfare, with phenotypes generally less severe than in EAE [72].

A separation between the demyelination and remyelination phases is another feature of the lysolecithin and EB models. While this does not recapitulate the natural history of diseases such as multiple sclerosis, it has allowed pathways involved in pathology and regeneration to be disentangled. A notable example was identifying the positive contribution of the innate immune system to remyelination [28], which is obscured in other models by its roles in myelin damage.

Toxin-based models have generally been viewed as reductionist tools, while EAE and viral models are used as our best approximations to human disease. However, there is much heterogeneity between lesions in multiple sclerosis and some may in fact be better modeled using toxin-based approaches. For example, cuprizone is thought to be one of the best experimental models for pattern III clinical demyelination [6]. It is clear that a continued synthesis of results from different approaches is necessary to advance the field of myelin research. Toxin-based models can be a valuable resource, so long as the strengths and limitations are considered and the choice of model is well suited to answer the question at hand.

2 Materials

Protocols for a typical paradigm using each of our three toxins will be described:

1. **Lysolecithin**: focal injection into mouse ventral spinal cord.
2. **Ethidium bromide**: focal injection into rat caudal cerebellar peduncle.
3. **Cuprizone**: dietary administration.

2.1 Focal Injection Models

Lysolecithin and ethidium bromide are both focally injected under anesthesia and much of the equipment required is common to both.

2.1.1 Lysolecithin Spinal Cord Lesions

1. 1% lysolecithin (*see* below)—at least 1 μL per lesion.
2. Sterile gown and gloves.
3. Anesthetic machine (isoflurane in oxygen). An injectable anesthetic may be used if an inhalation setup is not available.
4. Operating table with a stage for mouse and two surgical stands.
5. 3-way manipulator for Hamilton syringe.
6. Operating microscope.
7. Heat pad and rectal probe to monitor temperature.
8. Clippers.
9. Disposable drapes.

10. Gauze swabs.
11. Viscotears (Alcon).
12. Antiseptic solution, e.g., Betadine®.
13. Sterile saline.
14. Analgesic, e.g., buprenorphine, 0.05 mg/kg (Vetergesic (Champion Alstoe) contains 0.3 mg/mL buprenorphine, so can be diluted 1 in 6 in saline and used 100 μL/100 g).
15. Cotton wool or similar (for bleeding).
16. Hamilton syringe with glass tip for injecting lysolecithin.
17. 30G dental needle to cut dura.
18. 5 mL syringe and needle to administer saline.
19. 1 mL syringe and needle to administer analgesic.
20. Scalpel with small blade (e.g., No. 15).
21. Fine-tipped tweezers (2 sets).
22. Mosquito forceps.
23. Needle holder.
24. Scissors.
25. 6-0 sutures.
26. Recovery unit.
27. Lysolecithin should be prepared in advance. Add 10 mL sterile saline to 100 mg lysolecithin (Sigma L4129), dissolve, then make aliquots to freeze at −20 °C. On the day of lesioning, defrost aliquot and sonicate in a sonicating water bath for 20–30 min until clear.

2.1.2 Ethidium Bromide CCP Lesions

1. 0.01% ethidium bromide (*see* below)—at least 4 μL per lesion.
2. Sterile gown and gloves.
3. Anesthetic machine (isoflurane in oxygen). An injectable anesthetic may be used if an inhalation setup is not available.
4. Operating table with a small animal stereotactic frame.
5. Operating microscope.
6. Heat pad and rectal probe to monitor temperature.
7. Clippers.
8. Disposable drapes.
9. Gauze swabs.
10. Viscotears (Alcon).
11. Antiseptic solution, e.g., Betadine®.
12. Sterile saline.
13. Analgesic, e.g., buprenorphine, 0.05 mg/kg (Vetergesic (Champion Alstoe) contains 0.3 mg/mL buprenorphine, so can be diluted 1 in 6 in saline and used 100 μL/100 g).

14. Cotton wool or similar (for bleeding).
15. 10 mL Hamilton syringe with 26G cone tip needle for injecting EB.
16. 5 mL syringe and needle to administer saline.
17. 1 mL syringe and needle to administer analgesic.
18. Scalpel with small blade (e.g., No. 15).
19. Surgical drill.
20. Retractor.
21. Needle holder.
22. Scissors.
23. 5-0 sutures
24. Recovery unit.

Ethidium bromide can be purchased as a liquid and diluted in PBS to 0.01%.

2.2 Dietary Cuprizone Model

2.2.1 The Cuprizone Model

1. Cuprizone diet

   The easiest option is to purchase a diet with cuprizone incorporated at 0.2%. A number of suppliers are able to provide this, including Envigo (TD.140804) and SDS (as a "Custom Made Diet"). A common alternative is to buy cuprizone separately and mix it with powdered chow. The diet should not be autoclaved, as cuprizone is heat-sensitive, and it should be stored in the dark and refrigerated.
2. Biological safety cabinet to open cages.
3. Scales for weighing mice.

3 Methods

Surgical procedures should be carried out wearing a sterile gown and gloves, with good aseptic technique.

3.1 Lysolecithin Lesion in Mouse Ventral Spinal Cord

Spinal cord lesions are normally carried out on mice aged 8 weeks and onward. The procedure should take less than 30 min and sterile saline can be applied periodically to prevent the wound from drying out.

1. Prepare anesthetic machine, operating microscope, operating table with stage suitable for mice, and sterile surgical instruments.
2. Prepare the following syringes:
   - Glass-tipped Hamilton syringe with 1% lysolecithin—ensure there are no bubbles and attach this to a 3-way manipulator fitted to a stand.

- 1 mL syringe of analgesic, e.g., buprenorphine, 0.05 mg/kg.

3. Weigh mouse and transfer to anesthetic chamber to begin induction of anesthesia with isoflurane/oxygen. When effective, transfer mouse to operating stage equipped with mask for maintaining anesthesia throughout the procedure. Position mouse prone with head away from operator.
4. Inject analgesic subcutaneously. Position the temperature probe and apply one drop of Viscotears liquid gel to each eye.
5. Prepare skin by shaving the back around the natural "hump" and clean the shaved area with antiseptic solution. Do not treat excessively to prevent heat loss. Cut a small access window in a sterile drape and lay over mouse.
6. Begin operating after verifying the mouse has no response to pain. Start with a 0.5 cm midline incision at the top of the hump on the back (thoraco-lumbar region). Cut away the subcutaneous tissue to expose the muscles. Use a small scalpel blade to gently separate the muscles on both sides.
7. Under the microscope, find the space between two vertebrae using two pairs of fine-tipped tweezers and remove the soft tissues: first muscle and then the yellow ligament to expose the surface of the spinal cord. This should appear pale with the central vein visible in the midline. Care must be taken not to damage the spinal cord at this stage.
8. Use a pair of curved mosquito forceps to clamp the soft tissue above the exposed intervertebral space, then fix to a stand to minimize movement of the injection site from breathing. Use a very thin beveled needle (such as a 30G dental needle) to open the dura just to the right of the central vein (Fig. 3).

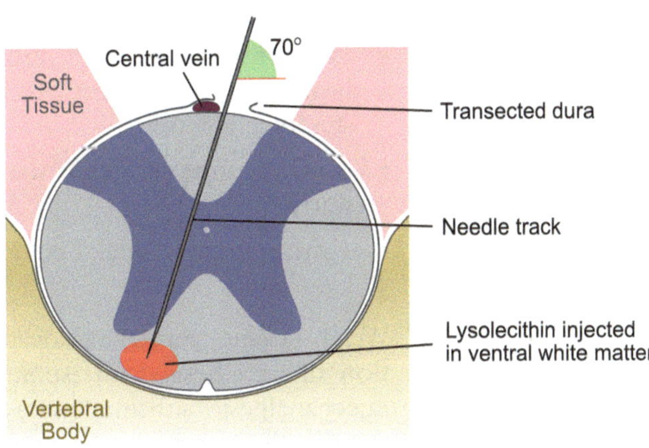

Fig. 3 Injection of lysolecithin into the ventral white matter of the spinal cord

9. Position the manipulator with the Hamilton syringe at a 70° angle relative to the horizontal plane. Gently adjust the tip of the syringe to touch the surface of the cord through the cut dura. Slowly lower the needle through the cord, until a slight deflection indicates contact with the vertebral bone underneath. Retract the needle fractionally so the deflected tip becomes straight.
10. Inject 0.5 μL of toxin in about 20 s, then retract the needle about 0.2–0.25 mm and inject another 0.5 μL. Keep the needle in place for 1 min.
    (a) Slowly raise the needle out of the spinal cord. And move away clamp and manipulator. Using a 6-0 suture, make one stitch to close the muscular layer, then 2–3 stitches for the skin. Lidocaine cream can be applied to reduce pain at the wound site.
11. Transfer the mouse to a warm chamber for recovery before returning it to the cage.

After the procedure, there should be no signs of distress or clinical abnormality. No special care is required and individual cages are not necessary. Occasionally a mouse may develop a unilateral hind leg impairment, which should recover in within 48 h and will not prevent animal accessing food or water. External sutures are often removed by the mouse, and may need to be replaced if this occurs before wound closure.

3.2 Ethidium Bromide Lesion in Rat Caudal Cerebellar Peduncle

Lesions of the CCP are typically carried out on rats aged 8 weeks and onward. The procedure is more complex than spinal cord lesions due to the need to drill through the skull and use precise stereotactic coordinates. In total, it should take around 45–60 min.

1. Prepare anesthetic machine, operating microscope, operating table with stereotactic frame, and sterile surgical instruments.
2. Prepare the following syringes:
    - Hamilton syringe (with 26G needle) with 0.01% ethidium bromide—ensure there are no bubbles and attach this to stereotactic frame.
    - 5 mL syringe of sterile saline—place in recovery unit to warm
    - 1 mL syringe of analgesic, e.g., buprenorphine, 0.05 mg/kg.
3. Weigh rat and transfer to anesthetic chamber to begin induction of anesthesia with isoflurane/oxygen. When effective, inject analgesic (subcutaneous) and 2.5–3 mL saline (intraperitoneal). Position the temperature probe.

4. Shave top of head from between the eyes to behind the ears. Scrub area with two gauze swabs wetted with antiseptic solution and apply one drop of Viscotears liquid gel to each eye.
5. Transfer rat to stereotactic frame equipped with mask for maintaining anesthesia throughout the procedure. Open the mouth and pull tongue outside if possible. Attach the ear bars carefully, holding the animal. Each bar should be in the same position and an eye-closing reflex can be seen when the bar is correctly located. The head of the animal should be positioned flat, which can be verified once the skull is exposed.
6. Apply skin prep to the shaved area skin. Cut a small access window in a sterile drape and lay over rat.
7. Using a scalpel, make a 1.5 cm midline incision at the back of the head. Remove the connective tissue with scissors and forceps. Cotton wool can be used to stop the bleeding. Scrape away tissue overlying the bone with the scalpel.
8. With the aid of the microscope, locate the bregma and note its coordinates, β (Fig. 4). Touching the bones to move them slightly can aid with this.
9. Locate the lambda, which is defined as the midpoint of the curve of best fit along the lambda suture (Fig. 4) rather than the intersection of the lambdoid and sagittal sutures [73]. Note its coordinates, λ.
10. Confirm that β and λ lie on the same horizontal plane, and adjust the position of the rat as necessary.
11. Calculate the stereotactic coordinates for the CCP using Table 2. These will depend on the size of the rat, and thus the β-λ distance.
12. Apply the retractor. Position tip of drill at the point of entry.
13. Drill through the skull, advancing slowly. The syringe can be used periodically to confirm the hole is in the correct location.

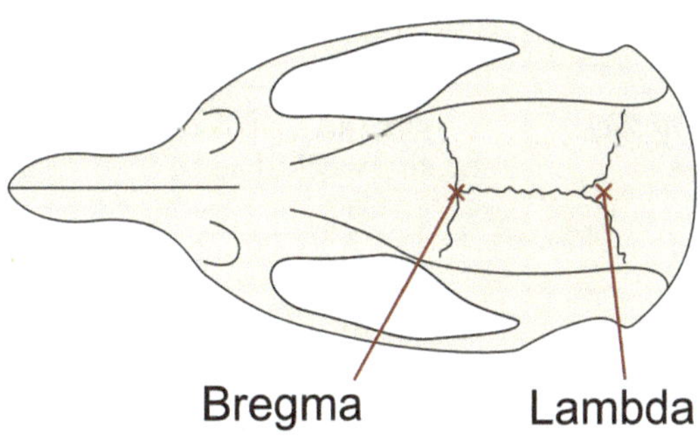

Fig. 4 The dorsal surface of the rat skull. The lamba, λ, is defined as the midpoint of the curve of best fit along the lambdoid suture, rather than the intersection between the lambdoid and sagittal sutures

Table 2
Guide for calculating the coordinates of CCP relative to lambda, λ

	β–λ (mm)	<8.0	8.0–8.2	8.2–8.5	8.5–9.0	9.0–9.5	>9.5
Distance to CCP (mm)	Anterior	2.1	2.2	2.2	2.2	2.3	2.4
	Ventral	7.1–7.2	7.2	7.2	7.2	7.3	7.4
	Lateral	2.5	2.5	2.5	2.5	2.6	2.65

14. When drilling is complete, move the syringe to the calculated coordinates for the CCP. The coordinates can be adjusted slightly if the needle catches on the edge of the burr hole. This is preferable to allowing the needle to deflect, which may result in the tip being several millimeters away from the intended site.
15. Inject 4 μL in 4 min, then wait another 4 min. Remove the syringe slowly, and stop any bleeding.
16. Remove the retractor and close the skin with 5-0 suture.
17. Transfer the rat to a warm chamber for recovery before returning it to the cage.

The surgery for CCP lesions involves more risk than for spinal cord lesions (e.g., hemorrhage) and postoperative complications are also more common. For the first 24 h after the procedure, the rat should be housed individually and provided with a liquid/gel diet *ad libitum* and a long spout water bottle. If showing signs of pain, another dose of analgesic can be administered several hours after surgery. Rats are often lethargic following the surgery but should regain their usual level of activity over 2–3 days. If a mild head tilt develops postoperatively, this should be monitored, with special feeding arrangements as required, and will usually resolve. In occasional cases, a rat may develop spontaneous rolling and will need to be culled.

3.3 Dietary Administration of Cuprizone to Mice

The effects of cuprizone can depend on the species, strain, age, and gender of the animals used. The literature standard is to use 8-week-old male C57BL/6 mice, and feed a diet of 0.2% cuprizone for 4–6 weeks [52].

1. For a period of 4–6 weeks, substitute the regular feed with a 0.2% cuprizone diet. This should be replaced every 2–3 days, due to degradation of the toxin. Food intake can be estimated as 4–5 g per mouse per day.
2. Cuprizone poses a safety risk through skin contact or inhalation. For this reason, mice should be housed in sealed cages and diet/bedding changes carried out in a biological safety cabinet with appropriate waste disposal.

3. Mouse weights should be monitored throughout the experiment. Weight loss often occurs during the first 1–2 weeks of administration, with subsequent weight gain.
4. Mice are often returned to their regular diet for 1–2 weeks for a period of remyelination without ongoing demyelination.

Under these conditions, substantial demyelination is achieved while minimizing clinical severity, which is frequent at higher doses [46, 74]. However, some weight loss is common and other signs have been reported including reduced social interaction [75] and impaired performance on a balance-beam task [76].

3.4 Conclusions

Toxin-based models can offer a valuable insight into the mechanisms of myelin diseases, and into broader aspects of regenerative medicine [84]. Predictable kinetics and a temporal dissociation between demyelination and remyelination are advantages of these paradigms that complement the findings of other model systems (*see* **Notes 1** and **2**).

4 Notes

1. Several methods can be used to visualize and quantify myelin, including immunohistochemistry for myelin protein, light microscopy dyes including luxol fast blue or lipophilic fluorophores such as FluroMyelin (ThermoFisher). Magnetic resonance imaging (MRI) can also be useful tool to map myelin across the entire brain. However, these techniques are unable to distinguish between the myelin produced during remyelination and that originating in development. Thus, it can be difficult to determine whether an intervention has truly enhanced remyelination, or if more of the original sheaths have survived.

   At a higher resolution it becomes clear that the dimensions of the myelin relative to its axon differs between myelination and remyelination. During development, myelin sheath thickness increases with larger diameter axons. In remyelination, this relationship is not maintained and thus a given diameter axon will be myelinated by a sheath which appears thinner and with shorter internodes than developmentally generated myelin (Fig. 5) [77]. To be able to visualize this, the ultrastructure of the tissue must be very well preserved, and this is best done by perfusion-fixing animals in glutaraldehyde before embedding regions of interest resin [78]. Myelin sheath thickness can then be measured using 1 μm sections stained with toluidine blue (with practice), or using electron microscopy. The thinner sheaths are less clear in regions of the CNS with smaller-

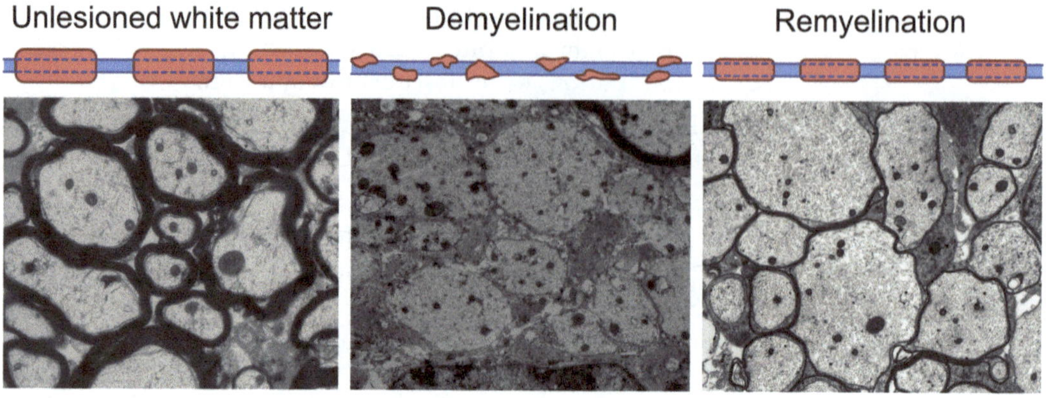

Fig. 5 Electron micrographs of unlesioned white matter, demyelination and remyelination, highlighting the thinner appearance of the remyelinated sheaths

diameter axons such as the corpus callosum, which can be problematic for interpreting electron microscopy in the cuprizone model [74].

2. Besides imaging myelin sheaths themselves, a wide variety of techniques have been applied to toxin-induced lesions to study the cells and molecular cues involved in demyelination and remyelination. To preserve the CNS for immunohistochemistry, a perfusion-fixation method using 4% paraformaldehyde in PBS is recommended. This delivers fixative rapidly and uniformly to all tissues, and is thus preferable to immersion-fixation [79]. In situ hybridization is a useful alternative to immunohistochemistry, particularly to visualize transcripts encoding secreted factors or myelin sheath proteins, which would be difficult to relate to a cell body.

   Flow cytometry has been applied to dissected lesion tissue with some success in the cuprizone model, although as the effected regions are relatively small this requires pooling of large numbers of mice [62]. In the lysolecithin model, inflammatory cells from the lesion and normal tissue have proved difficult to separate by flow cytometry, probably due to the small volume of affected tissue [80] relative to the more global inflammation which makes this possible in EAE [81]. In contrast to live cells, RNA and protein from a small region of CNS containing a lysolecithin lesion have been reliably used to investigate remyelination [82, 83]. Total RNA has been successfully extracted from EB lesions very selectively using laser capture microdissection of frozen sections [35], revealing general patterns in gene expression at different timepoints during remyelination.

Acknowledgments

Thanks to Natalia Murphy for consultation regarding the ethidium bromide protocol, and to Joseph Guy for performing the cuprizone brain MRI.

References

1. Morell P (1984) Myelin. Doi: https://doi.org/10.1007/978-1-4757-1830-0
2. Blakemore WF (1975) Remyelination by Schwann cells of axons demyelinated by intraspinal injection of 6-aminonicotinamide in the rat. J Neurocytol 4:745–757. https://doi.org/10.1007/BF01181634
3. Carroll WM, Jennings AR, Mastaglia FL (1984) Experimental demyelinating optic neuropathy induced by intra-neural injection of galactocerebroside antiserum. J Neurol Sci 65:125–135. https://doi.org/10.1016/0022-510X(84)90077-7
4. Felts PA, Woolston AM, Fernando HB et al (2005) Inflammation and primary demyelination induced by the intraspinal injection of lipopolysaccharide. Brain 128:1649–1666. https://doi.org/10.1093/brain/awh516
5. Bakker DA, Ludwin SK (1987) Blood-brain barrier permeability during Cuprizone-induced demyelination. Implications for the pathogenesis of immune-mediated demyelinating diseases. J Neurol Sci 78:125–137. https://doi.org/10.1016/0022-510X(87)90055-4
6. Torkildsen O, Brunborg LA, Myhr K-M, Bø L (2008) The cuprizone model for demyelination. Acta Neurol Scand Suppl 188:72–76. https://doi.org/10.1111/j.1600-0404.2008.01036.x
7. Nishimoto S, Tanaka H, Okamoto M et al (2015) Methylcobalamin promotes the differentiation of Schwann cells and remyelination in lysophosphatidylcholine-induced demyelination of the rat sciatic nerve. Front Cell Neurosci 9:298. https://doi.org/10.3389/fncel.2015.00298
8. Zhang F, Yao SY, Whetsell WO, Sriram S (2013) Astrogliopathy and oligodendrogliopathy are early events in CNS demyelination. Glia 61:1261–1273. https://doi.org/10.1002/glia.22513
9. Lauber K, Blumenthal SG, Waibel M, Wesselborg S (2004) Clearance of apoptotic cells: getting rid of the corpses. Mol Cell 14:277–287. https://doi.org/10.1016/S1097-2765(04)00237-0
10. Hall SM (1972) The effect of injections of lysophosphatidyl choline into white matter of the adult mouse spinal cord. J Cell Sci 10:535–546
11. Blakemore WF (1976) Invasion of Schwann cells into the spinal cord of the rat following local injections of lysolecithin. Neuropathol Appl Neurobiol 2:21–39. https://doi.org/10.1111/j.1365-2990.1976.tb00559.x
12. Blakemore WF (1978) Observations on remyelination in the rabbit spinal cord following demyelination induced by lysolecithin. Neuropathol Appl Neurobiol 4:47–59. https://doi.org/10.1111/j.1365-2990.1978.tb00528.x
13. Nait-Oumesmar B, Decker L, Lachapelle F et al (1999) Progenitor cells of the adult mouse subventricular zone proliferate, migrate and differentiate into oligodendrocytes after demyelination. Eur J Neurosci 11:4357–4366. https://doi.org/10.1046/j.1460-9568.1999.00873.x
14. Woodruff RH, Franklin RJM (1999) Demyelination and remyelination of the caudal cerebellar peduncle of adult rats following stereotaxic injections of lysolecithin, ethidium bromide, and complement/anti-galactocerebroside: a comparative study. Glia 25:216–228. https://doi.org/10.1002/(SICI)1098-1136(19990201)25:3<216::AID-GLIA2>3.0.CO;2-L
15. Arnett HA, Fancy SPJ, Alberta JA et al (2004) bHLH transcription factor Olig1 is required to repair demyelinated lesions in the CNS. Science 306:2111–2115. https://doi.org/10.1126/science.1103709
16. Jeffery ND, Blakemore WF (1995) Remyelination of mouse spinal cord axons demyelinated by local injection of lysolecithin. J Neurocytol 24:775–781. https://doi.org/10.1007/BF01191213
17. Gensert JM, Goldman JE (1997) Endogenous progenitors remyelinate demyelinated axons in the adult CNS. Neuron 19:197–203. https://doi.org/10.1016/S0896-6273(00)80359-1
18. Gilson J, Blakemore WF (1993) Failure of remyelination in areas of demyelination produced in the spinal cord of old rats. Neuropathol Appl Neurobiol 19:173–181

19. Weltzien HU (1979) Cytolytic and membrane-perturbing properties of lysophosphatidylcholine. BBA Rev Biomembr 559:259–287. https://doi.org/10.1016/0304-4157(79)90004-2
20. Gregson NA, Hall SM (1973) A quantitative analysis of the effects of the intraneural injection of lysophosphatidyl choline. J Cell Sci 13:257–277
21. Quinn MT, Parthasarathy S, Steinberg D (1988) Lysophosphatidylcholine: a chemotactic factor for human monocytes and its potential role in atherogenesis. Proc Natl Acad Sci U S A 85:2805–2809
22. McMurray HF, Parthasarathy S, Steinberg D (1993) Oxidatively Modified Low-Density-Lipoprotein Is a Chemoattractant for Human T-Lymphocytes. J Clin Invest 92:1004–1008. https://doi.org/10.1172/JCI116605
23. Ngwenya BZ, Yamamoto N (1985) Activation of peritoneal macrophages by lysophosphatidylcholine. Biochim Biophys Acta 839:9–15
24. Huang YH, Schäfer-Elinder L, Wu R et al (1999) Lysophosphatidylcholine (LPC) induces proinflammatory cytokines by a platelet-activating factor (PAF) receptor-dependent mechanism. Clin Exp Immunol 116:326–331. https://doi.org/10.1046/j.1365-2249.1999.00871.x
25. Schilling T, Lehmann F, Rückert B, Eder C (2004) Physiological mechanisms of lysophosphatidylcholine-induced de-ramification of murine microglia. J Physiol 557:105–120. https://doi.org/10.1113/jphysiol.2004.060632
26. Ousman SS, David S (2001) MIP-1alpha, MCP-1, GM-CSF, and TNF-alpha control the immune cell response that mediates rapid phagocytosis of myelin from the adult mouse spinal cord. J Neurosci 21:4649–4656
27. Ghasemlou N, Jeong SY, Lacroix S, David S (2007) T cells contribute to lysophosphatidylcholine-induced macrophage activation and demyelination in the CNS. Glia 55:294–302. https://doi.org/10.1002/glia.20449
28. Kotter MR, Setzu A, Sim FJ et al (2001) Macrophage depletion impairs oligodendrocyte remyelination following lysolecithin-induced demyelination. Glia 35:204–212. https://doi.org/10.1002/glia.1085
29. Bieber AJ, Kerr S, Rodriguez M (2003) Efficient central nervous system remyelination requires T cells. Ann Neurol 53:680–684. https://doi.org/10.1002/ana.10578
30. Miron V, Franklin R (2014) Macrophages and CNS remyelination. J Neurochem 130:1–7. https://doi.org/10.1111/jnc.12705
31. Liu L, Belkadi A, Darnall L et al (2010) CXCR2-positive neutrophils are essential for cuprizone-induced demyelination: relevance to multiple sclerosis. Nat Neurosci 13:319–326. https://doi.org/10.1038/nn.2491
32. Birgbauer E, Rao TS, Webb M (2004) Lysolecithin induces demyelination in vitro in a cerebellar slice culture system. J Neurosci Res 78:157–166. https://doi.org/10.1002/jnr.20248
33. Zhang H, Jarjour AA, Boyd A, Williams A (2011) Central nervous system remyelination in culture - A tool for multiple sclerosis research. Exp Neurol 230:138–148. https://doi.org/10.1016/j.expneurol.2011.04.009
34. Miron VE, Ludwin SK, Darlington PJ et al (2010) Fingolimod (FTY720) enhances remyelination following demyelination of organotypic cerebellar slices. Am J Pathol 176:2682–2694. https://doi.org/10.2353/ajpath.2010.091234
35. Huang JK, Jarjour AA, Nait Oumesmar B et al (2011) Retinoid X receptor gamma signaling accelerates CNS remyelination. Nat Neurosci 14:45–53. https://doi.org/10.1038/nn.2702
36. Waring MJ (1965) Complex Formation between Ethidium Bromide and Nucleic Acids. J Mol Biol 13:269–282. https://doi.org/10.1016/S0022-2836(65)80096-1
37. Chowdhury AR, Bakshi R, Wang J et al (2010) The killing of African trypanosomes by ethidium bromide. PLoS Pathog 6(12): e1001226. https://doi.org/10.1371/journal.ppat.1001226
38. Blakemore WF (1982) Ethidium bromide induced demyelination in the spinal cord of the cat. Neuropathol Appl Neurobiol 8:365–375
39. Goudarzvand M, Choopani S, Shams A et al (2016) Focal injection of ethidium bromide as a simple model to study cognitive deficit and its improvement. Basic Clin Neurosci J 7:63–73
40. Sim FJ, Zhao C, Penderis J, Franklin RJM (2002) The age-related decrease in CNS remyelination efficiency is attributable to an impairment of both oligodendrocyte progenitor recruitment and differentiation. J Neurosci 22:2451–2459. https://doi.org/10.1097/00041327-200306000-00025
41. Zawadzka M, Rivers LE, Fancy SPJ et al (2010) CNS-resident glial progenitor/stem cells produce Schwann cells as well as oligodendrocytes during repair of CNS demyelination. Cell Stem

Cell 6:578–590. https://doi.org/10.1016/j.stem.2010.04.002

42. Monteiro De Castro G, Deja NA, Ma D et al (2015) Astrocyte activation via Stat3 signaling determines the balance of oligodendrocyte versus Schwann cell remyelination. Am J Pathol 185:2431–2440. https://doi.org/10.1016/j.ajpath.2015.05.011
43. Graça DL, Blakemore WF (1986) Delayed remyelination in rat spinal cord following ethidium bromide injection. Neuropathol Appl Neurobiol 12:593–605
44. Kotter MR, Li W-W, Zhao C, Franklin RJM (2006) Myelin impairs CNS remyelination by inhibiting oligodendrocyte precursor cell differentiation. J Neurosci 26:328–332. https://doi.org/10.1523/JNEUROSCI.2615-05.2006
45. Carlton WW (1966) Response of mice to the chelating agents sodium diethyldithiocarbamate, alpha-benzoinoxime, and biscyclohexanone oxaldihydrazone. Toxicol Appl Pharmacol 8:512–521. https://doi.org/10.1016/0041-008X(66)90062-7
46. Hiremath MM, Saito Y, Knapp GW et al (1998) Microglial/macrophage accumulation during cuprizone-induced demyelination in C57BL/6 mice. J Neuroimmunol 92:38–49. https://doi.org/10.1016/S0165-5728(98)00168-4
47. Steelman AJ, Thompson JP, Li J (2012) Demyelination and remyelination in anatomically distinct regions of the corpus callosum following cuprizone intoxication. Neurosci Res 72:32–42. https://doi.org/10.1016/j.neures.2011.10.002
48. Mason JL, Jones JJ, Taniike M et al (2000) Mature oligodendrocyte apoptosis precedes IGF-1 production and oligodendrocyte progenitor accumulation and differentiation during demyelination/remyelination. J Neurosci Res 61:251–262. https://doi.org/10.1002/1097-4547(20000801)61:3<251::AID-JNR3>3.0.CO;2-W
49. Morell P, Barrett CV, Mason JL et al (1998) Gene expression in brain during cuprizone-induced demyelination and remyelination. Mol Cell Neurosci 12:220–227. https://doi.org/10.1006/mcne.1998.0715
50. Zoupi L, Markoullis K, Kleopa KA, Karagogeos D (2013) Alterations of juxtaparanodal domains in two rodent models of CNS demyelination. Glia 61:1236–1249. https://doi.org/10.1002/glia.22511
51. Manrique-Hoyos N, Jürgens T, Grønborg M et al (2012) Late motor decline after accomplished remyelination: impact for progressive multiple sclerosis. Ann Neurol 71:227–244. https://doi.org/10.1002/ana.22681
52. Praet J, Guglielmetti C, Berneman Z et al (2014) Cellular and molecular neuropathology of the cuprizone mouse model: clinical relevance for multiple sclerosis. Neurosci Biobehav Rev 47:485–505. https://doi.org/10.1016/j.neubiorev.2014.10.004
53. Venturini G (1973) Enzymic activities and sodium, potassium and copper concentrations in mouse brain and liver after cuprizone treatment in vivo. J Neurochem 21:1147–1151. https://doi.org/10.1111/j.1471-4159.1973.tb07569.x
54. Benetti F, Ventura M, Salmini B et al (2010) Cuprizone neurotoxicity, copper deficiency and neurodegeneration. Neurotoxicology 31:509–517. https://doi.org/10.1016/j.neuro.2010.05.008
55. Zatta P, Raso M, Zambenedetti P et al (2005) Copper and zinc dismetabolism in the mouse brain upon chronic cuprizone treatment. Cell Mol Life Sci 62:1502–1513. https://doi.org/10.1007/s00018-005-5073-8
56. Bénardais K, Kotsiari A, Škuljec J et al (2013) Cuprizone [bis(cyclohexylidenehydrazide)] is selectively toxic for mature oligodendrocytes. Neurotox Res 24:244–250. https://doi.org/10.1007/s12640-013-9380-9
57. Bradl M, Lassmann H (2010) Oligodendrocytes: biology and pathology. Acta Neuropathol 119:37–53. https://doi.org/10.1007/s00401-009-0601-5
58. Suzuki K (1969) Giant hepatic mitochondria: production in mice fed with cuprizone. Science 163:81–82. https://doi.org/10.1126/science.163.3862.81
59. Doan V, Kleindienst AM, Mcmahon EJ et al (2013) Abbreviated exposure to cuprizone is sufficient to induce demyelination and oligodendrocyte loss. J Neurosci Res 91:363–373. https://doi.org/10.1002/jnr.23174
60. Clarner T, Janssen K, Nellessen L et al (2015) CXCL10 triggers early microglial activation in the cuprizone model. J Immunol 194:3400–3413. https://doi.org/10.4049/jimmunol.1401459
61. Arnett HA, Mason J, Marino M et al (2001) TNF alpha promotes proliferation of oligodendrocyte progenitors and remyelination. Nat Neurosci 4:1116–1122. https://doi.org/10.1038/nn738
62. Voß EV, Škuljec J, Gudi V et al (2012) Characterisation of microglia during de- and remyelination: can they create a repair promoting environment? Neurobiol Dis 45:519–528. https://doi.org/10.1016/j.nbd.2011.09.008

63. Lampron A, Larochelle A, Laflamme N et al (2015) Inefficient clearance of myelin debris by microglia impairs remyelinating processes. J Exp Med 212:481–495. https://doi.org/10.1084/jem.20141656
64. Mason JL, Langaman C, Morell P et al (2001) Episodic demyelination and subsequent remyelination within the murine central nervous system: Changes in axonal calibre. Neuropathol Appl Neurobiol 27:50–58. https://doi.org/10.1046/j.0305-1846.2001.00301.x
65. Lindner M, Fokuhl J, Linsmeier F et al (2009) Chronic toxic demyelination in the central nervous system leads to axonal damage despite remyelination. Neurosci Lett 453:120–125. https://doi.org/10.1016/j.neulet.2009.02.004
66. Sachs HH, Bercury KK, Popescu DC et al (2014) A new model of cuprizone-mediated demyelination/remyelination. ASN Neuro 6:1–16. https://doi.org/10.1177/1759091414551955
67. Olitsky PK, Yager RH (1949) Experimental disseminated encephalomyelitis in white mice. J Exp Med 90:213–224. https://doi.org/10.1084/jem.90.3.213
68. Robinson AP, Harp CT, Noronha A, Miller SD (2014) The experimental autoimmune encephalomyelitis (EAE) model of MS: utility for understanding disease pathophysiology and treatment. Handb Clin Neurol 122:173–189. https://doi.org/10.1016/B978-0-444-52001-2.00008-X.The
69. Dal Canto MC, Lipton HL (1975) Primary demyelination in Theiler's virus infection: an ultrastructural study. Lab Investig 33:626–637
70. Mecha M, Carrillo-Salinas FJ, Mestre L et al (2013) Viral models of multiple sclerosis: neurodegeneration and demyelination in mice infected with Theiler's virus. Prog Neurobiol 101–102:46–64. https://doi.org/10.1016/j.pneurobio.2012.11.003
71. Griffiths IR, Schneider A, Anderson J, Nave KA (1995) Transgenic and natural mouse models of proteolipid protein (PLP)-related dysmyelination and demyelination. Brain Pathol 5:275–281
72. Wolfensohn S, Hawkins P, Lilley E et al (2013) Reducing suffering in experimental autoimmune encephalomyelitis (EAE). J Pharmacol Toxicol Methods 67:169–176. https://doi.org/10.1016/j.vascn.2013.01.009
73. Paxinos G, Watson C (2013) The rat brain in stereotaxic coordinates: hard cover edition. Elsevier Science, Amsterdam
74. Stidworthy MF, Genoud S, Suter U et al (2003) Quantifying the early stages of remyelination following cuprizone-induced demyelination. Brain Pathol 13:329–339. https://doi.org/10.1111/j.1750-3639.2003.tb00032.x
75. Xu H, Yang H, Clough R, Browning R (2009) Behavioral and neurobiological changes in C57BL / 6 mice exposed to cuprizone. Behav Neurosci 123:418–429. https://doi.org/10.1037/a0014477
76. Hagemeyer N, Boretius S, Ott C et al (2012) Erythropoietin attenuates neurological and histological consequences of toxic demyelination in mice. Mol Med 18:628–635. https://doi.org/10.2119/molmed.2011.00457
77. Franklin RJM, Zhao C, Lubetzki C, Ffrench-Constant C (2013) Endogenous remyelination in the CNS. In: Myelin repair neuroprotection mult. Scler. Springer, Boston, MA, pp 71–92
78. Blakemore WF, Franklin RJM (2008) Remyelination in experimental models of toxin-induced demyelination. Curr Top Microbiol Immunol 318:193–212. https://doi.org/10.1007/978-3-540-73677-6-8
79. Gage GJ, Kipke DR, Shain W (2012) Whole animal perfusion fixation for rodents. J Vis Exp 65:e3564. https://doi.org/10.3791/3564
80. Döring A, Sloka S, Lau L et al (2015) Stimulation of monocytes, macrophages, and microglia by amphotericin B and macrophage colony-stimulating factor promotes remyelination. J Neurosci 35:1136–1148. https://doi.org/10.1523/JNEUROSCI.1797-14.2015
81. Goncalves DaSilva A, Yong VW (2009) Matrix metalloproteinase-12 deficiency worsens relapsing-remitting experimental autoimmune encephalomyelitis in association with cytokine and chemokine dysregulation. Am J Pathol 174:898–909. https://doi.org/10.2353/ajpath.2009.080952
82. Zhao C, Li W-W, Franklin RJM (2006) Differences in the early inflammatory responses to toxin-induced demyelination are associated with the age-related decline in CNS remyelination. Neurobiol Aging 27:1298–1307. https://doi.org/10.1016/j.neurobiolaging.2005.06.008
83. Boyd A, Zhang H, Williams A (2013) Insufficient OPC migration into demyelinated lesions is a cause of poor remyelination in MS and mouse models. Acta Neuropathol 125:841–859. https://doi.org/10.1007/s00401-013-1112-y
84. Aurora AB, Olson EN (2014) Immune modulation of stem cells and regeneration. Cell Stem Cell 15:14–25. https://doi.org/10.1016/j.stem.2014.06.009

Chapter 22

Magnetic Resonance Techniques for Imaging White Matter

Cassandra Sampaio-Baptista, Kata Diosi, and Heidi Johansen-Berg

Abstract

The white matter is a complex network of brain fibers connecting different information processing regions in the brain. In recent years, the investigation of white matter in humans and in animal models has greatly benefitted from the introduction of in vivo noninvasive magnetic resonance imaging (MRI) techniques. MRI allows for multiple in vivo time-point whole-brain acquisition in the same subject, thus it can be used longitudinally to monitor white matter brain change, intervention effects, as well as disease progression. However, MRI has low spatial resolution compared to gold standard cellular techniques and MRI measures are sensitive to a number of tissue properties resulting in a lack of specificity.

The following chapter describes in simple technical terms to non-imaging experts some common MRI techniques that can be used to investigate white matter structure noninvasively, covering some of the advantages and pitfalls of each technique.

Key words MRI, White matter, Myelin, Image analysis

1 Introduction to MRI

MRI relies on nuclear magnetic resonance (NMR). Nuclei that contain an odd number of protons and/or neutrons spin around their own axis. The most commonly used nucleus in MRI is hydrogen (H), which is present in water molecules and is widely available in body tissue. In a free environment the spins will be randomly oriented, however when placed in a magnetic field, the spins will align with the field and precess around it. In MRI this field is generated by the scanner and is measured in tesla units. In MRI, the alignment to the main magnetic field (named *b0* and conventionally represented as the z axis) is perturbed by a perpendicular magnetic field that emits a brief radio frequency (RF) pulse. This causes the protons to flip their spin and precess in the x,y axis, producing a signal that can be detected as a current by a receiving coil. If the signal is acquired immediately after the RF pulse, most protons will be maximally aligned with the x,y axis and the resulting image will reflect proton density. On the other hand, if the signal acquisition is delayed, proton spins will start to realign with the

David A. Lyons and Linde Kegel (eds.), *Oligodendrocytes: Methods and Protocols*, Methods in Molecular Biology, vol. 1936, https://doi.org/10.1007/978-1-4939-9072-6_22,

main magnetic field, resulting in signal decay in the measurable x,y axis. The recovery time in the longitudinal (z) axis is known as *T1* and varies between different brain tissues which can be a source of tissue contrast. *T2* reflects the decaying time in the transverse plane (x,y) and also varies between tissue types. *T2* decay occurs faster than *T1* due to the small magnetic fields around each proton that affect the spin of the neighboring nuclei, causing them to dephase and rapidly decrease the signal in the transverse plane. MRI sequences can be tuned in order to maximize or minimize the influence of each type of recovery or decay time in order to obtain images with high contrast between tissues.

To be able to image the location of the protons, it is necessary to use an additional set of small magnetic field gradients distributed across the x,y,z axes. These gradients are rapidly turned on and off, after the RF pulse, to create small linear changes in the main magnetic field along the three orthogonal axes. As a result, protons at different spatial locations will experience slightly different magnetic fields. Since precession frequency depends on the magnetic field experienced by a proton, this means proton precession can be used to estimate the position of the nuclei. Fourier transforms are then used to reconstruct MR images from frequency space, conventionally known as k-space.

In the next sections some common MRI methods to assess myelin and white matter microstructure will be described (for a more extensive list of MRI aquisition methods see [1]).

2 *T1*-Weighted and *T2*-Weighted Imaging

T1-weighted (*T1*w) and *T2*-weighted (*T2*w) imaging are some of the earliest methods used in MR studies to provide image contrast between brain tissue and are sensitive indicators of anatomical structures. The images are generated by exciting and acquiring signal at different time intervals. *T1*-weighted images allow for a high contrast between white matter (WM), gray matter (GM), and cerebrospinal fluid (CSF), making it ideal for tissue segmentation and tissue volume estimates. The signal intensity from free hydrogen proton filled compartments (such as the CSF) appear attenuated and dark on *T1*-weighted images, whereas in highly structured materials, such as the lipid-rich myelin sheaths (where the macromolecule-bound protons proportion is higher), the *T1* signal appears brighter. While *T1*w and *T2*w images are sensitive indicators of anatomical structures and the *T1*w contrast is related to the presence of myelin, these methods are not capable of directly measuring myelin concentration in the brain [2, 3]. Additionally, individual experimental parameters will influence *T1*w and *T2*w contrast, precluding the possibility of direct comparisons of *T1*w and $T2$w values across patients and healthy controls, or across longitudinal time-points [4, 5].

An alternative approach is to use the ratio of $T1$w and $T2$w images to produce maps whose gray-scale intensities have been found to co-localize with myelin-stained histology from the cortical surface [6]. These structural images are routinely acquired in MR exams and thus eliminate the need of additional complex MR acquisitions and long scan times. Furthermore, they can be acquired at high resolution, which can be beneficial to examine cortical myelination, in a relatively short period of time. However, $T1$w and $T2$w are also related to the overall water content in tissues, and thus images individually have low specificity to myelin.

In order to overcome the comparability, reproducibility, and specificity issues of qualitative imaging, quantitative mapping that provides absolute values on voxel-by-voxel basis has also been developed [4, 5].

2.1 Quantitative T1 Mapping

Quantitative representation of $T1$ measures offers a more direct assessment of white matter tissue properties. $T1$ relaxometry refers to quantitative methods that measure and map relaxation times or susceptibility values within tissues. $T1$ contrast is highly sensitive to the glycolipids and cholesterol components of myelin sheath [7] and to a lesser degree, iron content, so a decrease in quantitative $T1$ values relates mainly to an increase in myelin [2].

An alternative approach to $T1$ relaxometry is to use the ratio of standard structural images to produce maps of myelination. $R1$ maps ($=1/T1$) positively correlates with the amount of myelination and has been shown to reflect myeloarchitecture in cortical areas [8, 9]. Postmortem ultra-high resolution $R1$ images have shown a good correspondence with histological measures of myelin in cortical areas [2].

Quantitative MRI measurements, including $T1$ and $R1$ maps, are theoretically independent of acquisition parameters; therefore, the generated information is comparable across healthy participants and pathological conditions, cortical regions, and over different time-points [4, 5]. However, long scan times are necessary, which might prevent these quantitative techniques from being routinely included in clinical studies.

3 Magnetization Transfer Imaging

Magnetization Transfer (MT) imaging has the ability to indirectly detect macromolecule-bound protons such as membrane lipids and myelin [7]. The MT effect relies on exchanges between the macromolecule-bound water and the free water. The protons in the free water have relatively long $T2$ relaxation times making it easy to detect by conventional MRI systems. By contrast, the magnetization from the restricted macromolecule-bound protons decays faster. In MT imaging these macromolecule-bound protons are selectively saturated without influencing the free water. This saturated magnetization is then transferred to the free pool by a continuous exchange of magnetization, which is referred to cross-relaxation or MT, and this effect is then measured.

MT ratio (MTR) imaging is a semiquantitative technique with the advantage of short acquisition times at relatively high resolution. To calculate MTR only two types of images are needed: one with MT saturation and one without. The resulting maps are found to correlate with myelin [10]. MTR maps have been used to assess changes in myelin content in Multiple Sclerosis (MS) lesions with studies showing higher MTR values in remyelinated WM lesions [10]. Despite the strong correlation between MTR and myelin, MTR lacks neuropathological specificity [11]. In other words, MTR values are influenced by other factors unrelated to myelin, such as axonal injury, inflammation, edema [12] and experimental parameters, such as direct saturation effects and pulse magnitudes [10, 13]. Furthermore, a lack of standardization across imaging centers makes it hard to compare results (for review *see* [14]).

Quantitative MT (qMT) is more sensitive and specific to myelin than MTR [11, 15, 16]. For instance, qMT is sensitive to regional differences in white matter of healthy participants that are related to tissue myelination [16]. Furthermore, qMT is able to detect aging-related changes in myelination in rodent models while MTR is not sufficiently sensitive [11]. However, this type of imaging requires more than the two images necessary for the MTR calculation; thus, longer scan times are needed for the calculation of qMT. The higher specificity and sensitivity therefore comes at the expense of resolution, coverage, and scan time [17–19].

4 Diffusion Imaging

The MRI technique most commonly used to probe white matter structure in humans is Diffusion-weighted imaging (DWI). DWI is sensitive to the local diffusion of water molecules. Conventional DWI estimates the overall diffusion at each voxel and can be useful to detect certain types of pathology. However, in many tissue types diffusion will vary along different axes and so an estimate of overall diffusion is less useful.

In the absence of any barriers, water diffusion is roughly uniform in all directions, known as isotropic diffusion. In biological tissue, however, many boundaries alter free water diffusivity; thus, it becomes more restricted, or in other words anisotropic. For instance, in the fibrous white matter tracts, the axonal membrane confines the movements of water; diffusion is restricted in the perpendicular direction, so water tends to diffuse along the axis of the tract but not across it. In such cases it is useful to estimate diffusion along different directions. The most common approach to do so is diffusion tensor imaging (DTI) [20]. In DTI imaging water diffusion is mathematically modeled by an ellipsoid, or tensor. The diffusion tensor is calculated on a voxel-by-voxel basis by applying a mathematical algorithm on multiple diffusion-weighted images, each sensitized to diffusion along a different direction. A minimum of six sampled directions are required to estimate the six parameters that define any given diffusion tensor: the length of the longest, middle, and shortest axes of the tensor (called eigenvalues) and their orientations in space (called eigenvectors). The larger the number of acquired directions the more accurate is the estimation of the tensor, so most DTI acquisitions sample more than six directions at the expense of scan time.

4.1 Imaging White Matter Microstructure and Integrity with DTI

Multiple parameters can be estimated from the diffusion tensor model. Fractional anisotropy (FA) is the most commonly used parameter to measure directional dependence of water diffusion. FA characterizes the shape of the tensor and provides a normalized value to the degree of anisotropy from zero (completely isotropic) to one (completely anisotropic) [21]. For instance, at the typical resolution of a DTI acquisition, the diffusion of water molecules shows little directional dependence in the gray matter and cerebrospinal fluid; thus, the value of FA in these tissues is close to zero, which is represented as a near-spherical tensor within each voxel. In contrast, in the highly organized white matter fiber bundle the diffusion has a preferred direction; thus, the FA value is closer to one and represented by an elongated diffusion ellipsoid.

FA is highly sensitive to various white matter features, such as myelination, fiber caliber and density, fiber organization and axonal degeneration [22]. For instance, FA is higher in white matter tracts that are highly myelinated. Other important measures calculated from the eigenvalues are mean diffusivity (MD), axial diffusivity (AD) also referred to as parallel diffusivity, radial diffusivity (RD), also known as perpendicular diffusivity. MD reflects tissue density, the greater the barriers to diffusion, the lower the value; MD is a commonly used measure in both GM and WM. AD is a measure of the diffusivity in the direction of maximum diffusion in a voxel and can be an indirect marker of axonal membrane integrity [23–25]. Whereas RD represents the average of the diffusivity values in the two axes perpendicular to the principle direction of diffusion and there is some evidence it might particularly relate to myelination [24, 26]. These measures allow for the indirect probing of white matter microstructure and integrity; however, there is not a one-to-one relationship between a particular DTI-derived metric and a specific cellular feature [27]. In other words, a decrease in FA, for instance, can be due to a number of causes such as a decrease in myelination, cell swelling, and axon degeneration, among others.

Although DTI measures are nonspecific, they are very sensitive to events that alter the microstructure of the brain; hence, this technique has been widely used. For instance, it has been used to investigate white matter microstructure in relation to development and aging [28, 29], brain plasticity during adulthood [30, 31] and brain disorders such as multiple sclerosis (MS) [32] or Alzheimer's disease [33]. FA has been found to increase during development well into adulthood, while MD and RD decrease, partly reflecting increases in myelination [28, 34]. FA decreases with aging while MD and RD have the opposite trajectory, which is associated with decreases in myelination as well as a loss of nerve fibers associated with normal age-related degeneration [28, 29, 35–37]. Naturally, diffusion imaging has been used to assess MS lesions during acute phases and it might be useful to monitor disease progression and treatment efficacy. Most studies have shown increased MD and decreased FA in lesions of MS patients when compared with normal appearing white matter or with healthy controls [32, 38, 39]. Diffusion imaging has made a large contribution in fostering interest in white matter neuroplasticity during adulthood. Neuroimaging studies have reported that cognitive and motor learning and practice results in changes in white matter structure [30, 31, 40, 41], which are related at least in part to myelination [42, 43].

4.2 Advanced Diffusion Analysis Methods

DTI and its estimated parameters give unprecedented insights into the architectural organization and integrity of the living brain but this simple model has several shortcomings (Assaf et al. 2008; Panagiotaki et al. 2012; De Santis et al. 2014). For instance, DTI does not account for multiple tissue types within a voxel, and a number of different white matter microstructural features (such as the cellular membranes of glial cells or the axon diameter and density) contribute to the obtained DT indices [44, 45]. However, the water residing in different tissue compartments has different diffusion patterns. For example, water molecules residing outside the axon (extra-axonal) has different diffusion characteristics than those confined inside an axon (intra-axonal) [44, 46, 47]. Hence, by estimating diffusion from the individual compartments, further information could be revealed about white matter microstructure and axonal morphology not provided by DTI. Different multi-compartment models have been proposed to complement DTI measurements and provide more specific estimates of brain microstructural properties.

One of the multi-compartment models is the *Composite hindered and restricted model of diffusion*, or CHARMED (Assaf & Basser 2005). CHARMED estimates and models the contribution of the extra-axonal and intra-axonal space diffusion compartments. Hindered diffusion characterizes the displacement of water in the extra-axonal space that includes extra- and intracellular spaces of glial cells [44]. In contrast, restricted diffusion refers to the distribution of water in the intra-axonal space bounded by compartments like axonal membranes [44]. Therefore, CHARMED is able to provide improved local fiber orientations and measures of axonal density.

Based on the two-compartment CHARMED framework, further methods have been developed. The *AxCaliber* model was proposed to estimate the axonal density, the extra-axonal diffusivity, and the mean axon diameter distribution by probing the entire white matter tissue continuously with multiple diffusion times and gradient strengths [48]. The diameter distribution of neuronal fibers is an important morphological parameter that may link to various functional properties of white matter fascicles, such as their conduction velocity [49, 50]. Additionally, *AxCaliber* is a very promising technique for future insights into the longitudinal changes of mean axonal diameter distributions in vivo during development, aging, and disease.

While most diffusion MRI techniques for the central nervous system have focused on the white matter, recent promising methods for estimating the distribution of orientations and density of neurites (the collective name for axons and dendrites) in either the white matter or gray matter have begun to emerge. Neurite morphology is an important marker of normal brain development, aging, brain connectivity, and numerous neurological disorders, such as Alzheimer's disease and Multiple Sclerosis [46, 51]. One neuroimaging technique to assess such information is the neurite orientation dispersion and density imaging (NODDI) and its recent extension Bingham-NODDI that aims to estimate the density and orientation dispersion of axons and dendrites in the living human brain in as little as 10 min on standard clinical MRI scanners [46, 51].

4.3 Tracing White Matter Tracts In Vivo

MRI can also be used to perform in vivo tract tracing. Currently, diffusion imaging is the only method that allows for the tracking of white matter fibers noninvasively. Tractography can reconstruct entire white matter pathways through the human brain by tracing and connecting voxels. The information provided by this method allows the visualization of the three-dimensional architecture of the white matter tracts. It can reveal patterns of anatomical connectivity between different regions, which in turn can provide information about the functional specialization of the brain region.

There are several tracing methods that rely on different assumptions to estimate fiber orientation and in some cases require advanced diffusion acquisition (for more information *see* [52]). The simplest method of fiber tracing known as deterministic tractography can use DTI data, and is based on the assumption that the orientation of the largest eigenvector of the fitted tensor corresponds to the direction of the main white matter tracts [20]. Probabilistic single-fiber tractography, on the other hand, requires sampling of a large number of diffusion directions, and additionally considers the uncertainty on the estimate of principal fiber direction [53]. Both these approaches assume a single fiber orientation at each voxel. This is often not the case in white matter voxels where crossing, diverging, or complex pathways mean that a single fiber orientation does not provide a useful representation of the true fiber architecture. Alternative methods, allowing for estimate of multiple fiber populations at each voxel, have therefore been proposed [54]. These make it possible to trace crossing fibers and provide a more accurate reconstruction of the underlying anatomy.

Although there has been encouraging evidence on the validation of diffusion tractography relative to gold standard methods [55–58], it is important to highlight that diffusion tractography does not trace axonal fibers directly like injected tracers, and tens of thousands of axons are present in a voxel. This method also does not differentiate between anterograde and retrograde connections; thus, it does not allow for an identification of the direction of the connection or presence of synapses. Furthermore, crossing-fiber regions, fiber fanning and axon geometry influence tractography performance and accuracy, even when more sophisticated crossing fiber models are used, as such the results should be carefully interpreted in light of these limitations.

5 Conclusions

In the last few decades we have seen a rapid development in MRI techniques that allow for probing of the microstructure of the living brain. This offers huge opportunities for understanding human brain development, aging, plasticity, and pathology. The rapid development of increasingly sophisticated MRI acquisitions and tissue modeling are already providing a better characterization of the microstructure of the brain and more specific markers of the underlying cellular structure. New technical developments (*see* review [1]) will continue to offer more specificity, higher sensitivity, and faster acquisitions, at a higher resolution. Furthermore, with faster acquisitions, several modalities can be acquired in a single scan session providing complementary information and helping to disentangle and to quantify the different cellular components, offering a more complete picture of the in vivo brain.

It is necessary to continue to validate these methods by means of animal models, where both MRI and histological measures can be acquired [42, 43, 55], and human postmortem tissue studies [2, 59]. Still, even with further development, MRI will never achieve the same resolution and specificity of gold standard microscopy—but the ability to measure in vivo and in humans offers the exciting opportunity to exploit the methods to understand the importance of WM structure to human development, behavior, and disease.

References

1. Heath F, Hurley SA, Johansen-Berg H, Sampaio-Baptista C (2017) Advances in non-invasive myelin imaging. Dev Neurobiol. 78(2):136–151
2. Stuber C et al (2014) Myelin and iron concentration in the human brain: a quantitative study of MRI contrast. NeuroImage 93 Pt 1:95–106
3. Tardif CL et al (2016) Advanced MRI techniques to improve our understanding of experience-induced neuroplasticity. NeuroImage 131:55–72
4. Deoni SC (2010) Quantitative relaxometry of the brain. Top Magn Reson Imaging 21:101–113
5. Weiskopf N et al (2013) Quantitative multi-parameter mapping of R1, PD(*), MT, and R2(*) at 3T: a multi-center validation. Front Neurosci 7:95
6. Glasser MF, Van Essen DC (2011) Mapping human cortical areas in vivo based on myelin content as revealed by T1- and T2-weighted MRI. J Neurosci 31:11597–11616
7. Kucharczyk W, Macdonald PM, Stanisz GJ, Henkelman RM (1994) Relaxivity and magnetization transfer of white matter lipids at MR imaging: importance of cerebrosides and pH. Radiology 192:521–529
8. Sereno MI, Lutti A, Weiskopf N, Dick F (2013) Mapping the human cortical surface by combining quantitative T(1) with retinotopy. Cereb Cortex 23:2261–2268
9. Lutti A, Dick F, Sereno MI, Weiskopf N (2014) Using high-resolution quantitative mapping of R1 as an index of cortical myelination. NeuroImage 93 Pt 2:176–188
10. Schmierer K, Scaravilli F, Altmann DR, Barker GJ, Miller DH (2004) Magnetization transfer ratio and myelin in postmortem multiple sclerosis brain. Ann Neurol 56:407–415
11. Samsonov A et al (2012) Quantitative MR imaging of two-pool magnetization transfer model parameters in myelin mutant shaking pup. NeuroImage 62:1390–1398
12. Vavasour IM, Laule C, Li DK, Traboulsee AL, MacKay AL (2011) Is the magnetization transfer ratio a marker for myelin in multiple sclerosis? J Magn Reson Imaging 33:713–718
13. Dortch RD et al (2013) Quantitative magnetization transfer imaging of human brain at 7 T. NeuroImage 64:640–649
14. Alexander AL et al (2011) Characterization of cerebral white matter properties using quantitative magnetic resonance imaging stains. Brain Connect 1:423–446
15. Schmierer K et al (2007) Quantitative magnetization transfer imaging in postmortem multiple sclerosis brain. J Magn Reson Imaging 26:41–51
16. Sled JG et al (2004) Regional variations in normal brain shown by quantitative magnetization transfer imaging. Magn Reson Med 51:299–303
17. Henkelman RM, Stanisz GJ, Graham SJ (2001) Magnetization transfer in MRI: a review. NMR Biomed 14:57–64
18. Schmierer K et al (2010) Effects of formalin fixation on magnetic resonance indices in multiple sclerosis cortical gray matter. J Magn Reson Imaging 32:1054–1060
19. Schmierer K et al (2008) Quantitative magnetic resonance of postmortem multiple sclerosis brain before and after fixation. Magn Reson Med 59:268–277
20. Basser PJ, Mattiello J, LeBihan D (1994) Estimation of the effective self-diffusion tensor from the NMR spin echo. J Magn Reson B 103:247–254
21. Pierpaoli C et al (2001) Water diffusion changes in Wallerian degeneration and their dependence on white matter architecture. NeuroImage 13:1174–1185
22. Beaulieu C (2009) The biological basis of diffusion anisotropy. In: Diffusion Mri: from quantitative measurement to in vivo neuroanatomy. Elsevier, Amsterdam, pp 105–126
23. Kinoshita Y, Ohnishi A, Kohshi K, Yokota A (1999) Apparent diffusion coefficient on rat brain and nerves intoxicated with methylmercury. Environ Res 80:348–354
24. Song SK et al (2003) Diffusion tensor imaging detects and differentiates axon and myelin degeneration in mouse optic nerve after retinal ischemia. NeuroImage 20:1714–1722
25. Kim JH et al (2006) Detecting axon damage in spinal cord from a mouse model of multiple sclerosis. Neurobiol Dis 21:626–632

26. Klawiter EC et al (2011) Radial diffusivity predicts demyelination in ex vivo multiple sclerosis spinal cords. NeuroImage 55:1454–1460
27. Zatorre RJ, Fields RD, Johansen-Berg H (2012) Plasticity in gray and white: neuroimaging changes in brain structure during learning. Nat Neurosci 15:528–536
28. Lebel C, Beaulieu C (2011) Longitudinal development of human brain wiring continues from childhood into adulthood. J Neurosci 31:10937–10947
29. Sexton CE et al (2014) Accelerated changes in white matter microstructure during aging: a longitudinal diffusion tensor imaging study. J Neurosci 34:15425–15436
30. Scholz J, Klein MC, Behrens TE, Johansen-Berg H (2009) Training induces changes in white-matter architecture. Nat Neurosci 12:1370–1371
31. Hofstetter S, Tavor I, Tzur Moryosef S, Assaf Y (2013) Short-term learning induces white matter plasticity in the fornix. J Neurosci 33:12844–12850
32. Larsson HB, Thomsen C, Frederiksen J, Stubgaard M, Henriksen O (1992) In vivo magnetic resonance diffusion measurement in the brain of patients with multiple sclerosis. Magn Reson Imaging 10:7–12
33. Pini L et al (2016) Brain atrophy in Alzheimer's disease and aging. Ageing Res Rev 30:25–48
34. Krogsrud SK et al (2016) Changes in white matter microstructure in the developing brain--A longitudinal diffusion tensor imaging study of children from 4 to 11years of age. NeuroImage 124:473–486
35. Lebel C et al (2012) Diffusion tensor imaging of white matter tract evolution over the lifespan. NeuroImage 60:340–352
36. Burzynska AZ et al (2010) Age-related differences in white matter microstructure: region-specific patterns of diffusivity. NeuroImage 49:2104–2112
37. Marner L, Nyengaard JR, Tang Y, Pakkenberg B (2003) Marked loss of myelinated nerve fibers in the human brain with age. J Comp Neurol 462:144–152
38. Sbardella E, Tona F, Petsas N, Pantano P (2013) DTI measurements in multiple sclerosis: evaluation of brain damage and clinical implications. Mult Scler Int 2013:671730
39. Filippi M, Cercignani M, Inglese M, Horsfield MA, Comi G (2001) Diffusion tensor magnetic resonance imaging in multiple sclerosis. Neurology 56:304–311
40. Taubert M et al (2010) Dynamic properties of human brain structure: learning-related changes in cortical areas and associated fiber connections. J Neurosci 30:11670–11677
41. Takeuchi H et al (2010) Training of working memory impacts structural connectivity. J Neurosci 30:3297–3303
42. Sampaio-Baptista C et al (2013) Motor skill learning induces changes in white matter microstructure and myelination. J Neurosci 33:19499–19503
43. Blumenfeld-Katzir T, Pasternak O, Dagan M, Assaf Y (2011) Diffusion MRI of structural brain plasticity induced by a learning and memory task. PLoS One 6:e20678
44. Assaf Y, Basser PJ (2005) Composite hindered and restricted model of diffusion (CHARMED) MR imaging of the human brain. NeuroImage 27:48–58
45. Alexander DC et al (2010) Orientationally invariant indices of axon diameter and density from diffusion MRI. NeuroImage 52:1374–1389
46. Zhang H, Schneider T, Wheeler-Kingshott CA, Alexander DC (2012) NODDI: practical in vivo neurite orientation dispersion and density imaging of the human brain. NeuroImage 61:1000–1016
47. Assaf Y, Cohen Y (2000) Assignment of the water slow-diffusing component in the central nervous system using q-space diffusion MRS: implications for fiber tract imaging. Magn Reson Med 43:191–199
48. Assaf Y, Blumenfeld-Katzir T, Yovel Y, Basser P (2008) AxCaliber: a method for measuring axon diameter distribution from diffusion MRI. Magn Reson Med 59:1347–1354
49. Barazany D, Basser PJ, Assaf Y (2009) In vivo measurement of axon diameter distribution in the corpus callosum of rat brain. Brain 132:1210–1220
50. Horowitz A et al (2015) In vivo correlation between axon diameter and conduction velocity in the human brain. Brain Struct Funct 220:1777–1788
51. Tariq M, Schneider T, Alexander DC, Gandini Wheeler-Kingshott CA, Zhang H (2016) Bingham-NODDI: mapping anisotropic orientation dispersion of neurites using diffusion MRI. NeuroImage 133:207–223
52. Behrens TEJ, Jbabdi S (2009) MR diffusion tractography. In: Diffusion Mri: from quantitative measurement to in vivo neuroanatomy. Elsevier, Amsterdam, pp 333–351

53. Behrens T et al (2003) Non-invasive mapping of connections between human thalamus and cortex using diffusion imaging. Nat Neurosci 6:750–757
54. Behrens TEJ, Berg HJ, Jbabdi S, Rushworth MFS, Woolrich MW (2007) Probabilistic diffusion tractography with multiple fibre orientations: what can we gain? NeuroImage 34:144–155
55. Jbabdi S, Lehman JF, Haber SN, Behrens TE (2013) Human and monkey ventral prefrontal fibers use the same organizational principles to reach their targets: tracing versus tractography. J Neurosci 33:3190–3201
56. Hubbard PL, Parker GJM (2009) Validation of Tractography. In: Diffusion Mri: From Quantitative Measurement to in Vivo Neuroanatomy. Elsevier, Amsterdam, pp 353–375
57. Azadbakht H et al (2015) Validation of high-resolution tractography against in vivo tracing in the macaque visual cortex. Cereb Cortex 25:4299–4309
58. Dyrby TB et al (2007) Validation of in vitro probabilistic tractography. NeuroImage 37:1267–1277
59. Kolasinski J et al (2012) A combined post-mortem magnetic resonance imaging and quantitative histological study of multiple sclerosis pathology. Brain 135:2938–2951

INDEX

David A. Lyons and Linde Kegel (eds.), *Oligodendrocytes: Methods and Protocols*, Methods in Molecular Biology, vol. 1936, https://doi.org/10.1007/978-1-4939-9072-6, © Springer Science+Business Media, LLC, part of Springer Nature 2019

W

Z

FSC
www.fsc.org
MIX
Papier | Fördert gute Waldnutzung
FSC® C083411

Kolibri 360 GmbH
Ferdinand-Jühlke-Straße 7
99095 Erfurt, Deutschland
produktsicherheit@kolibri360.de